I0072090

Soil Science: Conservation and Nutrient Management

Soil Science: Conservation and Nutrient Management

Editor: Donald Cronin

www.callistoreference.com

Callisto Reference,
118-35 Queens Blvd., Suite 400,
Forest Hills, NY 11375, USA

Visit us on the World Wide Web at:
www.callistoreference.com

© Callisto Reference, 2019

This book contains information obtained from authentic and highly regarded sources. Copyright for all individual chapters remain with the respective authors as indicated. All chapters are published with permission under the Creative Commons Attribution License or equivalent. A wide variety of references are listed. Permission and sources are indicated; for detailed attributions, please refer to the permissions page and list of contributors. Reasonable efforts have been made to publish reliable data and information, but the authors, editors and publisher cannot assume any responsibility for the validity of all materials or the consequences of their use.

ISBN: 978-1-64116-063-6 (Hardback)

Trademark Notice: Registered trademark of products or corporate names are used only for explanation and identification without intent to infringe.

Cataloging-in-Publication Data

Soil science : conservation and nutrient management / edited by Donald Cronin.
  p. cm.
Includes bibliographical references and index.
ISBN 978-1-64116-063-6
1. Soil science. 2. Soil conservation. 3. Soils and nutrition. 4. Soil management.
I. Cronin, Donald.
S591 .S65 2019
631.4--dc23

Table of Contents

Preface

I am honored to present to you this unique book which encompasses the most up-to-date data in the field. I was extremely pleased to get this opportunity of editing the work of experts from across the globe. I have also written papers in this field and researched the various aspects revolving around the progress of the discipline. I have tried to unify my knowledge along with that of stalwarts from every corner of the world, to produce a text which not only benefits the readers but also facilitates the growth of the field.

Soil science is the study of soil as a natural resource. It is concerned with various chemical, physical and biological phenomena associated with soil. The tools of soil science are significant to the theory and practice of many disciplines like agriculture, forestry, anthropology, geology, environmental science, etc. Nutrient management, on the other hand, deals with the optimization of nutrient use efficiency, crop yield and quality. This is achieved by combining soil, hydrologic and climatic factors with conservation practices associated with soil and water. These two fields are integral to each other and are highly significant for attaining better crop yield and developing new methods and techniques of crop production and management. The chapters in this book are compiled to provide detailed information about soil science, particularly on nutrient management and conservation. This book presents researches and studies performed by experts across the globe. It will serve as a valuable source of reference for graduate and postgraduate students, as well as experts working in this domain.

Finally, I would like to thank all the contributing authors for their valuable time and contributions. This book would not have been possible without their efforts. I would also like to thank my friends and family for their constant support.

Editor

An insight into pre-Columbian raised fields: the case of San Borja, Bolivian lowlands

Leonor Rodrigues[1], Umberto Lombardo[2], Mareike Trauerstein[1], Perrine Huber[*], Sandra Mohr[*], and Heinz Veit[1]

[1]Institute of Geography, University of Berne, Hallerstrasse 12, 3012 Bern, Switzerland
[2]University of Pompeu Fabra, Ramon Trias Fargas 25–27, Mercè Rodoreda, 08005 Barcelona, Spain
[*]Independent researcher, 3012 Bern, Switzerland

Correspondence to: Leonor Rodrigues (leonor.rodrigues@giub.unibe.ch)

Abstract. Pre-Columbian raised field agriculture in the tropical lowlands of South America has received increasing attention and been the focus of heated debates regarding its function, productivity, and role in the development of pre-Columbian societies. Even though raised fields are all associated to permanent or semi-permanent high water levels, they occur in different environmental contexts. Very few field-based studies on raised fields have been carried out in the tropical lowlands and little is known about their use and past management. Based on topographic surveying and mapping, soil physical and chemical analysis and OSL and radiocarbon dating, this paper provides insight into the morphology, functioning and time frame of the use of raised fields in the south-western Llanos de Moxos, Bolivian Amazon. We have studied raised fields of different sizes that were built in an area near the town of San Borja, with a complex fluvial history. The results show that differences in field size and height are the result of an adaptation to a site where soil properties vary significantly on a scale of tens to hundreds of metres. The analysis and dating of the raised fields sediments point towards an extensive and rather brief use of the raised fields, for about 100–200 years at the beginning of the 2nd millennium.

1 Introduction

The Llanos de Moxos (LM) is one of the largest floodplains in Latin America, characterized by a prolonged rainy season and a contrasting dry season, often resulting in either severe flooding or droughts (Hanagarth, 1993). Nowadays, the LM is sparsely inhabited, soils have been considered unsuitable for agriculture and the main economic activity is extensive cattle grazing (Erickson, 2003b). However, the presence of numerous pre-Columbian earthworks indicates that in the past, humans modified the landscape in several ways and cultivated crops (Denevan, 2001; Erickson, 2008; Lombardo et al., 2011b; Jaimes Betancourt, 2013; Prümers and Jaimes Betancourt, 2014; Carson et al., 2015). The number and variety of earthworks has made the area of the LM one of the most important examples of pre-Columbian anthropogenic landscapes in the Amazon Basin (Erickson, 2008). Raised fields are one of the most abundant and impressive

types of earthworks found in the LM. Pre-Columbian raised fields are elevated agricultural earth platforms. They are also found in a number of other Latin American countries. Distributed over a wide range of latitudes, spanning from almost 20° N to 40° S, they exist in very different environmental contexts: in seasonal inundated floodplains, around lake shores and in coastal zones; all of them associated to permanent or semi-permanent high water levels (Erickson, 1988; Kolata and Ortloff, 1989; Wilson et al., 2002; Dillehay et al., 2007; Denevan, 2001; Rostain, 2010). Since their discovery about 100 years ago by Erland Nordenskiöld (Denevan, 2009), increasing interest in raised field agriculture has led to a body of literature focused on understanding how raised fields worked and were managed and how pre-Columbian societies were sustained (Gliessmann et al., 1985; Kolata and Ortloff, 1989; Erickson, 2008; Iriarte et al., 2010; McKey et al., 2010; Rostain, 2010; Lombardo et al., 2011a; Renard

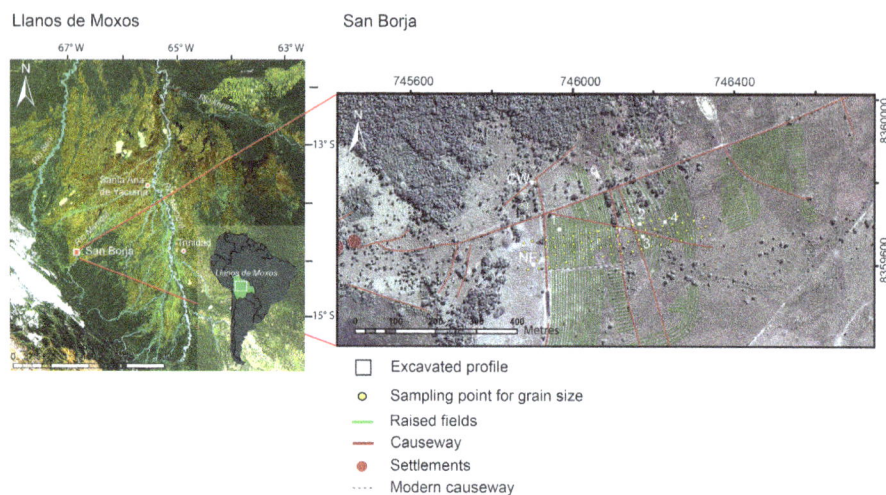

Figure 1. (a) MODIS image of the Llanos de Moxos in the Bolivian Lowlands including a small continent map of South America. **(b)** Zoom in to the study area in San Borja showing raised fields and causeways, including the location of the excavated profiles and locations sampled for grain size analysis. Image source: Google Earth.

et al., 2011; Whitney et al., 2014; Rodrigues et al., 2015; Walker, 2004). Nevertheless, there is still limited data and a lack of consensus about the time frame during which raised fields were in use, why they were built and how they were managed. Some studies have suggested that raised fields were highly productive and could have sustained dense populations in the LM (Erickson, 2008), and the technology of raised fields has been proposed as a model for sustainable agriculture today (Erickson, 2008; Renard et al., 2012; Whitney et al., 2014). In contrast, other studies suggest that raised fields in the LM were not more productive per se and were rather constructed to overcome past climate extremes (Lombardo et al., 2011a; Rodrigues et al., 2016).

Experimental studies have aimed to assess the productivity of raised fields and, because of their alleged high yields, they have been presented as a "sustainable agriculture" alternative for rural development projects (Erickson, 1992; Stab and Arce, 2000; Saavedra, 2009). However, most of these rehabilitation projects failed due to a number of reasons: overestimation of field productivity (Bandy, 2005), unfavourable structure of modern society (Erickson, 2003a), environmental conditions (Chapin, 1988), non-involvement of local communities due to a top-down approach (Renard et al., 2012), other methodological weaknesses (Lombardo et al., 2011a) and the generalisation of one single technology as a model for all raised fields regardless of their location (Baveye, 2013). As already stated, raised fields are highly diverse in design and exist in very different environmental contexts. Therefore, it has been proposed that the way they were built and their management was probably determined by geographical and/or environmental constraints (Denevan, 2001; McKey et al., 2010; Lombardo et al., 2011a; Baveye, 2013; Rostain, 2013; Rodrigues et al., 2016). However, the reasons behind the regional differences in raised field types

is still not clear, as differences in raised fields could also be related to cultural diversity (Erickson, 1995; Denevan, 2001; Walker, 2011; Rostain, 2013).

Understanding the link between environmental variables (e.g. soils, topography, and hydrology) and design of raised fields is key in order to better understand the reasons behind raised field diversity and to infer how they could have been managed in the past. Very little is known with regards to the environmental characteristics of the areas where raised fields were built in lowland South America. Only recently some studies have started to include detailed soil studies providing an insight to the internal structure of raised fields and their relation to the environment (Iriarte et al., 2010; McKey et al., 2010; Lombardo et al., 2011a; McMichael et al., 2014; Rodrigues et al., 2015).

The LM is very suitable for the study of raised fields, as they appear in a variety of forms (ridged fields, platform fields, mound fields, ditched fields) and patterns (e.g. parallel or randomly scattered, with or without embankment and causeways). These different types of fields tend to be present in different parts of the LM and do not normally coexist in the same site (Denevan, 2001; Lombardo et al., 2011b). Nevertheless, recently it has been shown that, in at least two cases in the LM, differences in shape, height and layout can vary considerably within the same cultural area, and are the result of an adaptation to the distinct local edaphology (Lombardo et al., 2011a; Rodrigues et al., 2016).

The present study explores the raised fields located in the savannah on the western bank of the Rio Maniqui, in the San Borja area (Fig. 1). Archaeological evidence of occupational sites and raised fields here were already described by Nordenskiöld (1916) at the beginning of the 20th century followed by Erickson (1980) and again recently by Iriarte and Dickau (2012). Up until now no data from archaeological excavations have been available for this area. Here, raised fields of different shapes co-exist. This study aims to further our understanding of pre-Columbian agricultural systems in the Bolivian lowlands. It uses topographic surveying and mapping, soil physical and/or chemical analysis and OSL and radiocarbon dating to address the following questions:

- In which environmental context can raised fields be found and what are their characteristics (e.g. morphology and soil properties)?

- Are there links between the dimension and/or shape of raised fields and soil properties?

- When were these raised fields in use?

2 Study area

2.1 Geography and environment

The study area is located near San Borja, a town situated in the south-western part of the Beni department, only a few kilometres away from the Andean foothills (Fig. 1). The Beni department almost completely overlaps with the Llanos de Moxos (LM), a seasonally inundated floodplain drained by three major rivers: Río Mamoré, Río Beni and Río Iténez. The diverse geomorphology of the LM is shaped by past and present fluvial dynamics such as alluvial deposition and erosion and river shifting (Hanagarth, 1993; Dumont and Fournier, 1994; May, 2011; Plotzki et al., 2011, 2013; Lombardo, 2016), as well as tectonics (Hanagarth, 1993; Dumont and Fournier, 1994; Lombardo, 2014). These processes are responsible for changes in the local topography, determining the flooding dynamics and, in turn, the forest-savannah ecotone (Mayle et al., 2007). The climate in the LM is controlled by the South American Summer Monsoon, leading to heavy convective rainfalls in austral summer and dry conditions in winter (Zhou and Lau, 1998; Garreaud et al., 2009). The mean annual temperature is 25.8 °C and is fairly stable year round (Navarro and Maldonado, 2002), although it can decrease considerably during the austral winter months with the arrival of cold southern winds locally called *surazos* (Espinoza et al., 2013). The mean annual precipitation in the region amounts to 1900 mm (Hijmanns et al., 2005); most of the precipitation falls during the austral summer, followed by prolonged flooding which covers an area ranging from 30 000 to > 80 000 km^2 (Hanagarth, 1993; Hamilton et al., 2004). In general floods are greater in the southern LM due to increased precipitation towards the Andes and higher groundwater tables (Hanagarth, 1993). The landscape around San Borja has been shaped by the fluvial history of the Río Maniqui, a meandering white water river which has shifted its course several times in the past. Hanagarth (1993) has distinguished nine phases of ancient river courses; the Maniqui has been shifting its course in both clockwise and anti-clockwise directions. The Río Maniqui is one of the most

dynamic rivers in the LM (Lombardo, 2016) with crevasses occurring every few years, leading to complete river avulsions on a sub-decadal time frame (a crevasse describes the process when a river breaks though its river levée; an avulsion is a natural change of river course that leads to the abandonment of the old channel and the establishment of a new one; Charlton, 2007). Throughout the Holocene, this river, with a high sedimentary load, has built a large interior delta (Hanagarth, 1993; Lombardo, 2014). An extensive forest has grown on this delta, which today forms part of the Biosphere "Estación Biológica del Beni" (Hanagarth, 1993).

Soils along the Río Maniqui are in general acidic, with a pH ranging from 3.86 to 5.11, but they are very heterogeneous in terms of plant available nutrients, mostly correlated to the particle size of the sediments (Guèze et al., 2013). Soils in the southern LM are generally loam or silty loam and silty clay loam (Boixadera et al., 2003); all soils are subject to hydromorphic processes and are mainly acidic (Boixadera et al., 2003; Rodrigues et al., 2015), with some exceptions of saline soils (pH of > 8) (Lombardo et al., 2015) and soils with accumulation of calcium carbonates in the subsoil (Boixadera et al., 2003).

2.2 The archaeology of raised fields in the Llanos de Moxos

Our knowledge about the chronology, complexity and evolution of pre-Columbian cultures in the LM is still in its early stages (Prümers and Jaimes Betancourt, 2014). The archaeological landscape is roughly divided by the Río Mamoré; on the western side no raised fields exist, while on the eastern side thousands of hectares of fields have been documented (Denevan, 2001; Lombardo et al., 2011b). Detailed archaeological research has mostly concentrated in the Monumental Mounds Region (MMR) (Prümers, 2008; Bruno, 2010; Dickau et al., 2012; Jaimes Betancourt, 2012; Lombardo and Prümers, 2010; Lombardo et al., 2013), in the south-western LM. Here, no raised fields are present. Unfortunately, almost no chronological data exist for the eastern Llanos, where fields are widespread. Up till now, habitational sites associated to raised fields have only been dated in four locations: the San Juan site (AD 446–613) and the Cerro site (AD 1300–1400), in the northern part of the LM (Walker, 2004), and the Moxitania site (AD 700–1000), Abularach and Carretera Santa Ana sites (AD 900–1100), close to San Ignacio de Moxos (Villalba et al., 2004). Raised fields in Bermeo, close to San Ignacio, have been dated; they were used intermittently from AD 570–770 up to the 14th century (Rodrigues et al., 2015).

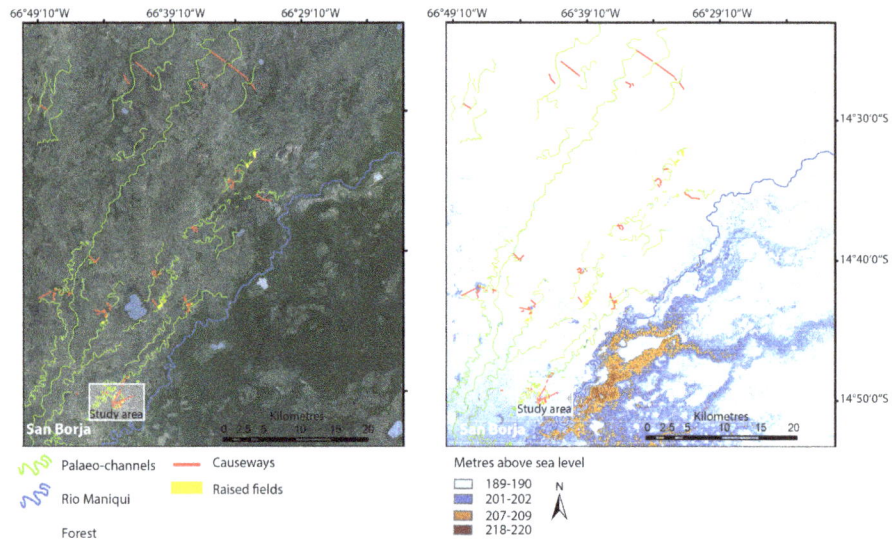

Figure 2. Left panel: Google earth image showing mapped palaeo-river features, causeways (red) and raised field areas (yellow). Right panel: digital elevation model (CGIAR_CSISRTM) including same mapped palaeo-river features, causeways (red) and raised field areas (yellow). Higher areas are relict sediment deposits from former rivers.

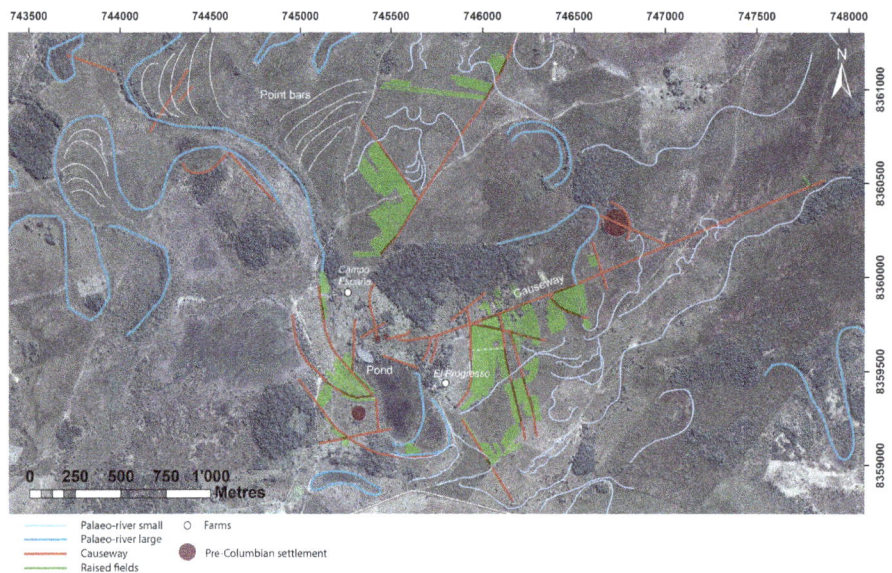

Figure 3. Study area including anthropogenic earthworks, natural geomorphological features and the two farms Campo España and El Progresso.

3 Methods

3.1 Mapping and field work

The location for the study of raised fields was first predefined with the help of Google Earth. An area of about 8 ha was selected, where raised fields of different shapes coexist (Fig. 1). ArcGIS was used to map natural and anthropogenic features at two different scales (Google Earth, 2002, 2011, provided by the ArcGIS Basmap extension). On a large scale, covering an area of 2500 km^2, paleo-channels and areas with

raised fields and causeways have been mapped (Fig. 2a), including topographic information using CGIAR-CSI SRTM (resolution 90 m) (Fig. 2b). On a smaller scale (12.3 km^2), in the area where field work was conducted, individual raised fields, causeways, ponds, settlement sites, as well as natural features like palaeo-channels and creeks, were mapped (Fig. 3).

Field work was conducted in August 2012 and again in 2013. The local relief was measured using a digital level Sokkia D50. A 480 m long topographic transect perpendicular to the fields was drawn based on measurements taken approximately every 1 m (Fig. 7). In addition, a specific area where higher and lower fields lie next to each other was selected for in depth morphological analysis: 2400 elevation points were measured and a digital elevation model (DEM) was generated using the 3-D analyst extension of ArcGis with natural neighbour interpolation (Fig. 7). Four fields were excavated; trenches were dug from the ridge to the canal (Fields 1–4). Two additional pits were dug; one in an area away from the raised fields (No Field, NF) and one in a causeway (CW) (Fig. 1). In total 10 stratigraphic profiles were prepared and sampled every 10 cm: the NF profile, the CW profile and two for each Field (1–4), one profile in the ridge and a second one in the adjoining canal. The description of the horizons and/or layers follow the guidelines of FAO (IUSS Working Group WRB, 2014). In addition, a virtual grid was applied onto an area of 450×64 m and samples for particle size analysis were taken every 16 m in the N–S direction and every 20 m in the E-W direction (Figs. 1 and 6). Particle size was analysed every 20 cm up to a depth of 100 cm.

3.2 Laboratory analysis

All the samples have been air-dried. The colour determination, however, was carried out on moist samples, using the Munsell soil colour charts (1994). For particle size distribution, organic matter was removed with 30 % H_2O_2 and afterwards measured with a laser diffraction instrument (Malvern Mastersizer Hydro 2000S). The pH was measured with a glass-electrode after mixing the sample with a 0.01 M $CaCl_2$ solution. C and N concentrations were analysed by dry combustion and gas chromatographic separation with a CNS analyser (vario El cube). Cation exchange capacity (CEC) was measured by means of extraction of the exchangeable Cations Ca_2^+, Mg_2^+, K^+, Na^+, Mn_2^+, and Al_3^+, with 1 M ammonium nitrate solution (NH_4NO_3). Concentrations were measured using an atomic absorption spectrometer of the type analytikjena ZEEnit 700P. Effective cation exchange capacity (CEC_{eff}) and base saturation (BS) were calculated as follows: $CEC_{eff} = \sum$ Cations (exchangeable Cation = Cation mg kg^{-1} molar mass mmol^{-1} L^{-1} · valence) and BS = ([Ca] + [Mg] + [K] + [Na]/CEC_{eff}) × 100. Available phosphorus (P_{av}) was measured with the Mehlich I extraction method (Mehlich, 1953), following the guidelines recommended for acidic soils in Sparks (1996) and standard analysis for tropical soils in Brazil (Solos, 2011). P_{av} content was determined by means of a spectrophotometer reacting with ammonium molybdate, following the slightly modified method of Murphy and Riley (1962) developed by Watanabe and Olsen (1965). Element analysis has been performed by x-ray fluorescence spectroscopy (XRF) and quantified by means of the *UniQuant* method.

3.3 Radiocarbon and optically stimulated luminescence (OSL) dating

AMS C analysis on three charcoal samples and two palaeo-soil samples (Table 1) was conducted at the Poznan Radiocarbon Laboratory and LARA AMS Laboratory in Bern, calibrated using Calib 7.1 (Stuiver and Reimer, 1993) and the SHcal13 calibration curve for the Southern Hemisphere (Hogg et al., 2013).

Samples for optically stimulated luminescence (OSL) dating were taken by pushing steel tubes into the exposed sediment. The concentration of dose rate relevant elements (Table S1 in the Supplement) was determined using high-resolution low-level gamma spectrometry, performed on bulk material from the surrounding sediment. Dose rates have been calculated assuming an average moisture content of 25–35 % and present-day sediment cover. For equivalent dose (D_e) determination, samples were dry sieved to separate the 100–150 µm particle size fraction, followed by HCl and H_2O_2 treatment. The quartz fraction was extracted using heavy liquids and etching in 40 % HF for 1 h. Luminescence measurements were carried out on a Risø DA-20 TL/OSL reader fitted with an internal ^{90}Sr/^{90}Y beta-source. Quartz signals were detected through a Hoya U340 detection filter. De measurements were performed on 48 small aliquots (2 mm) per sample applying the single-aliquot regenerative dose (SAR) protocol (Murray and Wintle, 2000) using a preheat of 230 °C for 10 s prior to all OSL measurements. The small aliquot OSL signals show no indication for feldspar contamination (IR depletion ratio > 0.8) and are dominated by the fast component. Small aliquot D_e distributions of sample SB1-C36 and SB1-C40 are slightly skewed, with some outliers at the upper end of the distribution, indicating partial bleaching. To exclude signal averaging, single grain measurements were additionally carried out on all samples using the same measurements parameters as for the small aliquots. For the two sediment samples (SB1 C60 and SB1 F140), single grain D_e distributions are symmetric and exhibit an overdispersion of 28 and 21 %, respectively. The resulting single grain CAM ages are consistent with the small aliquot ones (see Table S1). The single grain D_e distributions for the samples from the field and canal deposits (SB1-C36 and SB1-C40) are skewed and exhibit an overdispersion of 66 and 62 %, respectively. To calculate MAM ages (Galbraith et al., 1999) from the single grain data of these samples a σb value of 0.28 was applied. In the following discussion single grain ages of all samples are used.

4 Results

4.1 Field work and mapping of the study site

The area studied has been shaped by several shifting rivers, and more recently by anthropogenic earthworks. Natural features like palaeo-channels and oxbows, and anthropogenic

Table 1. Down-profile values of selected geochemical parameters and grain size of all Fields (ridges and canals).

Depth	Sand 63–2000 µm	Silt 4–63 µm	Clay < 4.00 µm	pH	Corg %	N %	CECeff mmolc kg^{-1} %	Bs %	Ca %	Al	Ca/Al ppm	Pav
						Soil geochemical and physical properties						
						Ridge 1						
0–10	52.10	39.73	8.17	4.1	0.68	0.08	19	54	32	39	0.82	3.7
10–20	44.58	44.97	10.46	4.1	0.4	0.03	20	40	20	54	0.37	1.0
20–30	47.42	42.97	9.62	3.9	0.23	0.03	18	65	30	31	0.98	1.2
30–40	49.72	41.44	8.84	4.2	0.3	0.034	26	72	34	26	1.30	0
40–50	45.78	42.65	11.58	4.4	0.22	0.03	27	84	46	14	3.40	0.2
50–60	41.36	47.14	11.51	4.4	0.13	0.02	29	82	46	17	2.75	12
60–70	37.99	49.23	12.77	4.3	0.14	0.03	46	92	40	8	5.30	18.7
70–80	43.04	45.42	11.54	4.4	–	–	100	96	53	3	15.98	16.6
80–90	37.79	50.18	12.04	4.5	0.19	0.03	67	95	37	4	8.47	20.6
90–100	27.16	57.09	15.76	4.5	0.16	0.03	56	100	39	0.00	–	–
						Canal 1						
0–10	43.94	47.41	8.65	3.9	0.87	0.10	30	60	46	6	1.36	3.4
10–20	43.69	46.21	10.11	3.9	0.59	0.06	30	58	43	4	1.14	1.2
20–30	39.07	48.59	12.34	4.1	0.49	0.05	4	55	35	2	0.83	2.3
30–40	30.91	55.35	13.74	4	0.30	0.04	45	62	39	4	1.05	6.8
40–50	19.71	62.59	17.70	4.2	0.19	0.04	71	70	38	4	1.29	0.5
50–60	33.49	53.97	12.55	4.2	0.14	0.03	52	88	42	0.3	3.53	8.5
80–90	46.47	44.87	8.66	4.3	0.08	0.02	52	93	32	0.2	4.30	15.2
90–100	45.47	44.85	9.68	4.6	0.10	0.02	74	96	34	0.3	9.06	26.8
						Ridge 2						
0–10	65.60	27.55	6.85	4.2	0.74	0.07	17	42	30	55	0.55	–
10–20	59.46	31.91	8.63	4.0	0.46	0.05	16	19	12	79	0.15	–
20–30	58.71	31.38	9.92	4.0	0.33	0.03	15	17	10	81	0.12	–
30–40	56.82	34.13	9.04	4.0	0.22	0.02	12	15	8	83	0.10	–
40–50	60.49	32.15	7.36	4.0	0.16	0.02	11	18	11	80	0.14	–
60–70	58.34	32.58	9.08	4.0	0.13	0.02	12	17	10	82	0.13	–
70–80	61.42	29.65	8.94	4.1	0.11	0.01	11	22	13	76	0.18	–
90–100	60.77	31.65	7.58	4.0	0.10	0.01	16	17	9	82	0.11	–
100–110	73.96	20.54	5.50	4.0	0.09	0.01	16	12	6	83	0.07	–
110–120	70.79	22.07	7.15	4.0	0.05	0.01	17	16	9	85	0.10	–
120–130	77.98	17.17	4.85	3.1	0.06	DL	–	–	–	–	–	–
130–140	84.61	12.21	3.18	4.0	0.04	DL	–	–	–	–	–	–
140–150	90.88	7.53	1.59	4.1	0.03	DL	–	–	–	–	–	–
						Canal 2						
0–10	43.93	39.41	16.66	4.0	1.13	0.11	20	33	9	66	0.14	–
15–25	37.08	49.54	13.38	4.0	0.39	0.04	23	38	13	60	0.22	–
30–40	50.00	36.70	13.31	4.1	0.30	0.03	25	23	8	76	0.10	–
40–50	41.37	41.55	17.08	4.0	0.24	0.03	33	25	10	74	0.13	–
50–60	46.85	37.63	15.52	3.9	0.21	0.03	42	22	7	77	0.10	–
60–70	63.81	25.98	10.21	4.0	0.01	0.02	31	26	11	74	0.15	–
70–80	55.77	32.88	11.35	4.0	0.10	0.01	31	31	16	68	0.23	–
80–90	50.84	37.44	11.72	4.0	0.08	0.01	22	40	21	60	0.36	–
100–110	59.80	30.08	10.13	4.0	0.06	0.01	25	44	24	56	0.42	–
110–120	67.82	25.56	6.63	4.0	0.07	0.01	22	39	15	6	0.25	–

Table 1. Continued.

Depth	Sand 63–2000 μm	Silt 4–63 μm	Clay < 4.00 μm	pH	Corg %	N %	CECeff mmolc kg^{-1} %	Bs %	Ca %	Al	Ca/Al ppm	Pav
						Soil geochemical and physical properties						
						Ridge 3						
10–25	65.75	29.77	4.47	4.0	0.37	0.05	12	13	6	87	0.07	1.16
30–40	65.80	29.96	4.15	4.0	0.14	0.02	11	20	13	77	0.17	0
40–50	68.94	27.12	3.94	4.1	0.13	0.02	10	24	13	73	0.18	0
50–60	70.06	26.21	3.73	4.0	0.10	0.02	12	17	9	80	0.11	0
60–70	73.48	23.07	3.45	4.0	0.08	0.02	14	18	9	80	0.11	0.84
70–80	76.51	21.01	2.48	4.0	0.06	0.01	12	28	17	71	0.23	2.6
90–100	88.02	10.71	1.27	4.0	0.04	0.01	11	18	9	81	0.11	2.56
100–110	88.59	10.12	1.29	4.1	0.03	0.01	12	22	11	77	0.14	–
120–130	89.02	9.65	1.33	4.1	0.02	0.01	13	32	18	68	0.26	–
						Canal 3						
15–25	53.5	40.22	6.28	3.9	0.60	0.07	20	26	12	71	0.16	5.32
25–35	58.32	36.11	5.58	4.0	0.35	0.04	18	40	17	56	0.30	10.48
35–45	64.56	31.02	4.42	4.0	0.03	0.03	20	34	11	63	0.18	9.8
45–55	67.62	28.39	3.99	4.0	0.18	0.03	15	32	10	65	0.15	12.6
55–65	79.50	18.32	2.18	4.0	0.06	0.01	13	32	7	67	0.10	20.96
65–75	82.20	15.99	1.81	4.0	0.05	0.01	13	23	8	77	0.10	17.6
75–85	89.06	9.92	1.01	4.0	0.03	0.01	18	21	8	78	0.10	14.8
85–90	80.06	17.46	2.48	4.0	0.04	0.01	12	21	6	79	0.08	11.2
90–100	90.80	8.44	0.77	4.0	0.02	0.01	14	16	4	84	0.05	7.6
						Ridge 4						
0–10	41.40	49.59	10.72	4	0.75	0.08	56	71	35	26	1.33	2.04
10–20	36.01	52.88	11.34	4.3	0.32	0.04	34	83	39	15	2.60	0
20–35	36.24	49.87	15.80	4.2	0.34	0.05	71	88	37	10	3.84	0.04
40–50	35.29	49.05	16.91	4.6	0.28	0.04	80	97	42	3	16.04	1.2
50–60	34.98	46.93	19.05	4.8	0.25	0.05	98	100	45	0.00	–	1.4
70–80	21.47	59.23	21.53	5.1	0.21	0.05	128	100	45	0.00	–	6.2
80–90	2.23	71.51	26.26	5.2	0.21	0.05	75	100	87	0.00	–	9.04
90–100	6.02	68.50	26.60	5.1	0.19	0.05	172	100	49	0.00	–	6.24
						Canal 4						
0–10	29.51	52.58	19.02	4	0.86	0.10	52	75	32	22	1.43	5.04
10–20	28.79	55.02	16.73	3.9	0.40	0.05	36	67	36	30	1.19	3.6
20–30	47.54	46.29	14.41	4	0.33	0.05	37	72	35	25	1.39	4.6
30–40	22.73	57.60	19.67	4	0.52	0.06	85	80	32	18	1.73	3.56
50–60	21.03	59.42	20.22	4.2	0.29	0.04	84	87	37	12	3.14	4.04
70–80	4.34	67.83	28.68	4.7	0.20	0.06	170	100	42	0.00	–	4.48
						No field REF						
0–10	18.09	62.02	19.89	3.9	1.13	0.13	32	67	31	28	1.11	2.6
10–20	15.70	62.65	21.65	4.0	0.70	0.09	38	66	29	32	0.92	0.88
20–30	16.41	60.18	23.41	4.0	0.52	0.08	75	43	19	56	0.34	0.8
30–40	22.42	56.75	20.83	4.1	0.38	0.07	82	47	20	52	0.39	0
40–50	39.24	46.42	14.33	4.2	0.31	0.06	75	52	22	48	0.46	0
50–60	4.12	63.34	32.54	4.3	0.20	0.04	51	63	25	37	0.69	0
60–70	13.24	65.31	21.45	7.9	0.34	0.06	121	100	50	0.00	–	0
70–80	11.55	65.47	23.02	4.7	0.23	0.05	71	97	42	8	14.70	3.8
80–90	5.11	69.30	25.58	4.7	0.23	0.06	110	100	43	0.5	87.41	2.8
						Causeway						
0–10	29.28	59.33	11.39	–	0.77	0.10	–	–	–	–	–	2.4
20–30	35.16	51.75	13.10	–	0.32	0.04	–	–	–	–	–	0
30–40	41.51	47.29	11.21	–	0.23	0.04	–	–	–	–	–	0
50–60	39.22	47.96	12.81	–	0.25	0.04	–	–	–	–	–	3.68
60–70	42.07	44.61	13.32	–	0.18	0.04	–	–	–	–	–	3.72
80–90	32.92	50.67	16.41	–	0.13	0.04	–	–	–	–	–	1.28
90–100	29.28	59.33	11.39	–	0.12	0.03	–	–	–	–	–	2.76

earthworks, including raised fields and causeways, are illustrated in Fig. 2. Several generations of palaeo-river channels were clearly identified; these share the direction of the modern Río Maniqui, from the south-east to the north-west. One paleo-river can be traced continuously while in the other cases only segments of paleo-channels can be recognized. Channels and oxbows often appear washed out due to enduring erosion, the superimposing of other channels and the construction of earthworks, mostly raised fields. Patches of gallery forest exist along the channels and are recognisable by the topographic differences in Fig. 2b. In total, 370 ha of raised fields were mapped, as well as 52.2 km of causeways (Fig. 2a). All the raised fields are found along palaeo-rivers, on alluvial deposits, and are associated to causeways, but not vice versa. Some causeways were built through the pampa, connecting higher laying areas covered by forest along palaeo rivers (Fig. 2).

On a smaller area, detailed structures like small creeks, point bars, anthropogenic features, including raised fields, causeways, settlement sites and ponds, were mapped (Fig. 3). The natural features can be categorised as continuous meander channels, creeks, oxbows and point bars. The assignment of the oxbows to the larger channels is not straight forward as they sometimes display similar channel width and sometimes they differ by several metres. It is therefore difficult to say if oxbows and larger channels have been formed by the same river. In some cases oxbows are dried out, but most of them exist as wetlands. Point bars are found next to the larger channels and oxbows. During the field survey we mapped four little earth mounds, locally called *lomas*, and one pond (Fig. 3). Lomas are anthropogenic earth mounds found in the lowlands of Bolivia, which could have served several purposes: as settlements, cemeteries, ritual sites and, as well, for agriculture (Erickson, 2000). The *lomas* surveyed are located on the most elevated parts of the study site, near two farms: *Campo España* and *El Progreso* (Fig. 3).

Within this area individual raised fields were mapped, as well as 14 577 m of causeways (Fig. 3). The longest causeway, crossing the whole area from the south-west to the north-east, is 2997 m long. From this major causeway several shorter causeways go north and south, always in connection with raised fields. The shape of the raised fields, as well as the causeways, are particularly interesting. Most of the fields (> 100 m) and causeways share a curved course, which is similar to the shape of the meanders and point bars of the palaeo-rivers. Some fields (< 100 m) were built perpendicular to a causeway, resembling a comb, a pattern already described for raised fields in the Titicaca Basin (Denevan, 2001). They are all elongated, with a mean length of 87 m; however, their length varies considerably, ranging from a minimum of 9 m to a maximum of 582 m.

4.2   Sedimentary and pedogenic characteristics

4.2.1   Individual profiles

As field work was conducted twice, in 2012 and 2013, and the rainy season during 2013 was much wetter, the depth of the groundwater table was significantly different each year: at the beginning of August 2012 the water table was 2 m below the surface, whereas at the end of July 2013 it was at a depth of 1 m. Detailed profile descriptions are summarized in Fig. 4.

All profiles share some pedogenic characteristics, but in general they differ remarkably in a number of aspects. All profiles show intense mottling, typical of hydromorphic processes. The iron mottles are soft, up to 1.5 cm in diameter and the colour varies between yellow and orange, which usually indicates the presence of goethite and lepidocrocite, typically formed in waterlogged soils (Cornell and Schwertmann, 2003). Black millimetre scale manganese concretions were also observed. Hydromorphy is present, on average, up to 35 cm below the top of the ridge, but there is no clear boundary. Manganese concretions tend to accumulate at the upper limits of the hydromorphic affected layers. The depth of this diffuse boundary slightly differs from field to field and is indicated with a blue line in Fig. 4. With regards to the canals, the profiles can generally be divided into two major layers: the infilling of the canal after the construction of the fields and the undisturbed sediments below. This sharp boundary can be clearly recognized due to the brown infilled canals contrasting with the yellowish sediments below. This boundary is always present around 30–50 cm depth (Fig. 4). With regards to the texture, all profiles differ remarkably (Fig. 5 and Table 1) and distinctive characteristics were observed for each field.

- *Field 1*: soil texture in Field 1 is mainly loam-silty loam. Particle size decreases towards the bottom, ranging from 52 % to only 27 % of sand. During the excavation two small pieces of burnt earth were found.

- *Fields 2 and 3*: the soil in these fields is relatively coarse in texture, ranging from sand to sandy loam, with 65–90 % sand and only 1.5–8 % clay. Particle size increases towards the bottom of the fields, from sand to sandy loam. Both profiles comprise a ferruginous yellow-orange continuous line (Fig. 4, Ridge 2). This line originates at the boundary of the infilling of the canal, following the topography of the field about 40 cm below the surface, and disappears towards the top/middle of the ridge. One charcoal piece (diameter 5 mm) was extracted from the ridge of Field 2 at a depth of 80 cm. Beneath the canal of Field 2 a palaeosol (Palaeosol I), recognisable by its dark brown colour, was found at a depth of 270 cm using a hand auger (Table 2).

- *Field 4*: the soil here has the finest texture of all raised fields, ranging from silt to silty loam. There is a signif-

Profile	Soil characteristics	Munsell color
Ridge 1	Silty, dull brown topsoil	7.5 YR 6/3
	Lighter dull orange layer with abundant manganese concretions 40 % (Ø 2 mm)	7.5 YR 6/6
	Orange matrix with some few iron and manganese concretions	7.5 YR 6/4
	Boundary with increasing orange iron hydromorphic mottling	7.5 YR 6/8
	Silty, hydromorphic orange mottles (30%) increasing towards the bottom	
	Pieces of burned earth	7.5 YR 6/8
	Water table at 110 cm (dry season 2013)	
Canal 1	Silty, brown organic rich layer	7.5 YR 4/3
	Orange mottles (5 %) and manganese concretions (Ø 2 mm)	
	Boundray of infilling, hydromorphic orange mottles (30 %)	7.5 YR 4/3
	Silty, hydromorphic layer, still partly saturated, reduced pale orange matrix with orange mottles	7.5 YR 7/3
		7.5 YR 4/3
	Water table at 95 cm (dry season 2013)	
Ridge 2	Sandy loam light grey top soil	10 YR 5/3
	Hydromorphic yellow orange mottling and iron and manganese concretions (5 %) in light grey matrix	7.5 YR 5/4
		7.5 YR 6/6
	Increasing amount of hydromorphic yellow-orange mottles and iron concretions (5 %)	7.5 YR 6/6
	Charcoal pieces	
	Hydromorphic yellow-orange staining of whole matrix with vertical bleaching structures	7.5 YR 6/4
	Water table below 2 m (dry season 2012)	7.5 YR 6/4
yellow line	Yellow oxidized line reffered as rust line in the text following the topography of the field present in Field 2 and 3	
Canal 2	Greyish yellow-brown, dense topsoil	10 YR 4/2
	Slightly lighter greyish yellow brown layer with very fine manganese and iron concretions (Ø 2mm)	10 YR 5/2
	Greyish yellow brown matrix with orange mottles (20%)	10 YR 6/2
		10 YR 6/6
	Boundary of infilling to light yellow orange matrix, abundant orange mottling (50 %)	7.5 YR 8/3
	Matrix orange 10 YR 6/4 with vertical bleaching structures (10 YR 8/3)	7.5 YR 6/6
		7.5 YR 8/3
	Very sandy almost uniform orange layer. Water table below 2 m (dry season 2012)	7.5 YR 6/6

icant increase in clay content from 10 to 26 % towards the bottom, whilst sand decreases from 41 to 6 %. A lot of natural and anthropogenic disturbance is evident. In the upper layers some clear signs of modern deformation by cattle trampling can be recognized (Fig. 4). Further down in the ridge profile desiccation cracks are filled with fine sand. The relative high amount of clay in the deeper layers (80–100 cm) is responsible for the cracks, which commonly develop by the expansion and shrinking of the clays due to repeated wetting and drying (Schachtschabel et al., 2002).

– *No Field Profile (NF)*: the soil of the profile from the area without raised fields is extremely dense and has the finest texture, with up to 32 % clay, ranging from

Profile	Soil characteristics	Munsell color
Ridge 3		
	Sandy loam, light grey, very thin top soil	10 YR 5/3
	Yellow-orange mottles, iron and manganese concretions (5 %) in light grey matrix	10 YR 6/4
	Increasing amount of yellow-orange mottles (30 %)	7.5 YR 6/4
	Dark brownish mottles	7.5 YR 4/6
	Dark brownish and orange mottles with vertical bleaching structures	7.5 YR 4/6 7.5 YR 6/4
	Very sandy uniform orange yellow layer Water table below 2 m (dry season 2012)	7.5 YR 6/4
Canal 3		10 YR 4/2
	Laomy, greyish yellow-brown dense topsoil	
	Slightly lighter greyish yellow-brown with very fine manganese and iron concretions (Ø 2 mm)	10 YR 5/2
	Greyish yellow-brown matrix with orange mottles	7.5 YR 8/3
	Boundary of infilling	7.5 YR 4/6
	Light yellow-orange matrix, abundant dark brownish mottles (50 %)	
	Orange matrix 10 YR 6/4 with vertical bleaching structures (10 YR 8/3)	10 YR 6/4 10 YR 8/3
	Very sandy almost homogenous orange layer Water table below 2 m (dry season 2012)	7.5 YR 6/6
Ridge 4	Silty, greyish brown topsoil, diffuse boundary	7.5 YR 6/2
	Silty, light brownish grey, hydromorphic mottling (5 %), manganese concretions (Ø 5 mm)	7.5 YR 7/2
	Sandy layer, hydromorphic mottling (50%) : small manganese concretions (Ø 5 mm), deformation by cattle steps (black dashed line)	
	Transition to light brownish-grey silty sediments, orange mottles (50%), diffuse boundary	7.5 YR 7/2 2.5 YR 6/8
	Increasing amount of clay and orange mottles (50%)	5.0 YR 6/8 5.0 YR 8/1
	Desiccation cracks (black dashed line), boundaries along cracks are filled with fine sand	5.0 YR 6/8
	Water table 100 cm (dry season 2013)	
Canal 4		7.5 YR 4/2
	Silty, brownish and organic rich topsoil	
	Silty, light brownish-grey matrix, manganese concretion (40 %), yellow-orange (7.5 YR 7/8) iron mottles (5 %)	7.5 YR 7/2
	Increasing amount of clay and hydromorphic orange mottles (50%)	5.0 YR 6/8 5.0 YR 8/1
	Silty clay, hydromorphic reddish-brown mottling 5 YR 4.8 > 50%	5.0 YR 4.8
	Water table 80 cm (dry season 2013)	

silt loam to silty clay loam. A palaeosol (Palaeosol II) was detected at a depth of 60–70 cm, recognisable by its darker brownish colour (Fig. 4).

– *Causeway (CW)*: the profile from the causeway differs in many aspects from the other profiles. The CW in general is much denser and was hard to excavate. The texture is loam (27–52 % sand and 8–15 % clay). Hydromorphic features are almost absent up to a depth of 50 cm, where density increases remarkably and there is a sharp colour change from light yellow-brown to darker brown. This change may be interpreted as the ancient surface upon which sediments were heaped up to construct the causeway. The bottom part of the profile,

starting at around 50 cm, is affected by hydromorphism, characterized by few manganese concretions and about 20 % of iron mottling. There are a striking amount of pieces of charcoal at a depth of 30–50 cm, which were almost completely absent in the other profiles.

4.2.2 Particle size distribution from grid sampling

Results from the grid sampling shows that most of the sediments contain a high percentage of silt, meanwhile the content of sand with respect to clay is very heterogeneous. The proportion of sand varies considerably from 2 to 41 % (compare with Fig. 5). The median particle size of each sample

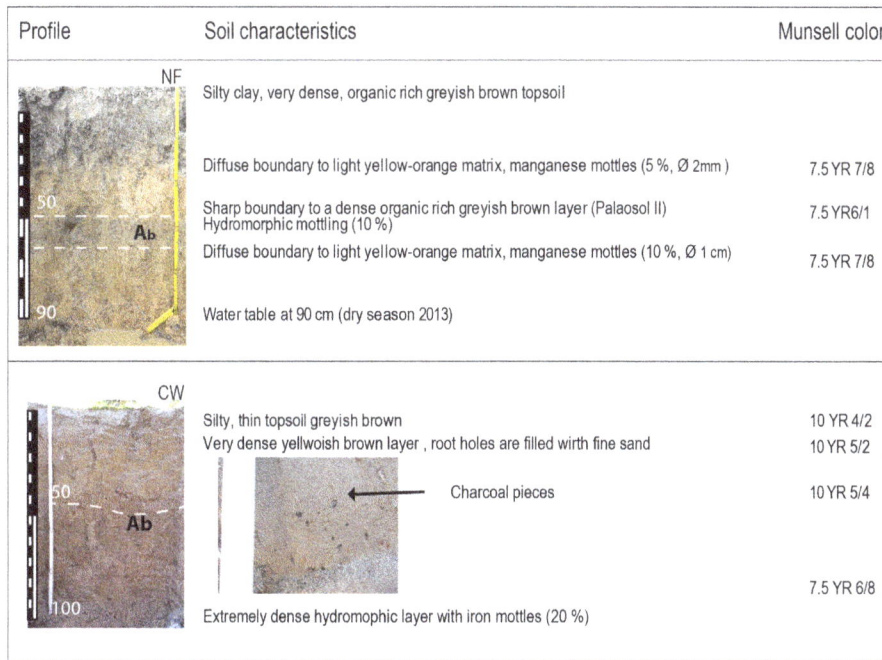

Figure 2. Profile description: Munsell colour signature is given for each layer, water table boundaries are illustrated as blue dashed lines, Ab = buried topsoil.

Table 2. AMS radiocarbon ages of charcoal and soil samples, given both as ^{14}C age BP and calibrated radiocarbon age in AD/BC format at two-sigma level. OSL ages are given as years before sampling (rounded to the next 5 years) and converted to AD/BC format to allow direct comparison with radiocarbon ages.

	Radiocarbon ages ^{14}C				
Profile	Palaeosol I Field 2	Causeway	Causeway	Ridge 2	Palaeosol II NF
Depth (cm)	270	60	60	80	65
Material	humin/no humates	charcoal	charcoal	charcoal	humin/humates
^{14}C age	6163 ± 41 BP	2590 ± 23 BP	2451 ± 30 BP	3139 ± 30 BP	7034 ± 31/2790 ± 28 BP
95.4 % (2σ) cal age ranges	5212–4940 BC	765–471 BC	566–398 BC	1438–1262 BC	5934–5775/976–816 BC
RAUPD	1	0.953	0.791	1	0.801/1
Lab Number	D_AMS-006318	BE-3265.1.1	BE-3266.1.1	D-AMS 002333	D-AMS 006330
Cal date	30 Nov 2015	5 Feb 2016	5 Feb 2016	5 Feb 2016	26 Jan 2016
	OSL ages				
Profile	Ridge 2	Ridge 2		Canal 2	Canal 2
Depth (cm)	40	140		36	60
OSL age	790 ± 70 (MAM)	5100 ± 410		635 ± 55 (MAM)	4510 ± 370
Converted to AD	AD 1150–1290	3500–2680 BC		AD 1320–1430	2870–2130 BC

(d50 µm) has been used to illustrate particle size distribution at each depth, as it has shown to best reflect the differences within the area (Fig. 6).

One south–north oriented sand structure which is partly buried by fine sediments is evident in column 13. Two additional sand structures, one in column 11 and the other in column 7, are visible at the border of the sampled area; they seem to have the same south–north orientation. The finest texture of silty clay loam can be found at both west and east ends, in columns 0–3 and 18–20 respectively, where no fields were built (Fig. 6).

For all the fields, down-profile particle size distribution was measured separately and the results are consistent with the results from the grid. The fields with a finer texture of silt loam (Fields 1 and 4) show a coarsening up profile, where in the upper 20 cm there is considerably more sand com-

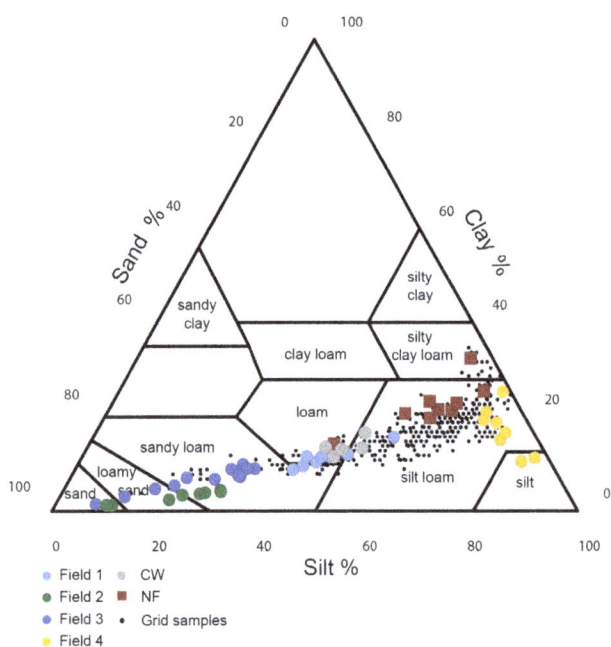

Figure 3. Soil texture triangle including all measured samples.

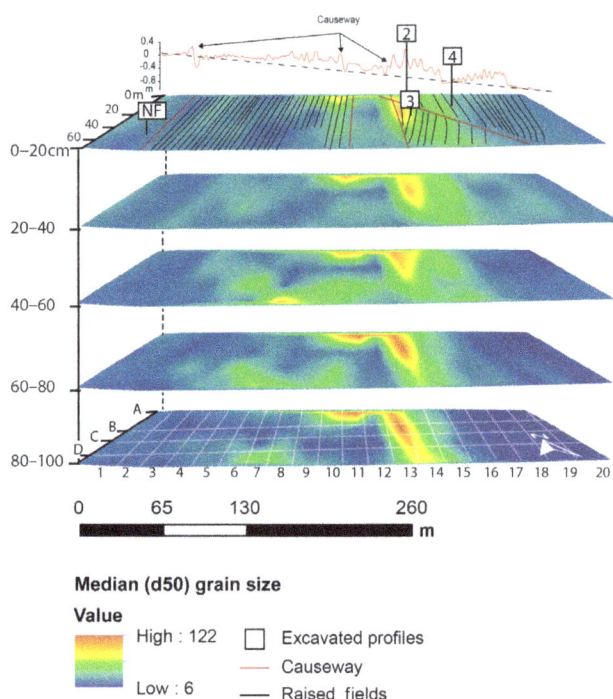

Figure 4. Interpolated grain size distribution of the sampled area using the median particle size of each sample (d50 µm) at five different depths. Top layer includes mapped fields and excavation sites. Bottom layer shows virtual grid, samples for grain size analysis were taken every 20 m along the east-west direction and every 16 m along the north-south direction.

pared to the bottom. In contrast, the profiles in the middle columns 10–13 show, on top of the sand structures, a fining upward sequence (Field 2 and 3). The highest fields are all built on the coarsest textures (row A, columns 8–15), while the lower smaller fields were mainly built on silty loam (row B–D, columns 3–18). It is striking that in the upper 60 cm of Field 3 the texture changes along its course towards the south-east, starting with sandy loam at the point of the excavation of the raised fields and ending up with silty loam towards the causeway.

Groups of parallel fields with similar heights were identified; these are always separated by a causeway (Fig. 7). The fields we excavated (Fields 1–4 in Fig. 7) were built on a slight slope. A topographic transect going from west to east reveals a downward trend, with a maximum difference of 90 cm (Figs. 6 and 7).

The highest points were measured on the ridge of Field 2 and the causeways (Fig. 6b). The difference between ridge and canal for most of the fields is around 25 cm, but in some fields it is up to 60 cm (Field 2). The DEM, which includes the higher and the lower fields, illustrates the difference in height and shows that the larger fields are, on average, 50 % taller than the smaller fields (Fig. 6c). The higher fields are much better preserved than the lower fields, because the latter have been partially destroyed by cattle trampling. There are several causeways; one major causeway going from the south-west to north-east, three perpendicular to it and one cutting all the others, going from the south-east to north-west. The latter separates the lower fields from the higher ones (Fig. 6c). The major causeway connects the area of the raised fields with a settlement area and the *lomas*, while the

other causeways are only linked to raised fields (compare with Fig. 3).

4.3 Geochemistry

The sediments elemental composition is poor in carbonates, with less than 0.1 % of CaO. The sediments are dominated by quartz (SiO_2), with a mean percentage of 76 % for Field 1, 85 % for Field 3, 70 % for Field 4 and 63 % for NF, followed by a mean percentage of aluminium (Al_2O_3) of 12, 7, 14 and 17 % respectively. For all the ridge profiles the more soluble elements (Na, K_2O, CaO and MgO as well as Fe_2O_3 and Al_2O_3 oxides) increase towards the bottom (Fig. 8). Some anomalies can be observed where Mn, P_2O_5 and Fe_2O_3 are more concentrated in specific layers. A clear enrichment of Mn always occurs at the top of all profiles, except for Canal 4 where Fe_2O_3 is accumulated instead. It is surprising that P_2O_5, which is always concentrated in the relatively upper part (20–45 cm) of the canals, in the fields the highest concentrations are found deeper in the profile (40–70 cm). This enrichment of PO_4 is clearly accompanied by higher Fe/Al and Fe/Ti, indicating a higher level of ferruginisation (relative accumulation of Iron) (McQueen, 2006) (Fig. 7 and Table S3). In NF there is no important accumulation of P_2O_5, however, it does increase towards the bottom. P_2O_5 in the

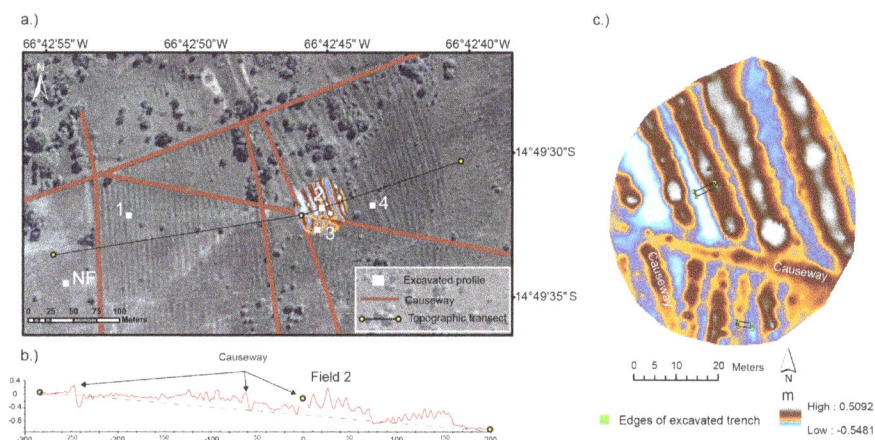

Figure 5. (a) Study area including causeways and excavation sites. **(b)** Topographic transect going from west to east. **(c)** Digital elevation model present-day morphology covering Field 2 and 3.

NF Profile is slightly lower, very high values of CaO and Na have been detected in Palaeosol II, at a depth of 60–70 cm (Fig. 6).

Further soil chemical properties (C_{org}, N, CEC_{eff}, pH, Bs, P_{av}) describing soil fertility are summarized in Table 1 and partly illustrated in Fig. 9. In all profiles C_{org} decreases with depth. Values in the ridges range from 0.74 to 0.01 % towards the bottom and in the canals from 1.13 to 0.01 %. In the ridges C_{org} decreases smoothly, whereas in the canals an abrupt change occurs always at the boundary of the in-filling. An exception is profile NF, where at a depth of 60–70 cm there is an increase of C_{org} that reaches 0.34 % due to the presence of the palaeosol (Palaeosol II). In general C_{org} values for the profile NF are high, going from 1.31 to 0.21 %. In all profiles the C/N values in the first centimetres range from 7.7 to 10.57, which are typical values for the tropics, where organic matter is mineralized fast (Schachtschabel et al., 2002). The exchangeable cation (CEC_{eff}) and pH for each profile are given in Table 1. Considerable differences are evident: according to the criteria of Hazelton and Murphy (2007), CEC_{eff} is very low in Fields 2 and 3, low in Field 1 and low to moderate in Field 4 and NF. Base saturation is very low for Fields 2 and 3, whereas levels are more favourable in 1, 4 and NF, with moderate (40–60 %) to very high (80–100 %) (Table 1 and Fig. 9).

Available phosphorous (P_{av}) is generally low, but amounts are highly variable going from below limit of detection to 23 ppm (Table 1 and Fig. 9). According to the criteria of Hazelton and Murphy (2007), P_{av} values in the upper parts of the ridges can be classified as low (< 5 ppm), but become moderate (10–17 ppm) towards the bottom. There are some very high values in Field 1, at the bottom of the ridge and the canal (20 ppm/26 ppm) and in the canal belonging to Field 3 (21 ppm) at a depth of 55-65 cm. Here, the available phosphorous (P_{av}) accounts for up to 1 % of the total phosphorous, whereas in the other samples P_{av} is < 0.5 %. To assess

the relationship of the parameters describing the fertility of the sediments a pearson correlation has been performed for all samples (Fig. 10).

CEC_{eff} and Bs values correlate with particle size, whereas P_{av} does not (Fig. 10). In general the P_{av} values are strongly associated to the total amount of P_2O_5 which is accumulated in a specific layer (compare Table 1 with Fig. 8). The pH is acidic in all profiles and is correlated to the total amount of Ca. This can be clearly seen in Field 4 and NF, where the pH increases together with CaO. The lowest PH values, which do not exceed 4.2, can be found in the sandiest Fields 2 and 3, with low Bs and a high percentage of exchangeable Al. In these fields the Ca / Al ratio is < 1.0 and aluminium toxicity can be a problem for plants (Cronan and Grigal, 1995). In contrast, in Field 4, which is composed of much finer sediments (silt-silt loam), the pH values reach 5.2, Bs is high, with low to no exchangeable Al and the Ca / Al ratio is > 1.0.

4.4 Chronological framework

A total of four OSL and five radiocarbon ages have been obtained (Table 2, Fig. 11). Two OSL ages were taken from the ridge profile 2 (40 and 140 cm) and two from the adjoining canal (36 and 60 cm). Samples for radiocarbon dating were taken from Palaeosol I, at a depth of 270 cm below Field 2 and from Palaeosol II, below NF, at a depth of 65 cm. The three remaining radiocarbon ages are from charcoal pieces: two extracted from the excavated CW and one from the ridge of Field 2. The oldest age, 5212–4940 cal BC, was found in the Palaeosol I, beneath the canal of Field 2 (270 cm depth). In this case only the humin fraction could be dated, as humates were missing. In contrast, both fractions could be dated in the case of Palaeosol II in the NF profile (65 cm). The difference between the age of the humins (5934–5775 cal BC) and the much younger age of the humates (976–816 cal BC) is significant and might point to-

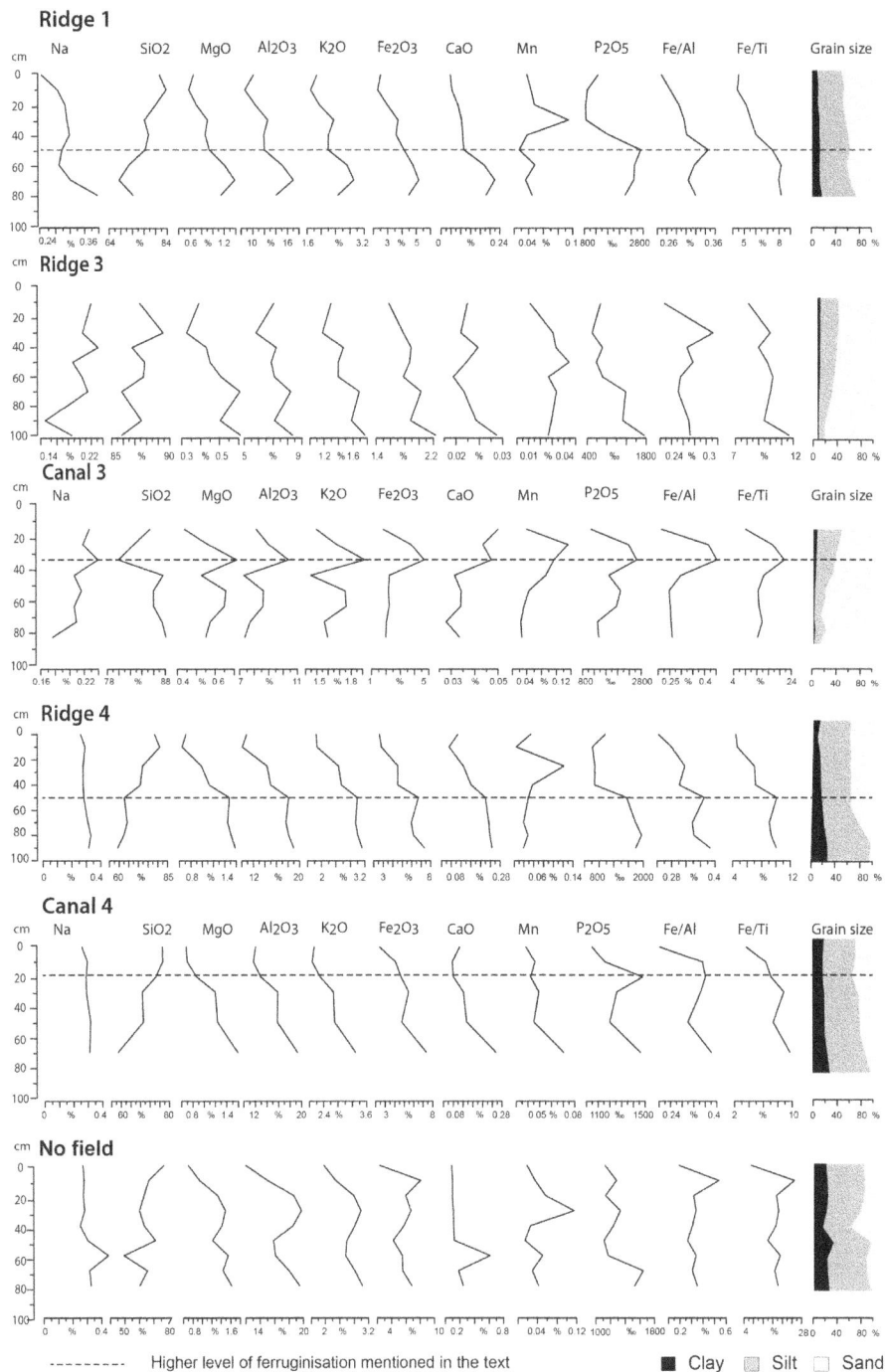

Figure 6. Down-profile variations of selected elements and grain size.

wards a contamination (Walker, 2005). The OSL ages are consistent with the stratigraphy, the two basal ages are 3500–2680 BC for the ridge and 2870–2130 BC for the canal and the top ages are AD 1150–1290 and AD 1320–1430, respectively. The radiocarbon ages of two charcoal pieces extracted from the CW were 797–751 BC and 743–687 cal BC and from Field 2 1438–1262 BC.

5 Discussion

All the fields mapped in the vicinity of San Borja show two essential characteristics which could help explain why they were built and their shape and size.

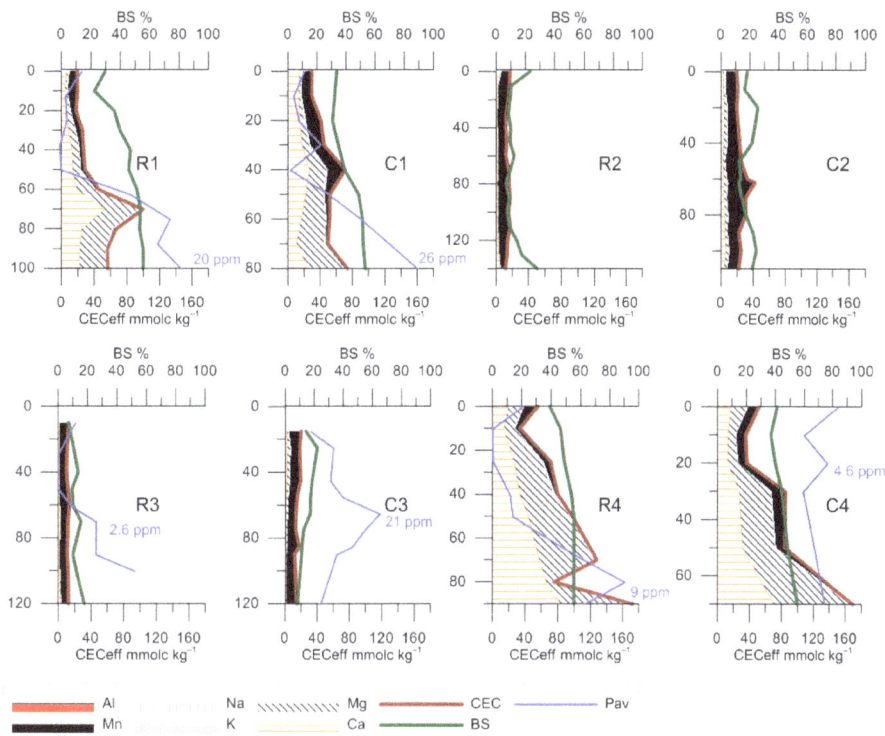

Figure 7. Down-profile variations of available cations (CEC), base saturation (BS) and phosphorous (Pav).

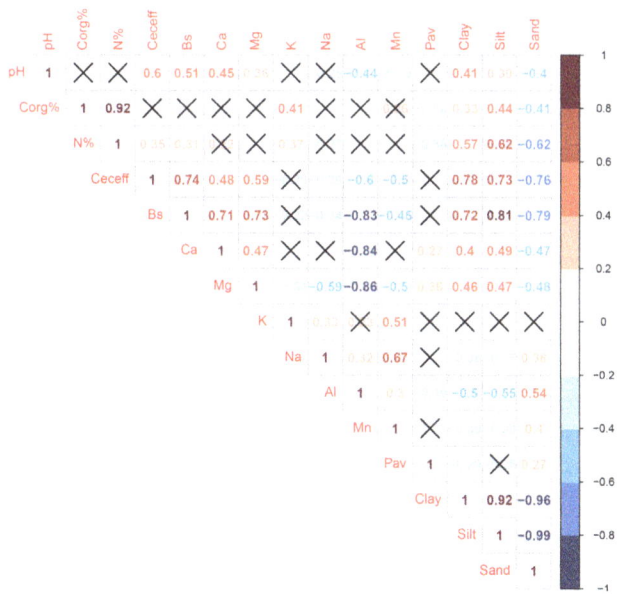

Figure 8. Pearson correlation matrix. Values with significance level $p > 0.05$ are crossed out.

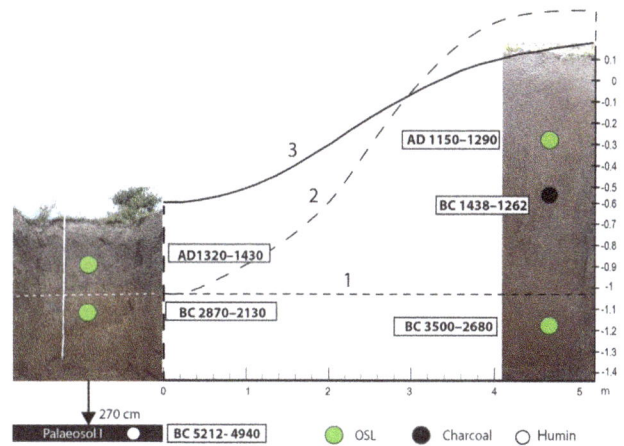

Figure 9. Reconstruction of Field 2 illustrating the three suggested different phases: (1) ancient surface before the construction of the fields (short dashed line), (2) original field profile (long dashed line) and (3) present-day profile of abandoned field (solid line).

Firstly, all the fields were constructed on fluvial deposits, which are naturally higher than the pampa, and are made of relatively coarse, hence better drained, sediments. Secondly, while field height seems to be related to sediment charac-teristics, distribution pattern seems to depend on the natural landscapes morphology.

5.1   Main characteristics of raised fields versus the landscape

Existing studies of raised fields in the LM have similarly shown that the majority of raised fields were built mainly on fluvial levees and on the naturally well-drained areas, often on silty to sandy sediments, with the aim of improving drainage (Walker, 2004; Lombardo et al., 2011a; Rodrigues et al., 2015). It seems that pre-Columbian people took advantage of the natural morphology of the rivers and built raised fields on the levees or point bars, where the coarser sandy sediments are deposited. Areas where surface sediments were too fine were avoided (Fig. 6). It should be noted that no connection exists between the height of the raised fields and the general topography, in fact, the lowest fields (for example Field 4) are located in the lowest lying areas. The soil properties of the raised fields studied differ considerably, probably due to the heterogeneity of the sediments in the area. Frequent changes in soil texture may be explained by the high frequency of crevasse splays and avulsions of the Maniqui River (Lombardo, 2016). The analysis of satellite image and particle size distribution shows that the landscape history of the area is complex, resulting from a combination of several palaeo-river generations of different dimensions and the depositional behaviour of the meandering rivers. On such diverse landscapes sediment properties can therefore vary naturally within the range of metres, in this case resulting in the great variability of sediments comprising the excavated fields.

5.2   Age and morphology of raised fields

5.2.1   OSL

OSL dating of Field 2 shows three major time phases (Fig. 11): phase 1 comprises the oldest ages, corresponding to the deposition of the fluvial sediments, phase 2 indicates the period during which the field was built/used and phase 3 marks the time when the raised field was abandoned.

The top layers include an age of AD 1150–1290 for the ridge of Field 2 and AD 1320–1430 for the adjoining Canal (Fig. 10, Table 2). The OSL age refers to the moment in which quartz grains were buried, therefore the age AD 1150–1290 probably indicates the moment in which Field 2 was built. In Canal 2, the age AD 1320–1430 marks the time of the canal infilling. This implies that there was no further digging of the canal afterwards, suggesting the abandonment of the fields. The estimated time of abandonment, AD 1320–1430, is consistent with the abandonment of the raised fields studied in Bermeo, around AD 1400 (Rodrigues et al., 2015). The exact time of construction and abandonment of the fields, however, cannot be deduced from these ages, as the fields could have been elevated more than once (Rodrigues et al., 2015). Furthermore, the arrangement of fields in groups separated by causeway indicates that fields were most probably built separately and could therefore have

different ages. However, the OSL ages show that the raised fields were in use during a rather short time span, about 100–200 years at the beginning of the 2nd millennium. This is consistent with the raised fields in Bermeo (Rodrigues et al., 2015), where fields were used for short periods, and with the relatively brief occupation of settlements associated to raised fields in the northern LM close to Santa Ana (Walker, 2004). The age below the infilling of Canal 2, 2870–2130 BC, is consistent with the age obtained for the base of the ridge profile (3500–2680 BC), indicating that the sediments below 60 cm in Canal 2 were not reworked for the construction of the raised fields.

The suggested original depth of Canal 2 is consistent with these results and it may be reasonably assumed that the canal was originally 50 cm deeper (Fig. 11). Hence, the difference between the canal and the ridge was at least 150 cm, without taking into account the eroded material from the ridge (Fig. 11). Most probably all fields were much steeper, as in the case of Bermeo (Rodrigues et al., 2015). Compared to the platform fields in the northern part of the Llanos de Moxos (Walker, 2004; Lombardo, 2010), the large fields of San Borja are about three times higher. In addition, the raised fields in this area are associated to causeways which most probably were needed to reach the fields during high water season. It is important to note that causeways do not embank the fields, providing open runoff. There is no evidence that causeways were built to hold back water in the raised field's area or to extend the time of flooding in the area, as has been proposed for canals in the Apere Region (Erickson and Walker, 2009). During the year 2012, the canals adjoining the causeways were already completely dried out in July. Most causeways found in the adjacent settlement area, away from raised fields, must have served as a form of transportation and communication between settlements during the wet season (Erickson and Walker, 2009), suggesting that even the most elevated parts got flooded (Fig. 3). Because of its location close to the Andean Piedmont, San Borja gets on average 400 mm more precipitation compared to the northern LM (Hijmanns et al., 2005) and ground water levels during the dry season are 2 to 3 m below the surface, up to 3 times higher compared to the northern LM (Hanagarth, 1993). Because of this, during the rainy season sediments get saturated quickly. This combination makes flooding in the southern LM much more pronounced compared to the northern part of the LM and may be an important reasons why fields in the southern part of the LM are commonly much higher.

5.2.2   Radiocarbon dating [14]C

Charcoal derived from the CW (765–471 BC; 566–998 BC) and from the ridge profile of Field 2 (1438–1262 BC) are much older than the construction of the fields, probably predating the occupation of the area. While in the CW charcoal was plentiful, in the area of raised fields only one single piece has been found. In comparison, charcoal in the raised fields of Bermeo was more abundant, found in the canals and specific layers in the elevated fields (Rodrigues et al., 2015). In Bermeo the raised fields are under dense forest and it has been shown that the fields were in use during at least two different periods. This suggests that fire could have been used to clear the forest between periods of field use. The use of fire for raised field management in the northern LM has been also suggested by Whitney et al. (2014) and Erickson and Balée (2006). The use of fire however stands in contrast to the fact that charcoal was not directly found in the fields in San Borja and there is no evidence of the fields having been used during several different periods and managed with fire.

The age of the Palaeosol I (5212–4940 cal BC) below Field 2 is consistent with the age of the sediments that cover it (3500–2680 BC) and with the general chronology. There is an important difference between the ages from the two fractions of Palaeosol II in the NF (5934–5775 cal BC for the humin fraction and 976–816 cal BC humate fraction), this might point towards significant contamination. In theory, the humin fraction (residues) is considered to be more stable and represents the oldest age, whereas the humate (humic acid) fraction gives crucial information about the degree of contamination and, consequently, its reliability (Pessenda et al., 2001). The fact that the sample in the Palaeosol II of the NF was taken relatively near the surface (at a depth of 65 cm) could point towards contamination of the humate fraction with modern carbon from the surface (Walker, 2005). In addition, the fact that the Palaeosol I is chronologically consistent with the OSL ages above it, suggests that the age of the humine fraction might be more reliable. Taking the humin ages of the two palaesols does suggest that these two could have belonged to the same palaeosurface. Thus, the topography was much steeper than today and the depressions were later filled up with sediments. Palaeosol I could have been covered by a crevasse splay, while Palaeosol II could have been covered more slowly, with fine flood sediments from the surrounding area or the overflow of a distal river.

5.3 Local soil properties versus raised fields

Hydromorphic characteristics present in all raised fields show that they are highly influenced by high water tables. The average depth of the water table can be derived from the Fe / Al ratio (McQueen, 2006), showing the in situ accumulation of iron which in some profiles is clearly expressed with a peak. Field observations show that for most profiles manganese tends to accumulate some centimetres above the iron oxides. This is common in hydromorphic soils, because of the different solubility of iron and manganese (Lindbo et al., 2010). The depth of these oxides and the Fe/Al ratio suggest that the depth of the modern water table in the ridge profiles of Fields 1 and 2 normally oscillates between 40 and 50 cm below the surface, while in the lower Fields 3 and 4 the depth of the water table is between 30 and 35 cm respectively. The hydromorphic features formed at the time of field construc-

tion have probably been erased, as hydromorphic features are forming continuously and the fields were not used for a very long period of time. As already mentioned, no connection exists between the height of the raised fields and the topography. For raised fields studied in the northern part of the LM it has been shown that these were built higher on finer sediments, as coarse sediments provided better drainage (Rodrigues et al., 2016). Surprisingly, in San Borja the opposite can be observed: smaller fields are built on finer sediments and the higher fields are built on the coarser sediments. There are several possible explanations for this apparent contradiction.

When comparing the relative depth of the water table below the ridge in the high fields vs. the low fields, we can see that there is a small difference of about 10–20 cm. Surprisingly, the water table below the ridge in Field 3, which has a similar height but much coarser sediments than Field 4, is almost at the same depth as in Field 4. This suggests that the drainage is not significantly better in the sandy area. This might be explained by the fact that the regional water table in San Borja is very close to the surface (Hanagarth, 1993; Miguez-Macho and Fan, 2012), hindering vertical drainage. Furthermore, as seen in Fig. 6, the sandy areas on which the fields were built seem to be enclosed by fine sediments, hence also hindering lateral water movement.

Besides the hydrology, another important factor determining the height of the fields seems to be the sedimentary characteristics of each field. In the case of the lower/smaller fields, sediments become finer towards the bottom, with a high percentage of clay in the lower layers (Fig. 4). The sediments in these deeper layers have a similar clay content (up to 28 %) to the sediments in the "no fields area" (NF). Such soils are generally avoided in agriculture because of their poor physical properties (e.g. low permeability and poor soil structure) and workability. It is therefore conceivable that due to the limited availability of coarser sediments in these deeper layers only the uppermost layers were used to raise the fields. In contrast, in the area of the large fields this impediment does not exist and fields could be built higher. In addition, coarser sediments are much easier to work, which could also explain why these fields are higher.

Taking into account the very different soil properties and fertility status of the sediments, people could have brought the more sandy sediments from the large fields area in order to improve the soil conditions of the area with clay-rich sediments, and vice versa. However, there is no evidence of this. On the contrary, as we can see in Field 3, the northern part of the field is composed of sandy loams, whilst further south the field is formed by silty loams and there is no evidence of attempting to improve the field's soil structure by mixing the finer and coarser sediments (Fig. 6).

In general, the geochemical and physical properties of the soils here are similar to those of other soils studied along the Maniqui (Guèze et al., 2013). This area has unfertile, acidic sandy-loamy soils. The exceptional high pH of Palaeosol II in profile NF is a result of the considerably higher amount of Ca and Na within the same layer (Fig. 6 and Table 1). Such saline soils, with accumulation of carbonates in the subsoil, have been described by other authors (Boixadera et al., 2003; Hanagarth, 1993). The Ca and Na could have been supplied through capillary rise (Boixadera et al., 2003). Similarly, as shown by Boixadera et al. (2003), the present profile is poorly drained and clay-rich. The layer just above Palaeosol II, with considerably more clay, could further prevent the outwash of the bases. Whether the Ca and Na come from capillary rise or are relict features from Palaeosol II, is beyond the scope of this study.

In general there is a clear relationship between particle size and CEC_{eff}. As expected, soil fertility in Fields 2 and 3, which have been built on sand, is extremely low. Due to their limited capacity to retain cations and their high water conductivity, sandy soils tend to quickly leach valuable nutrients like calcium and magnesium and, as a result, plant growth might be hindered (Schachtschabel et al., 2002). This is reflected in the elemental composition of Ridge 3, where the more soluble elements CaO, Na, K_2O and MgO are washed out from the upper parts of the profiles, showing higher values at the bottom (Fig. 8). The soil's acidity further increases the amount of exchangeable Al^{3+} and Mn, which can be toxic for most plants (Jones, 2012; Cronan and Grigal, 1995). Fields 1 and 4 have much finer sediments and are able to retain more cations, making them more fertile. Besides CEC_{eff} $P_{(av)}$ is one of the most important limiting nutrients for crops (Fageria et al., 2011). In general, $P_{(av)}$ values in all profiles are very low to low, with less than 5–10 ppm, constituting < 0.5 % of the total amount of phosphorous. Similar values (4.5 to 10 ppm), comprising the same region, were reported by Guèze et al. (2013). Low $P_{(av)}$ values seem to be common in acidic soils as phosphorous normally occurs bounded to other elements and is therefore not directly available to plants (Schachtschabel et al., 2002). The results, however, also show some quite high values of $P_{av} > 20$ ppm, which are related to the higher levels of total amount of P. The higher P values are related to the Fe/Al, which coincides with the average depth of the water table. Similar positive links between P_2O_5 and Fe/Al have been reported by other studies (Lopez, 2004; Lopez et al., 2006; Huang et al., 2005). The overall high values of total PO_4 can be explained by the relatively young age of the sediments and the fact that sediments coming from the Andes are high in phosphorous. Up to 2200 ppm (for P_2O_5) and up to 4300 ppm (for CaO) have been reported for river sediments coming from the Andes (Guyot, 1992). While CaO, Mn, MgO are leached relatively fast, P in acidic conditions is stabilised in the soil by iron and aluminium (Schachtschabel et al., 2002). Due to the seasonal floods and droughts, oxides (e.g. P_2O_5 and Fe_2O_3), along with other elements, are redistributed and accumulated in the soil (Scott and Pain, 2009). It seems, therefore, unlikely that the accumulation of P in specific layers is the result of an anthropogenic enrichment of the soil due to practices of intensive manuring, as reported for other anthropogenic soils (Holliday, 2004; Costa et al., 2013; Glaser and Woods, 2004).

In the ridge profile of the coarser Field 3, the leaching of cations, geochemical changes and enhanced hydromorphism all suggests that, in the long run, the construction of raised fields could accelerate soil weathering. These processes should be taken into account when considering raised fields as a model for sustainable agriculture today.

It has been suggested that manure or muck grown in the canal could have been used to fertilise the fields, allowing continuous production without the need of fallow periods (Erickson, 1994; Lee, 1997; Barba, 2003; Saavedra, 2006). Nevertheless, in order to produce green manure the canals would have had to retain water during the dry season, which is not the case in the San Borja fields nor in other fields studied in the LM (Lombardo et al., 2011a; Rodrigues et al., 2015). It is possible, however, that earth from the canals could have been reused to raise the fields. If we compare the soil properties from the infilling of the canal and the raised field, the former has slightly higher content of organic matter, P_{av} and CEC_{eff} than the latter. By adding sediment from the canal onto the elevated bed they could have improved the field's soil fertility. However, it has to be considered that the fertility of the soil from the canal that we see today is the result of at least 500 years of accumulation of sediments and organic matter. If, in pre-Columbian times, the fields had been in use continuously, there would not have been sufficient fertile sediments in the canals left to fertilise the fields after one growing season and manure would have had to be brought from somewhere else.

Up till now raised fields studied in the LM have not shown evidence of intensive manuring but rather of extensive agricultural practices (Lombardo, 2010; Lombardo et al., 2011a; Rodrigues et al., 2015' 2016). However comparing the fertility status of the raised fields studied here with the ones in the northern LM, the results show that the soils in San Borja are considerably more fertile.

The soils in the northern LM are much older and much more weathered (Rodrigues et al., 2015). On the other hand, density of fields in the northern LM is much higher than in San Borja. This high density of raised fields has been interpreted as the result of intermittent use by small groups of mobile people which were shifting their fields over a period of hundreds of years (Rodrigues et al., 2016).

Even though the fertility status of the soils in San Borja is better, raised fields here would similarly have needed fallow periods, especially those built on the sandy sediments. It is probable that the fields built on the more fertile soils could have been cultivated for longer periods, with shorter fallows. If we assume a similar scenario of small groups of mobile people in the present study area, the better soil quality could explain the much lower density of raised fields here compared to the northern LM.

Another reason for the lower density could be related to the widespread availability of more elevated well-drained areas. These areas normally do not get flooded and similarly, as shown for the region of Bermeo, raised fields could have been constructed to overcome periods of more severe and frequent flooding (Lombardo et al., 2011a; Rodrigues et al., 2015). On the contrary, the northern LM is affected by ponding water on a regular basis, because of relatively impermeable soils, and raised fields were needed to improve the drainage (Rodrigues et al., 2016). Consequently, the density of fields in the different parts of the LM could also be related to the frequency of use, with raised fields in the southern LM used only during periods of extreme events while in the northern part fields were used annually.

However, as almost no data exist about the size and timing of pre-Columbian occupations such scenarios are difficult to prove.

Furthermore, as argued by McKey and Rostain (2014), the raised fields may have possibly been complemented by other subsistence systems which also could have been different for each region.

Nevertheless, information about soil properties are important in order to understand the development of societies (McNeill and Winiwarter, 2006), differences in agricultural strategies and in distribution and density of people living in the LM. Up till now, raised fields studied in the LM all show that fields were constructed with the main purpose of drainage. The fact that some of the raised fields studied here were constructed on the highly unfertile sands supports the idea that drainage was the first priority. There is no clear evidence suggesting that raised fields were more productive compared with similar soils which are naturally drained. Hence, there is no indication which suggests that the construction of raised fields was a highly productive strategy which could sustain dense populations.

As already proposed with the first description of raised fields in 1916 by Erland Nordenskiöld (Denevan, 2009), they must have played a crucial role in protecting the crops from the floods. The abundant causeways, even on the more elevated area, further suggest that in San Borja flooding used to be a frequent problem.

As similarly suggested for the raised fields in Bermeo, the period of use in San Borja coincides with a period of higher ENSO activity, which has been reported to be an important factor responsible for extreme floods and droughts during the past 2500 years in South America (Meggers, 1994; Markgraf and Díaz, 2000; Moy et al., 2002; Rein et al., 2005). Some major floods in the LM have been associated with the negative ENSO phase La Niña, where rainfall is above normal in the Basin (Aalto et al., 2003). Moy et al. (2002) reported higher frequency of extreme ENSO events occurring between 1000 and 2000 years ago, with its maximum around AD 800, which coincides with the time when fields were in use. However, as reported for the extreme event in 2014, severe flooding can as well occur in absence of ENSO, as a result of tropical and subtropical changes in South Atlantic Sea Surface Temperature (SST) (Espinoza et al., 2014). To-

day losses of harvest due to flooding are frequently reported in the LM (UNDP, 2011).

6 Conclusion

We analyse raised fields of different sizes which were built in an area, near San Borja, with a complex fluvial history. Different generations of palaeo rivers, partly overlapping each other, coexist in the area, resulting in a heterogeneous depositional environment. This is reflected in the great variability of sediment particle size of the excavated raised fields. The results show that differences in field size and height are the result of an adaptation to this heterogeneous depositional environment. The dimension of the fields is related to particle size. Only coarse, silty to sandy sediments were used for the construction of the raised fields. The height of the fields depends on how deep the coarse sediments are: fields are relatively small where the coarse sediments are limited to the surface, whilst in areas where the subsoil was also made of coarse sediments these could be used to build larger fields. Areas with exclusively fine clay rich sediments were not used for the construction of raised fields. Raised fields were built by piling up the sediments taken from the excavation of the adjacent canal; there is no evidence of other agricultural strategies such as mixing of sediments or intensive manuring. Geochemical changes along the stratigraphic profiles show that the construction of fields might accelerate the weathering process in the long term, calling into question the idea of reintroducing raised fields as a very productive model of sustainable agriculture for today. The raised fields in the area are always associated to causeways. There is no evidence that causeways were built to manage the floodwaters; they were more likely used to reach the fields and to connect settlement areas. Although the construction of raised fields did not directly improve soil fertility, leading to higher productivity, it was of major importance to protect crops during severe flooding.

Author contributions. Leonor Rodrigues and Umberto Lombardo conceived and designed the study. Leonor Rodrigues, Umberto Lombardo, Perrine Huber and Heinz Veit performed field work. Leonor Rodrigues, Perrine Huber and Sandra Mohr carried out laboratory analyses. Mareike Trauerstein conducted OSL measurements. Heinz Veit secured funding. Leonor Rodrigues prepared the manuscript with contributions from all co-authors.

Acknowledgements. The present study has been funded by the Swiss National Science Foundation (SNSF), grant no. SNF 200020-141277/1, and performed under authorisation N_017/2012 issued by the Unidad de Arqueología y Museos (UDAM) del Estado Plurinacional de Bolivia. We thank M. R. Michel López from the Ministerio de Culturas and our Bolivian counterpart J. M. Capriles for their support. A special thanks to the owners of the farms El Progresso and Campo España and their workers for their logistical support in the field and for allowing us free access to their land. Fieldwork assistance by B. Vogt, L. M. Salazar and C. Welker is gratefully acknowledged. We thank D. Fischer for technical support in the laboratory. X-ray fluorescence spectroscopy (XRF) was measured at the Geological Institute of the University of Fribourg. A special thanks to E. Canal for improvement of the manuscript.

Edited by: O. Evrard

References

Aalto, R., Maurice-Bourgoin, L., Dunne, T., Montgomery, D. R., Nittrouer, C. A., and Guyot, J.-L.: Episodic sediment accumulation on Amazonian flood plains influenced by El Niño/Southern Oscillation, Nature, 425, 493–497, doi:10.1038/nature01990, 2003.

Bandy, M. S.: Energetic efficiency and political expediency in Titicaca Basin raised field agriculture, J. Anthropolog. Archaeol., 24, 271–296, 2005.

Barba, J.: Campos elevados, in: Moxos: una limnocultura, Gràfiques 92, CEAM, Barcelona, 89–92, 2003.

Baveye, P. C.: Comment on "Ecological engineers ahead of their time: The functioning of pre-Columbian raised-field agriculture and its potential contributions to sustainability today" by Delphine Renard et al., Ecol. Eng., 52, 224–227, 2013.

Boixadera, J., Poch, R. M., García-González, M. T., and Vizcayno, C.: Hydromorphic and clay-related processes in soils from the Llanos de Moxos (northern Bolivia), Catena, 54, 403–424, doi:10.1016/s0341-8162(03)00134-6, 2003.

Bruno, M.: Carbonized Plant Remains from Loma Salvatierra, Department of Beni, Bolivia, Z. Arch. Außereurop. Kult., 3, 151–206, 2010.

Carson, J. F., Watling, J., Mayle, F. E., Whitney, B. S., Iriarte, J., Prümers, H., and Soto, J. D.: Pre-Columbian land use in the ring-ditch region of the Bolivian Amazon, Holocene, 25, 1285–1300, doi:10.1177/0959683615581204, 2015.

Chapin, M.: The seduction of models, Grassroots Dev., 12, 8–17, 1988.

Charlton, R.: Fundamentals of fluvial geomorphology, Routledge, New York, 2007.

Cornell, R. M. and Schwertmann, U.: The iron oxides: Structure, properties, reactions, occurences and uses, 2., completely rev. and extended ed., Vol. XXXIX, Wiley-VCH, Weinheim, 664 pp., 2003.

Costa, J. A., Costa, M. L. D., and Kern, D. C.: Analysis of the spatial distribution of geochemical signatures for the identification of prehistoric settlement patterns in ADE and TMA sites in the lower Amazon Basin, J. Archaeol. Sci., 40, 2771–2782, doi:10.1016/j.jas.2012.12.027, 2013.

Cronan, C. S. and Grigal, D. F.: Use of calcium/aluminum ratios as indicators of stress in forest ecosystems, J. Environ. Qual., 24, 209–226, 1995.

Denevan, W. M.: Cultivated landscapes of native Amazonia and the Andes, Oxford University Press, Oxford, 2001.

Denevan, W. M.: Indian Adaptations in Flooded Regions of South America, J. Lat. Am. Geogr., 8, 209–224, 2009.

Dickau, R., Bruno, M. C., Iriarte, J., Prümers, H., Betancourt, C. J., Holst, I., and Mayle, F. E.: Diversity of cultivars and other plant resources used at habitation sites in the Llanos de Mojos, Beni, Bolivia: evidence from macrobotanical remains, starch grains, and phytoliths, J. Archaeol. Sci., 39, 357–370, doi:10.1016/j.jas.2011.09.021, 2012.

Dillehay, T. D., Quivira, M. P., Bonzani, R., Silva, C., Wallner, J., and Le Quesne, C.: Cultivated wetlands and emerging complexity in south-central Chile and long distance effects of climate change, Antiquity, 81, 949–960, 2007.

Dumont, J. F. and Fournier, M.: Geodynamic environment of Quaternary morphostructures of the subandean foreland basins of Peru and Bolivia: Characteristics and study methods, Quatern. Int., 21, 129–142, doi:10.1016/1040-6182(94)90027-2, 1994.

Erickson, C. L.: Sistemas agrícolas prehispánicas en los Llanos de Mojos, América Indígena, 40, 731-755, 1980.

Erickson, C. L.: Raised field agriculture in the Lake Titicaca Basin, Expedition, 30, 8–16, 1988.

Erickson, C. L.: Prehistoric landscape management in the Andean highlands: Raised field agriculture and its environmental impact, Popul. Environ., 13, 285–300, doi:10.1007/BF01271028, 1992.

Erickson, C. L.: Raised fields as a sustainable Agricultural System from Amazonia, Recovery of indigenous technology and resources in Bolivia – XVIII International Congress of the Latin American Studies Association, Atlanta, 1994.

Erickson, C. L.: Archaeological methods for the study of ancient landscapes of the Llanos de Mojos in the Bolivian Amazon, Cambridge University Press, Cambridge, 1995.

Erickson, C. L.: Lomas de ocupación en los Llanos de Moxos, in: Arqueologia de las tierras bajas, edited by: Coirolo, D., Bracco-Boksar, A., and Bracco-Boksar, R., Comision Nacional de Arqueologia, Montevideo, Uruguay, 207–226, 2000.

Erickson, C. L.: Agricultural landscapes as world heritage: Raised field agriculture in Bolivia, Peru, in: Managing Change: Sustainable Approaches to the Conservation of the Built Environment, edited by: Teutonico, J. M. and Matero, F., Getty Conservation Institute, Los Angeles, 181–204, 2003a.

Erickson, C. Historical ecology and future explorations, in: Amazonian Dark Earths – Origin, Properties, and Management, edited by: Lehmann, J., Kern, D., Glaser, B., and Woods, W., Kluwer Academic Publ., Dordrecht, Holland, 455–500, 2003b.

Erickson, C. L.: Amazonia: the historical ecology of a domesticated landscape, in: Handbook of South American archaeology, edited by: Silverman, H. and Isbell, W. H., Springer, Berlin, 157–183, 2008.

Erickson, C. L. and Balée, W.: The historical ecology of a complex landscape in Bolivia, in: Time and complexity in historical ecology, edited by: Balée, W., and Erickson, C. L., Columbia University Press, New York, 2006.

Erickson, C. L. and Walker, J. H.: Precolumbian causeways and canals as landesque capital, in: Landscapes of movement, edited by: Snead, J. E., Erickson, C. L., and Darling, J. A., University of Pennsylvania Museum of Archaeology and Anthropology, Philadelphia, 2009.

Espinoza, J. C., Ronchail, J., Lengaigne, M., Quispe, N., Silva, Y., Bettolli, M. L., Avalos, G., and Llacza, A: Revisiting wintertime cold air intrusions at the east of the Andes: propagating features from subtropical Argentina to Peruvian Amazon and relationship with large-scale circulation patterns, Clim. Dynam., 41, 1983–2002, 2013.

Espinoza, J. C., Marengo, J. A., Ronchail, J., Carpio, J. M., Flores, L. N., and Guyot, J. L.: The extreme 2014 flood in south-western Amazon basin: the role of tropical-subtropical South Atlantic SST gradient, Environ. Res. Lett., 9, 124007, doi:10.1088/1748-9326/9/12/124007, 2014.

Fageria, N. K., Baligar, V. C., and Jones, C. A.: Growth and Mineral Nutrition of Field Crops, CRC Press, Taylor and Francis Group, Boca Raton, Forida, 2011.

Galbraith, R. F., Roberts, R. G., Laslett, G. M., Yoshida, H., and Olley, J. M.: Optical dating of single and multiple grains of quartz from jinmium rock shelter, northern Australia, part 1, Experimental design and statistical models, Archaeometry, 41, 339–364, 1999.

Garreaud, R. D., Vuille, M., Compagnucci, R., and Marengo, J.: Present-day south american climate, Palaeogeogr. Palaeocl., 281, 180–195, 2009.

Glaser, B. and Woods, W. I.: Amazonian dark earths: explorations in space and time, Springer, Berlin, Germany, 2004.

Gliessmann, S. R., Turner II, B. L., Rosado May, F. J., and Amador Alarcîn, M. F.: Ancient raised field agriculture in the Maya lowlands of southeastern Mexico, in: Prehistoric Intensive Agriculture in the Tropics, BAR International Series 232, edited by: Farrington, I. S., Oxford, England, 97–111, 1985.

Guèze, M., Paneque-Gálvez, J., Luz, A. C., Pino, J., Orta-Martínez, M., Reyes-García, V., and Macía, M. J: Determinants of tree species turnover in a southern Amazonian rain forest, J. Veg. Sci., 24, 284–295, 2013.

Guyot, J. L.: Hydrogéochemie de fleuves de Amazonie Bolivienne, Doctoral thesis, Université de Bordeaux, Bordeaux, France, 1992.

Hamilton, S. K., Sippel, S. J., and Melack, J. M.: Seasonal inundation patterns in two large savanna floodplains of South America: the Llanos de Moxos (Bolivia) and the Llanos del Orinoco (Venezuela and Colombia), Hydrol. Process., 18, 2103–2116, doi:10.1002/hyp.5559, 2004.

Hanagarth, W.: Acerca de la geoecología de las sabanas del Beni en el noreste de Bolivia, Instituto de Ecología, La Paz, 1993.

Hazelton, P. A. and Murphy, B. W.: Interpreting soil test results: what do all the numbers mean?, CSIRO publishing, CSIRO Publishing, Collingwood Victoria, Australia, 2007.

Hijmanns, R. J., Cameron, S. E., Parra, J. L., Jones, P. G., and Jarvis, A.: Very high resolution interpolated climate surfaces for global land areas, Int. J. Climatol., 25, 1965–1978, doi:10.1002/joc.1276, 2005.

Hogg, A. G., Hua, Q., Blackwell, P. G., Niu, M., Buck, C. E., Guilderson, T. P., Heaton, T. J., Palmer, J. G., Reimer, P. J., and Reimer, R. W.: SHCal13 southern hemisphere calibration, 0–50,000 cal yr BP, Radiocarbon, 55, 1889–1903, 2013.

Holliday, V. T.: Soils in archaeological research, Oxford Univ. Press, New York, 448 pp., 2004.

Huang, Q.-H., Wang, Z.-J., Wang, D.-H., Wang, C.-X., Ma, M., and Jin, X.-C.: Origins and Mobility of Phosphorus Forms in the Sediments of Lakes Taihu and Chaohu, China, J. Environ. Sci. Health A, 40, 91–102, doi:10.1081/ESE-200033593, 2005.

Iriarte, J. and Dickau, R.: As culturas do Milho? Arquebontanica das sociedades hidraulicas das terras baixas Sul-Americanas, Amazônica – Revista de Antropologia, 4, 30–58, 2012.

Iriarte, J., Glaser, B., Watling, J., Wainwright, A., Birk, J. J., Renard, D., Rostain, S., and McKey, D.: Late Holocene Neotropical agricultural landscapes: phytolith and stable carbon isotope analysis of raised fields from French Guianan coastal savannahs, J. Archaeolog. Sci., 37, 2984–2994, doi:10.1016/j.jas.2010.06.016, 2010.

IUSS Working Group WRB: World Reference Base for Soil Resources: International soil classification system for naming soils and creating legends for soil maps, World soil resources reports, Online-Ressource, FAO, Rome, 2014.

Jaimes Betancourt, C.: La cerámica de dos montículos habitacionales en el área de Casarabe, Llanos de Moxos, The Past Ahead, in: Language, Culture, and Identity in the Neotropics, edited by: Isendahl, C., Acta Universitatis Upsaliensis, Studies in Global Archaeology 18, Uppsala, 260 pp., 2012.

Jaimes Betancourt, C.: Diversidad cultural en los Llanos de Mojos, in: Arqueología Amazonica; las civilizaciones ocultas del bosque tropical: Diversidad cultural en los Llanos de Mojos, edited by: Valdez, F., Instituto Francés de Estudios Andinos, UMIFRE, MAE-CNRS, Quito, Ecuador, 235–278, 2013.

Jones, J. B.: Plant nutrition and soil fertility manual: [how to make soil fertility; plant nutrition principles work], 2nd Edn., CRC Press, Boca Raton, 282 pp., 2012.

Kolata, A. L. and Ortloff, C. R.: Thermal analysis of Tiwanaku raised field systems in the Lake Titicaca basin of Bolivia, J. Archaeolog. Sci., 16, 233–263, 1989.

Lee, K.: Apuntes sobre las obras hidrauclicas prehispánicas de las llanuras de Moxos, Paititi, 11, 24–26, 1997.

Lindbo, D. L., Stolt, M. H., and Vepraskas, M. J.: Redoximorphic Features, in: Interpretation of Micromorphological Features of Soils and Regoliths, Elsevier, Oxford, Amsterdam, 129–147, 2010.

Lombardo, U.: Raised fields of northwestern Bolivia: a GIS based analysis, Z. Arch. Außereurop. Kult., 3, 127–149, 2010.

Lombardo, U.: Neotectonics, flooding patterns and landscape evolution in southern Amazonia, Earth Surf. Dynam., 2, 493–511, doi:10.5194/esurf-2-493-2014, 2014.

Lombardo, U.: Alluvial plain dynamics in the southern Amazonian foreland basin, Earth Syst. Dynam., 7, 453–467, doi:10.5194/esd-7-453-2016, 2016.

Lombardo, U. and Prümers, H.: Pre-Columbian human occupation patterns in the eastern plains of the Llanos de Moxos, Bolivian Amazonia, J. Archaeolog. Sci., 37, 1875–1885, 2010.

Lombardo, U., Canal-Beeby, E., Fehr, S., and Veit, H.: Raised fields in the Bolivian Amazonia: a prehistoric green revolution or a flood risk mitigation strategy?, J. Archaeolog. Sci., 38, 502–512, 2011a.

Lombardo, U., Canal-Beeby, E., and Veit, H.: Eco-archaeological regions in the Bolivian Amazon, Geogr. Helvet., 66, 173–182, 2011b.

Lombardo, U., Szabo, K., Capriles, J. M., May, J.-H., Amelung, W., Hutterer, R., Lehndorff, E., Plotzki, A., and Veit, H.: Early and middle holocene hunter-gatherer occupations in Western

Amazonia: the hidden shell middens, PloS One, 8, e72746, doi:10.1371/journal.pone.0072746, 2013.

Lombardo, U., Denier, S., and Veit, H.: Soil properties and pre-Columbian settlement patterns in the Monumental Mounds Region of the Llanos de Moxos, Bolivian Amazon, SOIL, 1, 65–81, doi:10.5194/soil-1-65-2015, 2015.

Lopez, P.: Spatial distribution of sedimentary P pools in a Mediterranean coastal lagoon 'Albufera d'es Grau' (Minorca Island, Spain), Mar. Geol., 203, 161–176, doi:10.1016/S0025-3227(03)00333-5, 2004.

Lopez, P., Navarro, E., Marce, R., Ordoñez, J., Caputo, L., and Armengol, J.: Elemental ratios in sediments as indicators of ecological processes in Spanish reservoirs, Asociación Española de Limnología, Madrid, Spain, 2006.

Markgraf, V. and Díaz, H. F.: The past ENSO record: a synthesis, El Niño: Historical and Paleoclimatic Aspects of the Southern Oscillation, Cambridge University Press, Cambridge, 465–488, 2000.

May, J. H.: The Río Parapetí – Holocene megafan formation in the southernmost Amazon basin, Geogr. Helvet., 66, 193–201, 2011.

Mayle, F. E., Langstroth, R. P., Fisher, R. A., and Meir, P.: Long-term forest-savannah dynamics in the Bolivian Amazon: implications for conservation, Philos. T. Roy. Soc. B, 362, 291–307, doi:10.1098/rstb.2006.1987, 2007.

McKey, D. and Rostain, S.: Farming Technology in Amazonia, in: Encyclopaedia of the History of Science, Technology, and Medicine in Non-Western Cultures, edited by: Selin, H., Springer Netherlands, Dordrecht, 1–14, 2014.

McKey, D., Rostain, S., Iriarte, J., Glaser, B., Birk, J. J., Holst, I., and Renard, D.: Pre-Columbian agricultural landscapes, ecosystem engineers, and self-organized patchiness in Amazonia, P. Natl. Acad. Sci. USA, 107, 7823–7828, doi:10.1073/pnas.0908925107, 2010.

McMichael, C. H., Palace, M. W., Bush, M. B., Braswell, B., Hagen, S., Neves, E. G., Silman Tamanaha, E. K., and Czarnecki, C.: Predicting pre-Columbian anthropogenic soils in Amazonia, P. Roy. Soc. B, 281, 20132475, doi:10.1098/rspb.2013.2475, 2014.

McNeill, J. R. and Winiwarter, V. (Eds.): Soils and societies: perspectives from environmental history, The White Horse Press, Isle of Harris, UK, 2006.

McQueen, K. (Ed.): Unravelling the regolith with geochemistry, in: Regolith 2006: Consolidation and Dispersion of Ideas, edited by: Fitzpatrick, R. W. and Shand, P., Proceedings of the CRC LEME Regolith Symposium Hahndorf Resort, South Australia, 230–235, 2006.

Meggers, B. J.: Archeological evidence for the impact of mega-Niño events on Amazonia during the past two millennia, Climatic Change, 28, 321–338, 1994.

Mehlich, A.: Determination of P, K, Na, Ca, Mg and NH_4, Soil Test Division Mimeo, North Carolina Department of Agriculture, Raleigh, NC, USA, 1953.

Miguez-Macho, G. and Fan, Y.: The role of groundwater in the Amazon water cycle: 1. Influence on seasonal streamflow, flooding and wetlands, J. Geophys. Res.-Atmos., 117, 1–30, doi:10.1029/2012JD017539, 2012.

Moy, C. M., Seltzer, G. O., Rodbell, D. T., and Anderson, D. M.: Variability of El Niño/Southern Oscillation activity at millennial timescales during the Holocene epoch, Nature, 420, 162–165, 2002.

Munsell soil color charts: revised edition, Munsell Color, New Windsor, 1994.

Murphy, J. and Riley, J. P.: A modified single solution method for the determination of phosphate in natural waters, Analyt. Chim. Ac., 27, 31–36, 1962.

Murray, A. S. and Wintle, A. G.: Luminescence dating of quartz using an improved single-aliquot regenerative-dose protocol, Radiat. Measure., 32, 57–73, 2000.

Navarro, G. and Maldonado, M.: Geografía ecológica de Bolivia: Vegetación y ambientes acuáticos, Quinta, Santa Cruz de la Sierra, Bolivia, 2002.

Nordenskiöld, E.: Die Anpassung der Indianer an die Verhältnisse in den Ueberschwemmungsgebieten in Südamerika, Ymer, Stockholm, 36, 55–135, 1916.

Pessenda, L., Gouveia, S., and Aravena, R.: Radiocarbon dating of total soil organic matter and humin fraction and its comparison with ^{14}C ages of fossil charcoal, Radiocarbon, 43, 595–601, 2001.

Plotzki, A., May, J. H., and Veit, H.: Past and recent fluvial dynamics in the Beni lowlands, NE Bolivia, Geogr. Helvet., 66, 164–172, 2011.

Plotzki, A., May, J.-H., Preusser, F., and Veit, H.: Geomorphological and sedimentary evidence for late Pleistocene to Holocene hydrological change along the Río Mamoré, Bolivian Amazon, J. S. Am. Earth Sci., 47, 230–242, doi:10.1016/j.jsames.2013.08.003, 2013.

Prümers, H.: Der Wall führt zum See. Die Ausgrabungen 2005–2006 in der Loma Salvatierra (Bolivien), Z. Arch. Außereurop. Kult., 2, 371–379, 2008.

Prümers, H. and Jaimes Betancourt, C.: 100 años de investigación arqueológica en los Llanos de Mojos, Arqueoantropológicas, Universidad Mayor de San Simón, Cochabamba, Bolivia, 11–53, 2014.

Rein, B., Lückge, A., Reinhardt, L., Sirocko, F., Wolf, A., and Dullo, W.: El Niño variability of Peru during the last 20,000 years, Paleoceanography, 20, 1–17, doi:10.1029/2004P, 2005.

Renard, D., Birk, J. J., Glaser, B., Iriarte, J., Grisard, G., Karl, J., and McKey, D.: Origin of mound-field landscapes: a multi-proxy approach combining contemporary vegetation, carbon stable isotopes and phytoliths, Plant Soil, 351, 337–353, doi:10.1007/s11104-011-0967-8, 2011.

Renard, D., Iriarte, J., Birk, J. J., Rostain, S., Glaser, B., and McKey, D.: Ecological engineers ahead of their time: The functioning of pre-Columbian raised-field agriculture and its potential contributions to sustainability today, Ecol. Eng., 45, 30–44, 2012.

Rodrigues, L., Lombardo, U., Fehr, S., Preusser, F., and Veit, H.: Pre-Columbian agriculture in the Bolivian Lowlands: Construction history and management of raised fields in Bermeo, Catena, 132, 126–138, 2015.

Rodrigues, L., Lombardo, U., Canal-Beeby, E., and Veit, H.: Linking soil properties and pre-Columbian agricultural strategies in the Bolivian Lowlands: The case of raised fields in Exaltación, Quatern. Int., doi:10.1016/j.quaint.2015.11.09, in press, 2016.

Rostain, S.: Pre-Columbian Earthworks in Coastal Amazonia, Diversity, 2, 331–352, doi:10.3390/d2030331, 2010.

Rostain, S.: Islands in the Rainforest: Landscape Management in Pre-Columbian Amazonia, New Frontiers in Historical Ecology, Left Coast Press, Walnut Creek, 278 pp., 2013.

Saavedra, O.: El sistema agrícola prehispánico de camellones en la Amazonía Boliviana, in: Agricultura ancestral. camellones y albarradas. Contexto social, usos y retos del pasado y del presente, edited by: Valdez, F., Abya Yala, Quito, 295–311, 2006.

Saavedra, O.: The lowlands, Rescuing the Past, in: Bolivia: climate change, poverty and adaptation, OXFAM, Oxfam International, La Paz, Bolivia, 45–51, 2009.

Schachtschabel, P., Blume, H. P., Brümmer, G., Hartge, K. H., and Schwertmann, U.: Scheffer/Schachtschabel, Lehrbuch der Bodenkunde, 15th Edn. Spektrum Akademischer Verlag, Heidelberg, 2002.

Scott, K. M. and Pain, C. F.: Regolith science, CSIRO Pub, Collingwood, Vic., 1st online resource, 461 pp., 2009.

Solos, E.: Manual de métodos de análise de solo, Embrapa Solos, Rio de Janeiro, 2011.

Sparks, D. L.: Methods of soil analysis, Soil Science Society of America book series, no. 5, Soil Science Society of America, American Society of Agronomy, Madison, Wisconsin, 1996.

Stab, S. and Arce, J.: Pre-Hispanic raised-field cultivation as an alternative to slash-and-burn agriculture in the Bolivian Amazon: agroecological evaluation of field experiments, in: Biodiversity, Conservation and Management in the Region of the Beni Biological Biosphere Reserve, Bolivia, edited by: Herrera-MacBryde, O., Dallmeier, F., MacBryde, B., Comiskey, J. A., and Miranda, C., SI/MAB Biodiversity Program, Smithsonian Institution, Washington, D.C., 317–327, 2000.

Stuiver, M. and Reimer, P. J.: Extended ^{14}C data base and revised CALIB 3.014 C age calibration program, Radiocarbon, 35, 215–230, 1993.

UNDP: Human Development Report 2011 Sustainability and Equity: A Better Future for All, United Nations Development Programme 1, UN Plaza, New York, USA, 2011.

Villalba, M. J., Alesán, A., Comas, M., Tresserras, J. J., SáeZ, J. L., Malgosa, A., Michel, M., and Playà, R.: Investigaciones arqueológicas en los Llanos de Moxos (Amazonía boliviana). Una aproximación al estudio de los sistemas de producción precolombinos, Revista Bienes Culturales IPCE, 3, 201–215, 2004.

Walker, J.: Agricultural Change in the Bolivian Amazon, Memoirs in Latin American Archaeology, University of Pittsburgh Latin American Archaeology Publications and Fundación Kenneth Lee, Trinidad, 2004.

Walker, J.: Social Implications from Agricultural Taskscapes in the Southwestern Amazon, Lat. Am. Antiquity, 22, 275–296, 2011.

Walker, M.: Quaternary dating methods, Wiley, Sussex, UK, 2005.

Watanabe, F. S. and Olsen, S. R.: Test of an ascorbic acid method for determining phosphorus in water and $NaHCO_3$ extracts from soil, Soil Sci. Soc. Am. J., 29, 677–678, 1965.

Whitney, B. S., Dickau, R., Mayle, F. E., Walker, J. H., Soto, J. D., and Iriarte, J.: Pre-Columbian raised-field agriculture and land use in the Bolivian Amazon, Holocene, 24, 231–241, doi:10.1177/0959683613517401, 2014.

Wilson, C., Simpson, I. A., and Currie, E. J.: Soil management in pre-Hispanic raised field systems: Micromorphological evidence from Hacienda Zuleta, Ecuador, Geoarchaeology, 17, 261–283, doi:10.1002/gea.10015, 2002.

Zhou, J. and Lau, K. M.: Does a monsoon climate exist over South America?, J. Climate, 11, 1020–1040, 1998.

2

Soil fauna: key to new carbon models

Juliane Filser[1], Jack H. Faber[2], Alexei V. Tiunov[3], Lijbert Brussaard[4], Jan Frouz[5], Gerlinde De Deyn[4], Alexei V. Uvarov[3], Matty P. Berg[6], Patrick Lavelle[7], Michel Loreau[8], Diana H. Wall[9], Pascal Querner[10], Herman Eijsackers[11], and Juan José Jiménez[12]

[1]Center for Environmental Research and Sustainable Technology, University of Bremen, General and Theoretical Ecology, Leobener Str. – UFT, 28359 Bremen, Germany
[2]Wageningen Environmental Research (Alterra), P.O. Box 47, 6700 AA Wageningen, the Netherlands
[3]Laboratory of Soil Zoology, Institute of Ecology & Evolution, Russian Academy of Sciences, Leninsky prospekt 33, 119071 Moscow, Russia
[4]Dept. of Soil Quality, Wageningen University, P.O. Box 47, 6700 AA Wageningen, the Netherlands
[5]Institute for Environmental Studies, Charles University in Prague, Faculty of Science, Benátská 2, 128 43 Praha 2, Czech Republic
[6]Vrije Universiteit Amsterdam, Department of Ecological Science, De Boelelaan 1085, 1081 HV Amsterdam, the Netherlands
[7]Université Pierre et Marie Curie, Centre IRD Ile de France, 32, rue H. Varagnat, 93143 Bondy CEDEX, France
[8]Centre for Biodiversity Theory and Modelling, Station d'Ecologie Théorique et Expérimentale, UMR5321 – CNRS & Université Paul Sabatier, 2, route du CNRS, 09200 Moulis, France
[9]School of Global Environmental Sustainability & Dept. Biology, Colorado State University, Fort Collins, CO 80523-1036, USA
[10]University of Natural Resources and Life Sciences, Department of Integrated Biology and Biodiversity Research, Institute of Zoology, Gregor-Mendel-Straße 33, 1180 Vienna, Austria
[11]Wageningen University and Research Centre, P.O. Box 9101, 6700 HB Wageningen, the Netherlands
[12]ARAID, Soil Ecology Unit, Department of Biodiversity Conservation and Ecosystem Restoration, IPE-CSIC, Avda. Llano de la Victoria s/n, 22700 Jaca (Huesca), Spain

Correspondence to: Juliane Filser (filser@uni-bremen.de)

Abstract. Soil organic matter (SOM) is key to maintaining soil fertility, mitigating climate change, combatting land degradation, and conserving above- and below-ground biodiversity and associated soil processes and ecosystem services. In order to derive management options for maintaining these essential services provided by soils, policy makers depend on robust, predictive models identifying key drivers of SOM dynamics. Existing SOM models and suggested guidelines for future SOM modelling are defined mostly in terms of plant residue quality and input and microbial decomposition, overlooking the significant regulation provided by soil fauna. The fauna controls almost any aspect of organic matter turnover, foremost by regulating the activity and functional composition of soil microorganisms and their physical–chemical connectivity with soil organic matter. We demonstrate a very strong impact of soil animals on carbon turnover, increasing or decreasing it by several dozen percent, sometimes even turning C sinks into C sources or vice versa. This is demonstrated not only for earthworms and other larger invertebrates but also for smaller fauna such as Collembola. We suggest that inclusion of soil animal activities (plant residue consumption and bioturbation altering the formation, depth, hydraulic properties and physical heterogeneity of soils) can fundamentally affect the predictive outcome of SOM models. Understanding direct and indirect impacts of soil fauna on nutrient availability, carbon sequestration, greenhouse gas emissions and plant growth is key to the understanding of SOM dynamics in the context of global carbon cycling models. We argue that explicit consideration of soil fauna is essential to make realistic modelling predictions on SOM dynamics and to detect expected non-linear responses of SOM dynamics to global change. We present a decision framework, to be further developed through the activities of KEYSOM, a European COST Action, for when mechanistic SOM models include soil fauna. The research activities of KEYSOM, such as field experiments and literature reviews, together with dialogue between empiricists and modellers, will inform how this is to be done.

1   Introduction

Despite continuous refinement over the past decades, estimates of the global carbon cycle still show large discrepancies between potential and observed carbon fluxes (Ballantyne et al., 2012; Schmitz et al., 2014). Soils contain more carbon than the atmosphere and above-ground vegetation together and play an important role for many of the recently adopted UN Sustainable Development Goals. Therefore, soil organic matter (SOM) modelling is key to understanding and predicting changes in global carbon cycling and soil fertility in a changing environment. SOM models can facilitate a better understanding of the factors that underlie the regulation of carbon cycling and the persistence of SOM. The predictive power of current global SOM models is, however, limited, as the majority rely on a relatively restricted set of input parameters such as climate, land use, vegetation, pedological characteristics and microbial biomass (Davidson and Janssens, 2006). Other parameters, such as the leaching of organic matter or soil erosion of organic matter, have been suggested for improving model predictions, and recent research has demonstrated what drastic effects, for example, living roots (Lindén et al., 2014) and soil fungi (Clemmensen et al., 2013) exert on SOM persistence. In an overview of the performance of SOM models, none of 11 tested models could predict global soil carbon accurately, nor were 26 regional models able to assess gross primary productivity across the USA and Canada (Luo et al., 2015).

Some years ago Schmidt et al. (2011) proposed eight "key insights" to enrich model predictions on the persistence of SOM. However, they ignored a major component of SOM dynamics, soil fauna, which plays a fundamental role in most of the insights they propose (e.g. Fox et al., 2006; Jiménez et al., 2006; Osler and Sommerkorn, 2007; De Deyn et al., 2008; Wilkinson et al., 2009). By moving through and reworking soil, feeding on living plant roots, detritus and all types of microorganisms growing on these, soil animals are intimately involved in every step of SOM turnover. Omission of soil fauna from SOM models will, therefore, hamper the potential predictive power of these models.

In a review focusing mostly on large mammals, terrestrial herbivores and aquatic ecosystems, Schmitz et al. (2014) recently called for "animating the carbon cycle". Bardgett et al. (2013) argued that differential responses of various trophic groups of above-ground and below-ground organisms to global change can result in a decoupling of plant–soil interactions, with potentially irreversible consequences for carbon cycling. A correlative large-scale field study has suggested that including soil animal activities could help clarify discrepancies in existing carbon models (de Vries et al., 2013). Similar attempts to connect animal activity to carbon cycling have occurred in the past (e.g. Lavelle and Martin, 1992; Lavelle et al., 1998; Lavelle and Spain, 2006; Osler and Sommerkorn, 2007; Brussaard et al., 2007; Sanders et al., 2014), without any further change in the structure of carbon models. This was partly due to a lack of communication between modellers and experimenters, but also because the magnitude of animal effects on SOM dynamics remains poorly quantified (Schmitz et al., 2014).

Here we use the "key insights" proposed by Schmidt et al. (2011) as a basis to review current evidence and to identify research needs on the relationship of soil fauna to SOM dynamics. Our review justifies the relevance of incorporating the soil fauna into SOM models. How important animal activities are for manifold geological and pedological processes has been reviewed repeatedly (e.g. Swift et al., 1979; Wilkinson et al., 2009), but carbon turnover – which is highly dynamic and both directly and indirectly affected by animals – had never been the focus. Due to their prime role in most processes in soil (Briones, 2014), we mostly focus on earthworms, but also give examples for other groups of soil fauna whose role in C turnover appears to be much more relevant than thought thus far (e.g. David, 2014). We point out regional differences in climate, soils and land use with respect to soil fauna composition, abundance and activity and derive implications for SOM modelling. Finally, we introduce a new COST Action (ES 1406) that is working on the implementation of soil fauna into SOM models, also exploring the pros and caveats in such a process.

2   Key insights

The eight "key insights" compiled by Schmidt et al. (2011) are shown in Fig. 1, together with the most important activities of soil animals affecting them. As many animal-mediated processes are tightly interconnected, they also matter for most of these insights. For instance, aggregate formation in faeces simultaneously affects molecular structure, humic substances, physical heterogeneity and soil microorganisms. In the following text we briefly summarise the role of animal activities for each of the "key insights". As a more detailed example of animal impacts on SOM turnover, we consider their role on soil aggregate formation in a separate section.

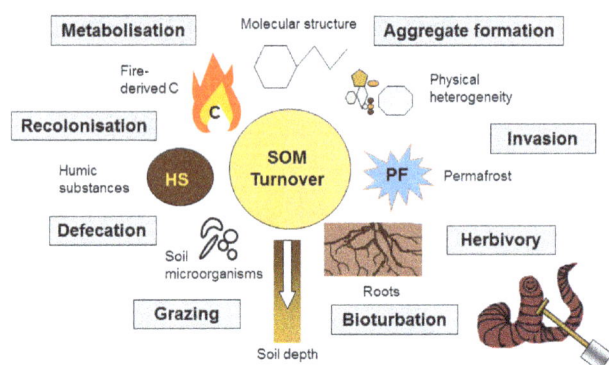

Figure 1. Main animal-mediated processes (boxes) affecting the eight insights (symbols) identified by Schmidt et al. (2011) that should be considered for improving SOM models.

2.1 Molecular structure

The molecular structure of root exudates and dead organic matter is modified during metabolisation, decomposition and associated food web transfer, both by microorganisms and soil fauna. Prominent examples are the release of ammonium by bacterivorous protozoans and nematodes, due to their higher C : N ratio compared to their bacterial prey (Osler and Sommerkorn, 2007), or the strong mediation of the direction and rate of humus formation by soil animals (see Sect. 2.2). Recently, the significant impact of eight different species of ants over 25 years on mineral dissolution and accumulation of calcium carbonate has even been discussed in the context of geoengineering and carbon sequestration (Dorn, 2014).

Many soil animals ingest and process SOM (and accompanying microorganisms) in their gut system, where it is partly assimilated with the help of mutualistic gut microflora and partly egested. Metabolisation alters the chemical structure of ingested SOM (Jiménez and Lal, 2006; Hedde et al., 2005; Coulis et al., 2009; Frouz et al., 2015b; Schmitz et al., 2014) and, consequently, the decomposition dynamics of animal faeces, which can be a substantial component of SOM (humus). Humification as such renders SOM less decomposable (Blume et al., 2009; Dickinson, 2012), whereas the alkaline milieu in invertebrate midguts accelerates mineralisation (e.g. Li and Brune, 2007).

For instance, earthworm casts have species-specific near-infrared spectral signatures, indicating presence of specific organic compounds (Hedde et al., 2005). Under grass/legume pasture they are characterised by significant enrichment of slightly altered plant residues in the sand particle size (> 53 µm). CPMAS 13C NMR (cross-polarisation magic-angle spinning carbon-13 nuclear magnetic resonance) spectra showed that earthworm casts and surrounding soil were dominated by carbohydrates, with a decrease in O-alkyl C and an increase in alkyl C with decreasing particle size (Guggenberger et al., 1996). Moreover, earthworms likely possess a unique capability of neutralising plant polyphenols

that otherwise strongly decrease decomposition rates of fresh plant litter (Liebeke et al., 2015). Micro- and mesofauna excrete ammonium or dissolved organic carbon (Filser, 2002; Fox et al., 2006; Osler and Sommerkorn, 2007) and affect the quantity of microbial metabolites (Bonkowski et al., 2009). Gut passage, defecation, and excretion, together with bioturbation by macro- and mesofauna, facilitate humification and decomposition, also altering nutrient stoichiometry (Bohlen et al., 2004). These modifications in the molecular structure of SOM due to soil fauna activity have significant effects on its dynamics (Swift et al., 1979; Guggenberger et al., 1995; Blume et al., 2009; Dickinson, 2012; and other references related to points 1 and 2 in Table 1).

2.2 Humic substances

As stated above, humification increases SOM stability. The term "humic substances" (here defined as very large and highly complex, poorly degradable organic molecules with manifold aromatic rings; Lehmann and Kleber, 2015) may be considered problematic by part of the scientific community: neither is the concept itself clear, nor is there any evidence that the often mentioned highly complex large organic molecules play any relevant role in organic matter stabilisation under natural conditions (Schmidt et al., 2011; Lehmann and Kleber, 2015). However, here we stick to it when referring to the "insights", simply for reasons of consistency with the article our argumentation is based on Schmidt et al. (2011). We acknowledge that "humus" or "humic substances" represent a continuum of more or less decomposed dead organic matter of which energy content and molecule size mostly should decrease over time, and that water solubility, sorption to the mineral matrix and accessibility for microorganisms are highly relevant for OM turnover (Lehmann and Kleber, 2015).

Humic substances are formed during the gut passage: organic matter in young soils and humic horizons almost completely consists of soil animal faeces (Lavelle, 1988; Martin and Marinissen, 1993; Brussaard and Juma, 1996). Humus forms mainly comprise animal casts, e.g. casts of ants, isopods, millipedes, beetle larvae or termites in deadwood; casts of insect larvae and spiders in leaf litter; or casts of collembolans, mites and enchytraeids in raw humus. In his review, David (2014) considered macroarthropod casts being a factor of partial SOM stabilisation, rather than hotspots of microbial activity. The dark colour of casts (compared to the ingested organic material) visually demonstrates the strong chemical OM modification in animal guts, which is accompanied by a substantial physical modification. Clay–humus complexes, physically protecting organic matter (Jiménez and Lal, 2006), are mainly faeces of earthworms and diplopods (see also Sect. 2.5 on physical heterogeneity). Due to differences in feeding preferences, gut microflora, SOM quantities consumed, etc. of soil animals, their faeces vary in size, shape and quality not only between

Table 1. Quantitative examples of the impact of earthworms and selected groups of other soil fauna on soil properties and processes involved in soil organic matter (SOM) turnover. If not mentioned otherwise, any numbers or percentages refer to the control without fauna. Selected, particularly striking examples are printed in bold.

Insight*	Examples	Source
	Earthworms	
1. Molecular structure	An indicator of lignin degradation in earthworm casts was twice that of the surrounding soil.	Guggenberger et al. (1995)
2. Humic substances	Introduced earthworms can double microaggregate formation and the stabilisation of new C in the topsoil.	Marashi and Scullion (2003), Six et al. (2004)
	C protection is promoted by microaggregates within large macroaggregates, and earthworms can add 22 % anew to this C pool	Bossuyt et al. (2005)
	Exclusion of earthworms reduced SOC accumulation by 0 (at 0–10 cm depth) to 75 % (at 30–40 cm depth), associated with a decrease in percentage of water-stable aggregates.	Albrecht et al. (2004), cited in Schmidt et al. (2011)
	In organic layers of a Canadian aspen forest, in locations with earthworms, N (1.5–0.8 %) and especially C concentrations (25.3–9.8 %) were strongly reduced, together with C/N ratio (16.7–13.2) and soil pH (6.5–6.1); in brackets: control values vs. values with earthworms. **This suggests a shift towards a faster cycling system, resulting in a net loss of C from the soil and turning northern temperate forests from C sinks into C sources.**	Eisenhauer et al. (2007)
3. Fire-derived carbon	Small charcoal particles from burned plots after 1 year increased by 21 % at 0–1 cm depth. One year later they were concentrated in earthworm casts at the soil surface, after 6.5 years such casts were found at 8 cm depth.	Eckmeier et al. (2007)
4. Roots	**Presence of earthworms in a continuous maize plot in Peruvian Amazonia increased the organic C input from roots by 50 %.**	Jiménez and Lal (2006)
5. Physical heterogeneity (see also insights no. 2, 3, 6 and 7)	Up to 50 % of soil aggregates in the surface layer of temperate pastures are earthworm casts.	van de Westeringh (1972)
	Mull-type forest soil top layers and wooded savanna soils consist almost entirely of earthworm casts.	Kubiena (1953), Lavelle (1978)
	Earthworm inoculation in pastures on young polder soils completely removed the organic surface layer within 8–10 years and incorporated it into deeper layers, creating an A horizon. This affected manifold measures, increasing, for example, grass yield by 10 %, root content in 0–15 % from 0.38 to 1.31 g dm^{-3}, C content at 0–20 cm from 1.78 to 16.9 kg C $\cdot$ 10^3 ha^{-1}, and water infiltration capacity from 0.039 to 4.6 m 24 h^{-1}. In turn, penetration resistance at 15 cm depth decreased from 35 to 22 kg cm^{-2}.	Hoogerkamp et al. (1983)
	In average temperature pasture and grasslands, earthworms cast 40–50 t ha^{-1} yr^{-1} on the surface and even more below surface.	Lee (1985)
	Passage of a tropical soil through the gut of the invading earthworm *Pontoscolex corethrurus* reduced macroporosity from 21.7 to 1.6 cm^3 g^{-1}, which exceeded the effect of mechanically compacting the same soil at 10^3 kPa (resulting macroporosity: 3 cm^3 g^{-1}).	Wilkinson et al. (2009)
	After invasion of European earthworms into a Canadian aspen forest a thick layer of their cast material (thickness up to 4 cm) on top of organic layers was developed.	Eisenhauer et al. (2007)
6. Soil depth	Burrows of anecic earthworms are up to several metres deep and last for many years.	Edwards and Bohlen (1996)
7. Permafrost and boreal areas	**Earthworm invasions in boreal forests completely transformed mor to mull soils and significantly altered the entire plant community.**	Frelich et al. (2006)
8. Soil microorganisms	Earthworms may lower actual microbial activity (by 11–23 %) but markedly (by 13–19 %) optimise microbial resource utilisation.	Scheu et al. (2002)

Table 1. Continued.

Insight*	Examples	Source
Ants and termites		
2. Humic substances	**In a degraded marsh in NE China, ant mounds were CH_4 sinks, contrary to the control soils, which were CH_4 sources (-0.39 to -0.19 mg vs. 0.13–0.76 m^{-2} h^{-1}).**	Wu et al. (2013)
5. Physical heterogeneity	Ant and termite mounds can occupy up to 25 % of the land surface	Bottinelli et al. (2015)
5. Physical heterogeneity and 6. Soil depth	**Underground nests of leafcutter ants (e.g. *Atta* spp.) can cover up to 250 m^2 and extend down to 8 m, which is associated with a massive impact on forest vegetation.**	Corréa et al. (2010)
Collembola		
8. Soil microorganisms	Grazing by Collembola affected community composition of ectomycorrhizal fungi and on average reduced $^{14}CO_2$ efflux from their mycelia by 14 %.	Kanters et al. (2015)
	Grazing by *Protaphorura armata* at natural densities on AM fungi disrupted carbon flow from plants to mycorrhiza and its surrounding soil by 32 %.	Johnson et al. (2005)
	The presence of a single Collembola species may enhance microbial biomass by 56 %.	Filser (2002)
	At elevated temperature, litter decay rates were up to 30 % higher due to Collembola grazing.	A'Bear et al. (2012)
Various or mixed groups		
1. Molecular structure	Microbial grazing by Collembola or enchytraeids alone enhanced leaching of NH_4^+ or DOC by up to 20 %.	Filser (2002)
	Feeding by millipedes and snails reduced the content of condensed tannins in three Mediterranean litter species from 9–188 mg g^{-1} dry matter to almost zero.	Coulis et al. (2009)
	Long-term mineralisation of fauna faeces may be slower than the mineralisation of litter from which the faeces were produced. This decrease in decomposition rate corresponds to a decrease in the C : N ratio and in the content of soluble phenols.	Frouz et al. (2015a, b)
	Due to stoichiometric constraints, soil animals tend to reduce the C concentration of SOM but increase N and P availability. About 1.5 % of the total N and P in the ingested soil was mineralised during gut passage in humivorous larvae of the scarabaeid beetle *Pachnoda ephippiata*. In *Cubitermes ugandensis* termites, the ammonia content of the nest material was about 300-fold higher than that of the parent soil.	Li et al. (2006), Li and Brune (2007), Ji and Brune (2006)
2. Humic substances	In a laboratory experiment, activity of earthworms, Collembola, enchytraeids and nematodes in coarse sand liberated > 40 % from the insoluble C pool as compared to the control.	Fox et al. (2006)
	Radiolabelled proteins and phenolic compounds in litter are transformed faster to humic acids (as revealed by alkaline extraction and acid precipitation) via faeces of Bibionidae (Diptera) than from litter not eaten by fauna.	Frouz et al. (2011)
	The quantitative contribution of invertebrates (mainly beetles and termites) to wood decomposition ranges between 10 and 20 %.	Ulyshen (2016)
	Depending on fungal and animal species (Collembola, isopods and nematodes), grazing on fungi colonising wood blocks altered (mostly decreased) their decay rates by more than 100 %. Isopods and nematodes had opposite effects in this study.	Crowther et al. (2011)

Table 1. Continued.

Insight*	Examples	Source
	Various or mixed groups	
2. Humic substances (continued)	Carbon and nitrogen losses from soil followed by drought and rewetting were substantially affected by microarthropod richness, which explained 42 % of the residual variance.	de Vries et al. (2012)
5. Physical heterogeneity	**Bioturbation rates of soil animal groups typically range between 1 and 5 Mg ha^{-1} yr^{-1} but may reach up to 10 (crayfish, termites), 20 (vertebrates), 50 (earthworms) and $>$ 100 Mg ha^{-1} yr^{-1} (earthworms in some tropical sites), which is equivalent to maximum rates of tectonic uplift.**	Wilkinson et al. (2009)
8. Soil microorganisms	In the course of a 2.5-year succession, fauna activities (especially of nematodes and mesofauna during the first year, and later of earthworms) accelerated microbial decomposition of clover remains in an arable soil by 43 %.	Uvarov (1987)
	Depending on vegetation, animal group and climate, soil animals directly or indirectly increased C mineralisation between 1 and 32 %. However, intensive grazing by fungal feeders may even reduce C mineralisation.	Persson (1989)

* According to Schmidt et al. (2011).

fauna groups but also between species within one group (see Sect. 3 on aggregate formation). Discovering the important role of animal faeces in humification is essential to improve our understanding of carbon dynamics in soil.

2.3 Fire-derived carbon

Fire-derived carbon is chemically highly condensed and thus often hardly degradable. However, its stability in soil is variable and still poorly understood (Schmidt et al., 2011; Wang et al., 2016). Two of the factors identified by a meta-analysis on the stability of biochar in soil were association with aggregates and translocation in the soil profile (Wang et al., 2016), which are both strongly affected by soil fauna (see Sects. 2.5, 2.6 and 3). Microbial recolonisation of burned sites is mediated by wind and soil animals that survived in soil or emigrated from neighbouring areas, e.g. by macro- and mesofauna, birds and mice (Malmström, 2012; Zaitsev et al., 2014). Moreover, soil fauna also ingest the charcoal particles (Eckmeier et al., 2007; see Table 1). Due to animal activity, charcoal is sorted by size and translocated down the soil profile. Mice and earthworms (Eckmeier et al., 2007) and the tropical earthworm *Pontoscolex corethrurus* (Topoliantz and Ponge, 2003; Topoliantz et al., 2006) had been suggested as responsible for rapid incorporation of charcoal into the soil. Quantitative data are, however, scarce (Table 1). In spite of potentially great importance, the effect of soil animals on the fate of the "black carbon" in soil remains practically unknown (Ameloot et al., 2013).

2.4 Roots

Roots not only represent a major input pathway of carbon into soil, but together with associated microflora they also have a large influence on the turnover dynamics of existing soil carbon (Finzi et al., 2015). Roots preferably grow in existing soil cavities (Jiménez and Lal, 2006), mostly formed by soil fauna (Wilkinson et al., 2009). Both burrowing and non-burrowing soil animals have a strong impact on root growth, allocation, length and density (Brown et al., 1994; Bonkowski et al., 2009; Arnone and Zaller, 2014). Animal grazing of root bacteria and mycorrhiza affects their activity and community composition, and animal excreta are enriched in micronutrients and selectively affect plant nutrition (Brown, 1995; Filser, 2002; Brussaard et al., 2007). Root herbivores and rhizosphere grazers affect C allocation of roots (Wardle et al., 2004) and largely regulate nutrient acquisition and plant productivity (Bonkowski et al., 2009). Not only root herbivores but also saprotrophic/microbivorous soil animals may obtain a significant proportion of energy from plant roots (Pollierer et al., 2007). This suggests an animal-mediated regulatory loop that connects plant roots and SOM.

2.5 Physical heterogeneity

Schmidt et al. (2011) considered the physical disconnection between decomposers and organic matter to be one reason for SOM persistence in deep soil. However, physical heterogeneity in soils occurs at all spatial scales, and animals play a fundamental role in the distribution of organic matter and associated microorganisms. According to body size, decomposers act at various spatial scales, from micro-aggregates to landscapes (Ettema and Wardle, 2002;

Jouquet et al., 2006). They fragment organic residues, perform bioturbation, distribute dead organic matter and generate smaller and larger organic matter hotspots (e.g. faecal pellets, ant and termite mounds). Mounds and burrows are obvious signs of physical heterogeneity created by ecosystem engineers (Meysmann et al., 2006; Wilkinson et al., 2009; Sanders et al., 2014). These structures significantly affect microorganisms and plants (Chauvel et al., 1999; Frelich et al., 2006) and associated soil properties such as aggregate stability (Bossuyt et al., 2005, 2006) and hydraulic properties (Bottinelli et al., 2015; Andriuzzi et al., 2015). This has consequences for the sorption and degradation (Edwards et al., 1992; Bolduan and Zehe, 2006) and for C emissions (Wu et al., 2013; Lopes de Gerenyu et al., 2015). Earthworms in particular feed on organic and mineral parts of the soil and mix them (Eckmeier et al., 2007; Wilkinson et al., 2009). The resulting clay–organic matter complexes considerably increase SOM retention of soils (Jiménez and Lal, 2006; Fox et al., 2006; Brussaard et al., 2007), although C loss from fresh casts is much higher than from surrounding soil (Zangerlé et al., 2014). The impact on soil processes and physical heterogeneity varies considerably between different groups of ecosystem engineers (Jouquet et al., 2006; Bottinelli et al., 2015). For instance, some earthworm species strongly affect their physical environment, while others are more linked to the soil organic matter content (Jiménez et al., 2012).

2.6 Soil depth

In most soil types, pore volume, carbon content, associated biotic processes and temperature variability strongly decrease with depth, whereas other parameters such as bulk density and water content increase – all of which significantly affect SOM turnover rates. The depth of organic horizons varies with soil type, from almost zero to several metres. Thus, Schmidt et al. (2011) identified soil depth as another "key insight". Nevertheless, digging animals play a key role in the development of soil depth. Animal burrows, which can reach several metres deep, are a considerable part of physical heterogeneity. Bioturbation (e.g. by earthworms, termites, ants, beetle and Diptera larvae, spiders, solitary bees and wasps, snails, isopods and amphipods, puffins, lizards, porcupines, pigs, moles, voles, rabbits, foxes, or badgers) is a key process to the formation of soil depth, soil structure and associated C translocation, as shown by several examples in Table 1 and reviewed, for example, by Wilkinson et al. (2009).

2.7 Permafrost

In permafrost soil up to 1672×10^{15} g carbon is stored (Tarnocai et al., 2009). Organism activity is mostly restricted to the short periods of time when the upper centimetres of the soil are thawed. Due to unfavourable environmental conditions (resulting in low animal biomass, activity and diver-

sity), there is only a minor impact of fauna in permafrost soils (De Deyn et al., 2008). However, fauna invasions, especially of the above-mentioned soil engineers, due to soil melting in tundra and boreal forests are likely to have drastic effects (Frelich et al., 2006; van Geffen et al., 2011). Data on earthworm invasions in North American forests (Bohlen et al., 2004; Frelich et al., 2006; Eisenhauer et al., 2007) show that they must be taken into consideration in carbon-rich soils, particularly in melting permafrost soils (Frelich et al., 2006; Schmidt et al., 2011), where they may affect many soil functions.

2.8 Soil microorganisms

After roots, microorganisms constitute by far the largest share of biomass in soil biota. Accordingly, they have a crucial role in SOM turnover. They consume root exudates and dead organic matter, attack plants and animals as pathogens, or support them as mutualists. Finally, microorganisms are the most important food source for the majority of soil animals, and also to a considerable extent for above-ground insects and vertebrates. Soil fauna comprises ecosystem engineers as well as an armada of mobile actors connecting elements of the soil system, thus mediating microbial processes (Briones, 2014). Countless isopods, ants, termites, enchytraeids, microarthropods, nematodes or protozoans make large contributions to SOM turnover underground (Persson, 1989; Filser, 2002; Wardle et al., 2004; Fox et al., 2006; Osler and Sommerkorn, 2007; Wilkinson et al., 2009; Wu et al., 2013). They affect the activity and community composition of soil microorganisms in multiple ways such as feeding and burrowing, facilitating the coexistence of different fungal species (Crowther et al., 2011) or by modifying micro-habitat conditions. Litter comminution by detritivores increases SOM accessibility for microorganisms, and propagules are dispersed with body surface and casts. The gut environment provides protected microsites with modified biotic and abiotic conditions, which increase bacterial abundance substantially – e.g. by three orders of magnitude in earthworm guts (Edwards and Fletcher, 1988). Grazing affects microbial biomass, activity and community composition, and animal excreta modify nutrient availability for microorganisms (Brown, 1995; Filser, 2002).

Table 1 contains quantitative examples of animal activity taken from different biomes and land-use types, showing that earthworms alone strongly affect each of the "key insights". However, much smaller soil animals can also have substantial effects (Table 1). It has to be kept in mind that the separation of animals' effects according to the insights is somewhat arbitrary as the associated soil processes are often interconnected. This is particularly obvious for molecular structure, humic substances, roots, physical heterogeneity, soil depth and microorganisms: metabolisation implies by definition an alteration of the molecular structure, often associated with the formation of humic substances. The stability of the latter has a very strong association with physical protection, and whether metabolisation of dead organic matter occurs at all depends on its horizontal and vertical distribution. For instance, earthworms will (a) translocate dead organic matter both vertically and horizontally; (b) transform part of it via metabolisation; (c) mix ingested OM with minerals, thus affecting its physical protection; (d) increase and alter the microbial community; and (e) affect hydraulic properties and aeration of the soil through digging and tunnelling, which has an immediate impact on the activity of microorganisms and on root growth.

As this example illustrated only the most important aspects of interacting processes, the next section provides a more elaborate overview of aggregate formation.

3 Aggregate formation

The modern view on the stability of organic matter in soils requires a thorough understanding of aggregate structure and formation, including the role of soil biota (Lehmann and Kleber, 2015). Soil aggregation is the process by which aggregates of different sizes are joined and held together by different organic and inorganic materials. Thus, it includes the processes of formation and stabilisation that occur more or less continuously and can act at the same time. With clay flocculation being a prerequisite for soil aggregation, the formation of aggregates mainly occurs as a result of physical forces, while their stabilisation results from a number of factors, depending in particular on the quantity and quality of inorganic and organic stabilising agents (Amézketa, 1999).

By bioturbation, feeding and dispersal of microbial propagules soil animals regulate all of the above forces and agents and are therefore a crucial factor in the formation and stabilisation of soil aggregates. Earthworms, many insect larvae and other larger fauna may stabilise aggregate structure by ingesting soil and mixing it intimately with humified organic materials in their guts and then egesting it as casts or pellets (Tisdall and Oades, 1982; Oades, 1993).

Earthworms have a direct and fast impact on microaggregate formation and the stabilisation of new C within these microaggregates (Bossuyt et al., 2005) (Table 1). There are several mechanisms to explain the increase in micro- and macroaggregate stability by earthworms, but no mechanism has been quantified in relation to population size yet. Effects are related to ecological groups of earthworms, associated with feeding habit, microhabitat in the soil profile, and burrow morphology. However, irrespective of this classification, species may enhance or mitigate soil compaction (Blanchart et al., 1997; Guéi et al., 2012). The tensile strength of casts (roughly defined as the force required to crush dried aggregates, i.e. an indirect measure of physical SOM protection) appears to be species-dependent: for example, the casts of *Dendrobaena octaedra* have a lower tensile strength compared to those of *L. terrestris* (Flegel et al., 1998). Similarly, organic carbon and water-stable aggregation was significantly higher in casts of *L. terrestris* than in casts of *A. caliginosa* (Schrader and Zhang, 1997).

Some research, however, suggests that earthworm activity can also evoke soil degradation. Shipitalo and Protz (1988) proposed that ingestion of soil by earthworms results in disruption of some existing bonds within micro-aggregates and realignment of clay domains. Therefore, fresh casts are more dispersible than uningested soil, contributing to soil erosion and crusting. Significant improvement in the water stability of fresh, moist casts only occurs when incorporated organic debris from the food sources is present and when moist casts are aged or dried. Nevertheless, in the long term, casting activity enhances soil aggregate stability.

However, our understanding of the contribution of soil fauna to aggregate formation and stabilisation is limited, and mostly qualitative in nature. Different methodologies complicate the comparison among aggregate stability data (Amézketa, 1999). Data in terms of functional response to density are limited as many studies have been conducted in arable systems, where the diversity and abundance of soil animals are reduced as a consequence of tillage, mineral fertilisers and pesticide use. Recently, some studies on these topics have emerged. A negative correlation between earthworm abundance and total macroaggregates and microaggregates within macroaggregates in arable treatments without organic amendments could be linked to the presence of high numbers of *Nematogenia lacuum*, an endogeic species that feeds on excrements of other larger epigeic worms and produces small excrements (Ayuke et al., 2011). Under the conditions studied, differences in earthworm abundance, biomass and diversity were more important drivers of management-induced changes in aggregate stability and soil C and N pools than differences in termite populations. Another study highlighted that in fields converted to no-tillage management, earthworms incorporated C recently fixed by plants and moved C from soil fragments and plant residues to soil aggregates of > 1 mm (Arai et al., 2013). Thus, soil management practices altering fauna activities may have a significant effect on the redistribution of soil organic matter in water-stable aggregates, impacting agronomically favourable size fractions of water-stable macro-aggregates, and water-stable micro-aggregates, which are the most important source of carbon sequestration (Šimanský and Kováčik, 2014).

4 Regional differences in climate, soils and land use

In a global meta-analysis spanning several continents, García-Palacios et al. (2013) show that, across biomes and scales, the presence of soil fauna contributes on average 27 % to litter decomposition. Depending on the situation, this contribution can be substantially lower or higher. For instance, the authors report an average increase in decomposi-

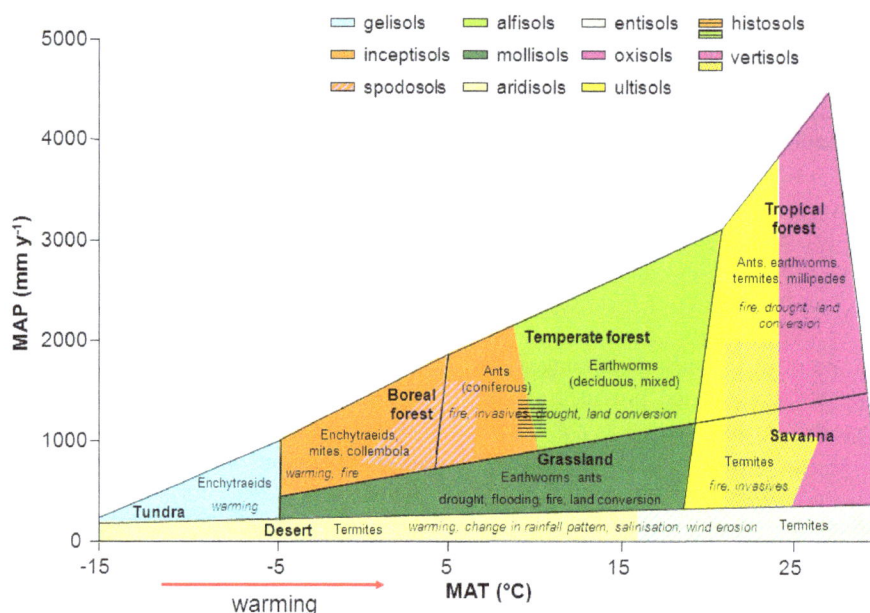

Figure 2. Dominant soil types and characteristic soil forming invertebrates across biomes (major global change threats are shown in italics). MAT: mean annual temperature; MAP: mean annual precipitation. For data and biome sources, see Brussaard et al. (2012). ©John Wiley and Sons. Reprint (slightly modified) by kind permission from John Wiley and Sons and Oxford University Press.

tion rates of 47 % in humid grasslands, whereas in coniferous forests this figure amounts to only 13 %. The high impact of soil fauna in humid grasslands is all the more important as such grasslands are among those ecosystems that are most severely affected by global environmental change (Chmura et al., 2003; Davidson and Janssens, 2006).

Many of our examples refer to earthworms and temperate regions as they have been studied most intensively. However, we suggest that any dominant group of soil fauna, irrespective of body size or the ability to create larger soil structures, may substantially affect carbon dynamics. Table 1 gives a number of respective case studies. The key players and specific effects of soil animals vary across space (Fig. 2), with increasing importance for SOM dynamics in humid-warm and nutrient-limited conditions (Persson, 1989; Filser, 2002; Wardle et al., 2004; Fox et al., 2006; Osler and Sommerkorn, 2007; De Deyn et al., 2008; Briones, 2014). Once key players in a given ecosystem have been identified as relevant for being included in SOM models (see Sect. 6 and Fig. 3), more detailed information on their biology is required, in particular on their activity, ecological niche and corresponding tolerance limits. All this varies with species, and often extremely within one systematic group. Variation in drought or soil temperature towards limiting conditions will first increase (stress response, e.g. downward migration) and then strongly decrease activity (mortality or transition to inactive resting stage). Some key players will exhibit high abundance and be extremely active throughout the year (Wilkinson et al., 2009), while others might only be moderately relevant dur-

ing a short period of time; the contribution of a third group might be considered insignificant.

Ecosystem engineers also differ between soil types, biomes and land-use types, from rodents and ants in dry areas to termites, earthworms and millipedes in tropical rainforests. They consume different types of organic matter, make deep or shallow, narrow or wide burrows, and differ in aggregation behaviour (e.g. more or less regularly distributed earthworms versus distinct ant nests and termite mounds). Accordingly, their role in SOM redistribution and turnover differs as well.

In cold ecosystems – where, together with wetlands and peatlands, the majority of terrestrial carbon is stored (Davidson and Janssens, 2006) – the response of detritivores to climatic change is expected to be most pronounced (Blankinship et al., 2011). Melting of permafrost soil might lead to northward expansion of soil macro-invertebrates, associated with accelerated decomposition rates (van Geffen et al., 2011). Further examples are shown in Table 1.

More information is needed on how existing abiotic and biotic constraints to SOM decomposition will vary with changing climate and in different regions (Davidson and Janssens, 2006). Finally, human activity comes into play: any significant land use change, particularly soil sealing and conversion of native forest to agricultural land, has dramatic consequences for abundances and species composition of soil communities. The same holds true for management intensity and pollution (Filser et al., 1995, 2002; Filser and Prasse, 2008; de Vries et al., 2012). However, even seemingly harmless activities can be significant, as we will show for the case

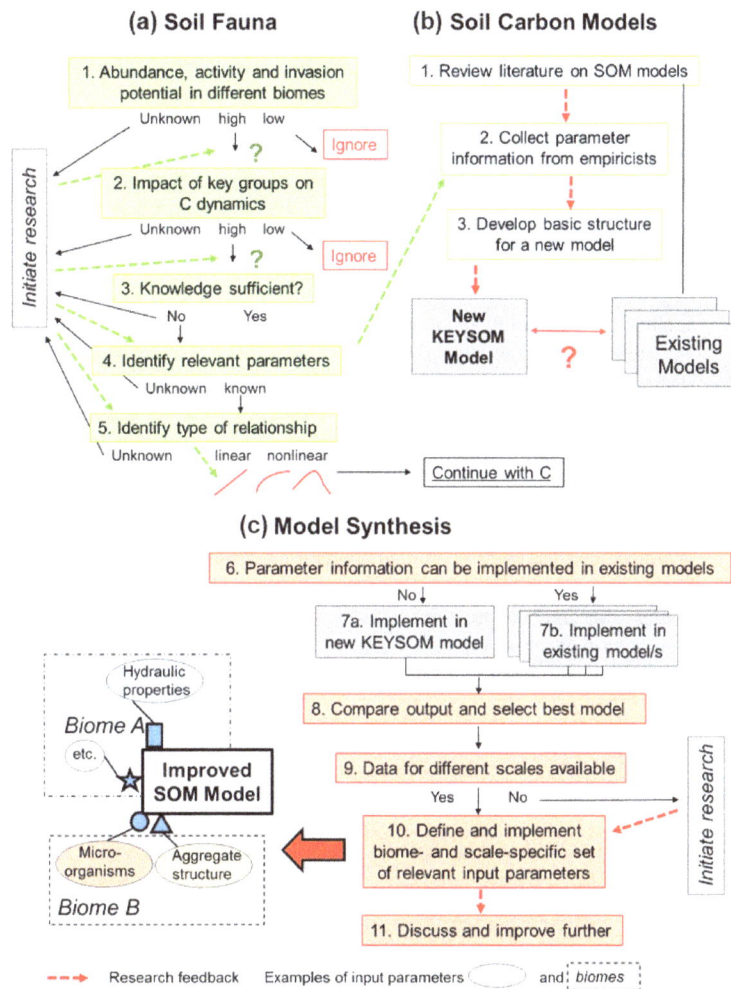

Figure 3. Flow scheme for an improved understanding of the role of soil fauna for soil organic matter (SOM) turnover. This scheme is basically followed within the COST Action ES 1406 (KEYSOM). Activities in **(a)** and **(b)** run parallel, followed by **(c)**, which ends with an improved SOM model. Example scenarios are shown for two biomes. For further explanations, see text.

of fishing at the end of Sect. 5 – pointing out the relevance of human activities for soil fauna beyond impact on global warming and land use change. How we address all this biogeographical and ecological variation is shown in Sects. 5 and 6.

5 Implications for modelling

As there is no unambiguous scientific support for the widespread belief in "humic substances", the question of how long organic carbon remains in soil is largely related to (a) physical protection and (b) how often the once photosynthesised dead organic matter is recycled in the soil food web. For both processes soil animals are of great importance, as we have shown above. Biomass and abundance of soil animals are generally constrained by temperature, humidity and food (living or dead organic matter). However, the effects of these constraints on their activity are not simply additive, nor

is there any simple relation between biomass and activity. For example, despite overall unfavourable conditions for the majority of soil organisms, burrowing activity in deserts can be extremely high (Filser and Prasse, 2008). Moreover, there is increasing evidence that fauna effects on energy and nutrient flow can be at least partly decoupled from other abiotic and biotic factors (Frouz et al., 2013). De Vries et al. (2013) even concluded that "soil food web properties strongly and consistently predicted processes of C and N cycling across land use systems and geographic locations, and they were a better predictor of these processes than land use". This implies that knowledge of fauna may increase our prediction power. The thermodynamic viewpoint makes the issue even more relevant: reaction speed increases with temperature, but most soil organisms are rather adapted to relatively cool conditions and might thus be pushed beyond their niche limits – with eventually negative consequences on their activity (see Sect. 4).

Table 2. "Insights" (compiled after Schmidt et al., 2011) for future soil organic matter models and recommendations for further improvements by implementing effects of soil fauna.

SOM modelling element ("insight")	Recommendations*
1. Molecular structure	Incorporate the knowledge on the structure of organic substances and element concentrations in faunal casts and excreta in SOM decay rate models. Consider linkage between C and N cycling mediated by fauna. See 8.
2. Humic substances	Add physical and chemical stability of casts, patterns of their microbial colonisation and degradation dynamics. See 1, 3, 5, 6, 7, 8.
3. Fire-derived carbon	Include recolonisation and inoculation potential of surviving soil fauna and adjacent fauna. Initiate studies on the impact of fauna on the fate of black carbon (fragmentation, gut, casts, decomposition, and recolonisation).
4. Roots	Add activity of bioturbators, rhizosphere microbial grazers and root herbivores. See 1, 5, 6, 8.
5. Physical heterogeneity	Consider spatial and physicochemical heterogeneity created by soil fauna, including consequences of soil aggregation and dis-aggregation (e.g. bulk density, infiltration rate, preferential flow, casts). See 1, 2, 6, 8.
6. Soil depth	Incorporate burrowing depth and annual transport rates of bioturbators and animal-induced spatial heterogeneity of old and young carbon in the deep soil. See 5.
7. Permafrost	For warming scenarios, take into account short- and long-term invasion effects, particularly of earthworms and enchytraeids.
8. Soil microorganisms	Add microbial grazer effects, effects on microorganisms during gut passage and faunal impact on C and N coupling. See 1–7.

* Recommendations refer to site-specific keystone groups of animals (dominating in terms of biomass or impact; see Fig. 2). Their prevalence is determined by climate, bedrock and land use (e.g. rodents or ants in deserts, earthworms in temperate grasslands, or microarthropods and enchytraeids in acidic northern forests).

Changes in climate (Blankinship et al., 2011), land use (Filser et al., 2002; Tsiafouli et al., 2015), resource availability and biotic interactions (de Vries et al., 2012; see Table 2) alter the distribution, community composition, activity and associated impact of soil animals on distribution and turnover rate of SOM (Wall et al., 2008) to the extent that underlying assumptions of SOM models may no longer be valid (Swift et al., 1998; Bardgett et al., 2013; Schmitz et al., 2014). Therefore, it is opportune to include approaches that have been developed during the past decades (Filser, 2002; Jiménez and Lal, 2006; Osler and Sommerkorn, 2007; Brussaard et al., 2007; Meysmann et al., 2006; Wall et al., 2008; Sanders et al., 2014). For instance, Lavelle et al. (2004) implemented earthworm activity in the CENTURY model. For this purpose, observations on long-term incubated earthworm casts and sieved control had been used as a reference. Afterwards, earthworm activity was simulated with CENTURY by replacing the active and slow soil C decomposition rates of the model with those obtained by calibration with the control soil. The simulations revealed a 10 % loss of the slow C pool within 35 years compared to the original model without earthworms.

Without considering the role of animals, models are less accurate: in a field study spanning four countries from Sweden to Greece, soil food web properties were equally important as abiotic factors and predicted C and N cycling processes better than patterns of land use (de Vries et al., 2013). In their study, earthworms enhanced CO_2 production, whereas Collembola and bacterivorous nematodes increased leaching of dissolved organic carbon. Mechanistic experiments confirm that earthworms have a detrimental effect on the greenhouse gas balance under nitrogen-rich conditions (Lubbers et al., 2013) and under no-till management (Lubbers et al., 2015). Inclusion of group-specific diversity of mesofauna in models of global-scale decomposition rates increased explained variance from 70 to 77 % over abiotic factors alone (Wall et al., 2008). Also, García-Palacios et al. (2013) provide additional evidence on the argument that soil fauna activity is not merely a product of climate, soil properties and land use but an independent parameter. These examples indicate that the actors that play an important role in SOM dynamics should be considered in SOM models.

Model parameters are often measured in situ at relatively large spatial scales – at least compared with the size or activity range of most soil animals. As a result, the fauna effect is de facto included, although not appreciated (Swift et al., 1998). However, in many cases parameters are measured or extrapolated by combining in situ methods (e.g. monitoring of gas flux or litterbag experiments) and ex situ techniques such as laboratory experiments at controlled, highly simplified conditions. Especially the results of the latter may be sensitive to neglecting soil fauna. A relationship between animal activity and C turnover may vary with scale, for instance when soil properties or animal abundance differ at larger distance. However, as data are often insufficient, it will be context-dependent whether the inclusion of fauna is sensible or not (see Sect. 6). On the other hand, not taking explicitly into account the spatial heterogeneity created by soil fauna in field measurements might lead to substantial errors in calculating carbon budgets (Wu et al., 2013; Lopes de Gerenyu et al., 2015). It is thus crucial to develop sound (and biome-specific) strategies for combining in and ex situ measurements as parameters in more realistic SOM models.

Next to space, scale effects also apply to temporal patterns – which poses a great challenge for SOM modelling as most studies refer to rather short periods of time. We illustrate this by the comparatively well studied impact of invasive earthworms. The meta-analysis of Lubbers et al. (2013) suggests that the effect of earthworms on total soil organic carbon (SOC) contents is on average relatively small. In contrast, in certain situations earthworms can strongly affect greenhouse gas emission. These data were, however, mainly obtained in relatively short-term experiments. Over a period of months to years and even decades, earthworms can reduce C decomposition by physical protection of C in ageing casts (Six et al., 2004; see Table 1).

Thus, long-lasting effects of invasive earthworms on the total SOC storage cannot be determined with certainty in short-term experiments, whereas field observations are rather controversial. For instance, Wironen and Moore (2006) reported ca. 30 % increase in the total soil C storage in the earthworm-invaded sites of an old-growth beech–maple forest in Quebec. Other studies (e.g. Sackett et al., 2013; Resner et al., 2014) suggest a decrease in C storage. Zhang et al. (2013) introduced the sequestration quotient concept to predict the overall effect of earthworms on the C balance in soils differing in fertility, but the question remains strongly understudied.

These well-documented examples of the impact of earthworms on soil C storage are related to invasive species. The presence of these species cannot be inferred directly from the climatic, soil and vegetation properties. The distributions of European invasive earthworms in North America, northern European forests or South Africa are largely driven by human activity. Often fishing (due to lost bait), imported plants or potting material of colonising farmers (Reinecke, 1983) is more important for these than habitat transformation – without human help earthworms are not active invaders (Stoscheck et al., 2012; Tiunov et al., 2006; Wironen and Moore, 2006). Thus, the presence of earthworms can be an environment-independent parameter of SOM dynamics.

Another fundamental issue in the large-scale approach is often neglected: when including the effects of the soil fauna implicitly, this assumes that the soil fauna will always have the same effects under the same conditions and hence that the soil fauna are essentially static. This assumption is increasingly unrealistic in a fast-changing world where both biodiversity and the climate are changing at accelerated paces, and where we are likely to witness major reorganisations of plant, animal and microbial communities. Therefore, explicit representation of the soil fauna, where possible, should increase the predictive ability of SOM models.

Given the fact that this issue had been raised decades back (see above), it appears somewhat astonishing that attempts to pursue it have not yet made any significant progress. We believe there are mainly three reasons for this: (a) missing information; (b) too much detail, irrespective of spatial scale; and (c) too little communication between empiricists and modellers. This is why we decided to bring into life a COST Action as an appropriate instrument to bridge these gaps. The next section gives an overview of it.

6 Ways to proceed: COST Action ES 1406

Based on the arguments compiled here, a COST Action entitled "Soil fauna – Key to Soil Organic Matter Dynamics and Modelling (KEYSOM)" was launched in March 2015 (http://www.cost.eu/COST_Actions/essem/ES1406). An interdisciplinary consortium of soil biologists and biogeochemists, experimenters and modellers from 23 European countries plus the Russian Federation and the USA cooperates to implement soil fauna in improved SOM models as a basis for sustainable soil management. The main aim of KEYSOM is to test the hypothesis that the inclusion of soil fauna activities into SOM models will result in a better mechanistic understanding of SOM turnover and in more precise process descriptions and output predictions of soil processes, at least locally. A number of workshops address key challenges in experimentation and modelling of SOM and soil fauna and support research exchange and access to experimental data. Special attention is given to the education of young scientists. The action comprises four working groups (WGs) with the following topics:

1. knowledge gap analysis of SOM–soil fauna interactions;

2. potentials and limitations for inclusion of soil fauna effects in SOM modelling;

3. data assemblage and data sharing;

4. knowledge management and advocacy training.

After an intensive and enthusiastic workshop held in Osijek, Croatia, in October 2015, first activities included compilation of literature as well as setting up and keeping a website constantly up to date (http://keysom.eu/). Meanwhile, short-term scientific missions for early-career scientists have been launched (http://keysom.eu/stsm/KEYSOM-STSMs-are-open-for-application), aiming for complementing the action's activities. The second workshop was held in Prague in April 2016.

Next to a first compilation of knowledge gaps in this article, present activities of KEYSOM involve

– a literature review on biome-specific effects of soil fauna impact on SOM turnover;

- a literature review on the impact of soil fauna other than earthworms on SOM turnover;

- a compilation of the potentials and limitations of existing SOM models;

- the development of a simple SOM model that also explicitly incorporates soil animals and associated processes in it, based on the current state of knowledge exchange between empiricists and modellers within KEYSOM;

- the preparation of a common Europe-wide field study into the impact of soil fauna composition and abundance on SOM breakdown, distribution and aggregate formation, which will start in autumn 2016;

- the preparation of a summer school, to be held in early October 2016 in Coimbra, Portugal.

Figure 3 illustrates the present state of our interdisciplinary discussions, providing a roadmap for how SOM models could be supplemented with the effects of soil fauna. In the first phase, empiricists (Fig. 3a) and modellers (Fig. 3b) work in parallel. Mutual exchange between these groups is guaranteed by the regular workshop meetings such as in Osijek and Prague.

The stepwise approach functions like a decision tree, with various feedback loops and options at every step if and how known effects of soil fauna could be implemented into SOM. It also identifies under which circumstances additional research (literature review or experimental studies) needs to be initiated before proceeding further. Like many existing models, the new model should also have a modular structure so that different modules can be used and combined according to the respective biome- and scale-specific scenario (Fig. 3c). It can also be seen that we do not aim to include every detail everywhere: in some situations (Fig. 3a) the impact of soil fauna on SOM dynamics might be too small (or existing information too scanty) to be included, and not all input parameters will be feasible or relevant at each scale (miniature in Fig. 3c). This keeps the model manageable, and also flexible enough to allow for more precise predictions in critical scenarios, like in the case of earthworm invasions outlined in Sect. 5. We generally think that focusing on such critical scenarios (analogous to, for example, global biodiversity hotspots) is a crucial precondition for well-informed management decisions, one of the final aims of KEYSOM.

As an example, box no. 1 in Fig. 3a represents the first literature review in the above list. Depending on the outcome, for each biome a decision will be made if the impact of fauna on SOM turnover is unknown, relevant or low. In the first case, more research is needed; in the last case, the faunal effect can be ignored. Depending on the outcome of additional research, the knowledge base will be improved and the decision between ignoring and proceeding further can be made

anew. If a strong effect is expected, the next question (box no. 2 in Fig. 3a) will be addressed and so forth.

Once the procedure in Fig. 3a has reached box no. 4, intensive exchange with modellers (Fig. 3b) is mandatory to identify the relevant model parameters and the type of functional relationship (box 5). Mechanistic aspects (such as chemical transformation in the gut, physical protection within aggregates or impact on hydraulic soil properties via digging) are of prime importance here as each of these examples may have different effects on C turnover. Effects of fauna abundance or biomass (in comparison to presence–absence) on the shape of the function will be addressed as well. Note, however, that to date necessary data for such an approach appear to be limited (García-Palacios et al., 2013). In the meantime, the modellers will have developed a basic model structure and compared it with the structures of existing SOM models concerning potentials and limitations for including fauna effects (Fig. 3b).

The second phase (Fig. 3c) starts with the practical tests of the collected model parameters (boxes 6 and 7), using data that have been compiled by then by WG 3, allowing for selecting the best model (box 8). At this point, spatial scale comes into play, which is likely to be the most critical issue: as we have also seen while preparing this article, existing data on the impact of soil fauna on SOM turnover are highly diverse, from short-term and often highly artificial experiments at controlled conditions to large-scale correlative field studies in all kinds of different environments (and with a strong bias with respect to certain biomes). The type of relationship between faunal abundance and SOM turnover will in most cases vary with scale. If data for different scales are not available (box 9), further research is needed. In the second case, one can proceed with boxes 10 and 11.

Importantly, the idea is not to include the fauna in every situation everywhere. Rather, we aim at identifying critical hotspots and scenarios (see above) where faunal activities play a crucial role in SOM turnover, as demonstrated in Sect. 5. Due to the above-mentioned differences between biomes and scale effects, these scenarios will be biome- and scale-specific. An example is shown in the lower left corner of Fig. 3c. For Biome A, hydraulic properties have been identified to be crucial for SOM dynamics. Thus, data are needed on animals that affect these, such as digging earthworms or rodents. Instead, the analyses for Biome B have revealed aggregate structure and microorganisms being most relevant – requiring respective data at the small scale. On a larger scale such data for microorganisms might not be available, which implies proceeding with aggregate structure alone.

Overall, the whole approach requires a modular model structure, allowing for using different models according to the respective situation and data availability. This is what WG 2 is currently developing. Certainly all the research outlined here cannot be done within one single COST Action. Based on the outcome of our work, we hope to come up with a more detailed roadmap for how to further proceed to improve SOM modelling. This roadmap, together with what could be achieved with the limited resources of KEYSOM, will provide information material, decision tools and management options for decision makers and politicians (WG 4).

7 Conclusions and outlook

Understanding and modelling SOM is essential for managing the greenhouse gas balance of the soil, for land restoration from desertification, for sustaining food production and for the conservation of above- and below-ground biodiversity and associated ecosystem services (Nielsen et al., 2015). Soil animal abundance, biodiversity, species traits and interactions are crucial for SOM turnover (Chauvel et al., 1999; Bohlen et al., 2004; Wardle et al., 2004; Wall et al., 2008; Uvarov, 2009). In Table 2 we give recommendations on how the known impact of soil fauna on SOM turnover could be used for improving carbon models. Due to the pronounced differences with respect to climate, soil and land use outlined above, it is important that these recommendations are considered region- and scale-specific, taking into account the key players and their specific activities in the respective area.

Author contributions. Juliane Filser wrote the article, prepared Figs. 1 and 3 and the tables and compiled the contributions from all co-authors, who are listed according to their quantitative and qualitative impact on the manuscript, except for Juan José Jiménez, who was placed last as he is the chair of COST Action ES 1406 (KEYSOM). Lijbert Brussaard suggested including Fig. 2.

Acknowledgements. The two anonymous referees and O. Schmitz are acknowledged for their critical comments, which significantly contributed to the revision of the original manuscript. We thank Antje Mathews for compiling the references and editing the manuscript. Many thanks to Karin Nitsch for linguistic proofreading. Oxford University Press and Wiley and Sons are acknowledged for the permission to include Fig. 2. This paper is a contribution to the COST Action ES1406 (KEYSOM) lead by the first and last author. A lot of the writing was inspired by the lively discussions within the workshop meetings of KEYSOM – thanks to all contributors! We thank the COST Association for financially supporting collaboration and networking activities across Europe.

Edited by: M. Muñoz-Rojas

References

A'Bear, A. D., Boddy, L., and Jones, T. H.: Impacts of elevated temperature on the growth and functioning of decomposer fungi are influenced by grazing collembola, Global Change Biol., 18, 1823–1832, 2012.

Ameloot, N., Graber, E. R., Verheijen, F. G., and De Neve, S.: Interactions between biochar stability and soil organisms: review and research needs, Eur. J. Soil Sci., 64, 379–390, 2013.

Amézketa, E.: Soil aggregate stability: a review, J. Sustain. Agricult., 14, 83–151, 1999.

Andriuzzi, W. S., Pulleman, M. M., Schmidt, O., Faber, J. H., and Brussaard, L.: Anecic earthworms (Lumbricus terrestris) alleviate negative effects of extreme rainfall events on soil and plants in field mesocosms, Plant Soil, 397, 103–113, doi:10.1007/s11104-015-2604-4, 2015.

Arai, M., Tayasu, I., Komatsuzaki, M., Uchida, M., Shibata, Y., and Kaneko, N.: Changes in soil aggregate carbon dynamics under no-tillage with respect to earthworm biomass revealed by radiocarbon analysis, Soil Till. Res., 126, 42–49, 2013.

Arnone, J. A. and Zaller, J. G.: Earthworm effects on native grassland root system dynamics under natural and increased rainfall, Front. Plant Sci., 5, 1–8, 2014.

Ayuke, F. O., Brussaard, L., Vanlauwe, B., Six, J., Lelei, D. K., Kibunja, C. N., and Pulleman, M. M.: Soil fertility management: Impacts on soil macrofauna, soil aggregation and soil organic matter allocation, Appl. Soil Ecol., 48, 53–62, 2011.

Ballantyne, A. P., Alden, C. B., Miller, J. B., Tans, P. P., and White, J. W. C.: Increase in observed net carbon dioxide uptake by land and oceans during the past 50 years, Nature, 488, 70–72, 2012.

Bardgett, R. D., Manning, P., Morrien, E., and de Vries, F. T.: Hierarchical responses of plant-soil interactions to climate change: consequences for the global carbon cycle, J. Ecol., 101, 334–343, 2013.

Blanchart, E., Lavelle, P., Bruadeau, E., Le Bissonnais, Y., and Valentin, C.: Regulation of soil structure by geophagous earthworm activities in humid savannas of Côte d'Ivoire, Soil Biol. Biochem., 29, 431–439, 1997.

Blankinship, J. C., Niklaus, P. A., and Hungate, B. A.: A meta-analysis of responses of soil biota to global change, Oecologia, 165, 553–565, 2011.

Blume, H.-P., Brümmer, G. W., Horn, R., Kandeler, E., Kögel-Knabner, I., Kretzschmar, R., Stahr, K., and Wilke, B. M.: Scheffer/Schachtschabel: Lehrbuch der Bodenkunde, Springer, Heidelberg, 2009.

Bohlen, P. J., Groffmann, P. M., Fahey, T. J., Fisk, M. C., Suárez, E., Pelletier, D. M., and Fahey, R. T.: Ecosystem Consequences of Exotic Earthworm Invasion of North Temperate Forests, Ecosystems, 7, 1–12, 2004.

Bolduan, R. and Zehe, E.: Abbau von Isoproturon in Regenwurm-Makroporen und in der Unterbodenmatrix – Eine Feldstudie, J. Plant Nutr. Soil Sci., 169, 87–94, doi:10.1002/jpln.200521754, 2006.

Bonkowski, M., Villenave, C., and Griffiths, B.: Rhizosphere fauna: the functional and structural diversity of intimate interactions of soil fauna with plant roots, Plant Soil, 321, 213–233, 2009.

Bossuyt, H., Six, J., and Hendrix, P. F.: Protection of soil carbon by microaggregates within earthworm casts, Soil Biol. Biochem., 37, 251–258, 2005.

Bossuyt, H., Six, J., and Hendrix, P. F.: Interactive effects of functionally different earthworm species on aggregation and incorporation and decomposition of newly added residue carbon, Geoderma, 130, 14–25, 2006.

Bottinelli, N., Jouquet, P., Capowiez, Y., Podwojewski, P., Grimaldi, M., and Peng, X.: Why is the influence of soil macrofauna on soil structure only considered by soil ecologists?, Soil Till. Res., 146, 118–124, 2015.

Briones, M. J. I.: Soil fauna and soil functions: a jigsaw puzzle, Front. Environ. Sci., 2, 1–22, 2014.

Brown, G. G.: How do earthworms affect microfloral and faunal community diversity?, Plant Soil, 170, 209–231, 1995.

Brown, G. G., Edwards, C. A., and Brussaard, L.: How Earthworms Affect Plant Growth: Burrowing into the Mechanisms, in: Earthworm Ecology, 2nd Edn., CRC Press, Boca Raton, 13–49, 1994.

Brussaard, L. and Juma, N. G.: Organisms and humus in soils, in: Humic substances in terrestrial ecosystems, edited by: Piccolo, A., Elsevier, Amsterdam, 329–359, 1996.

Brussaard, L., Pulleman, M. M., Ouédraogo, E., Mando, A., and Six, J.: Soil fauna and soil function in the fabric of the food web, Pedobiologia, 50, 447–462, 2007.

Brussaard, L., Aanen, D. K., Briones, M. J. I., Decaëns, T., De Deyn, G. B., Fayle, T. M., James, S. W., and Nobre, T.: Biogeography and Phylogenetic Community Structure of Soil Invertebrate Ecosystem Engineers: Global to Local Patterns, Implications for Ecosystem Functioning and Services and Global Environmental Change Impacts, Soil Ecol. Ecosyst. Serv., 201–232, 2012.

Chauvel, A., Grimaldi, M., Barros, E., Blanchart, E., Deshardins, T., and Lavelle, P.: Pasture damage by an Amazonian earthworm, Nature, 398, 32–33, 1999.

Chmura, G. L., Anisfeld, S. C., Cahoon, D. R., and Lynch, J. C.: Global carbon sequestration in tidal, saline wetland soils, Global Biogeochem. Cy., 17, 1–12, 2003.

Clemmensen, K. E., Bahr, A., Ovaskainen,, O., Dahlberg, A., Ekblad, A., Wallander, H., Stenlid, J., Finlay, R. D., Wardle, D. A., and Lindahl, B. D.: Roots and Associated Fungi Drive Long-Term Carbon Sequestration in Boreal Forest, Science, 339, 1615–1618, 2013.

Corréa, M. M., Silva, P. S. D., Wirth, R., Tabarelli, M., and Leal, I. R.: How leaf-cutting ants impact forests: drastic nest eVects on light environment and plant assemblages, Oecologia, 162, 103–115, 2010.

Coulis, M., Hättenschwiler, S., Rapior, S., and Coq, S.: The fate of condensed tannins during litter consumption by soil animals, Soil Biol. Biochem., 41, 2573–2578, 2009.

Crowther, T. W., Boddy, L., and Jones, T. H.: Outcomes of fungal interactions are determined by soil invertebrate grazers, Ecol. Lett., 14, 1134–1142, 2011.

David, J. F.: The role of litter-feeding macroarthropods in decomposition processes: A reappraisal of common views, Soil Biol. Biochem., 76, 109–118, 2014.

Davidson, E. A. and Janssens, I. A.: Temperature sensitivity of soil carbon decomposition and feedbacks to climate change, Nature, 440, 165–173, 2006.

De Deyn, G. B., Cornelissen, J. H. C., and Bardgett, R. D.: Plant functional traits and soil carbon sequestration in contrasting biomes, Ecol. Lett., 11, 516–531, 2008.

de Vries, F. T., Liiri, M. E., Bjørnlund, L., Bowker, M. A, Christensen, S., Setälä, H. M., and Bardgett, R. D.: Land use alters the resistance and resilience of soil food webs to drought, Nat. Clim. Change, 2, 276–280, 2012.

de Vries, F. T., Thébault, E., Liiri, M., Birkhofer, K., Tsiafouli, M. A., Bjørnlund, L, Bracht Jørgensen, H., Brady, M. V., Christensen, S., De Ruiter, P., d'Hertefeld, T., Frouz, J., Hedlund, K., Hemerik, L., Hol, W. H. G., Hotes, S., Mortimer, S. R., Setälä, H., Sgardelis, S. P., Uteseny, K., Van der Putten, W. H., Wolters,

V., and Bardgett, R. D.: Soil food web properties explain ecosystem services across European land use systems, P. Natl. Acad. Sci. USA, 110, 14296–14301, 2013.

Dickinson, C. H. and Pugh, G. J. F.: Biology of plant litter decomposition, Vol. 2, Academic Press, London, New York, 2012.

Dorn, R. I.: Ants as a powerful biotic agent of olivine and plagioclase dissolution, Geology, 42, 771–774, 2014.

Eckmeier, E., Gerlach, R., Skjemstad, J. O., Ehrmann, O., and Schmidt, M. W. I.: Minor changes in soil organic carbon and charcoal concentrations detected in a temperate deciduous forest a year after an experimental slash-and-burn, Biogeosciences, 4, 377–383, doi:10.5194/bg-4-377-2007, 2007.

Edwards, C. A. and Bohlen, P.: Biology and ecology of earthworms, Chapman & Hall, London, 1–426, 1996.

Edwards, C. A. and Fletcher, K. E.: Interactions between Earthworms and Micro-organisms in Organic-matter Breakdown, Agr. Ecosyst. Environ., 24, 235–247, 1988.

Edwards, W. M., Shipitalo, M. J., Traina, S. J., Edwards, C. A., and Owens, L. B.: Role of lumbrlcus terrestris (l.) burrows on quality of infiltrating, Soil Biol. Biochem., 24, 1555–1561, 1992.

Eisenhauer, N., Partsch, S., Parkinson, D., and Scheu, S.: Invasion of a deciduous forest by earthworms: Changes in soil chemistry, microflora, microarthropods and vegetation, Soil Biol. Biochem., 39, 1099–1110, 2007.

Ettema, C. H. and Wardle, D. A.: Spatial soil ecology, Trends Ecol. Evol., 17, 177–183, 2002.

Filser, J.: The role of Collembola in carbon and nitrogen cycling in soil, Pedobiologia, 46, 234–245, 2002.

Filser, J. and Prasse, R.: A glance on the fauna of Nizzana, in: A Sandy Ecosystem at the Desert Fringe, edited by: Yair, A., Veste, M., and Breckle, S.-W., Springer, Heidelberg, 125–147, 2008.

Filser, J., Fromm, H., Nagel, R., and Winter, K.: Effects of previous intensive agricultural management on microorganisms and the biodiversity of soil fauna, Plant Soil, 170, 123–129, 1995.

Filser, J., Mebes, K.-H., Winter, K., Lang, A., and Kampichler, C.: Long-term dynamics and interrelationships of soil Collembola and microorganisms in an arable landscape following land use change, Geoderma, 105, 201–221, 2002.

Finzi, A. C., Abramov, R. Z., Spiller, K. S., Brzostek, E. R., Darby, B. A., Kramer, M. A., and Phillips, R. P.: Rhizosphere processes are quantitatively important components of terrestrial carbon and nutrient cycles, Global Change Biol., 21, 2082–2094, 2015.

Flegel, M., Schrader, S., and Zhang, H.: Influence of food quality on the physical and chemical properties of detritivorous earthworm casts, Appl. Soil Ecol., 9, 263–269, 1998.

Fox, O., Vetter, S., Ekschmitt, K., and Wolters, V.: Soil fauna modifies the recalcitrance-persistence relationship of soil carbon pools, Soil Biol. Biochem., 38, 1353–1363, 2006.

Frelich, L. E., Hale, C. M., Scheu, S., Holdsworth, A. R., Heneghan, L., Bohlen, P. J., and Reich, P. B.: Earthworm invasion into previously earthworm-free temperate and boreal forests, Biol. Invasions, 8, 1235–1245, 2006.

Frouz, J., Li, X., Brune, A., Pizl, V., and Abakumov, E. V.: Effect of Soil Invertebrates on the Formation of Humic Substances under Laboratory Conditions, Euras. Soil Sci., 44, 893–896, 2011.

Frouz, J., Livecková, M., Albrechtová, J., Chronaková, A., Cajthaml, T., Pizl, V., Hánel, L., Stary, J., Baldrian, P., Lhotáková, Z., Simácková, H., and Cepáková, S.: Is the effect of trees on soil properties mediated by soil fauna? A case study from post-mining sites, Forest Ecol. Manage., 309, 87–95, 2013.

Frouz, J., Roubicková, A., Hedenec, P., and Tajovský, K.: Do soil fauna really hasten litter decomposition? A meta-analysis of enclosure studies, Eur. J. Soil Biol., 68, 18–24, 2015a.

Frouz, J., Spaldonová, A., Lhotáková, Z., and Cajthaml, T.: Major mechanisms contributing to the macrofauna-mediated slow down of litter decomposition, Soil Biology Biochem., 91, 23–31, 2015b.

García-Palacios, P., Maestre, F. T., Kattge, J., and Wall, D. H.: Climate and litter quality differently modulate the effects of soil fauna on litter decomposition across biomes, Ecol. Lett., 16, 1045–1053, 2013.

Guéi, A. M., Baidai, Y., Tondoh, J. E., and Huising, J.: Functional attributes: Compacting vs decompacting earthworms and influence on soil structure, Curr. Zool., 58, 556–565, 2012.

Guggenberger, G., Zech, W., and Thomas, R. J.: Lignin and carbohydrate alteration in particle-size separates of an oxisol under tropical pastures following native savanna, Soil Biol. Biochem., 27, 1629–1638, 1995.

Guggenberger, G., Thomas, R. J., and Zech, W.: Soil organic matter within earthworm casts of an anecic-endogeic tropical pasture community, Colombia, Appl. Soil Ecol., 3, 263–274, 1996.

Hedde, M., Lavelle, P., Joffre, R., Jiménez, J. J., and Decaens, T.: Specific functional signature in soil macro-invertebrate biostructures, Funct. Ecol., 19, 785–793, 2005.

Hoogerkamp, M., Rogaar, H., and Eijsackers, H. J. P.: Effect of earthworms on grassland on recently reclaimed polder soils in the Netherlands, in: Earthworm Ecology, edited by: Satchell, J. E., Chapman and Hall, London, New York, 85–105, 1983.

Ji, R. and Brune, A.: Nitrogen mineralization, ammonia accumulation, and emission of gaseous NH_3 by soil-feeding termites, Biogeochemistry, 78, 267–283, 2006.

Jiménez, J. J. and Lal, R.: Mechanisms of C Sequestration in Soils of Latin America, Crit. Rev. Plant Sci., 25, 337–365, 2006.

Jiménez, J. J., Decaëns, T., and Rossi, J. P.: Soil environmental heterogeneity allows spatial co-occurrence of competitor earthworm species in a gallery forest of the Colombian "Llanos", Oikos, 121, 915–926, doi:10.1111/j.1600-0706.2012.20428.x, 2012.

Johnson, D., Krsek, M., Wellington, E. M. H., Stott, A. W., Cole, L., Bardgett, R. D., Read, D. J., and Leake, J. R.: Soil Invertebrates Disrupt Carbon Flow Through Fungal Networks, Science, 309, 1047, 2005.

Jouquet, P., Dauber, J., Lagerlöf, J., Lavelle, P., and Lepage, M.: Soil invertebrates as ecosystem engineers: Intended and accidental effects on soil and feedback loops, Appl. Soil Ecol., 32, 153–164, 2006.

Kanters, C., Anderson, I. C., and Johnson, D.: Chewing up the Wood-Wide Web: Selective Grazing on Ectomycorrhizal Fungi by Collembola, Forests, 6, 2560–2570, 2015.

Kubiena, W. L.: Soils of Europe, 1st Edn. (December 1953), Thomas Murby & Co., London, 1–318, 1953.

Lavelle, P.: Les vers de terre de la savane de Lamto, Côte d'Ivoire: peuplements, populations et fonctions dans l'écosysteme, Publ. Lab. Zool. E.N.S., Ecole Normale Supérieure de Paris, Paris, 1–301, 1978.

Lavelle, P.: Earthworm activities and the soil system, Biol. Fert. Soils, 6, 237–251, 1988.

Lavelle, P. and Martin, A.: Small-Scale and Large-Scale Effects of Endogenic Earthworms on Soil Organic-Matter Dynamics in Soils of the Humid Tropics, Soil Biol. Biochem., 24, 1491–1498, 1992.

Lavelle, P. and Spain, A. V.: Soil Ecology, 2nd Edn., Kluwer Scientific Publications, Amsterdam, 2006.

Lavelle, P., Pashanasi, B., Charpentier, F., Gilot, C., Rossi, J.-P., Derouard, L., André, J., Ponge, J.-F., and Bernier, N.: Large-scale effects of earthworms on soil organic matter and nutrient dynamics, in: Earthworm Ecology, edited by: Edwards, C. A., St. Lucies Press, Boca Raton, 103–122, 1998.

Lavelle, P., Charpentier, F., Villenave, C., Rossi, J.-P., Derouard, L., Pashanasi, B., André, J., Ponge, J.-F., and Bernier, N.: Effects of Earthworms on Soil Organic Matter and Nutrient Dynamics at a Landscape Scale over Decades, in: Earthworm Ecology, edited by: Edwards, C. A., CRC Press, Boca Raton, 145–160, 2004.

Lee, K. E.: Earthworms: their ecology and relationships with soils and land use, Academic Press, Sydney, 1–654, 1985.

Lehmann, J. and Kleber, M.: The contentious nature of soil organic matter, Nature, 528, 60–68, doi:10.1038/nature16069, 2015.

Li, X. and Brune, A.: Transformation and mineralization of soil organic nitrogen by the humivorous larva of Pachnoda ephippiata (Coleoptera: Scarabaeidae), Plant Soil, 301, 233–244, 2007.

Li, X., Ji, R., Schäffer, A., and Brune, A.: Mobilization of soil phosphorus during passage through the gut of larvae of Pachnoda ephippiata (Coleoptera: Scarabaeidae), Plant Soil, 288, 263–270, 2006.

Liebeke, M., Strittmatter, N., Fearn, S., Morgan, J., Kille, P., Fuchser, J., Wallis, D., Palchykov, V., Robertson, J., Lahive, E., Spurgeon, D. J., McPhail, D., Takáts, Z., and Bundy, J. G.: Unique metabolites protect earthworms against plant polyphenols, Nat. Commun., 6, 1–7, 2015.

Lindén, A., Heinonsalo, J., Buchmann, N., Oinonen, M., Sonninen, E., Hilasvuori, E., and Pumpanen, J.: Contrasting effects of increased carbon input on boreal SOM decomposition with and without presence of living root system of Pinus sylvestris L., Plant Soil, 377, 145–158, 2014.

Lopes de Gerenyu, V. O., Anichkin, A. E., Avilov, V. K., Kuznetsov, A. N., and Kurganova, I. N.: Termites as a Factor of Spatial Differentiation of CO_2 Fluxes from the Soils of Monsoon Tropical Forests in Southern Vietnam, Euras. Soil Sci., 48, 208–217, 2015.

Lubbers, I. M., van Groenigen, K. J., Fonte, S. J., Brussaard, L., Six, J., and van Groenigen, J. W.: Greenhouse-gas emissions from soils increased by earthworms, Nat. Clim. Change, 3, 187–194, 2013.

Lubbers, I. M., van Groenigen, K. J., Brussaard, L., and van Groenigen, J. W.: Reduced greenhouse gas mitigation potential of no-tillage soils through earthworm activity, Nature, 5, 13787, doi:10.1038/srep13787, 2015.

Luo, Y., Keenan, T. F., and Smith, M.: Predictability of the terrestrial carbon cycle, Global Change Biol., 21, 1737–1751, 2015.

Malmström, A.: Life-history traits predict recovery patterns in Collembola' species after fire: A 10 year study, Appl. Soil Ecol., 56, 35–42, 2012.

Marashi, A. R. A. and Scullion, J.: Earthworm casts form stable aggregates in physically degraded soils, Biol. Fert. Soils, 37, 375–380, 2003.

Martin, A. and Marinissen, J. C. Y.: Biological and physico-chemical processes in excrements of soil animals, Geoderma, 56, 331–347, 1993.

Meysmann, F. J. R., Middelburg, J., and Heip, C. H. R.: Bioturbation: a fresh look at Darwin's last idea, Trends Ecol. Evol., 21, 688–695, 2006.

Nielsen, U. N., Wall, D. H., and Six, J.: Soil Biodiversity and the Environment, Annu. Rev. Environ. Resour., 40, 63–90, 2015.

Oades, J. M.: The role of biology in the formation, stabilization and degradation of soil structure, Geoderma, 56, 377–400, 1993.

Osler, G. H. R. and Sommerkorn, M.: Toward a Complete Soil C and N Cycle: Incorporating the Soil Fauna, Ecology, 88, 1611–1621, 2007.

Persson, T.: Role of soil animals in C and N mineralisation, in: Ecology of arable land, edited by: Clarholm, M. and Bergström, L., Kluwer Academic Publisher, Dordrecht, the Netherlands, 185–189, 1989.

Pollierer, M., Langel, R., Körner, C., Maraun, M., and Scheu, S.: The underestimated importance of belowground carbon input for forest soil animal food webs, Ecol. Lett., 10, 729–736, 2007.

Reinecke, A. J.: The ecology of earthworms in Southern Africa, in: Earthworm ecology, from Darwin to Vermiculture, edited by: Satchell, J. A., Chapman and Hall, London, 195–207, 1983.

Resner, K., Yoo, K., Sebestyen, S. D., Aufdenkampe, A., Hale, C., Lyttle, A., and Blum, A.: Invasive Earthworms Deplete Key Soil Inorganic Nutrients (Ca, Mg, K, and P) in a Northern Hardwood Forest, Ecosystems, 18, 89–102, 2014.

Sackett, T. E., Smith, S. M., and Basiliko, N.: Soil Biology & Biochemistry Indirect and direct effects of exotic earthworms on soil nutrient and carbon pools in North American temperate forests, Soil Biol. Biochem., 57, 459–467, 2013.

Sanders, D., Jones, C. G., Thébault, E., Bouma, T. J., van der Heide, T., van Belzen, J., and Barot, S.: Integrating ecosystem engineering and food webs, Oikos, 123, 513–524, 2014.

Scheu, S., Schlitt, N., Tiunov, A. V., Newington, J. E., and Jones, T. H.: Effects of the Presence and Community Composition of Earthworms on Microbial Community Functioning, Oecologia, 133, 254–260, 2002.

Schmidt, M. W., Torn, M. S., Abiven, S., Dittmar, T., Guggenberger, G., Janssens, I. A., Kleber, M., Kögel-Knabner, I., Lehmann, J., Manning, D. A. C., Nannipieri, P., Rasse, D. P., Weiner, S., and Trumbore, S. E.: Persistence of soil organic matter as an ecosystem property, Nature, 478, 49–56, 2011.

Schmitz, O. J., Raymond, P. A., Estes, J. A., Kurz, W. A., Holtgrieve, G. W., Ritchie, M. E., Schindler, D. E., Spivak, A. C., Wilson, R. W., Bradford, M. A., Christensen, V., Deegan, L., Smetacek, V., Vanni, M. J., and Wilmers, C. C.: Animating the Carbon Cycle, Ecosystems, 17, 344–359, 2014.

Schrader, S. and Zhang, H.: Earthworm Casting: Stabilization or Destabilization of Soil Structure?, Soil Biol. Biochem., 29, 469–475, 1997.

Shipitalo, M. J. and Protz, R.: Factors influencing the dispersibility of clay in worm casts, Soil Sci. Soc. Am. J., 52, 764–769, doi:10.2136/sssaj1988.03615995005200030030x, 1988.

Šimanský, V. and Kováčik, P.: Carbon Sequestration and its Dynamics in water-stable Aggregates, Agriculture, 60, 1–9, 2014.

Six, J., Bossuyt, H., Degryze, S., and Denef, K.: A history of research on the link between (micro)aggregates, soil biota, and soil organic matter dynamics, Soil Till. Res., 79, 7–31, 2004.

Stoscheck, L. M., Sherman, R. E., Suarez, E. R. and Fahey, T. J.: Exotic earthworm distributions did not expand over a decade in a hardwood forest in New York state, Appl. Soil Ecol., 62, 124–130, 2012.

Swift, M. J., Heal, O. W., and Anderson, J. M.: Decomposition in terrestrial ecosystems, Blackwell Scientific Publications, Oxford, 1–372, 1979.

Swift, M. J., Andrén, O., Brussaard, L., Briones, M., Couteaux, M.-M., Ekschmitt, K., Kjoller, A., Loiseau, P., and Smith, P.: Global change, soil biodiversity, and nitrogen cycling in terrestrial ecosystems: three case studies, Global Change Biol., 4, 729–743, 1998.

Tarnocai, C., Canadell, J. G., Schuur, E. A. G., Kuhry, P., Mazhitova, G., and Zimov, S.: Soil organic carbon pools in the northern circumpolar permafrost region, Global Biogeochem. Cy., 23, GB2023, doi:10.1029/2008GB003327, 2009.

Tisdall, J. M. and Oades, J. M.: Organic matter and water-stable aggregates in soils, J. Soil Sci., 33, 141–163, 1982.

Tiunov, A. V., Hale, C. M., Holdsworth, A. R., and Vsevolodova-Perel, T. S.: Invasion patterns of Lumbricidae into the previously earthworm-free areas of northeastern Europe and the western Great Lakes region of North America, Biol. Invasions, 8, 1223–1234, 2006.

Topoliantz, S. and Ponge, J.-F.: Burrowing activity of the geophagous earthworm Pontoscolex corethrurus (Oligochaeta: Glossoscolecidae) in the presence of charcoal, Appl. Soil Ecol., 23, 267–271, 2003.

Topoliantz, S., Ponge, J.-F., and Lavelle, P.: Humus components and biogenic structures under tropical slash-and-burn agriculture, Eur. J. Soil Sci., 57, 269–278, 2006.

Tsiafouli, M. A., Thébault, E., Sgardelis, S. P., De Ruiter, P. C., Van der Putten, W. H., Birkhofer, K., Hemerik, L., de Vries, F. T., Bardgett, R. D., Brady, M. V., Bjørnlund, L., Bracht Jørgensen, H., Christensen, S., d'Hertefeld, T., Hotes, S., Hol, W. H. G., Frouz, J., Liiri, M., Mortimer, S. R., Setälä, H., Tzanopoulos, J., Uteseny, K., Pizl, V., Stary, J., Wolters, V., and Hedlund, K.: Intensive agriculture reduces soil biodiversity across Europe, Global Change Biol., 21, 973–985, 2015.

Ulyshen, M. D.: Wood decomposition as influenced by invertebrates, Biol. Rev., 91, 70–85, 2016.

Uvarov, A. V.: Energetical evaluation of the role of soil invertebrates in the process of plant remains decomposition, in: Soil Fauna and Soil Fertility, Nauka Sci. Publ., edited by: Striganova, B. R., Proceedings of the 9th International Colloquium on Soil Zoology, August 1985, Moscow, 143–150, 1987.

Uvarov, A. V.: Inter- and intraspecific interactions in lumbricid earthworms: Their role for earthworm performance and ecosystem functioning, Pedobiologia, 53, 1–27, 2009.

van de Westeringh, W.: Deterioration of soil structure in worm free orchards, Pedobiologia, 12, 6–15, 1972.

van Geffen, K. G., Berg, M. P., and Aerts, R.: Potential macro-detritivore range expansion into the subarctic stimulates litter decomposition: a new positive feedback mechanism to climate change?, Oecologia, 167, 1163–1175, 2011.

Wall, D. H., Bradford, M. A., St. John, M. G., Trofymow, J. A., Behan-Pelletier, V., Bignell, D. E., Dangerfield, J. M., Parton, W. J., Rusek, J., Voigt, W., Wolters, V., Gardel, H. Z., Ayuke, F. O., Bashford, R., Beljakova, O. I., Bohlen, P. J., Brauman, A., Flemming, S., Henschel, J. R., Johnson, D. L., Jones, T. H., Kovarova, M., Kranabetter, J. M., Kutny, L., Lin, K.-C., Maryati, M., Masse, D., Pokarzhevskii, A., Rahman, H., Sabará, M. G., Salamon, J.-A., Swift, M. J., Varela, A., Vasconcelos, H. L., White, D., and Zou, X.: Global decomposition experiment shows soil animal impacts on decomposition are climate-dependent, Global Change Biol., 14, 2661–2677, 2008.

Wang, J., Xiong, Z., and Kuzyakov, Y.: Biochar stability in soil: meta-analysis of decomposition and priming effects, GCB Bioenergy, 8, 512–523, 2016.

Wardle, D. A., Bardgett, R. D., Klironomos, J. N., Setälä, H., Van der Putten, W. H., and Wall, D. H.: Ecological Linkages Between Aboveground and Belowground Biota, Science, 304, 1629–1633, 2004.

Wilkinson, M. T., Richards, P. J., and Humphreys, G. S.: Breaking ground: Pedological, geological, and ecological implications of soil bioturbation, Earth-Sci. Rev., 97, 257–272, 2009.

Wironen, M. and Moore, T. R.: Exotic earthworm invasion increases soil carbon and nitrogen in an old-growth forest in southern Quebec, Can. J. Forest Res., 36, 845–854, 2006.

Wu, H., Lu, X., Wu, D., Song, L., Yan, X., and Liu, J.: Ant mounds alter spatial and temporal patterns of CO_2, CH_4 and N_2O emissions from a marsh soil, Soil Biol. Biochem., 57, 884–889, 2013.

Zaitsev, A. S., Gongalsky, K. B., Persson, T., and Bengtsson, J.: Connectivity of litter islands remaining after a fire and unburnt forestdetermines the recovery of soil fauna, Appl. Soil Ecol., 83, 101–108, 2014.

Zangerlé, A., Hissler, C., Blouin, M., and Lavelle, P.: Near infrared spectroscopy (NIRS) to estimate earthworm cast age, Soil Biol. Biochem., 70, 47–53, 2014.

Zhang, W., Hendrix, P. F., Dame, L. E., Burke, R. A., Wu, J., Neher, D. A., Li, J., Shao, Y., and Fu, S.: Earthworms facilitate carbon sequestration through unequal amplification of carbon stabilization compared with mineralization, Nat. Commun., 4, 2576, doi:10.1038/ncomms3576, 2013.

3

Soil microbial biomass and function are altered by 12 years of crop rotation

Marshall D. McDaniel[1,a] and A. Stuart Grandy[1]

[1]Department of Natural Resources and the Environment, University of New Hampshire, Durham, NH, USA
[a]current address: Department of Agronomy, Iowa State University, 2517 Agronomy Hall,
716 Farm House Lane, Ames, IA 50011, USA

Correspondence to: Marshall D. McDaniel (marsh@iastate.edu)

Abstract. Declines in plant diversity will likely reduce soil microbial biomass, alter microbial functions, and threaten the provisioning of soil ecosystem services. We examined whether increasing temporal plant biodiversity in agroecosystems (by rotating crops) can partially reverse these trends and enhance soil microbial biomass and function. We quantified seasonal patterns in soil microbial biomass, respiration rates, extracellular enzyme activity, and catabolic potential three times over one growing season in a 12-year crop rotation study at the W. K. Kellogg Biological Station LTER. Rotation treatments varied from one to five crops in a 3-year rotation cycle, but all soils were sampled under a corn year. We hypothesized that crop diversity would increase microbial biomass, activity, and catabolic evenness (a measure of functional diversity). Inorganic N, the stoichiometry of microbial biomass and dissolved organic C and N varied seasonally, likely reflecting fluctuations in soil resources during the growing season. Soils from biodiverse cropping systems increased microbial biomass C by 28–112 % and N by 18–58 % compared to low-diversity systems. Rotations increased potential C mineralization by as much as 53 %, and potential N mineralization by 72 %, and both were related to substantially higher hydrolase and lower oxidase enzyme activities. The catabolic potential of the soil microbial community showed no, or slightly lower, catabolic evenness in more diverse rotations. However, the catabolic potential indicated that soil microbial communities were functionally distinct, and microbes from monoculture corn preferentially used simple substrates like carboxylic acids, relative to more diverse cropping systems. By isolating plant biodiversity from differences in fertilization and tillage, our study illustrates that crop biodiversity has overarching effects on soil microbial biomass and function that last throughout the growing season. In simplified agricultural systems, relatively small increases in crop diversity can have large impacts on microbial community size and function, with cover crops appearing to facilitate the largest increases.

1 Introduction

Research manipulating aboveground biodiversity in grasslands has shown a strong link between plant species richness and soil functions (Tilman et al., 1997; Zak et al., 2003; Eisenhauer et al., 2010; Mueller et al., 2013). While this research has contributed to our understanding of aboveground–belowground biodiversity in natural ecosystems, it fails to capture the biodiversity dynamics in agroecosystems, where crop rotations can be used to substitute temporal for spatial biodiversity. Given that species richness at any given time

in a rotated cropping system is 1 (excluding any weeds), the aboveground–belowground relationships dependent on diversity in agroecosystems and spatially diverse ecosystems (e.g., grasslands) may not be the same.

Crop rotations have been shown to have large positive effects on soil C, N, and microbial biomass (McDaniel et al., 2014a), plant pathogen suppression (Krupinsky et al., 2002), and yields (Smith et al., 2008; Riedell et al., 2009). These positive effects on crop production have been colloquially referred to as the "rotation effect". However, the mechanistic processes that link aboveground crop rotational diversity and belowground soil processes and contribute to the "rotation effect" remain elusive. One hypothesis explaining the benefits of crop rotations is that greater diversity of plant inputs to soil organic matter (SOM) over time enhances belowground biodiversity and soil ecosystem functioning (Hooper et al., 2000; Waldrop et al., 2006; Grandy and Robertson, 2007). Despite being low in spatial diversity, crop rotations have been shown to increase soil microbial and faunal biodiversity (Ryszkowski et al., 1998; Wu et al., 2008; Tiemann et al., 2015) and increase microbial carbon use efficiency (Kallenbach et al., 2015).

One essential function of soil microbial communities is the catabolism of newly added substrates from crops. The range and efficiency of microbial catabolism has great implications for ecosystem services such as sequestering C and soil fertility (Carpenter-Boggs et al., 2000; Kallenbach et al., 2015), as well as for ecosystem "dis-services" such as emission of soil-to-atmosphere greenhouse gases (McDaniel et al., 2014b). Furthermore, the partitioning of resources used in catabolism of residue and formation of SOM will affect long-term soil fertility (Lange et al., 2015; Kallenbach et al., 2015).

Soil microbial catabolism can be assessed using many different methods. The two most common measures are soil extracellular enzyme activities, microbe-produced catalysts for catabolism of soil substrates, and respiration response when supplying microbes with a source of C. The latter method, when multiple C compounds are added to the same soil, is commonly referred to as community-level physiological profiles (CLPPs), or as catabolic response profiles. The basic method for measuring soil CLPP involves adding a suite of C substrates to soils and measuring the catabolic response as CO_2 production or O_2 consumption with redox indicators (e.g., Biolog; Guckert et al., 1996). These C substrates are typically ecologically relevant compounds found in soils, and are intended to represent root exudates, microbial or plant cell structures, or other more-processed soil organic molecules. Other studies have used CLPPs to establish a catabolic "fingerprint" to distinguish soil microbial communities from one another by how they utilize different C substrates (Lupwayi et al., 1998; McDaniel et al., 2014b). The CLPP data can also be used to derive measures of metabolic diversity including substrate-use richness or catabolic evenness.

What can catabolic potential, and even catabolic evenness, tell us about soil microbial functioning in agroecosystems? Previous studies have shown that these metabolic diversity measures are increased with agroecosystem management practices that also increase soil health, e.g., reduced tillage or crop rotations (Lupwayi et al., 1998; Degens et al., 2000). In other words, soil microbial catabolism may be a good proxy for long-term consequences of agroecosystem management practices. Given that soil microorganisms, and the resources available to them in the soil, regulate many critical processes in agroecosystems, CLPPs can provide an inte-

grated measure of how management practices alter microbes and substrates available to them. Modern agriculture's use of monocultures could have unknown consequences for soil microbial catabolism, as well as related processes such as SOM mineralization, but to date the effect of rotation practices and crop diversity on soil microbial functioning remains poorly understood.

Considering a lack of understanding of how soil microbial functions are influenced by crop rotations, we sought to examine the rotation effects on soil microbial biomass and function. We measured soil microbial catabolic potential, C and N mineralization, extracellular enzyme activities, and microbial biomass three times over one growing season in a long-term crop rotation experiment at the W. K. Kellogg Biological Station (established 2000). All soils were collected during the same crop phase, allowing us to separate historical rotation from current crop effects. We hypothesized that soils under more diverse crop rotations would show greater catabolic diversity and have higher measures of soil function (enzyme activities, soil microbial biomass, potentially mineralizable C and N). In addition, we hypothesized that crop rotation effects would vary seasonally, being greatest in the spring and lessening over the growing season with the emerging influence of the current crop. The rationale for this second hypothesis is that early in the season all soils are coming out of different crops from the previous year, but over the growing season under corn the soils will become more functionally similar as the immediate crop has greater influence. Alternatively, significant rotation × season interactions on soil microbial functioning that do not converge over the growing season point to historical effects of rotations on differences in soil microbial communities and SOM.

2 Materials and methods

This study was conducted in the Cropping Biodiversity Gradient Experiment (CBGE) at the W. K. Kellogg Biological Station Long-term Ecological Research site (42°24' N, 85°24' W). The CBGE was established in 2000 and consists of crop rotations ranging from monocultures to a five-species rotation (http://lter.kbs.msu.edu/research/long-term-experiments/biodiversity-gradient/). The crop rotations were repeated but with different rotation phases within all four blocks. For example, the corn–soy–wheat rotation is replicated three times within each block, but these replicates are planted to a different crop each year. The plot dimensions were 9.1 m × 27.4 m and received the same chisel plow tillage to a depth of approximately 15 cm, and received no inputs (e.g., pesticides or fertilizers) that would have confounded the treatment effects of rotation diversity (Smith et al., 2008). Mean annual temperature and precipitation at the site are 9.7 °C and 890 mm. The two main soil series located at the site are Kalamazoo, a fine-loamy, mixed, mesic Typic Hapludalf, and Oshtemo, a coarse-loamy, mixed, mesic Typic Hapludalf (KBS, 2015). Soil pH in the top 10 cm ranges from 4.9 to 6.1 (1 : 1 w of 0.01 M $CaCl_2$).

Soils were collected from the following cropping systems: monoculture corn (*Zea mays* L., mC), corn–soy (*Glycine max*, CS), corn–soy–wheat (*Triticum aestivum*, CSW), corn–soy–wheat with red clover cover crop (*Trifolium pratense*, CSW1), and corn–soy–wheat with red clover + rye cover crops (*Secale cereale*, CSW2). Most of the year there was just one crop per plot, except when red clover cover crops were inter-seeded, and thus overlapped, with the cash crop at the end of the growing season, ca. October (Fig. S1 in the Supplement; Smith et al., 2008). Soil sampling took place on 27 April 2012, 19 July 2012, and 1 November 2012 – hereafter referred to as spring, summer, and autumn. Corn was planted in all plots on 11 June 2012. Three 5 cm diameter soil cores (0–10 cm deep) were collected between rows from each plot, homogenized in the field, and then put on ice and shipped to the University of New Hampshire. In the lab, field-moist soils were immediately sieved using a 2 mm sieve. A subsample was taken from sieved soil and dried at 105 °C to determine gravimetric water content. Water-holding capacity was determined as the water content after soils were saturated and drained for 6 h.

2.1 Soil carbon and nitrogen parameters

Five g of field-moist soil were extracted for inorganic N with 40 mL of 0.5 M K_2SO_4. The soil slurries were shaken for 1 h before the extracts were filtered on Whatman GF/C (5) filters and filtrate frozen and stored until analysis. Soil nitrate (NO_3^-) and ammonium (NH_4^+) were measured using the methods detailed in McDaniel et al. (2014c). We also used the same extracts to measure dissolved organic C and N (DOC and DON). The extracts were run on a TOC-TN analyzer (TOC-V-CPN; Shimadzu Scientific Instruments Inc., Columbia, MD, USA). Total C and N were analyzed by sieving soils through 2 mm sieve, grinding and analyzing on an ECS 4010 CHNSO elemental analyzer (Costech Analytical Technologies, Inc., Valencia, CA).

Potential mineralization rates of C (PMC) and net N (or PMN) estimate the quantity of potentially mineralizable SOM at an optimal temperature and soil moisture, and reflect both the activity of the microbial community and availability of SOM (Paul et al., 1999; Robertson et al., 1999). These mineralization assays provide a good indicator of the potential for a soil to provide plants with N (Stanford and Smith, 1972; Robertson et al., 1999). Both PMC and PMN were measured on 10 g of air-dried soils in Wheaton serum vials and brought to 50 % water-holding capacity, which is near optimal water content for respiration in these soils (Grandy and Robertson, 2007), and incubated for 4 months. During this 4-month period, CO_2 efflux was measured on a LI-820 infrared gas analyzer (LI-COR, Lincoln, NE). Efflux was measured using the change in headspace CO_2 concentration

measured between two time points. Each soil efflux measurement began by aerating jars, capping, and injecting a time-zero sample and then a second sample between 5 h and 2 days later. Efflux was calculated as the difference in CO_2 concentration between the two time points divided by time. Measurements of PMC occurred more frequently at the beginning of the experiment (daily), and became less frequent toward the end (once every other week), for a total of 19 sampling events over 120 days. High-frequency measurements are required during the beginning of these incubations, when respiration rates are high, to prevent build-up of CO_2 (and lack of O_2). The PMN was assessed by extracting the inorganic N ($NH_4^+ + NO_3^-$) produced at the end of the incubation, measuring it with the methods described above, then subtracting this final value from the initial inorganic N extracted before the incubation began.

2.2 Soil microbial parameters

Soil microbial biomass C (MBC) and N (MBN) were determined using the modified chloroform fumigation and extraction method (Vance et al., 1987), but modified for extraction in individual test tubes (McDaniel et al., 2014c). Briefly, two sets of fresh, sieved soil (5 g) were placed in 50 mL test tubes, and 1 mL of chloroform was added to one set of tubes and capped. The tubes sat overnight (24 h) and were then uncapped and exposed to open air in a fume hood to allow chloroform to evaporate. Soils were then extracted in the tubes with 25 ml of 0.5 M K_2SO_4. The chloroform-fumigated and non-fumigated extracts were run on a TOC-TN analyzer (TOC-V-CPN; Shimadzu Scientific Instruments Inc., Columbia, MD, USA). We used 0.45 (Joergensen, 1996) and 0.54 (Brookes et al., 1985) for the C and N extraction efficiencies.

Soils were analyzed for eight extracellular enzyme activities (EEAs): β-1,4-glucosidase (BG), β-D-1,4-cellobiohydrolase (CBH), β-1,4-N-acetyl glucosaminidase (NAG), acid phosphatase (PHOS), tyrosine aminopeptidase (TAP), leucine aminopeptidase (LAP), polyphenol oxidase (PO), and peroxidase (PER). Given the large number of samples (60) and variety of measurements made at each of three sampling dates, soil EEAs were conducted on frozen samples within 4 weeks of sampling. While some studies show freezing has minor effects on EEAs (Peoples and Koide, 2012), others show no effects (Lee et al., 2007; DeForest, 2009), and we assume that any effects of freezing will be consistent among treatments. Extracellular enzyme activity assays were carried out following previously published protocols (Saiya-Cork et al., 2002; German et al., 2011), but with some modifications. Briefly, 1 g of soil was homogenized with a blender in 80 mL of sodium acetate buffer at pH 5.6 (the average pH at the site). Soil slurries were pipetted into 96-well plates and then analyzed on a Synergy 2 plate reader (BioTek Instruments, Inc., Winooski, VT). For oxidoreductase enzymes, the supernatant from the slurry plates were pipetted into a clean plate to avoid interference with soil particles. Hydrolase assays were read at 360/40 and 460/40 fluorescence and oxidoreductases at 450 nm absorbance. For more details on the extracellular enzyme methods see McDaniel et al. (2014c).

Community-level physiological profiles (CLPPs) were conducted using the MicroResp^{TM} system (Chapman et al., 2007; Zhou et al., 2012; McDaniel et al., 2014b). The MicroResp^{TM} system allows for high-throughput measurement of soil catabolic responses to multiple C substrates. Each soil was loaded into 96 deep-well plates using the MicroResp^{TM} soil dispenser, and then brought to 50 % water-holding capacity. Thirty-one substrates were used at concentrations ranging from 7.5 to 30 mg C per gram of soil H_2O, as recommended by the MicroResp^{TM} manual (Table S1 in the Supplement). Soil and substrates were combined in analytical triplicates and a CO_2 detection plate (agar-containing creosol red) was immediately placed onto the deep-well plate with an air tight seal provided by the MicroResp^{TM} kit. The soil and substrates were incubated in the dark for 6 h at 25 °C. The detector plate absorbencies were read at times 0 and 6 h at 540 nm on a Synergy 2 plate reader (BioTek Instruments, Inc., Winooski, VT). Absorbance data were normalized and converted to a CO_2 efflux rate ($\mu g \, CO_2 - C \, g \, soil^{-1} \, h^{-1}$), according to the MicroResp^{TM} procedure (Chapman et al., 2007).

2.3 Data analyses

Cumulative potentially mineralizable C and N were calculated in SigmaPlot v12.5 (Systat Software, Inc., San Jose, CA) using the integration macro *area below curves*. Data not conforming to ANOVA assumptions of homogeneity of variances and normality were transformed before analyses (Zuur et al., 2010). Catabolic evenness (CE), a measure of substrate diversity, was calculated using the Simpson–Yule index, $CE = 1/\Sigma p_i^2$, where p_i is the proportion of a substrate respiration response to the total response induced from all substrates (Degens et al., 2000; Magurran, 2004). Metabolic quotient (qCO_2) was calculated simply as the basal respiration over 6 h (determined in the MicroResp^{TM} method) divided by the MBC. Almost all the soil data were non-normal, including DOC, DON, PMC, PMN, microbial biomass, enzymes, and catabolic evenness. All these data were lognormally transformed, except for catabolic evenness, which was square-root-transformed to meet normality requirements.

Response variables were analyzed using a two-way analysis of variance (ANOVA), with season and rotation as main effects. The ANOVAs were conducted in SAS 9.3 (SAS Institute, Cary, NC) using the *proc mixed* function, and post hoc t tests were used to determine significant differences among means using *ls means*. Block was assigned as a random effect variable within the model. Correlations between variables were made using *proc corr*, and Pearson's corre-

lation coefficients are reported. Model effects were deemed significant if $\alpha < 0.05$.

All multivariate data analyses were performed with R software v. 3.0.0 (the R Foundation for Statistical Computing, Vienna, Austria). CLPP data were checked to ensure they conformed to principal components analysis assumptions. The *prcomp* function in the *vegan* package (Oksanen et al., 2016) was used for PCA of CLPP data. In order to correlate environmental variables with the multivariate CLPP data we used the *envirfit* function.

3 Results

It was a relatively dry year at the KBS-LTER in 2012, which had an annual precipitation of 742 mm, compared to the historical mean of 870 mm (Hamilton et al., 2015). There was also an anomalous warm spell in mid- to late March (Fig. S2). After harvest, the corn yield ($kg \, ha^{-1} \pm SE$) in each treatment was as follows: mC = 2846 ± 152, CS = 4208 ± 575, CSW = 4107 ± 220, CSW1 = 4015 ± 187, and CSW2 = 5219 ± 1180 (KBS, 2015).

3.1 Soil C and N biogeochemistry

There were few significant rotation or season effects on total soil C and N, except that CSW1 had greater N than CSW ($P = 0.040$), although both soil C and N tended to increase with the number of crops in rotation (Table 1). Seasonal soil NO_3^--N concentrations were highest in summer (10.33 ± 2.71), followed by spring (2.98 ± 0.69) and autumn (1.28 ± 0.20 mg kg^{-1}). Soil NH_4^+-N was generally low, but summer had more than twice the concentrations of spring and autumn. Dissolved organic C (DOC) and N (DON) were very dynamic over the year. The DOC was highest in the autumn, while DON was over 6 times greater in the summer than the other seasons ($P < 0.001$). The mean DOC : DON in autumn was 17.4 ± 5.9, 5 times higher than spring and 13 times higher than summer. Soil NO_3^--N was the only variable that showed a significant season × rotation interaction ($P < 0.001$). There were significant main effects of crop rotation on DOC and DON (Table 1). During the summer the two cover crop treatments had the highest NO_3^--N concentrations (16.68 ± 0.87 and 12.14 ± 4.03 mg kg^{-1}), which was 67 % greater than CSW and CS treatments and 158 % greater than mC. The CSW1 treatment had 112 % greater DOC concentrations than mC ($P < 0.001$), and two cover crop treatments had 107 % greater DON than non-cover-crop treatments and 211 % more than the mC treatment.

The potentially mineralizable pools of C and N showed significant main effects of both season and rotation ($P < 0.03$), but no interactions. The PMC was highest during the autumn (636 ± 105 $\mu g \, CO_2 - C \, g \, soil^{-1}$), while PMN was highest during the summer (89 ± 105 $\mu g \, NH_4^+ + NO_3^- \, g \, soil^{-1}$). Generally, both PMC

Table 1. Soil carbon (C) and nitrogen (N) pools by season and crop rotation.

Season	Crop rotation	Total organic C	Total N	NO_3^--N	NH_4^+-N	DOC	DON	C:N	DOC:DON
		g kg^{-1}		mg kg^{-1}					
Spring									
	mC	8.1 (0.8)	0.8 (0.1)ab	2.66 (0.79)	0.06 (0.01)B	14 (4)bB	5 (1)bB	9.8 (0.3)	2.8 (0.2)B
	CS	7.8 (1.2)	0.8 (0.1)ab	2.97 (1.13)	0.06 (0.01)B	11 (1)abB	5 (1)bB	10.3 (0.4)	2.1 (0.2)B
	CSW	7.0 (0.6)	0.7 (0.1)b	2.67 (0.39)	0.10 (0.02)B	21 (8)abB	6 (1)abB	10.4 (0.4)	4.2 (1.9)B
	CSW1	8.7 (0.4)	0.9 (0.1)a	3.10 (0.66)	0.10 (0.02)B	44 (18)aB	8 (1)aB	9.6 (0.2)	5.4 (2.6)B
	CSW2	8.2 (1.4)	0.8 (0.1)ab	3.49 (0.62)	0.12 (0.03)B	26 (7)abB	8 (2)aB	10.2 (0.2)	3.3 (0.4)B
Summer									
	mC	7.9 (0.8)	0.8 (0.1)ab	5.58 (0.67)c	0.08 (0.02)A	35 (4)bB	18 (1)bA	10.2 (0.4)	2.0 (0.1)C
	CS	7.6 (0.9)	0.8 (0.1)ab	9.47 (1.96)b	0.08 (0.01)A	32 (4)bB	33 (7)bA	9.8 (0.1)	1.0 (0.1)C
	CSW	7.6 (0.7)	0.8 (0.0)b	7.76 (0.75)b	0.08 (0.01)A	43 (7)bB	28 (4)abA	9.7 (0.3)	1.6 (0.3)C
	CSW1	8.1 (0.8)	0.9 (0.1)a	16.68 (0.87)a	0.37 (0.22)A	88 (32)aB	76 (8)aA	9.0 (0.2)	1.2 (0.4)C
	CSW2	8.7 (1.1)	0.9 (0.1)ab	12.14 (4.03)ab	0.34 (0.12)A	54 (7)bB	68 (13)aA	9.5 (0.1)	0.8 (0.1)C
Autumn									
	mC	8.1 (0.6)	0.7 (0.1)ab	1.31 (0.15)	0.07 (0.02)B	58 (21)bA	5 (1)bB	11.4 (0.3)	14.3 (7.3)A
	CS	7.7 (1.1)	0.7 (0.1)ab	1.44 (0.28)	0.06 (0.01)B	46 (15)abA	5 (1)bB	10.9 (1.0)	9.6 (3.2)A
	CSW	7.4 (0.8)	0.7 (0.1)b	1.28 (0.30)	0.08 (0.02)B	117(77)abA	6 (2)abB	10.6 (0.6)	15.6 (5.2)A
	CSW1	9.6 (0.6)	0.9 (0.0)a	1.41 (0.06)	0.05 (0.01)B	102 (27)aA	7 (1)aB	10.6 (0.5)	17.1 (7.2)A
	CSW2	8.9 (0.9)	0.9 (0.1)ab	0.96 (0.15)	0.05 (0.01)B	190 (42)abA	6 (1)aB	10.4 (0.4)	30.4 (4.0)A
ANOVA factor				*P* values					
Season		0.756	0.769	**< 0.001**	**0.004**	**< 0.001**	**< 0.001**	0.213	**< 0.001**
Crop rotation		0.298	**0.040**	**< 0.001**	0.084	**0.038**	**< 0.001**	0.223	0.947
Season × rotation		0.994	0.928	**< 0.001**	0.071	0.965	0.221	0.746	0.192

Note: crop rotation abbreviations are monoculture corn (mC), corn–soy (CS), corn–soy–wheat (CSW), corn–soy–wheat with red clover cover crop (CSW1), and corn–soy–wheat with red clover + rye cover crops (CSW2). Means ($n = 4$) are shown with standard errors in parentheses. Significant comparisons (*P* values in bold) are shown among rotations (lowercase) and season (capital) with letters.

and PMN increased with increasing number of crops in rotation (Fig. 1), and the incorporation of cover crops appeared important in regulating both PMC and PMN. For example, the PMC averages of both cover crop treatments (CSW1 and CSW2) were 53 and 41 % greater than mC and CS treatments ($P < 0.042$), respectively. The PMN average from the cover crop treatments was 36, 48, and 72 % greater than the mC, CS, and CSW treatments, respectively ($P < 0.015$). The potentially mineralizable C-to-N ratio (PMC:PMN), considered an index of the quality of accessible SOM (Schimel et al., 1985; Clein and Schimel, 1995), showed a significant season × rotation interaction ($P = 0.045$, Fig. S3). The PMC:PMN was markedly higher in the autumn than in summer and spring, indicating a greater demand for N in autumn. For summer and spring more diverse rotations had less CO_2 produced per unit of net inorganic N mineralized. However, in the autumn, after harvest, the crop rotation effects on the PMC:PMN were reversed, meaning that the more diverse crop rotations had greater CO_2 mineralized per unit of available N (Fig. S3).

3.2 Soil microbial dynamics

The range in soil MBC was 60–1661 µg C g soil^{-1} across all seasons and crop rotations, but both season ($P < 0.001$)

and rotation ($P = 0.008$) had significant effects on MBC (Fig. 2). Soils collected in autumn had more than twice the MBC than those collected in spring and summer. Generally, microbial biomass C was increased by increasing crop diversity across all seasons (Fig. 2), but only CSW1 was 112 and 28 % significantly greater than mC and CS, respectively ($P = 0.023$). Microbial biomass N ranged from 6 to 61 µg N g soil^{-1} and also showed both season ($P < 0.001$) and rotation ($P = 0.005$) effects, but no interaction. Once again, MBN generally increased with crop diversity, with the CSW (57 %), CSW1 (54 %), and CSW2 (50 %) significantly greater than the mC treatment ($P < 0.037$). Microbial biomass C:N showed a significant interaction ($P = 0.013$), with more diverse cropping systems having greater MBC:MBN in summer, but not in the spring or autumn. The metabolic quotient (qCO_2) is often used as a proxy for microbial respiration efficiency (Anderson and Domsch, 1990, 2010; Wardle and Ghani, 1995). Season ($P < 0.001$) and rotation ($P = 0.024$) both influenced qCO_2, with summer showing the greatest qCO_2 (0.11 ± 0.3) and autumn the lowest (0.04 ± 0.1) qCO_2. Crop diversity significantly decreased the qCO_2 in the CSW1 by 40 and 48 % compared to mC and CS.

Soil extracellular enzymes were very dynamic over the three seasons, as evidenced by radar plots in which the

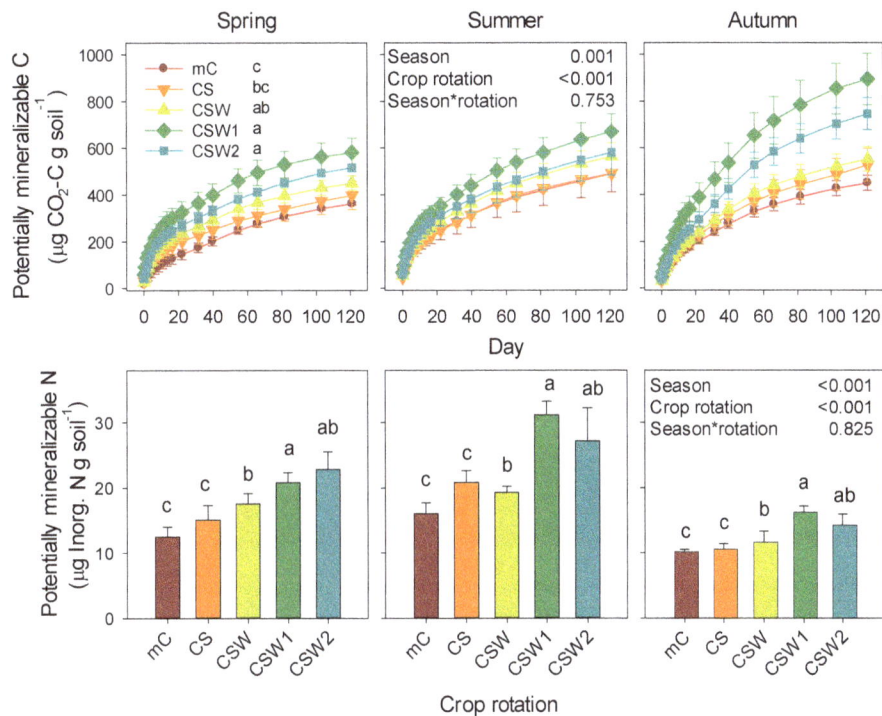

Figure 1. Potentially mineralizable carbon (top row panels) and potentially mineralizable nitrogen (bottom row panels). Crop rotation abbreviations are: monoculture corn (mC), corn–soy (CS), corn–soy–wheat (CSW), corn–soy–wheat with red clover cover crop (CSW1), and corn–soy–wheat with red clover + rye cover crops (CSW2). Means are shown and error bars are standard errors ($n = 4$). P values from ANOVA results are shown for each variable with the main effects (season and crop rotation) and the interaction, as well as significant differences from post hoc results shown as lowercase letters.

area and shape for each treatment change drastically over the growing season (Fig. 3). A MANOVA with all eight EEAs showed significant season ($P < 0.001$) and rotation ($P < 0.001$) main effects, but no interaction. Most individual enzymes showed only significant rotation effects except for PO, which also showed a significant season effect, with autumn greater than the other seasons (Table 2). The soil enzyme responsible for cleaving a glucosamine from chitin (NAG) and the lignin-reducing enzyme that uses peroxide (PER) were the only enzymes that showed a significant season × rotation interaction ($P < 0.001$). Spring had the greatest activities of LAP, 175 % greater than the average of the other seasons (Fig. 3, Table 2). In summer, we see a shift to the highest PHOS activity: 25 % greater than autumn and 99 % greater than spring. There were no main effects of season on BG or CBH, but rotation main effects were significant, with the CSW1 treatment having an average of 42 and 50 % higher BG and CBH activity than CS and mC soils, respectively. The majority of the hydrolase enzymes were higher in the cover crop treatments compared to that of the non-cover-crop treatments, especially mC (Table 2, Fig. 3). The two oxidoreductase enzymes (PO and PER) decreased with crop diversity. There were no significant main effects on the enzyme ratio used to assess C-versus-N demand (BG to NAG + LAP).

The CLPP, a catabolic profile of the soil microbial communities, showed both significant season ($P < 0.001$) and rotation ($P = 0.003$) main effects (Figs. 4 and S4; Table 3). A principal components analysis of the CLPP data showed that the summer soils corresponded with highest carboxylic acid utilization (Fig. 4), as season was the strongest discriminating factor along principal component 1 (PC1, Table 3). However, when rotating and examining PC2 and PC3, there was a strong treatment gradient from the bottom-right to upper-left quadrants of the graph (Fig. 4, right panel). The lower-diversity treatments corresponded with greater use of carboxylic acid substrates. Across seasons, summer exhibited the lowest catabolic evenness (12.9 ± 1.4), but there was no crop rotation effect on catabolic evenness using all substrates (i.e., Full, Table 4).

Due to the overwhelming influence of carboxylic acids in the PCA variation, and their possible role in abiotic reactions leading to CO_2 emissions (Maire et al., 2013; Pietravalle and Aspray, 2013), we split the 31 substrates into two sets to analyze separately: (1) non-carboxylic acid substrates (a total of 21 substrates) and (2) carboxylic acids by themselves (10 substrates). Season, again, was a dominant significant effect on the MANOVAs in both groups of substrates (P values < 0.001, Fig. S5, Tables S2 and S3). The non-carboxylic acid CLPP showed a significant treatment effect

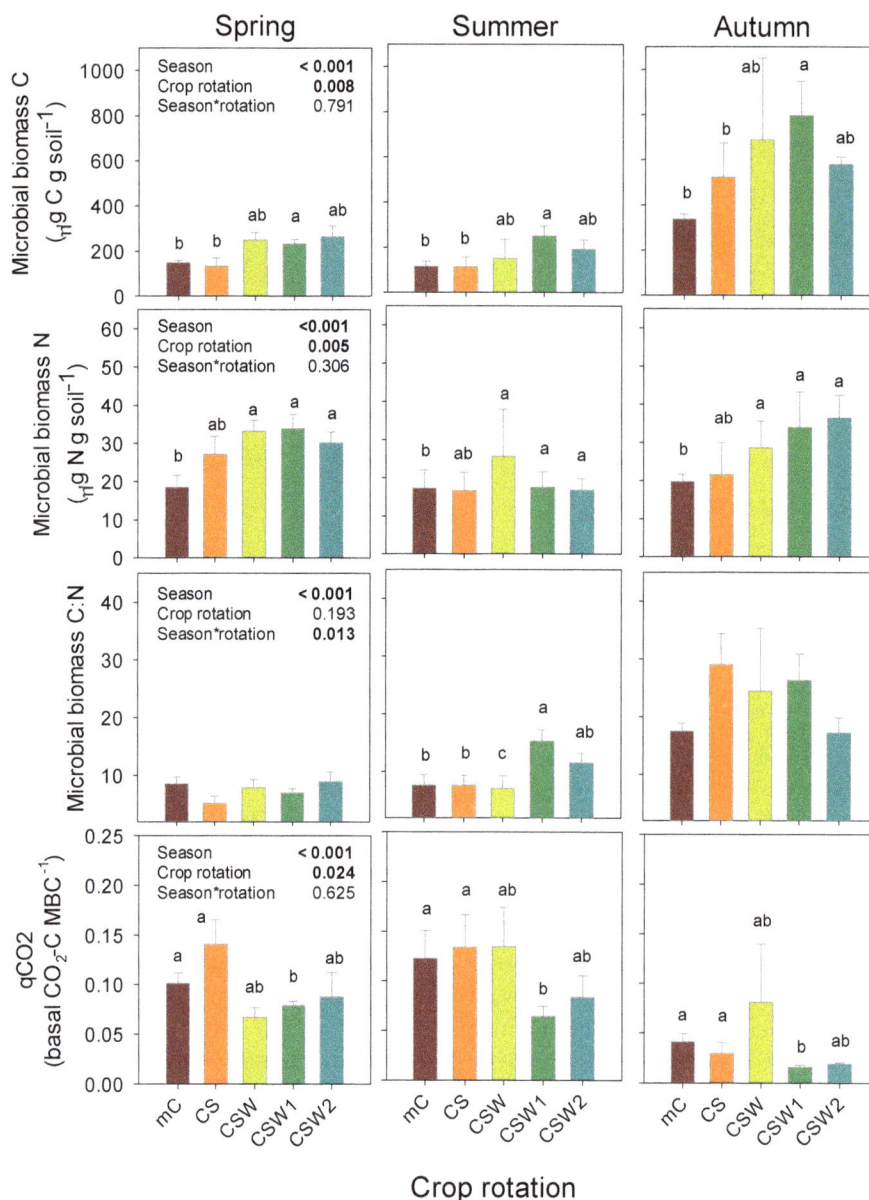

Figure 2. Soil microbial biomass parameters by season and crop rotation. See Fig. 1 for crop rotation abbreviations. Means are shown and error bars are standard errors ($n = 4$). P values from ANOVA results are shown for each variable with the main effects (season and crop rotation) and the interaction, as well as significant differences from post hoc results shown as lowercase letters.

with PC1 and PC2, and clear separation between low- and high-diversity cropping systems ($P = 0.012$, Fig. S4). The monoculture corn, as well as lower-diversity treatments, were associated with more complex substrates. In the carboxylic acid CLPP there was also a significant treatment effect, but with PC2 and PC3, and clear separation between low- and high-diversity cropping systems along PC3 ($P = 0.035$, Fig. S5). The low-diversity treatments (especially monoculture corn) were more associated with simple (lower molecular weight) carboxylic acids (Cit, Mlo, and Mli) on the positive half of PC3. When carboxylic acids were split from the

substrates, crop rotation had a significant effect on catabolic evenness – decreasing the catabolic evenness both within non-carboxylic acids and carboxylic acids by as much as 4 and 13 %, respectively (Table 4).

3.3   Relationships between soil biogeochemical factors, microbial functioning and yield

Over the three seasons many soil biogeochemical factors correlated with microbial catabolic potential, both with individual C substrate guilds and catabolic evenness (Table 5). Abiotic factors such as pH and sand content correlated with the

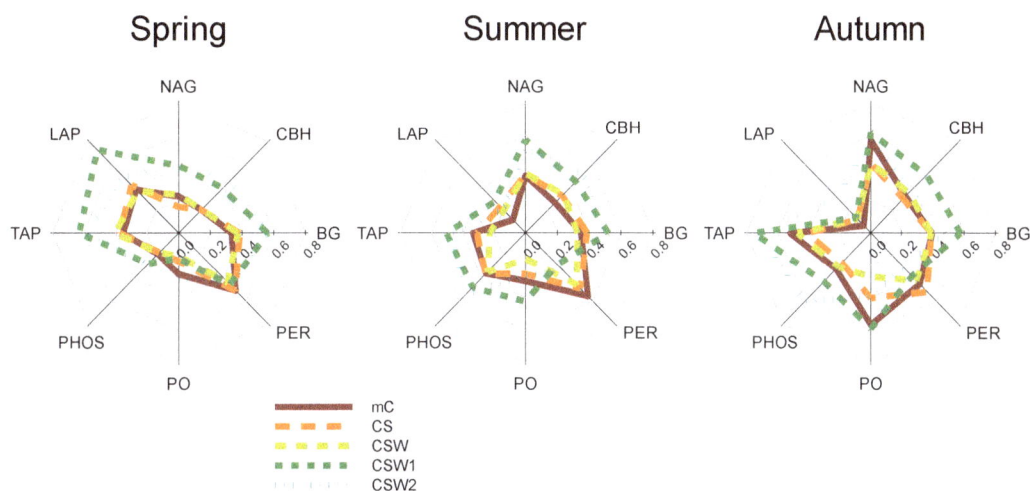

Figure 3. Mean extracellular enzyme activities (EEAs) normalized for the maximum value across all seasons. EEA abbreviations are β-1,4,-glucosidase (BG), β-D-1,4-cellobiohydrolase (CBH), β-1,4,-N-acetyl glucosaminidase (NAG), acid phosphatase (PHOS), tyrosine aminopeptidase (TAP), leucine aminopeptidase (LAP), phenol oxidase (PO), and peroxidase (PER). See Fig. 1 for crop rotation abbreviations.

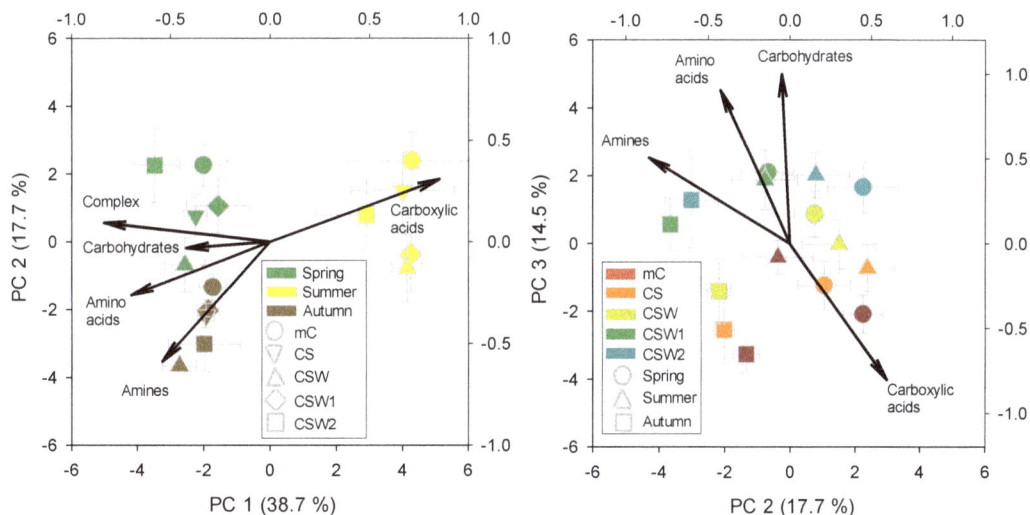

Figure 4. Principal components analysis (PCA) on the respiration response to all 31 substrates. Substrates were grouped into guilds (Table S1) which are the vectors. Left panel: principal components 1 and 2, where season is dominant discriminating factor ($P < 0.001$); right panel: principal components 2 and 3, where rotation is highlighted as a dominant discriminating factor. See also Table 5 for PCA and ANOVA results. Means are shown and error bars are standard errors ($n = 4$). See Fig. 1 for crop rotation abbreviations.

use of particular guilds of substrates. Soil pH positively correlated with N-containing and complex substrates but negatively with carboxylic acids. Sand content negatively correlated with amino acids and carbohydrates but positively with carboxylic acids. The microbial response to amino acids and amines correlated best with NO_3^--N (Table 5) and many of the specific enzyme activities, showing negative relationships which indicated a linkage between demand for N and usage of N-bearing substrates. Soil NO_3^--N was also significantly negatively correlated with catabolic evenness.

We used the soil microbial responses of EEA and the CLPP because we assumed they would be complementary. For example, adding N-acetyl glucosamine in the CLPP should be related to β-1,4-N-acetyl glucosaminidase (NAG) enzyme activity. Indeed, this was the case. Measuring NAG enzyme and adding the NAG amine to the soils showed a somewhat tight relationship, but this changed during autumn (Fig. S6). Additionally, when the CLPP substrates were grouped by guild they were significantly correlated with EEAs (Fig. S7). For instance, total amino acid catabolic response positively correlated well with LAP + TAP enzymes

Table 2. Soil extracellular enzyme activities (EEA) expressed as nanomoles of product per hour per gram of dry soil.

Season	Rotation	BGase	CBHase	LAPase	NAGase	PHOSase	TAPase	PPOase	PERase
						nmol h^{-1} g^{-1} soil			
Spring									
	mC	94 (8)b	27 (2)b	24 (4)bA	27 (2)ab	133 (19)bC	10 (1)abA	140 (47)B	614 (12)a
	CS	107 (18)b	28 (5)b	28 (4)abA	20 (2)b	129 (20)bC	11 (0)abA	100 (30)B	634 (53)a
	CSW	118 (12)ab	31 (4)ab	26 (8)abA	33 (2)ab	152 (7)abC	12 (2)bA	92 (27)B	602 (59)ab
	CSW1	148 (5)a	50 (5)a	43 (5)abA	47 (3)a	188 (17)aC	16 (1)aA	87 (13)B	516 (24)b
	CSW2	153(13)ab	56 (12)ab	33 (5)aA	48 (5)a	208 (8)aC	16 (1)aA	137 (61)B	562 (24)b
Summer									
	mC	100 (5)b	37 (3)b	7 (2)bB	43 (4)	270 (42)bA	9 (2)abB	174 (67)B	676 (88)a
	CS	111 (17)b	43 (10)b	14 (3)abB	44 (7)	291 (25)bA	9 (1)abB	140 (50)B	580 (124)b
	CSW	102 (7)ab	47 (12)ab	14 (2)abB	47 (3)	280 (13)abA	7 (2)bB	96 (29)B	578 (68)b
	CSW1	146 (12)a	61 (10)a	20 (3)abB	69 (10)	370 (45)aA	14 (1)aB	236 (91)B	317 (144)bc
	CSW2	132 (17)ab	62 (14)ab	13 (4)aB	59 (9)	400 (56)aA	12 (1)aB	126 (73)B	392 (97)c
Autumn									
	mC	111 (9)b	44 (6)b	5 (3)bB	67 (13)	238 (57)bB	14 (3)abA	330 (77)A	543 (113)a
	CS	110 (17)b	42 (8)b	8 (1)abB	55 (7)	209 (36)bB	11 (2)abA	234 (64)A	461 (103)bc
	CSW	115 (19)ab	49 (15)ab	9 (2)abB	54 (9)	245 (34)abB	14 (2)bA	176 (18)A	517 (150)b
	CSW1	138 (10)a	59 (6)a	8 (1)abB	63 (13)	277 (42)aB	18 (2)aA	300 (30)A	396 (76)c
	CSW2	117 (15)ab	46 (8)ab	17 (3)aB	63 (2)	308 (24)aB	18 (2)aA	202 (51)A	336 (49)c
ANOVA factor						*P* values			
Season		0.775	0.063	**< 0.001**	**< 0.001**	**< 0.001**	**0.003**	**< 0.001**	**< 0.001**
Crop rotation		**0.017**	**0.006**	**0.007**	**< 0.001**	**0.003**	**0.002**	0.224	**< 0.001**
Season × rotation		0.852	0.839	0.314	**< 0.001**	0.967	0.647	0.837	**< 0.001**

Note: see Table 1 for crop rotation abbreviations. Means ($n = 4$) are shown with standard errors in parentheses. Significant comparisons (*P* values in bold) are shown among rotations (lowercase) and season (capital) with letters.

($r^2 = 0.35$, $P < 0.001$), meaning that high activity of these enzymes in soils corresponded to high relative use of these substrates when added to soils, compared to other substrates added to the soil. This suggests that the LAP and TAP enzymes strongly reflect demand for N-bearing amino acids in soils. However, the catabolic response of the "complex" guild was negatively correlated with PO ($r^2 = 0.29$, $P < 0.001$). Soil PMN was better correlated with crop yields ($r^2 = 0.61$, $P < 0.001$) than NO_3^- in early spring (Fig. S8), highlighting the importance of PMN-like measurements being used as soil fertility tests.

4   Discussion

Increasing biodiversity in this long-term crop rotation experiment has altered the soil microbial dynamics across an entire growing season. This occurred even though the soils in our study were all in the same crop phase (corn) for the season, indicating that observed differences among soils reflect long-term rotation effects. Microbial biomass C, N, potential mineralization, and catabolic potential were all altered by crop rotations, although the rotation effect for some of these indicators of microbial functioning also depends upon the season. Soil microbial biomass and activity are now widely

recognized as pillars of soil health (Doran and Zeiss, 2000). Our results clearly indicate that diversifying agroecosystems (through crop rotations) enhances this aspect of soil health, and is also likely linked to changes in SOM dynamics (Tiemann et al., 2015) as well as the observed differences in yield among crop rotations (Smith et al., 2008; Fig. S8).

4.1   Crop biodiversity and soil microbial functioning

Both soil microbial biomass and functioning were strongly affected by increased crop diversity through rotation. This rotation effect was largely independent of the season, as indicated by the limited number of observed season × rotation interactions. The exception to this was microbial biomass C / N ratio (Fig. 2), potentially mineralizable C-to-N ratio (Figs. 1 and S3), and two extracellular enzyme activities (NAG and PER, Table 2), which together are likely indicative of the enhanced ability of soil microbes under diverse rotations to process, provision, and retain soil N. The stoichiometric shifts in microbial biomass and potentially mineralizable SOM suggest seasonal changes in microbial communities and/or how microbes shift between C and N resources among crop rotations. For instance, the MBC : MBN ratio is only significantly wider in the two cover crop treatments than those without

Table 3. Analysis of variance of results from the principal component analysis of community-level physiological profile (Fig. 4).

ANOVA[1] parameter	PC1		PC2		PC3		PC4		PC5		MANOVA (total)	
Proportion of variance	38.7		17.7		14.5		9		3.8		83.7	
ANOVA factor	F	P value	F	P value	F	P value	F	P value	F	P value	F	P value
Season	**64.02**	**< 0.001**	**22.57**	**< 0.001**	**5.4**	**0.008**	0.68	0.510	**10.33**	**< 0.001**	**33.28**	**< 0.001**
Crop rotation	0.69	0.605	**3.03**	**0.028**	**12.82**	**< 0.001**	0.36	0.834	1.81	0.146	**2.19**	**0.003**
Season × rotation	0.16	0.995	1.22	0.311	0.55	0.81	0.88	0.544	0.27	0.973	0.65	0.949
Significant comparisons[2]	1 = 3 ≠ 2		1 = 2 ≠ 3, CS ≠ CSW1		1 = 2 ≠ 3, mC = CS ≠ CSW = CSW2				1 ≠ 2 = 3			

[1] Degrees of freedom: season, 2; crop rotation, 4; season × rotation, 8. [2] Significant comparison abbreviations: 1. spring; 2, summer; 3, autumn. Note: see Table 1 for crop rotation abbreviations. Significant comparisons are in bold.

during the summer, when inorganic N was plentiful and labile C might have been limiting. On the other hand, during the autumn, when the soils were most N-limited, the potentially mineralizable C-to-N ratio widened in all treatments but was widest among diverse crop rotations (Figs. 1 and S3). Together these findings suggest that labile C might be a major regulating factor of soil N cycling, and that crop rotations change these dynamics.

With regard to provisioning of N, the PMN, MBN, and NAG enzyme activity were greater in soils under more diverse crop rotations during the spring (Figs. 1 and 2, Table 2). NAG has been shown to be strongly related to net N mineralization (Ekenler and Tabatabai, 2002); therefore, the alignment between these two measures of microbial function was not surprising. Taken together, though, these data indicate that soil microbes from diverse rotations might be able to better supply crops with N via mineralization, at this critical stage when corn crop N demand is high (Blackmer et al., 1989). Thus, in this severely N-limited cropping system, it makes sense that spring PMN was better related to yield than soil inorganic N concentrations because these crops are relying almost exclusively on SOM-derived N. Most importantly, it also suggests that the greater provisioning of N from SOM to plants in more diverse cropping systems is a likely factor for the higher yields in our study (Fig. S8). These findings are consistent with plant biodiversity studies that find increased aboveground diversity enhances soil microbial biomass and functioning in natural (Stephan et al., 2000; Zak et al., 2003; Lange et al., 2015) and agricultural ecosystems (Lupwayi et al., 1998; Xuan et al., 2012; McDaniel et al., 2014c).

While there were some significant differences in soil microbial dynamics between the non-cover-crop rotations (CS and CSW) and monoculture corn (Table 1, Figs. 1 and 2), the largest differences were between the two cover crop treatments and monoculture. This was particularly the case for the red-clover-only cover crop treatment (CSW1). A growing number of other studies show the large positive impact cover crops have on soil microbes and their activity (Mendes et al., 1999; Kabir and Koide, 2000; McDaniel et al., 2014c; Mbuthia et al., 2015). The reason cover crops consistently increase soil microbial biomass and activity is likely due to the increased quantity and quality of crop residue inputs, but cover crops have also been shown to improve soil physical properties that enhance biological activity (Williams and Weil, 2004; Schipanski et al., 2014). Another contributing feature of crop diversity via rotation is a greater likelihood of including "keystone" species, such as legumes like soy and red clover used in this study, which may have disproportionally large effects on soils (Wardle, 1999). While total soil N differences are largely undetectable, these legumes in diverse rotations are adding labile residues (including more N) to these N-limited soils, which could also be reflected in the enhanced soil microbial biomass and activity.

We hypothesized that increasing crop diversity through rotation would result in soil microbial communities that are

Table 4. Catabolic evenness by season and crop rotation (showing full suite of C substrates, no carboxylic acids, and carboxylic acids only).

Season	Rotation	Catabolic evenness		
		Full	No carboxylic acids	Carboxylic acids only
Spring				
	mC	24.37 (0.79)A	20.20 (0.05)aA	7.60 (0.23)aB
	CS	23.79 (0.91)A	19.80 (0.15)aA	7.21 (0.13)abB
	CSW	22.98 (0.63)A	19.65 (0.15)bA	6.56 (0.35)bB
	CSW1	24.28 (0.44)A	18.95 (0.19)abA	6.91 (0.12)abB
	CSW2	24.52 (0.72)A	19.75 (0.24)bA	6.90 (0.31)bB
Summer				
	mC	14.99 (1.61)B	18.95 (0.59)aa	4.91 (0.54)aC
	CS	12.86 (1.77)B	20.20 (0.18)aA	4.32 (0.38)abC
	CSW	12.10 (1.02)B	19.82 (0.54)bA	3.93 (0.20)bC
	CSW1	13.83 (1.65)B	18.59 (0.83)abA	4.34 (0.50)abC
	CSW2	12.78 (0.92)B	19.24 (0.51)bA	3.75 (0.11)bC
Autumn				
	mC	25.81 (0.79)A	19.62 (0.16)aB	8.47 (0.24)aA
	CS	25.82 (0.55)A	19.11 (0.22)aB	8.41 (0.22)abA
	CSW	25.71 (0.74)A	18.98 (0.28)bB	8.12 (0.61)bA
	CSW1	27.41 (0.63)A	18.63 (0.12)abB	8.90 (0.24)abA
	CSW2	26.08 (0.67)A	18.17 (0.28)bB	8.11 (0.08)bA
ANOVA factor				
Season		**< 0.001**	**0.002**	**< 0.001**
Crop rotation		0.357	**0.035**	**0.028**
Season × rotation		0.928	0.058	0.807

Note: see Table 1 for crop rotation abbreviations. Means ($n = 4$) are shown with standard errors in parentheses. Significant comparisons (P values in bold) are shown among rotations (lowercase) and season (capital) with letters.

more diverse and thus would more evenly use added C substrates (i.e., increase catabolic evenness, or decrease the variation in use among substrates). This hypothesis stems from arguments that soil community and functional biodiversity is linked to plant biodiversity, mostly through the diversity of plant inputs to SOM (Lodge, 1997; Hooper et al., 2000; Waldrop et al., 2006; Korboulewsky et al., 2016). However, in our study, we found no evidence that crop rotational diversity increased overall soil catabolic evenness (Table 4). There is some evidence that crop rotations can alter soil bacterial catabolic diversity or the ability to use different C substrates (Lupwayi et al., 1998; Larkin, 2003; Govaerts et al., 2007); however, all of these studies used Biolog, which has several limitations (Preston-Mafham et al., 2002). The MicroResp[TM] system's main benefit is that it adds C substrates directly to the soil instead of transferring an inoculum from a soil slurry. The discrepancy between our study and these other studies may be due to methodological differences between Biolog and MicroResp[TM]. Our lack of evidence for an aboveground–belowground link to catabolic potential aligns with findings from other studies that have found functional diversity measures of soil microbes are not related

to plant diversity (Bartelt-Ryser et al., 2005; Jiang et al., 2012), nor plant species in general (McIntosh et al., 2013). Both Jiang et al. (2012) and McIntosh et al. (2013) used the same MicroResp[TM] method used in this paper, while Bartelt-Ryser et al. (2005) used Biolog.

In our study, when a subset of the C substrates were analyzed (all non-carboxylic acids, or carboxylic acids only), we found that increased crop diversity decreased catabolic evenness (Table 4). This is unusual considering that soils from this same study, but collected a year prior, showed increases in soil biodiversity (Shannon–Weiner index or H') with increased crop diversity when measuring phospholipid fatty acids (Tiemann et al., 2015); in addition, diversity has been found to be strongly, positively related to species evenness in plants and animals (Stirling and Wilsey, 2001). In this study, our findings of a lack of an effect (or even a negative effect) of crop biodiversity on catabolic evenness is also contradictory to the findings of Degens et al. (2000), who showed that management practices that decreased soil C are associated with low catabolic evenness. However, evidence from these same soil samples showed that crop diversity significantly decreased H' for bacterial 16S rRNA by as much as 5 %

Table 5. Pearson correlation coefficients between soil properties and community-level physiological profile (CLPP) parameters.

Soil variable	Substrate guilds					Catabolic evenness		
	Amino acids	Amine	Carboxylic acids	Carbohydrates	Complex	Full	No carboxylic acids	Only carboxylic acids
Water content	ns	ns	ns	ns	ns	0.40	ns	0.52
pH	0.27	**0.43**	**−0.41**	ns	**0.53**	**0.68**	ns	**0.74**
Sand	**−0.36**	ns	0.28	−0.27	ns	ns	ns	ns
Silt	0.30	ns	ns	ns	ns	ns	ns	ns
Clay	ns	ns	ns	ns	ns	ns	−0.33	ns
Total C	ns	ns	ns	ns	ns	ns	**−0.40**	ns
Total N	ns	ns	ns	ns	ns	ns	**−0.40**	ns
C-to-N ratio	ns	0.27	ns	ns	0.30	**0.45**	ns	**0.53**
NH_4^+	ns	−0.31	0.33	ns	**−0.37**	**−0.40**	ns	**−0.38**
NO_3^-	**−0.58**	**−0.55**	**0.66**	−0.30	**−0.72**	**−0.74**	ns	**−0.70**
PMC	ns	0.29	ns	ns	ns	ns	**−0.63**	ns
PMN	ns	−0.27	0.32	ns	**−0.55**	**−0.49**	ns	**−0.52**
MBC	0.31	**0.49**	−0.37	ns	ns	**0.41**	−0.38	0.47
MBN	**0.36**	0.34	−0.37	**0.42**	ns	**0.36**	ns	0.31
MBC : MBN	ns	**0.40**	ns	ns	ns	0.31	−0.34	**0.40**
BGase	ns	**−0.43**	0.30	ns	ns	−0.29	0.32	−0.28
CBHase	−0.32	**−0.47**	**0.39**	−0.27	ns	−0.33	ns	−0.28
LAPase	ns	−0.29	ns	ns	ns	ns	0.49	ns
TAPase	ns	**−0.37**	ns	ns	ns	ns	ns	0.37
NAGase	**−0.35**	**−0.56**	**0.47**	**−0.39**	−0.29	**−0.46**	0.29	**−0.41**
PHOSase	**−0.45**	**−0.66**	**0.56**	**−0.46**	**−0.34**	**−0.63**	0.34	**−0.60**
PPOase	−0.38	−0.33	0.37	−0.31	ns	ns	ns	ns
PERase	−0.40	**−0.54**	0.42	−0.37	ns	−0.30	0.43	ns

Note: only significant correlations are shown (*P* values < 0.05); bold values are *P* < 0.01; ns = non-significant.

compared to monoculture corn (Peralta et al., 2016). Taken together, the decrease in functional and structural diversity of soil bacteria with crop diversity indicates that crop diversity might decrease bacterial diversity in this crop rotation experiment. Nevertheless, a recent meta-analysis showed that crop rotations tend to increase soil biodiversity by 3 % and richness by 15 % (Venter et al., 2016), but there was large variability around these estimates. Regardless of aboveground–belowground diversity trends, crop rotations did create functionally distinct microbial communities in our study (Fig. 4). We still do not have a good understanding of how crop rotations alter soil microbial dynamics, nor (arguably more importantly) how these changes in belowground communities might provide beneficial soil ecosystem services like increasing soil C or mineralizing more N to increase crop yields.

One trend that emerges across the suite of 31 C substrates is that crop rotations altered soil microbial preference for C substrates (i.e., complex versus simple C substrates). The soils from monoculture corn corresponded to greater use of simple C substrates (especially carboxylic acids), and showed less response to the suite of N-containing and complex substrates (Fig. 4). This finding corroborates a previous study we conducted using whole-plant residues, in which we showed diverse crop rotations resulted in greater decomposition of low-quality crop residues (e.g., corn and wheat; McDaniel et al., 2014c). Further, when looking only within the relatively labile carboxylic acid substrates, microbial communities in the less diverse crop rotations (mC, and to a lesser extent CS) responded to more labile, low-molecular-weight carboxylic acids (e.g., citric, malonic, and malic acid), while soil microbes from more diverse crop rotations responded more to complex, higher-molecular-weight carboxylic acids (e.g., caffeic, tartaric, and vanillic acids – Fig. S5d). The strong effects of crop diversity on catabolism of carboxylic acids is not surprising due to the small, yet dynamic, pool of these compounds in soil (Strobel, 2001). Since soil microbial function (as measured by CLPP) is an aggregate measure of both the community composition and available resources, it is impossible to tease out which (or both) have changed due to increased crop biodiversity. However, our overall findings indicate that increased aboveground biodiversity through crop rotations and cover crops appears to facilitate soil microbial communities' use of complex C substrates relative to simple ones.

4.2   Seasonal dynamics and N limitation

Season strongly influenced the measured pools of labile C and N (Table 1), as well as the microbial biomass size and functioning within this agroecosystem (Figs. 1–4). We hypothesized that soil microbial function would converge over the growing season, as the current crop exerted greater influence over soil microbes. We did find some support for this hypothesis. Both multivariate measures of extracellular enzyme activities and CLPP showed treatments becoming more similar over the growing season (Figs. 3 and 4). This is based on three time points, however, and we do not know for sure whether this convergence was due to the influence of the corn crop or other factors (like microclimatic). Some studies have shown that the current plant species identity often trumps biodiversity legacy in controlling belowground microbial structure and functioning (Stephan et al., 2000; Wardle et al., 2003; Bartelt-Ryser et al., 2005). Conversely, several studies have pointed to weak or no influence of current plant species on soil microbial structure and functioning (Costa et al., 2006; Kielak et al., 2008). The question of whether plant species identity versus spatial and temporal diversity has a stronger control on soil biota remains a critical question in terrestrial ecology.

The greatest microbial biomass and activity occurred in autumn, but potential N mineralization peaked in summer. In perennial and annual cropping systems in Iowa, potentially mineralizable N declined from spring to late summer; in addition, extracellular enzyme activities peaked in July, but there was little effect of the cropping system (Hargreaves and Hofmockel, 2014). In another study, season was shown to affect microbial biomass and potentially mineralizable C and N pools in a wheat–sorghum–soybean rotation in south-central Texas (Franzluebbers et al., 1994, 1995; Franzluebbers, 2002), but timing for peak values differed depending on the study and cropping systems, likely reflecting different climates and soil types. The frequently observed late-summer spike in microbial biomass and activity may be related to higher temperatures during this time period; however, even within agroecosystems, the timing for maximal microbial biomass varies substantially, although few microbial biomass maxima are reported in winter (Wardle, 1992). Our findings highlight the dynamic nature of soil microbial biomass and activity, especially with regard to the supply and demand of N (e.g., microbial C : N, substrate utilization, and extracellular enzyme activities), which is likely a limiting nutrient in these agroecosystems that are receiving no exogenous N inputs.

The summer warrants discussion because the sample was collected after a prolonged period of hot and dry days, but right after a large rainfall event. This rainfall event ($> 18 \, \text{mm} \, \text{day}^{-1}$, Fig. S2) increased the volumetric water content in the 0–10 cm of a nearby soil by over 54 % from the lowest value of the year (0.1 m m^{-3}, data shared from Hamilton et al., 2015), and we know from previous research

that drying–wetting cycles are important soil biogeochemical drivers (Borken and Matzner, 2009) and can alter microbial structure and functioning (Fierer et al., 2003; Schimel et al., 2007; Tiemann and Billings, 2011; McDaniel et al., 2014b). Indeed, the summer showed several signs of the soil microbial community being impacted by a rapid dry–wet event: lower overall microbial biomass C, high NO$_3^-$-N concentrations (Table 1), high potential N mineralization (Fig. 1), high extracellular enzyme activities per unit of microbial biomass (Fig. S9, presumably a result of lysed intracellular enzymes; Burns et al., 2013), and the particularly strong response of the summer soils to carboxylic acids (a highly labile class of compounds used by fast-growing, opportunistic microbes that would be found after a disturbance such as a dry–wet event, Figs. 4 and S3). Dry–wet cycles may drive microbial C and N to be reallocated to stress-response compounds instead of growth or reproduction, making C and N more vulnerable to loss from soils (Schimel et al., 2007). We captured one of these dry–wet events during one of the driest summers in the Kellogg Biological Station LTER's history and we show high soil inorganic N concentrations and altered microbial dynamics relative to the other dates. Climate change may increase the frequency and magnitude of these rapid dry–wet cycles (Groffman et al., 2001; McDaniel et al., 2014d) and thus may have long-term impacts on soil microbial functioning and biogeochemistry.

In the autumn we found several lines of evidence that indicate soil microbes are N, rather than C, limited. These lines of evidence include: lowest soil inorganic N concentrations, low potentially mineralizable N, high-microbial-biomass C : N and DOC : DON ratios, high TAP and NAG enzymes relative to other enzymes (although interestingly not LAP), and finally strong respiration response to the addition of amines and amino acids (Fig. 4). The unusually wide microbial biomass C : N in autumn was very surprising (mean of 24 vs. 10 and 8 in spring and summer, respectively), but microbial biomass C : N has been known to be as high as 30 in laboratory conditions (Schimel et al., 1989). Additionally, the few days before and after the collection of the autumn sample were unusually cold (Fig. S2), and cold temperatures and freezing can cause accumulation of carbohydrates in fungi (Tibbett et al., 2002), which could also widen t he microbial C : N ratio. While environmental conditions may be a factor in the microbial biomass C : N, it is likely that N limitation is a major factor in these long-term, unfertilized, agroecosystems.

5   Conclusions

As the growing population is increasingly reliant on soils for food, fiber, and fuel, we will either need to consume less, put more land into production, or better use the land we already have in production. Putting more land in production will likely result in declines in local and global biodiversity. Thus, it is critical to incorporate biodiversity through any means possible into the existing managed ecosystems – even including biodiversity through time as with crop rotations. Here we show that both microbial biomass and function are strongly influenced by cropping diversity. In fact, the influence of crop rotations on soil microbes and functioning lasts over an entire growing season and even when all soils are under the same crop. Crop rotations clearly enhance soil microbial biomass and activity, which are now considered a pillar of soil health, and it appears from our study that rotations also facilitate microbes in supplying more soil N to crops (Fig. S8). Overall, our study highlights the importance of incorporating biodiversity into agroecosystems by including more crops in rotation, especially cover crops, to enhance beneficial soil processes controlled by soil microbes.

6 Data availability

Core data will be made available at the Kellogg Biological Station Long-term Ecological Research website (http://lter.kbs.msu.edu/datatables). Otherwise, interested parties may email the corresponding author for data sets.

Acknowledgements. Support for this research was also provided by the NSF Long-Term Ecological Research Program (DEB 1027253) at the Kellogg Biological Station and by Michigan State University AgBioResearch. We are grateful for financial support from the United States Department of Agriculture (USDA) Soil Processes Program, grant #2009-65107-05961. Also, financial support came from USDA grant #2015-42247-519119. We would like to acknowledge both Kay Gross and Phil Robertson, who originally established these sites and have kindly provided our research team with access to them. Thanks to Stephen Hamilton and co-authors who provided soil microclimate data from a nearby experiment. Also, we would like to thank Serita Frey for helpful advice dealing with the CLPP data, and Christopher Fernandez for giving feedback on an early draft of the manuscript. Finally, we would like to thank the three anonymous reviewers, whose very valuable feedback improved the quality of this manuscript.

Edited by: E. Bach

References

Anderson, T. H. and Domsch, K. H.: Application of eco-physiological quotients (qCO$_2$ and qD) on microbial biomasses from soils of different cropping histories, Soil Biol. Biochem., 22, 251–255, 1990.

Anderson, T. H. and Domsch, K. H.: Soil microbial biomass: The eco-physiological approach, Soil Biol. Biochem., 42, 2039–2043, 2010.

Bartelt-Ryser, J., Joshi, J., Schmid, B., Brandl, H., and Balser, T.: Soil feedbacks of plant diversity on soil microbial communities and subsequent plant growth, Perspect. Plant Ecol. 7, 27–49, 2005.

Blackmer, A. M., Pottker, D., Cerrato, M. E., and Webb, J.: Correlations between soil nitrate concentrations in late spring and corn yields in Iowa, J. Prod. Agr., 2, 103–109, 1989.

Borken, W. and Matzner, E.: Reappraisal of drying and wetting effects on C and N mineralization and fluxes in soils, Global Change Biol., 15, 808–824, 2009.

Brookes, P. C., Landman, A., Pruden, G., and Jenkinson, D. S.: Chloroform fumigation and the release of soil nitrogen: A rapid direct extraction method to measure microbial biomass nitrogen in soil, Soil Biol. Biochem., 17, 837–842, 1985.

Burns, R. G., DeForest, J. L., Marxsen, J., Sinsabaugh, R. L., Stromberger, M. E., Wallenstein, M. D., Weintraub, M. N., and Zoppini, A.: Soil enzymes in a changing environment: Current knowledge and future directions, Soil Biol. Biochem., 58, 216–234, 2013.

Carpenter-Boggs, L., Picul Jr., J. L., Vigil, M. F., and Riedell, W. E.: Soil nitrogen mineralization influenced by crop rotation and nitrogen fertilization, Soil Sci. Soc. Am. J., 64, 2038–2045, 2000.

Chapman, S., Campbell, C., and Artz, R.: Assessing CLPPs using MicroResp^{TM}, J. Soils Sediments, 7, 406–410, 2007.

Clein, J. S. and Schimel, J. P.: Nitrogen turnover and availability during succession from alder to poplar in Alaskan taiga forests, Soil Biol. Biochem., 27, 743–752, 1995.

Costa, R., Götz, M., Mrotzek, N., Lottmann, J., Berg, G., and Smalla, K.: Effects of site and plant species on rhizosphere community structure as revealed by molecular analysis of microbial guilds, FEMS Microbiol. Ecol., 56, 236–249, 2006.

DeForest, J. L.: The influence of time, storage temperature, and substrate age on potential soil enzyme activity in acidic forest soils using MUB-linked substrates and l-DOPA, Soil Biol. Biochem., 41, 1180–1186, 2009.

Degens, B. P., Schipper, L. A., Sparling, G. P., and Vojvodic-Vukovic, M.: Decreases in organic C reserves in soils can reduce the catabolic diversity of soil microbial communities, Soil Biol. Biochem., 32, 189–196, 2000.

Doran, J. W. and Zeiss, M. R.: Soil health and sustainability: managing the biotic component of soil quality, Appl. Soil Ecol., 15, 3–11, 2000.

Eisenhauer, N., Beßler, H., Engels, C., Gleixner, G., Habekost, M., Milcu, A., Partsch, S., Sabais, A. C. W., Scherber, C., Steinbeiss, S., Weigelt, A., Weisser, W. W., and Scheu, S.: Plant diversity effects on soil microorganisms support the singular hypothesis, Ecology, 91, 485–496, 2010.

Ekenler, M. and Tabatabai, M. A.: ß-Glucosaminidase activity of soils: effect of cropping systems and its relationship to nitrogen mineralization, Biol. Fert. Soils, 36, 367–376, 2002.

Fierer, N., Schimel, J. P., and Holden, P. A.: Influence of drying-rewetting frequency on soil bacterial community structure, Microb. Ecol., 45, 63–71, 2003.

Franzluebbers, A. J.: Soil organic matter stratification ratio as an indicator of soil quality, Soil Till. Res., 66, 95–106, 2002.

Franzluebbers, A. J., Hons, F. M., and Zuberer, D. A.: Seasonal changes in soil microbial biomass and mineralizable C and N in wheat management systems, Soil Biol. Biochem., 26, 1469–1475, 1994.

Franzluebbers, A. J., Hons, F. M., and Zuberer, D. A.: Soil organic carbon, microbial biomass, and mineralizable carbon and nitrogen in sorghum, Soil Sci. Soc. Am. J., 59, 460–466, 1995.

German, D. P., Weintraub, M. N., Grandy, A. S., Lauber, C. L., Rinkes, Z. L., and Allison, S. D.: Optimization of hydrolytic and oxidative enzyme methods for ecosystem studies, Soil Biol. Biochem., 43, 1387–1397, 2011.

Govaerts, B., Mezzalama, M., Unno, Y., Sayre, K. D., Luna-Guido, M., Vanherck, K., Dendooven, L., and Deckers, J.: Influence of tillage, residue management, and crop rotation on soil microbial biomass and catabolic diversity, Appl. Soil Ecol., 37, 18–30, 2007.

Grandy, A. S. and Robertson, G. P.: Land-use intensity effects on soil organic carbon accumulation rates and mechanisms, Ecosystems 10, 58–73, 2007.

Groffman, P., Driscoll, C., Fahey, T., Hardy, J., Fitzhugh, R., and Tierney, G.: Effects of mild winter freezing on soil nitrogen and carbon dynamics in a northern hardwood forest, Biogeochemistry, 56, 191–213, 2001.

Guckert, J. B., Carr, G. J., Johnson, T. D., Hamm, B. G., Davidson, D. H., Kumagai, Y.: Community analysis by Biolog: curve integrationf ro statistical analysis of activated slude microbial habitats, J. Microb. Meth., 27, 183–197, 1996.

Hamilton, S. K., Hussain, M. Z., Bhardwaj, A. K., Basso, B., and Robertson, G. P.: Comparative water use by maize, errennial crops, restored prairie, and poplar trees in the US Midwest, Environ. Res. Lett., 10, 064015, doi:10.1088/1748-9326/10/6/064015, 2015.

Hargreaves, S. K. and Hofmockel, K. S.: Physiological shifts in the microbial community drive changes in enzyme activity in a perennial agroecosystem, Biogeochemistry, 117, 67–79, 2014.

Hooper, D. U., Bignell, D. E., Brown, V. K., Brussard, L., Dangerfield, M. J., Wall, D. H., Wardle, D. A., Coleman, D. C., Giller, K. E., Lavelle, P., Van Der Putten, W. H., De Ruiter, P. C., Rusek, J., Silver, W. L., Tiedje, J. M., and Wolters, V.: Interactions between aboveground and belowground biodiversity in terrestrial ecosystems: Patterns, mechanisms, and feedbacks, BioScience, 50, 1049–1061, 2000.

Jiang, Y., Chen, C., Xu, Z., and Liu, Y.: Effects of single and mixed species forest ecosystems on diversity and function of soil microbial community in subtropical China, J. Soils Sediments, 12, 228–240, 2012.

Joergensen, R. G.: The fumigation-extraction method to estimate soil microbial biomass: Calibration of the kEC value, Soil Biol. Biochem., 28, 25–31, 1996.

Kabir, Z. and Koide, R. T.: The effect of dandelion or cover crop on mycorrhiza inoculum potential, soil agregation and yield of maize, Agr. Ecosyst. Environ., 78, 167–174, 2000.

Kallenbach, C. M., Grandy, A. S., Frey, S. D., and Diefendorf, A. F.: Microbial physiology and necromass regulate agricultural soil carbon accumulation, Soil Biol. Biochem., 91, 279–290, 2015.

KBS – Kellogg Biological Station Long-term Ecological Research: http://lter.kbs.msu.edu/, http://lter.kbs.msu.edu/datatables/75, last access: May 2015.

Kielak, A., Pijl, A. S., Van Veen, J. A., and Kowalchuk, G. A.: Differences in vegetation composition and plant species identity lead to only minor changes in soil-borne microbial communities in a former arable field, FEMS Microb. Ecol., 63, 372–382, 2008.

Korboulewsky, N., Perez, G., and Chauvat, M.: How tree diversity affects soil fauna diversity: A review, Soil Biol. Biochem., 94, 94–106, 2016.

Krupinsky, J. M., Bailey, K. L., McMullen, M. P., Gossen, B. D., and Turkington, T. K.: Managing plant disease risk in diversified cropping systems, Agron. J., 94, 198–209, 2002.

Lange, M., Eisenhauer, N., Sierra, C. A., Bessler, H., Engels, C., Griffiths, R. I., Mellado-Vázquez, P. G., Malik, A. A., Roy, J., Scheu, S., and Steinbeiss, S.: Plant diversity increases soil microbial activity and soil carbon storage, Nat. Comm., 6, 6707, doi:10.1038/ncomms7707, 2015.

Larkin, R. P.: Characterization of soil microbial communities under different potato cropping systems by microbial population dynamics, substrate utilization, and fatty acid profiles, Soil Biol. Biochem., 35, 1451–1466, 2003.

Lee, Y. B., Lorenz, N., Dick, L. K., and Dick, R. P.: Cold storage and pretreatment incubation effects on soil microbial properties, Soil Sci. Soc. Am. J., 71, 1299–1305, 2007.

Lodge, D. J.: Factors related to diversity of decomposer fungi in tropical forests, Biodivers. Conserv., 6, 681–688, 1997.

Lupwayi, N. Z., Rice, W. A., and Clayton, G. W.: Soil microbial diversity and community structure under wheat as influenced by tillage and crop rotation, Soil Biol. Biochem., 30, 1733–1741, 1998.

Magurran, A. E.: Measuring Biological Diversity, John Wiley & Sons, Malden, MA, 2004.

Maire, V., Alvarez, G., Colombet, J., Comby, A., Despinasse, R., Dubreucq, E., Joly, M., Lehours, A.-C., Perrier, V., Shahzad, T., and Fontaine, S.: An unknown oxidative metabolism substantially contributes to soil CO_2 emissions, Biogeosciences, 10, 1155–1167, doi:10.5194/bg-10-1155-2013, 2013.

Mbuthia, L. W., Acosta-Martínez, V., DeBruyn, J., Schaeffer, S., Tyler, D., Odoi, E., Mpheshea, M., Walker, F., and Eash, N.: Long term tillage, cover crop, and fertilization effects on microbial community structure, activity: Implications for soil quality, Soil Biol. Biochem., 89, 24–34, 2015.

McDaniel, M. D., Grandy, A. S., Tiemann, L. K., and Weintraub, M. N.: Crop rotation complexity regulates the decomposition of high and low quality residues, Soil Biol. Biochem., 78, 243–254, 2014a.

McDaniel, M. D., Kaye, J. P., Kaye, M. W., and Bruns, M. A.: Climate change interactions affect soil carbon dioxide efflux and microbial functioning in a post-harvest forest, Oecologia, 174, 1437–1448, 2014b.

McDaniel, M. D., Tiemann, L. K., and Grandy, A. S.: Does agricultural crop diversity enhance soil microbial biomass and organic matter dynamics? a meta-analysis, Ecol. Appl., 24, 560–570, 2014c.

McDaniel, M. D., Wagner, R. J., Rollinson, C. R., Kimball, B. A., Kaye, M. W., and Kaye, J. P.: Microclimate and ecological threshold responses in a warming and wetting experiment following whole tree harvest, Theor. Appl. Climatol., 116, 287–299, 2014d.

McIntosh, A. C. S., Macdonald, S. E., and Quideau, S. A.: Linkages between the forest floor microbial community and resource heterogeneity within mature lodgepole pine forests, Soil Biol. Biochem., 63, 61–72, 2013.

Mendes, I. C., Bandick, A. K., Dick, R. P., and Bottomley, P. J.: Microbial biomass and activities in soil aggregates affected by winter cover crops, Soil Sci. Soc. Am. J., 63, 873–881, 1999.

Mueller, K. E., Hobbie, S. E., Tilman, D., and Reich, P. B.: Effects of plant diversity, N fertilization, and elevated carbon dioxide on grassland soil N cycling in a long-term experiment, Global Change Biol., 19, 1249–1261, 2013.

Oksanen, J, Blanchet, F. G., Kindt, R., Legendre, P., Minchin, P. R., O'hara, R. B., Simpson, G. L., Solymos, P. M., Stevens, H., and Wagner, H.: Vegan: Community Ecology Package, R package version 2.3-3, https://CRAN.R-project.org/package=vegan, last access: 26 January 2016.

Paul, E. A., Harris, D., Collins, H. P., Schulthess, U., Robertson, G. P.: Evolution of CO_2 and soil carbon dynamics in biologically managed, row-crop agroecosystems, Appl. Soil Ecol., 11, 53–65, 1999.

Peoples, M. S. and Koide, R. T.: Cosiderations in the storage of soil samples for enzyme activity analysis, Appl. Soil Ecol., 62, 98–102, 2012.

Peralta, A. L., Sun, Y., Brewer, M. S., McDaniel, M. D., and Lennon, J. T.: Crop diversity enhances disease suppressive potential in soils, Soil Biol. Biochem., in review, 2016.

Pietravalle, S. and Aspray, T. J.: CO_2 and O_2 respiration kinetics in hydrocarbon contaminated soils amended with organic carbon sources used to determine catabolic diversity, Environ. Poll., 176, 42–47, 2013.

Preston-Mafham, J., Boddy, L., and Randerson, P. F.: Analysis of microbial community functional diversity using sole-carbon-source utilisation profiles – a critique, FEMS Microbiol. Ecol., 42, 1–14, 2002.

Riedell, W. E., Pikul, J. L., Jaradat, A. A., and Schumacher, T. E.: Crop rotation and nitrogen input effects on soil fertility, maize mineral nutrition, yield, and seed composition, Agron. J., 101, 870–879, 2009.

Robertson, G. P., Coleman, D. C., Bledsoe, C. S., and Sollins, P.: Standard Soil Methods for Long-Term Ecological Research, Oxford University Press, New York, 1999.

Ryszkowski, L., Szajdak, L., and Karg, J.: Effects of continuous cropping of rye on soil biota and biochemistry, CRC CR Rev. Plant Sci.,17, 225–244, 1998.

Saiya-Cork, K. R., Sinsabaugh, R. L., and Zak, D. R.: The effects of long term nitrogen deposition on extracellular enzyme activity in an *Acer saccharum* forest soil, Soil Biol. Biochem., 34, 1309–1315, 2002.

Schimel, D. S., Coleman, D. C., and Horton, K. A.: Soil organic matter dynamics in paired rangeland and cropland toposequences in North Dakota, Geoderma, 36, 201–214, 1985.

Schimel, J. P., Scott, W. J., and Killham, K.: Changes in cytoplasmic carbon and nitrogen pools in a soil bacterium and a fungus in response to salt stress, Appl. Environ. Microbiol., 55, 1635–1637, 1989.

Schimel, J. P., Balser, T. C., and Wallenstein, M. D.: Microbial stress-response physiology and its implications for ecosystem function, Ecology, 88, 1386–94, 2007.

Schipankski, M. E., Barbercheck, M., Douglas, M. R., Finney, D. M., Haider, K., Kaye, J. P., Kemanian, A. R., Mortensen, D. A., Ryan, M. R., Tooker, J., and White, C.: A framework for evaluating ecosystem services provided by cover crops in agroecosystems, Agr. Syst., 125, 12–22, 2014.

Smith, R. G., Gross, K. L., and Robertson, G. P.: Effects of crop diversity on agroecosystem function: Crop yield response, Ecosystems, 11, 355–366, 2008.

Stanford, G. and Smith, S. J.: Nitrogen mineralization potentials of soils, Soil Sci. Soc. Am. J., 36, 465–472, 1972.

Stephan, A., Meyer, A. H., and Schmid, B.: Plant diversity affects culturable soil bacteria in experimental grassland communities, J. Ecol., 88, 988–998, 2000.

Stirling, G. and Wilsey, B.: Empirical relationships between species richness, evenness, and proportional diversity, Am. Nat., 158, 286–299, 2001.

Strobel, B. W.: Influence of vegetation on low-molecular-weight carboxylic acids in soil solution – a review, Geoderma, 99, 169–198, 2001.

Tibbett, M., Sanders, F. E., and Cairney, J. W. G.: Low-temperature-induced changes in trehalose, mannitol and arabitol associated with enhanced tolerance to freezing in ectomycorrhizal basidiomycetes (*Hebeloma* spp.), Mycorrhiza, 12, 249–255, 2002.

Tiemann, L. K. and Billings, S. A.: Changes in variability of soil moisture alter microbial community C and N resource use, Soil Biol. Biochem., 43, 1837–1847, 2011.

Tiemann, L. K., Grandy, A. S., Atkinson, E. E., Marin-Spiotta, E., and McDaniel, M. D.: Crop rotational diversity enhances belowground communities and functions in an agroecosystem, Ecol. Lett., 18, 761–771, 2015.

Tilman, D., Knops, J., Wedin, D., Reich, P., Ritchie, M., and Siemann, E.: The influence of functional diversity and composition on ecosystem processes, Science, 277, 1300–1302, 1997.

Vance, E. D., Brookes, P. C., and Jenkinson, D. S.: An extraction method for measuring soil microbial biomass C, Soil Biol. Biochem., 19, 703–707, 1987.

Venter, Z. S., Jacobs, K., and Hawkins, H.-J.: The impact of crop rotation on soil microbial diversity: A meta-analysis, Pedobiologia, 59, 215–233, 2016.

Waldrop, M. P., Zak, D. R., Blackwood, C. B., Curtis, C. D., and Tilman, D.: Resource availability controls fungal diversity across a plant diversity gradient, Ecol. Lett., 9, 1127–1135, 2006.

Wardle, D. A.: A comparative assessment of factors which influence microbial biomass carbon and nitrogen levels in soil, Biol. Rev., 67, 321–358, 1992.

Wardle, D. A.: Is "sampling effect" a problem for experiments investigating biodiversity-ecosystem function relationships?, Oikos, 87, 403–407, 1999.

Wardle, D. A. and Ghani, A.: A critique of the microbial metabolic quotient (qCO_2) as a bioindicator of disturbance and ecosystem development, Soil Biol. Biochem., 12, 1601–1610, 1995.

Wardle, D. A., Yeates, G. W., Williamson, W., and Bonner, K. I.: The response of a three trophic level soil food web to the identity and diversity of plant species and functional groups, Oikos, 102, 45–56, 2003.

Williams, S. M. and Weil, R. R.: Crop cover root channels may alleviate soil compaction effects on soybean crop, Soil Sci. Soc. Am. J., 68, 1403–1409, 2004.

Wu, T., Chellemi, D., Graham, J., Martin, K., and Rosskopf, E.: Comparison of soil bacterial communities under diverse agricultural land management and crop production practices, Microb. Ecol., 55, 293–310, 2008.

Xuan, D., Guong, V., Rosling, A., Alström, S., Chai, B., and Högberg, N.: Different crop rotation systems as drivers of change in soil bacterial community structure and yield of rice, *Oryza sativa*, Biol. Fert. Soils, 48, 217–225, 2012.

Zak, D. R., Holmes, W. E., White, D. C., Peacock, A. D., and Tilman, D.: Plant diversity, soil microbial communities, and ecosystem function: Are there any links?, Ecology, 84, 2042–2050, 2003.

Zhou, X., Wu, H., Koetz, E., Xu, Z., and Chen, C.: Soil labile carbon and nitrogen pools and microbial metabolic diversity under winter crops in an arid environment, Appl. Soil Ecol., 53, 49–55, 2012.

Zuur, A. F., Ieno, E. N., and Elphick, C. S.: A protocol for data exploration to avoid common statistical problems, Meth. Ecol. Evol., 1, 3–14, 2010.

4

A probabilistic approach to quantifying soil physical properties via time-integrated energy and mass input

Christopher Shepard[1], Marcel G. Schaap[1], Jon D. Pelletier[2], and Craig Rasmussen[1]

[1]Department of Soil, Water and Environmental Science, The University of Arizona, Tucson, AZ 85721-0038, USA
[2]Department of Geosciences, The University of Arizona, Tucson, AZ 85721-0077, USA

Correspondence to: Christopher Shepard (cbs8h@email.arizona.edu)

Abstract. Soils form as the result of a complex suite of biogeochemical and physical processes; however, effective modeling of soil property change and variability is still limited and does not yield widely applicable results. We suggest that predicting a distribution of probable values based upon the soil-forming state factors is more effective and applicable than predicting discrete values. Here we present a probabilistic approach for quantifying soil property variability through integrating energy and mass inputs over time. We analyzed changes in the distributions of soil texture and solum thickness as a function of increasing time and pedogenic energy (effective energy and mass transfer, EEMT) using soil chronosequence data compiled from the literature. Bivariate normal probability distributions of soil properties were parameterized using the chronosequence data; from the bivariate distributions, conditional univariate distributions based on the age and flux of matter and energy into the soil were calculated and probable ranges of each soil property determined. We tested the ability of this approach to predict the soil properties of the original soil chronosequence database and soil properties in complex terrain at several Critical Zone Observatories in the US. The presented probabilistic framework has the potential to greatly inform our understanding of soil evolution over geologic timescales. Considering soils probabilistically captures soil variability across multiple scales and explicitly quantifies uncertainty in soil property change with time.

1 Introduction

Pedogenic models that can be widely applied and easily utilized are paramount for understanding soil-landscape evolution, soil property change with time, and predicting future soil conditions. A mathematically simple, easily parameterized approach has yet to be developed that is capable of predicting current soil properties or recreating potential soil evolution with time. Here we address this knowledge gap through the development of a probabilistic model of soil property change capable of predicting soil properties across a wide range of terrains, climates, and ecosystems.

The state-factor approach has been one of the primary pedogenic models since its development in the late 1800s and early 1900s (Dokuchaev, 1883; Jenny, 1941). The soil state-factor approach (Jenny, 1941) assumes that the state of the soil system or specific soil properties (S) may be described as a function of the external environment, represented by climate (cl), biology (o), relief (r), parent material (p), and time (t): $S = f(\text{cl}, o, r, p, t)$. This approach increased our understanding of soil variation across each factor, but more complex, multivariate approaches are generally not possible or difficult to derive from this formulation (Yaalon, 1975). From the original state-factor model have evolved pedogenic models that include functional (Jenny, 1961), energetic (Rasmussen and Tabor, 2007; Rasmussen et al., 2005, 2011; Runge, 1973; Smeck et al., 1983; Volobuyev, 1964), and mechanistic approaches (Finke, 2012; Minasny and McBratney, 1999; Salvador-Blanes et al., 2007; Vanwalleghem et al., 2013). However, many of these approaches are either limited to a site-specific basis, require a high degree of parameterization, or lack wide-scale applicability.

Here we develop a simple probabilistic approach to predict soil physical properties using a large dataset of chronosequence studies. The model compresses state-factor variability into two key components (parent material and total pedogenic energy, defined in Sect. 1.1) that were parameterized and calibrated using the chronosequence database. We hypothesized that a probabilistic approach predicts accurate ranges of soil physical properties based on the soil-forming environment. Additionally, we modified the model to include soil depth to capture the influence of redistributive hillslope processes to predict soil properties. We hypothesized that by including soil depth, the model would effectively predict the clay content in an independent dataset synthesizing soil and landscape variability in complex, hilly terrain from a wide range of environments.

Probabilistic model of soil property change

The model presented here is based on a reformulated state-factor model, where a location has a probability of displaying a range of differing soil morphologies and properties based upon the state factors, with some range of values more probable than others, meaning that the state-factor model (Jenny, 1941) may be restated as

$$P(s_1 \leq S \leq s_2) = f(\text{cl}, o, r, p, t), \tag{1}$$

where the left-hand side of the equation, $P(s_1 \leq S \leq s_2)$, represents the probability that a given soil will have a value located between a lower limit (s_1) and an upper limit (s_2) (Phillips, 1993b). Equation (1) can be restated more simply as

$$P(s_1 \leq S \leq s_2) = f(L_o, P_x, t), \tag{2}$$

where the original soil-forming state factors have been simplified to represent the fluxes of matter and energy into the soil system (P_x), incorporating the influence of climate and biology, and the initial state of the soil-forming conditions (L_o), incorporating the influence of the initial topography and original soil parent material and time or age of the soil system (t) (Jenny, 1961).

Equation (2) was further simplified to make the approach operational. A quantitative measure of climate and biology was needed to represent the influence of P_x on soil formation. We used a quantification of P_x calculated from effective precipitation and biological productivity, termed effective energy and mass transfer (EEMT, $\text{J m}^{-2}\,\text{yr}^{-1}$) (Rasmussen and Tabor, 2007; Rasmussen et al., 2005, 2011). EEMT provides a measure of the energy transferred to the subsurface, in the form of reduced carbon from primary productivity and heat transfer from effective precipitation, which has the potential to perform pedogenic work, e.g., chemical weathering and carbon cycling. Using EEMT as a simplification of P_x, Eq. (2) was restated as (Rasmussen et al., 2011)

$$P(s_1 \leq S \leq s_2) = f(L_o, \text{EEMT}, t). \tag{3}$$

We further simplified Eq. (3) by combining the flux term EEMT and the age of the soil system (t). EEMT multiplied by the age of the soil system, i.e., $\text{EEMT} \times t$, provides an estimate of the total energy transferred to the soil system over the course of its evolution, referred to here as total pedogenic energy (TPE, J m^{-2}). The TPE provides an estimate of P_x that incorporates soil age; thus, Eq. (3) may be restated as

$$P(s_1 \leq S \leq s_2) = f(L_o, \text{TPE}), \tag{4}$$

where at a certain point in time the probability of a soil property existing between s_1 and s_2 is a function of L_o and TPE. L_o controls the spread or variation of the probability distribution $P(s_1 \leq S \leq s_2)$ over time and the potential observable soil states, whereas TPE is proportional to the internal soil state at a given time (Jenny, 1961). Explicitly including time in Eq. (4) through TPE partially captures variation in soil property change attributable to topography and parent material. Soil residence time may be directly related to landscape position through topographic control on soil production and sediment transport and deposition (Heimsath et al., 1997, 2002; Yoo et al., 2007). Additionally, parent material modulates soil residence time through control on soil depth (Heckman and Rasmussen, 2011; Rasmussen et al., 2005), soil production, and sediment transport rates (Andre and Anderson, 1961; Portenga and Bierman, 2011). The initial conditions of the soil-forming system (L_o) are never fully known; however, representing the state of the soil system as a probable distribution of values, implicitly accounting for soil age, and not constraining the initial soil-forming conditions, the influence of initial conditions can be partially ignored, and hence we herein focus on modeling soil properties using only TPE.

Quantitatively realizing Eq. (4) required the use of predetermined joint probability density functions parameterized with TPE and a selected soil physical property. Bivariate normal density functions were calculated to determine the probability of a soil property range given a TPE value. The bivariate density function was selected due to its simplicity and ease of parameterization; other bivariate density functions are available that may better fit the selected soil property data but are not considered here. The bivariate normal density distribution (Ugarte et al., 2008) was calculated as

$$f(x, y) = \frac{1}{2\pi \sigma_x \sigma_y \sqrt{1 - \rho^2}} \exp\left(-\frac{1}{2(1 - \rho^2)} \left[\frac{(x - \mu_x)^2}{\sigma_x^2} \right.\right.$$
$$\left.\left. + \frac{(y - \mu_y)^2}{\sigma_y^2} - \frac{2\rho(x - \mu_x)(y - \mu_y)}{\sigma_x \sigma_y} \right]\right), \tag{5}$$

where ρ represents the Pearson correlation coefficient, μ_x is the mean of TPE, μ_y is the mean of the selected soil physical property, σ_x is the standard deviation of TPE, and σ_y is the standard deviation of the selected soil physical property. Using the bivariate normal density functions, conditional mean and variance values were calculated given a value of TPE; the conditional means and variances parameterized conditional univariate normal distributions for the selected soil physical properties. The conditional mean (Ugarte et al., 2008) was calculated as

$$\mu_{Y|X=x} = \mu_y + \rho \frac{\sigma_y}{\sigma_x}(x - \mu_x), \tag{6}$$

where $\mu_{Y|X=x}$ is the conditional mean soil property value given a value for TPE. The conditional variance (Ugarte et al., 2008) was calculated as

$$\sigma^2_{Y|X=x} = \sigma^2_y \left(1 - \rho^2\right), \tag{7}$$

where $\sigma^2_{Y|X=x}$ is the conditional variance of the soil property given a value of TPE.

Applying this approach required certain assumptions and simplifications. The model assumes that climate was constant over the entire duration of pedogenesis. The model makes no assumptions about the progressive and regressive processes that drive pedogenesis; by weighting all profiles equally, the net effects of both progressive (e.g., horizonation, clay accumulation, reddening) and regressive (e.g., haplodization, erosion, pedoturbation) pedogenic processes (Johnson and Watson-Stegner, 1987; Phillips, 1993a) are captured in the model structure. The model also does not consider the net effect of progressive and regressive pedogenic processes on the distribution of selected soil properties with depth. The model makes no assumptions about the initial soil-forming system, and we did not constrain the model to any particular initial condition for either parent material or geomorphic landform; the model simply describes the probability of a location exhibiting a range of soil properties based on TPE. The model assumes that all changes in soil physical properties are due to pedogenic processes. We used a bivariate normal distribution; consequently the model assumes that the data conform to a normal distribution.

2 Methods

2.1 Data collection and preparation

The probability distributions were parameterized using an extensive literature review of chronosequence studies. More than 140 chronosequence publications were identified using Google Scholar (www.scholar.google.com) and Thomson Reuters Web of Science (www.webofknowledge.com), 44 of which contained the required data. Inclusion within the present study required the following: profile descriptions with horizon-level clay, sand, and silt content and soil depth; well-defined ages of the soil-geomorphic surfaces; and geographic coordinates or maps showing locations of the described profiles. The chronosequences spanned a wide range of geographic locations, ecosystems, climates, rock types, and geomorphic landforms (Fig. 1, Table S1 in the Supplement). The chronosequence soils spanned ages from 10 years

Figure 1. Map of study sites. Yellow points indicate location of chronosequences, and red triangles indicate location of soils in complex terrain.

to 4.35 Myr and depth ranges from 3.0 to 1460 cm, with mean annual temperature (MAT) and precipitation (MAP) ranging from -11.2 to $28.0\,°C$ and 3.0 to $400\,cm\,yr^{-1}$, respectively. We were limited in site selection by the available data; as such we could not control for any bias that may exist with regard to site selection and reported soil property values.

2.2 Total pedogenic energy

The influence of both climate and vegetation at the locations of each soil profile was determined using effective energy and mass transfer (EEMT) (Rasmussen and Tabor, 2007; Rasmussen et al., 2005). EEMT quantifies the heat and chemical energy from effective precipitation and net primary productivity added to the soil system (Rasmussen and Tabor, 2007; Rasmussen et al., 2005, 2011). EEMT describes the energy added to the soil system that can perform pedogenic work, such as chemical weathering and carbon cycling. EEMT is adaptable to include specific energetic inputs to the soil system based upon the prevailing soil-forming environment, e.g., the energetics from added fertilizer in an agriculture field or the impact of human-induced erosion (Rasmussen et al., 2011). The EEMT values for each soil profile were extracted from a global map of EEMT derived from the monthly global climate dataset of New et al. (1999) at $0.5° \times 0.5°$ resolution using ArcMap 10.1 (ESRI, Redlands, CA) (Rasmussen et al., 2011). For the chronosequence soils, EEMT values ranged from 2235 to $> 200\,000\,kJ\,m^{-2}\,yr^{-1}$. Total pedogenic energy (TPE, $J\,m^{-2}$) was derived simply by multiplying EEMT ($J\,m^{-2}\,yr^{-1}$) for each soil profile by its reported age (yr). TPE was used because it was a better predictor of soil physical properties relative to mean annual temperature, mean annual precipitation, or net primary productivity (NPP) (Table 3).

2.3   Application to chronosequence data

The chronosequence database included 44 distinct chronosequences representing 405 different soil profiles. We focused here on changes in sand, silt, and clay content and solum thickness as examples of soil property change with time. We tested the approach on depth-weighted (DWT) sand, silt, and clay content (reported as weight %), as well as the maximum measured value of sand, silt, and clay content within each soil profile. Buried horizons were removed from the soil profiles before either the maximum or DWT content values were calculated. Solum thickness was extracted for each profile, defined as the thickness of the horizons influenced by pedogenic processes or the depth to C horizons (Schaetzl and Anderson, 2005). The site RW-14 from McFadden and Weldon (1987) was not included in the solum thickness model calculations; the measured solum thickness of RW-14 was 1460 cm, 1 order of magnitude greater than all other soil profiles included in the study. Four hundred and five profiles reported clay content data, only 387 profiles reported sand and silt content, and 399 soil profiles contained a developed solum. We classified the soil profiles by parent material in terms of igneous, metamorphic, or sedimentary material and by geomorphic landform, e.g., alluvial surface, marine terrace, or moraine (Shoeneberger et al., 2012); for example, if a soil was formed on an alluvial fan from granitic parent material, it would be defined as alluvial and igneous.

Using the soil data, we calculated bivariate normal probability distributions using TPE and the soil physical properties (Eq. 5). The soil data were transformed using logarithmic and square root transformations when appropriate to meet the normality assumption of the bivariate normal probability distribution. Conditional univariate normal distributions (Eqs. 6, 7) were calculated to approximate probable ranges of soil properties using leave-one-out cross validation (LOOCV). Each of the soil chronosequences was removed from the model dataset, with the all remaining chronosequence data used to calculate the parameters of the bivariate and conditional univariate normal distributions. The conditional univariate normal distributions were calculated using the TPE values for the profiles within the left-out chronosequence.

2.4   Application to complex terrain

By design, soil chronosequences are generally sited on gentle, low, sloping terrain to minimize the influence of topography and erosion/deposition on soil formation (Harden, 1982). However, much of the Earth's surface is characterized by complex topography with high relief, steep slopes, and differences in slope aspect. Any predictive soil model or approach must be effective in both simple and complex terrain. To test the ability of the model to predict soil properties in complex terrain, we compiled data from upland catchments with variable parent material and topography from the literature, as well as data available from the US NSF Critical Zone Observatory Network (CZO, www.criticalzone.org) (Table 1) (Bacon et al., 2012; Dethier et al., 2012; Foster et al., 2015; Holleran et al., 2015; Lybrand and Rasmussen, 2015; Rasmussen, 2008; West et al., 2013). Data from several additional studies from complex terrain were also included to test the model (Table 1) (Dixon et al., 2009; Yoo et al., 2007). These data were accessed from www.criticalzone.org or Google Scholar (www.scholar.google.com). These studies were included because they all contained horizon-level soil texture data, soil depth, percent volume rock fragment data, and ^{10}Be or U series measures of soil erosion rates or residence time, where mean residence time (MRT) was calculated as $MRT = h/E$, where h is soil depth (m) and E is erosion rate (m yr^{-1}) (Pelletier and Rasmussen, 2009b). We used published coordinates to extract EEMT values, calculated from New et al. (1999), for each soil profile using ArcGIS 10.1 and used EEMT and MRT to calculate TPE. It should be noted the coarse resolution of New et al. (1999) EEMT values does not account for local-scale variation in water redistribution and primary productivity that can lead to significant topographic variation in EEMT (Rasmussen et al., 2015). Using Eq. (5) and the parameters generated from the chronosequence database, conditional mean depth-weighted clay content was calculated for each profile.

Due to the influence of redistributive hillslope processes on soil development (Yoo et al., 2007), soil depth varies systematically across hillslopes (Heimsath et al., 1997); thus, soil depth can be used to incorporate information about these processes within the model calculations. We calculated the mass per area clay content of these profiles using soil depth to incorporate this variation, as

$$\text{mass per area clay } \left(\text{kg m}^{-2} \right) \tag{8}$$
$$= (\rho_\text{b})(h) \left(\frac{\mu_{Y|X=x,\,\text{DWT CLAY}}}{100} \right) \left(1 - \left(\frac{\text{RF\%}}{100} \right) \right),$$

where ρ_b is the soil bulk density assumed to be 1500 kg m^{-3} for all soil profiles, $\mu_{Y|X=x,\,\text{DWT CLAY}}$ is the predicted conditional mean for depth-weighted clay content (DWT CLAY) using Eq. (6), RF% is the measured depth-weighted percent volume rock fragments within the soil (when no RF% data were available we assumed a value of 41.7 %, which was the average RF% for profiles with reported values), and h is the soil depth in meters. Using Eq. (8), mass per area clay was calculated for each soil profile. Further, we examined the impact of depth, rock fragment percentage, and predicted conditional mean DWT clay on the predicted mass per area clay predictions using multiple linear regression.

Coupling geomorphic model with probabilistic model

Additionally, we applied the probabilistic model independent of measured soil data, across a small complex catchment in

Table 1. Complex terrain study sites and characteristics.

Site	Study	Number of sites	Elevation (m)	MAP (cm)	MAT (°C)	Parent material	Slope	Aspect	Vegetation
Marshall Gulch granite sub-catchment, Arizona, USA	Holleran et al. (2015)	24	2300–2500	85–90	10	Granite, amphibolite, quartzite	45 %	North	*Pinus ponderosa, Pseudotsuga menziesii, Abies concolor*
Frog's Hollow, New South Wales, Australia	Yoo et al. (2007)	2	930	55–75	~16	Granodiorite	–	–	*Eucalyptus* grassland savannah
Cross Keys, South Carolina, USA	Bacon et al. (2012)	1	–	115–140	14–18	Granitic gneiss	<2 %	–	*Quercus, Carya*
Gordon Gulch, Colorado, USA	Foster et al. (2015)	9	2440–2740	52	5	Gneiss, quartz monzonite, granodiorite	15–28°	North and south	*Pinus ponderosa, Pinus contorta*
Rincon Mountains, Arizona, USA	Rasmussen (2008)	11	1050–2500	<40–80	10–18	Granodiorite	–	–	Oak grass woodland, piñon–juniper woodland, mixed conifer
Jemez Mountains, New Mexico, USA	Huckle et al. (2016)	4	2990–3100	~50	4	Rhyolite, tuff	–	West and east	*Pseudotsuga menziesii, Abies concolor*, Picea pungens, *Populus tremuloides*
Shale Hills, Pennsylvania, USA	West et al. (2013); Ma et al. (unpublished)	6	260–280	100	–	Shale, sandstone	15–20°	North and south	–
Sierra Nevada California, USA	Dixon et al. (2009)	5	216–2991	37–106	3.9–16.6	Tonalite, granodiorite	–	–	Oak-grass woodland, mixed conifer, subalpine

the Santa Catalina Mountains (Catalina-Jemez CZO, Fig. 2a–b, Table 1) (Holleran et al., 2015; Lybrand and Rasmussen, 2015). The ~6 ha catchment is located at an elevation between 2300 and 2500 m with mixed conifer vegetation, approximately 30 km northeast of Tucson, AZ (Fig. 2, Table 1). The approach utilized soil depth and residence time output from a process-based numerical soil depth model (Pelletier and Rasmussen, 2009a). The model used high-resolution lidar-derived topographic data to estimate 2 m pixel resolution soil depth and erosion rates (Fig. 2c) (Pelletier and Rasmussen, 2009a). These data were coupled with topographically resolved EEMT values that accounted for local hillslope-scale variation in water redistribution and primary productivity at a 10 m pixel resolution (Rasmussen et al., 2015) (Fig. 2d). We used calculated TPE from the topographically resolved EEMT and soil residence time values to predict DWT clay and coupled predicted DWT clay values with modeled depth from Pelletier and Rasmussen (2009a) in Eq. (8) to predict mass per area clay at a 2 m pixel resolution; the data processing and model apparatus are shown in Fig. 3. We assumed a constant 50 % rock fragment value for each location. The coupled geomorphic–TPE model outputs were compared with point measures of mass per area clay from

Holleran et al. (2015) and Lybrand and Rasmussen (2015). Model data were completely independent of the Holleran et al. (2015) and Lybrand and Rasmussen (2015) datasets such that they served as validation data for the modeled output.

2.5 Model domain

The model was parameterized using chronosequence studies; as such, the model is best suited for generally low, sloping terrain. The model was extended to complex terrain using the described correction above (Sect. 2.4), widening the model domain to steeply sloping terrain. The model does not consider human activities or aeolian additions and should not be extended to soils significantly impacted by either humans or dust. The model was trained on a diverse array of parent materials and ecosystems and could be utilized in climates with MAT ranging from -10 to $28\,°C$ and MAP ranging from 3 to $400\,\mathrm{cm\,yr^{-1}}$. The model could be utilized on soils spanning multiple magnitudes in age, from 10 years to greater than 4 Myr.

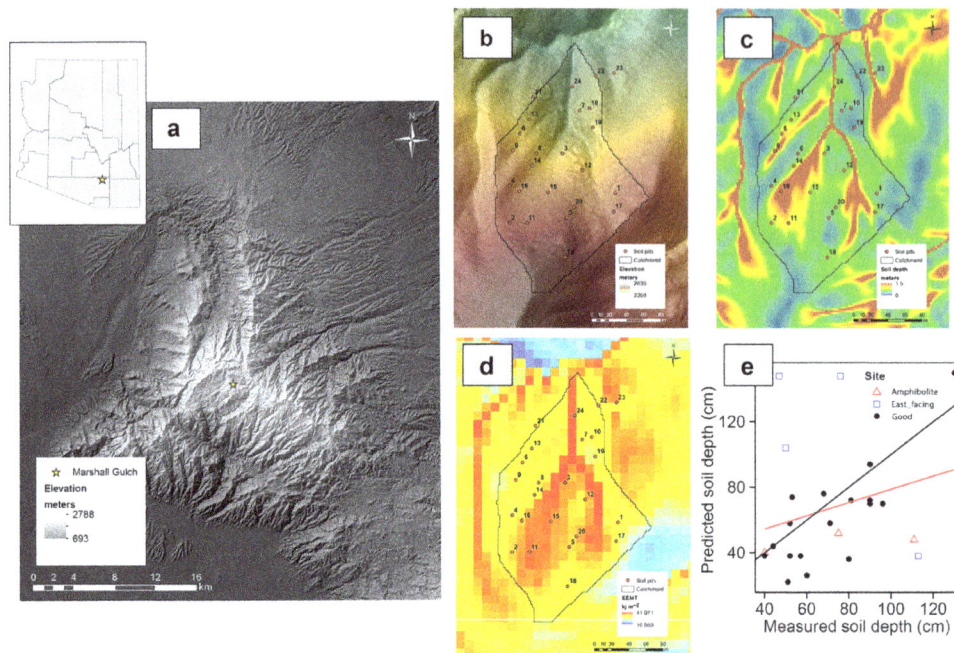

Figure 2. Marshall Gulch study site. **(a)** Location of the Santa Catalina Mountains and the Marshall Gulch catchment within Arizona, USA; **(b)** elevation of the granite sub-catchment of Marshall Gulch; **(c)** predicted soil depth in the granite sub-catchment (Pelletier and Rasmussen, 2009a); **(d)** EEMTv2.0 in the granite sub-catchment (Rasmussen et al., 2015); **(e)** mismatch between the measured soil depths and predicted soil depths.

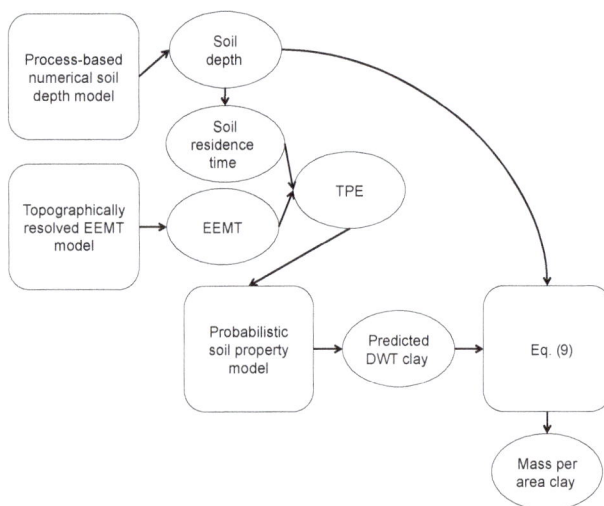

Figure 3. Coupled geomorphic–probabilistic model apparatus. The process-based numerical soil depth model is used to predict soil depth, which is used to predict soil residence time. The topographically resolved EEMT model is used to calculate TPE using the soil residence time and EEMT values. The probabilistic model is used to calculate DWT clay contents using the TPE values, and mass per area clay is calculated using predicted DWT clay and predicted soil depth values.

3   Results

3.1   Application and parameterization to chronosequences

The relationships between TPE and soil texture and solum thickness were used to calculate the bivariate probability distributions. The bivariate probability distributions (Eq. 5) were parameterized using the means, standard deviations, and Pearson's correlation from the chronosequence database (Table 2). Furthermore, the relationship between TPE and the soil properties was stronger than just using age, NPP, MAP, or MAT alone (Table 3). Age was expected to strongly correlate to the soil properties due to the design of chronosequence studies; however, comparing age and TPE separately, the percent increase in Spearman rank correlations (r) ranged from 8.7 % (DWT silt) to 25.6 % (max sand). Maximum and depth-weighted silt content were weakly correlated to both age and TPE and exhibited only a minimal change in Spearman's rank correlation with TPE relative to age.

The correlation between TPE and maximum clay content (Fig. 4, Pearson's $\rho = 0.78$, $r^2 = 0.62$, $\sqrt{\text{Max Clay}} = -7.38 + 1.37 \cdot \log(\text{TPE})$, df $= 403$) was highly significant and presented the strongest probabilistic relationship determined between TPE and the soil properties. The bivariate probability surface displayed the greatest probability around the joint means between TPE and maximum clay content (Fig. 4). Solum thickness and TPE were also strongly re-

Table 2. Parameters for the bivariate normal probability distributions for the soil physical properties and TPE; n is number of profiles; μ is mean; σ is standard deviation; and ρ is Pearson's correlation between soil variables and total pedogenic energy.

Variable	Soil property parameters			
	n	μ	σ	ρ
Max sand	387	70.97	25.55	−0.48
Max silt	387	34.27	18.32	0.32
Max clay[a]	405	4.52	2.26	0.78
DWT sand	387	59.47	26.22	−0.57
DWT silt[a]	387	4.50	1.66	0.26
DWT clay[a]	405	3.66	2.12	0.73
Solum thickness[b]	399	1.77	0.53	0.65
TPE[b]	405[c]	8.69	1.30	–
	387[d]	8.70	1.29	–
	399[e]	8.72	1.27	–

[a] Square root transformed. [b] Log10 transformed. [c] For clay variables. [d] For sand and silt variables. [e] For solum thickness, max indicates maximum content; DWT indicates depth-weighted average content.

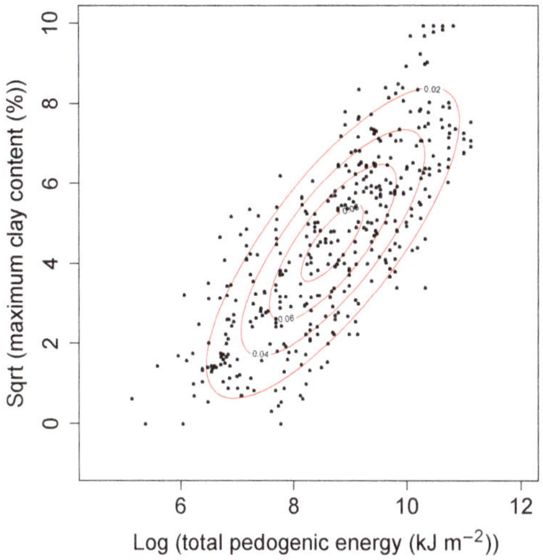

Figure 4. Bivariate normal distribution between TPE and max clay content. The points indicate individual soils. The red ellipses represent lines of equal probability, which corresponds to a three-dimensional probability distribution. From this relationship the conditional mean and variances for the soil physical properties were calculated.

lated, but weaker relative to the maximum clay–TPE relationship (Fig. S1 in the Supplement, Pearson's $\rho = 0.65$, $r^2 = 0.42 \times \log(\text{solum thickness}) = -0.58 + 0.27 \cdot \log(\text{TPE})$, df = 397). The relationships between TPE and max sand (Fig. S2) and silt (Fig. S3) contents were generally weaker, relative to clay and solum thickness, with little to no relationship between TPE and silt content.

The conditional univariate normal distribution parameters were determined for the soil physical properties from the bivariate distribution and using Eqs. (6) and (7). The bivariate normal distribution effectively predicted maximum clay content (Fig. 5) with an $r^2 = 0.54$ (RMSE = 14.8 %) between the measured maximum clay content and predicted conditional mean maximum clay content (Eq. 6) across all sites based on LOOCV (Fig. 5d). The model effectively predicted maximum clay content regardless of parent material with r^2 of 0.61 (RMSE = 14.4 %), 0.56 (RMSE = 12.0 %), and 0.59 (RMSE = 16.8 %), for igneous, metamorphic, and sedimentary parent materials, respectively. The r^2 between the measured values and predicted values for solum thickness, max sand, and max silt were 0.28 (RMSE = 101.0 cm, Fig. S4), 0.17 (RMSE = 23.4 %, Fig. S5), and 0.04 (RMSE = 18.0 %, Fig. S6), respectively.

The relationship of predicted to actual maximum clay content varied significantly across individual studies. The predicted values represent the predicted conditional means (Eq. 6) bounded by the conditional standard deviation (Eq. 7), which approximates a 50 % probability that the measured maximum clay content will be within 1 standard deviation of the conditional mean (Fig. 6). The individual studies presented in Fig. 6 were selected to represent a broad

range of climates and landforms and demonstrate both the strengths and weaknesses of the model. For Harden (1987) (Fig. 6a, $r^2 = 0.88$, $p < 0.0001$, df = 20, RMSE = 9.4 %) and Howard et al. (1993) (Fig. 6b, $r^2 = 0.86$, $p < 0.001$, df = 6, RMSE = 10.2 %), the model was generally successful at predicting the maximum clay content values; both the Harden (1987) and Howard et al. (1993) sequences were located in alluvial deposits but in vastly different climates: xeric (winter-dominated annual rainfall regime) vs. udic (evenly distributed annual rainfall regime), respectively. The model was capable of predicting maximum clay content values for glacial moraine deposits, in a frigid climate (Fig. 6c, $r^2 = 0.87$, $p < 0.0001$, df = 12, RMSE = 6.0 % Birkeland, 1984) and on marine terraces in northern California with a xeric climate (Fig. 6f, $r^2 = 0.98$, $p < 0.001$, df = 4, RMSE = 8.9 %; Merritts et al., 1991). The model was incapable of predicting clay accumulation on marine terraces in hot, wet climates in Barbados (Fig. 6d, $r^2 = 0.31$, $p = 0.08$, df = 9, RMSE = 44.9 % Muhs, 2001) or Taiwan (Fig. 6e, $r^2 = 0.67$, $p < 0.001$, df = 11, RMSE = 23.1 %, Huang et al., 2010).

3.2 Application in complex terrain

The model was much less effective in complex terrain and highly overpredicted DWT clay contents in soils located in complex landscapes (Fig. 7a, $r^2 = 0.26$, $y = 0.39x + 7.36$, $p < 0.0001$, RMSE = 5.4 %). The model highly overpredicted the clay content of the South Carolina site and the Gordon Gulch soils and underpredicted the clay content of the Rincon, Santa Catalina, and Jemez sites.

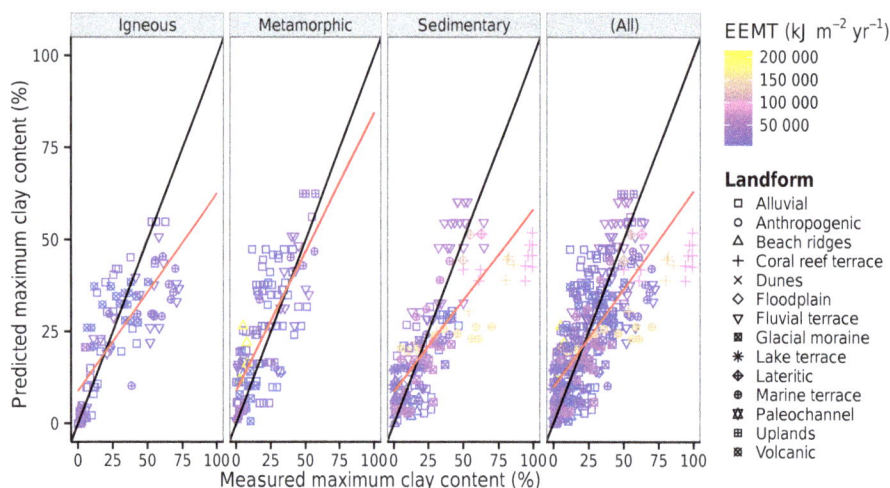

Figure 5. LOOCV results for max clay content. The results were subdivided by general soil parent material: igneous, metamorphic, and sedimentary; the points represent the geomorphic surface each soil formed on, and the colors represent the EEMT value for the location of each soil. Using LOOCV, where one chronosequence was removed from the model dataset and the remaining datasets were used to predict the parameters of the bivariate distributions, the conditional means of the left-out chronosequence was determined. The model was effectively able to predict the conditional mean values of the max clay contents with an $r^2 = 0.54$ (RMSE = 14.8 %). The model was least capable of predicting the clay contents on coral reef terraces (+) and appeared the most effective for alluvial surfaces (□).

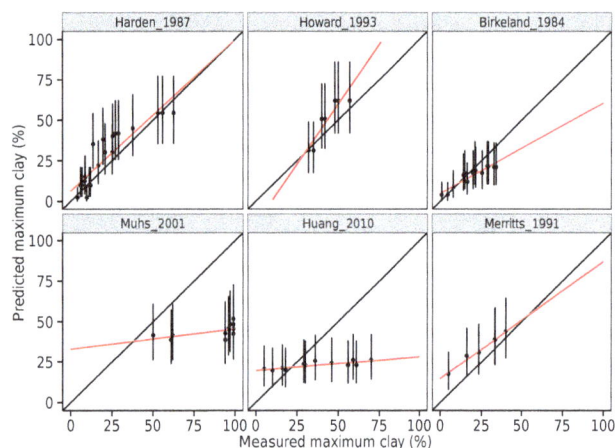

Figure 6. Selected relationships between the measured maximum clay content and predicted maximum clay content. **(a)** Harden (1987), **(b)** Howard et al. (1993), **(c)** Birkeland (1984), **(d)** Muhs (2001), **(e)** Huang et al. (2010), and **(f)** Merritts et al. (1991). The errors represent the conditional standard deviations around the mean, which correspond to a probability of 50 %. The model effectively predicted clay content across a diverse range of climates, landforms, and parent materials. The model was the least effective at predicting the clay content of soils in tropical climates and soils forming on coral reef terraces.

When correcting for the influence of hillslope processes by explicitly including soil depth and calculating mass per area clay, the approach effectively predicted clay content, with an $r^2 = 0.81$ (Fig. 7b, $y = 1.58x - 15.5$, $p < 0.0001$, RMSE = 86.4 kg clay m^{-2}), only slightly overpredicting clay content, with a regression slope of 1.58. Soil depth was the strongest contributing factor to the mass per area clay prediction with the greatest sums of squares in a simple multiple linear regression including depth, RF%, and DWT clay% (Table 4); predicted conditional mean clay content percentage was the second strongest contributing factor to the mass per area clay prediction. Rock fragment percentage did not influence the mass per area clay content prediction.

3.3 Coupled geomorphic–TPE model

The coupled geomorphic–TPE model effectively predicted mass per area clay for the majority of soils located within the Marshall Gulch sub-catchment with an $r^2 = 0.74$ (Fig. 8a, $y = 0.86x - 5.06$, $p < 0.0001$, RMSE = 17.7 kg clay m^{-2}). For a subset of soils, the model did not effectively predict mass per area clay, and this was excluded from the regression in Fig. 8a; four of these soils were located on the east-facing ridge of the catchment, and an additional two soils were formed on amphibolite rather than the granite or quartzite materials that all of the other soils in the catchment were derived from. All of these locations also exhibited a poor fit between modeled and measured soil depth (Fig. 2e). The spatial distribution of mass per area clay was also predicted across the catchment (Fig. 8b), independently of measured data, and generally conformed to previously predicted spatial distribution of clay stocks in the Marshall Gulch catchment (Holleran et al., 2015).

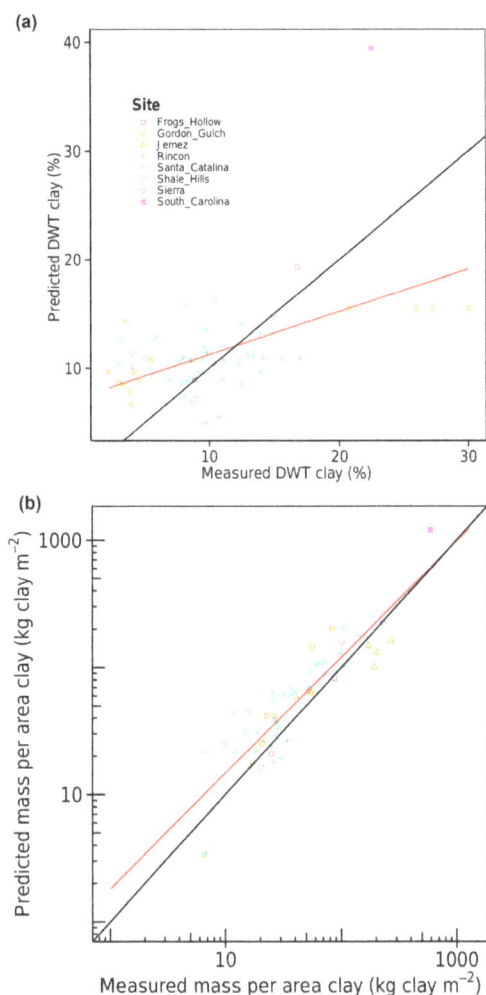

Figure 7. Model results in complex terrain. **(a)** Prediction of depth-weighted (DWT) clay contents; **(b)** prediction of mass per area clay using Eq. (9). The model was incapable of directly predicting DWT clay for the soils in complex terrain due to redistributive hillslope processes; $r^2 = 0.26$ between measured and predicted conditional mean DWT clay **(a)**. By including information about soil depth and percent volume rock fragment and converting DWT clay to mass per area clay, the model was significantly more effective at predicting clay contents for these soils $r^2 = 0.81$.

4 Discussion

4.1 Model effectiveness

4.1.1 Model results for chronosequences

The model predicted maximum clay content across a diverse range of lithologies, climates, and landforms. Weathering and clay production are primary pedogenic processes (Birkeland, 1999; Schaetzl and Anderson, 2005), and because the model assumed that all changes in the soil profile are due to these processes and TPE is closely related to degree of weathering, the model was the most effective at predicting clay content. For initial soil states that begin

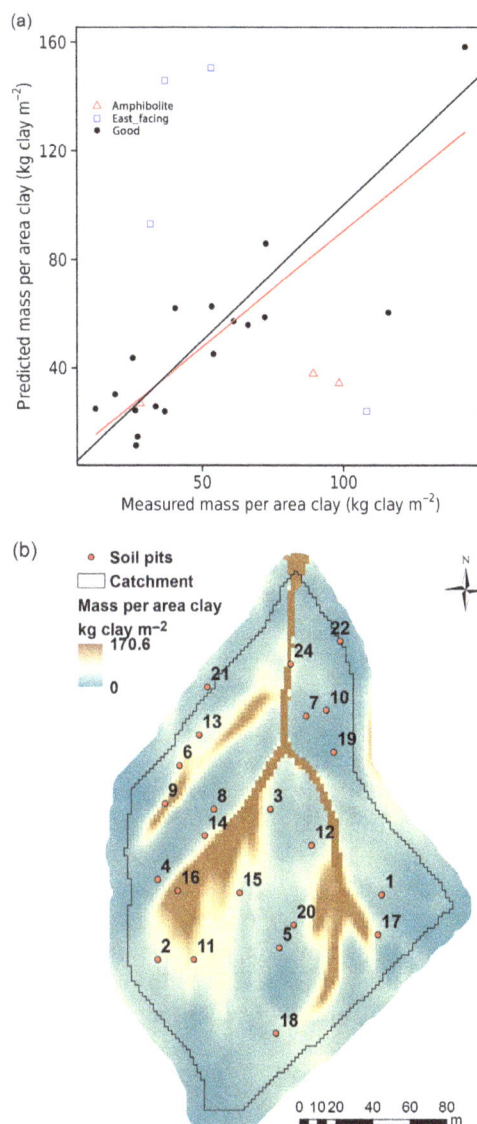

Figure 8. Model results of coupled geomorphic–EEMT–TPE model in Marshall Gulch granite sub-catchment. **(a)** Prediction of mass per area clay for sites from Holleran et al. (2015) and Lybrand and Rasmussen et al. (2015); **(b)** spatial prediction of mass per area clay. When combining the present approach, with a geomorphic-based soil depth model, the combined models together were highly effective at predicting the clay contents for a majority of soils in the Santa Catalina Mountains (Catalina-Jemez CZO), $r^2 = 0.74$.

pedogenesis with a potentially significant amount of clay-sized particles the model was much less effective. The soils of the Taiwanese chronosequence formed from conglomerates (Huang et al., 2010); conglomerates are typically poorly sorted, such that these soils initially formed with high clay contents slowing clay accumulation, limiting the effectiveness of the model to predict clay contents in these soils. Additionally, the model highly underestimated the clay content of soils located on coral reef terraces in tropical environments (Maejima et al., 2005; Muhs, 2001). Coral reef terraces represent a relatively unique landform that weathers rapidly to fine-sized particles, especially under tropical climates, and generally have complicated parent material compositions (Muhs et al., 1987). The combination of these factors limited the ability of the model to predict the soil properties on these surfaces.

Sand and silt displayed weaker relationships with increasing total pedogenic energy. The lack of correlation of sand and silt to TPE may result in part from the definitions of the particle size classes. Sand-sized particles span a difference in particle size of several orders of magnitude, ranging from particles of 2 to 0.05 mm (Soil Survey Staff, 2010), whereas clays are constrained to a particle size less than 0.002 mm. The sequential weathering of rock fragments and coarse sand to fine and very fine sands therefore is not reflected in total sand content and likely diminishes the relationship between sand content and total pedogenic energy and time (Pye and Sperling, 1983; Pye, 1983; Sharmeen and Willgoose, 2006). The relationship between silt content and pedogenic energy was the weakest of the three broad particles size classes (Tables 2, 3). Similar to sand, the silt size fractions span 1 order of magnitude in particle size ranging from 0.05 to 0.002 mm in diameter. Further, the sand and silt fractions are dominated by resistant primary minerals (Pye, 1983) and would not change greatly in response to increased TPE or weathering, which may partly account for the weaker correlations with TPE. Additionally, the silt fraction may also be heavily influenced by the deposition of eolian material and thereby introduce an additional mass of silt that was not derived from the direct weathering of the initial soil-forming system (McFadden et al., 1987) effectively uncoupling silt content from total pedogenic energy.

Solum thickness displayed a relatively strong relationship with increasing pedogenic energy, with TPE explaining up to 42 % of the variance in solum thickness (Tables 2, 3). Soil production is related to climatic variation (Amundson et al., 2015), with this variation partly captured by EEMT and TPE, leading to the slightly stronger predictive power of the model. However, soil production is also highly influenced by redistributive hillslope process, chemical and physical weathering, and tectonic uplift (Heimsath et al., 1997; Riebe et al., 2004; Yoo and Mudd, 2008b) and can be a highly nonlinear process (Pelletier and Rasmussen, 2009a). These factors were not directly accounted for in this study in that topography was not a quantified factor, which likely represents a large proportion of the remaining unexplained variance in solum thickness.

4.1.2 Model results in complex terrain

Due to using soil chronosequence data to parameterize the approach, the influence of redistributive hillslope processes was not captured. Additionally, in the amount of time required to transport soil across a hillslope, chemical and phys-

ical alterations of the soil particles are possible and may not be reflected in mean residence time calculations (Yoo and Mudd, 2008a; Yoo et al., 2007). Soil thickness is highly dependent upon hillslope position and landscape morphology (Dietrich et al., 2003; Heimsath et al., 1997; Pelletier and Rasmussen, 2009a). By using soil thickness as a proxy for the strength of these redistributive hillslope processes and converting the predicted conditional mean clay content value to a mass per area basis, the model was able to capture differences in clay content across complex terrain for a variety of lithologies and climates. The differing lithologies, climates, or vegetation types did not appear to impact the ability of the model to predict clay contents, likely because local variation in soil depth accounts for many of these controls. Parent material and climate influence the weathering process and production of clay in soils (Harden and Taylor, 1983; Muhs et al., 2001); however, these factors are collinear with soil depth (Heckman and Rasmussen, 2011; Lybrand and Rasmussen, 2015; Pelletier and Rasmussen, 2009a), such that by including soil depth, differences due to lithology or climate were partly incorporated in the model prediction.

4.1.3 Results from coupled geomorphic–TPE model

For the majority of sites in the Marshall Gulch sub-catchment, the coupled geomorphic–TPE model was highly effective at predicting clay content and the spatial distribution of clay stocks. Large differences were found for four soils located on the east-facing ridge of the catchment underlain by granite, with the model generally overpredicting soil depth and clay content. Discrepancies between the modeled and measured depths were likely the primary sources of error within the mass per area clay predictions for the four east-facing ridge soils (Fig. 2e). The geomorphic model predicted deeper soil depths due to the presence of an apparent convergent zone on the east-facing ridge of the sub-catchment; however, this convergent zone is only a small feeder tributary to the larger catchment drainage. The inability of the model to effectively predict clay contents and the mismatch between modeled and actual soil depths in the four soils located on the east-facing ridge is likely due to this local, fine-scale topographic variation. The fine-scale topographic variation may indicate that the scale of soil property predictions is important in achieving accurate predictions. Fine spatial scales match the scale of local soil-landscape variation and processes, but fine-scale variation in weathering rates and lithology is also required to better predict soil depth within the catchment (McKenzie and Ryan, 1999).

Error in predicted soil depths due to fine-scale differences in lithology within the Marshall Gulch sub-catchment partly explains the discrepancies between measured and predicted mass per area clay contents. For two amphibolite-derived soils, the model greatly underestimated mass per area clay. The geomorphic soil depth model assumed a uniform weathering rate based on the granitic soils (Pelletier and Ras-

Table 3. Spearman rank correlations between soil physical properties and TPE and age.

Variable	Spearman rank correlations						
	NPP	MAP	MAT	TPE	Age	% increase*	n
Max sand	−0.34	−0.15	−0.23	−0.46	−0.36	25.6	387
Max silt	0.00	−0.11	0.05	0.31	0.32	−1.1	387
Max clay	0.16	−0.01	0.37	0.80	0.73	8.8	405
DWT sand	−0.25	−0.07	−0.27	−0.57	−0.50	15.2	387
DWT silt	0.11	−0.01	0.02	0.23	0.21	8.7	387
DWT clay	0.22	0.02	0.40	0.75	0.67	11.7	405
Solum thickness	0.12	0.07	0.22	0.63	0.58	9.9	399

Max indicates maximum content; DWT indicates depth-weighted average content; * percent increase in Spearman rank correlation between TPE and age.

Table 4. Sensitivity analysis of model prediction in complex terrain.

Effects	Sensitivity analysis of model prediction in complex terrain				
	DF	Sums of squares	Mean sums of squares	F value	p
Depth, h (cm)	1	1 158 897	1 158 897	472.9	< 0.0001
CM DWT clay, $\mu_{Y|X=x}$ (%)	1	148 896	148 896	60.8	< 0.0001
Rock fragment, RF% (%)	1	1563	1563	0.6	0.428
Residuals	58	142 140	2451		

mussen, 2009a); due to differences in primary mineral assemblage, the amphibolite materials are likely weathering at a faster rate compared to the granite-derived soils (White et al., 2001; Wilson, 2004), resulting in greater clay production and likely explaining the underestimated clay contents. Inclusion of differential weathering rates for varying lithologies within the geomorphic model would likely lead to better prediction of clay contents, but in areas of complex lithology this would require detailed information about distributions of differing lithologies. With these adjustments, the coupled geomorphic–TPE model represents an effective, independent prediction of clay stocks.

4.2 Advantages of probabilistic approach

Simplifying and representing the soil-forming factors as multivariate distributions and probabilities has the potential to quantitatively represent the general state-factor model, making the approach universally applicable. The initial state of the soil can likely never be fully known, leading to variability in soil properties over time that cannot necessarily, or ever, be attributed to any external factor (Phillips, 1989, 1993b). A probabilistic approach utilizes that variability to drive predictions and understanding of these systems. Similar to the approach taken here, building distributions of the soil-forming state factors that are associated with distributions of particular soil properties could yield probabilistic predictions of soil formation and change. We selected to use a representation of climate and biology (EEMT). However, depend-

ing on the soil property of interest, the variables needed to parameterize the distributions would likely change; for example, if interested in organic matter content, aboveground net primary productivity or the normalized difference vegetation index may be better predictors of organic matter accumulation. The strength of this approach lies in the fact that no assumptions are made about the initial conditions of the soil-forming system or the specific soil-forming processes. Predicting probable distributions of soil physical properties implicitly acknowledges that our understanding of any system is incomplete but explicitly quantifies uncertainty in predictions and constrains the potential observable values to a predicted range. Utilizing this approach will require the necessary data to build distributions that are widely representative and applicable to most locations (Yaalon, 1975). With wide accessibility to large databases of soil information, such as the US National Soil Information System (NASIS) and the FAO Harmonized World Soil Database, access to the required amount and quality of data may be possible. Similar to the present study, simple bivariate distributions could be solved to calculate conditional distributions based on the soil-forming state factors, effectively producing quantitative probabilistic representations of Jenny's original equation (Jenny, 1941).

The simplicity of the present approach allows easy integration into preexisting geomorphic models of landscape evolution. Past approaches that have combined pedogenic and landscape evolution models have generally focused on producing hypothetical soil-landscape relationships that progress forward through time (Minasny and McBratney, 2001; Vanwalleghem et al., 2013) or have focused on idealized landscapes (Temme and Vanwalleghem, 2015). However, by combining probabilistic approaches parameterized using known landscapes and geomorphically based landscape evolution models, predictions of the current state of the soil landscape can be investigated. As was demonstrated in Fig. 7b, combining the present approach with geomorphically based soil depth models generated from DEMs has great potential to predict soil properties across a diverse range of environments, without needing prior knowledge of the landscape other than topography and climate. Further, potential soil landscapes can be investigated by updating EEMT values to incorporate future climate scenarios available from predictive climate models (Gent et al., 2011; Taylor et al., 2012) and topographic and hydrological impacts due to changes in topography over time (Rasmussen et al., 2015).

4.3 Limitations and potential refinements

There are obvious limitations within the current model: a lack of consideration of parent material influences, topographic variation, human impacts, internal soil feedbacks and thresholds, determination of landscape and soil age, and differences in paleoclimate variation. Parent material control on the relative proportion of weatherable minerals and mineral weathering rates (Jackson et al., 1948) can manifest itself as vastly different soil morphologies and rates of pedogenesis when controlling for other soil-forming factors or even without controlling for other factors (Heckman and Rasmussen, 2011; Parsons and Herriman, 1975; Phillips, 1993b). The current approach implicitly assumes no information about the initial conditions, only that all clay production is a pedogenic process. The application of this approach to parent materials, where a large fraction of clay-sized particles formed through non-pedogenic processes, is thus limited and may explain why the model was ineffective for some soils. Refining the current approach would require normalization of soil to the particle size distribution of the soil parent material. Past studies have utilized highly characterized parent material data to model soil property change with time (Chadwick et al., 1990; Harden, 1982), but these data are generally difficult to obtain and often not reported in the available chronosequence literature.

Topography dictates soil chemical and physical properties and residence times, especially in complex terrain (Almond et al., 2007; Egli et al., 2008; Lybrand and Rasmussen, 2015), where nonlinear diffusive hillslope processes control the fluxes of matter and energy into and out of the soil system (Heimsath et al., 1997; Pelletier and Rasmussen, 2009a; Rasmussen et al., 2015; Yoo and Mudd, 2008b; Yoo et al., 2007). Using earlier versions of EEMT (Rasmussen and Tabor, 2007; Rasmussen et al., 2005), the current formulation

of the model and TPE does not explicitly quantify topographic variation, which may account for error within current soil property distributions and predictions. With the inclusion of topographic variation in EEMT (Rasmussen et al., 2015) and topographic control of soil residence times (Foster et al., 2015; West et al., 2013), we were able to correct this error with the present approach and effectively predicted clay stocks in complex terrain.

Human activities significantly alter soil physical properties (Grieve, 2001; Neff et al., 2005; Pouyat et al., 2007). For example, differences in land use and increased grazing activity can alter soil physical properties such as clay and sand content across landscapes (Neff et al., 2005; Pouyat et al., 2007) or compaction from farming equipment leading to increased bulk density and increased erosion rates (Fullen, 1985; Hamza and Anderson, 2005). Human impacts on soil physical properties were not included in the presented model. The energetic contributions due to human impacts can be incorporated within the EEMT apparatus, and adjusted model parameters can be calculated (Rasmussen et al., 2011). Human impacts on soil physical properties may be locally important, but for the majority of locations, human energetic contributions to the soil system are generally orders of magnitude smaller compared to the energetic inputs from solar radiation, precipitation, or primary productivity.

Internal or intrinsic feedbacks and thresholds within the soil system drive pedogenic development without changes in the external state factors (Chadwick and Chorover, 2001; Muhs, 1984). For example, greater chemical weathering and clay production due to increased water residence time caused by argillic horizon development is the result of an internal feedback that is independent of the external climatic and biological system (Schaetzl and Anderson, 2005). These thresholds can operate as progressive or regressive processes, driving soil formation forward or hindering further development (Johnson and Watson-Stegner, 1987; Phillips, 1993a). Internal soil development feedbacks were not explicitly considered in the present model formulation. The presence of these internal feedbacks may partially explain error within the model predictions. Changes in EEMT would not explain all observed differences in soil properties over the age of the soil. However, if these feedbacks were operating in the included soils, the influence of intrinsic thresholds was implicitly captured within the probability distributions, partially accounting for the role of internal soil development feedbacks on soil formation.

Soil age is typically unmeasured in most geomorphological and pedological studies, limiting the applicability of the current model. Numerical age dating, e.g., cosmogenic radionuclides or optically stimulated luminescence, is expensive and requires time-consuming preparation to be broadly utilized and can be complicated by transport and burial histories of soil and sediment (Anderson et al., 1996; Bierman, 1994; Gosse and Phillips, 2001; Granger and Muzikar, 2001; Schaetzl and Anderson, 2005). Fortunately, relative-age dating methods using landscape position are easily utilized and can provide the necessary age constraint needed to make model predictions (Burke and Birkeland, 1979; Favilli et al., 2009; Huggett, 1998; Matthews and Shakesby, 1984; Nicholas and Butler, 1996; Schaetzl and Anderson, 2005). Age constraint may also be achieved using landscape or hillslope morphology derived from elevation transects or digital elevation models to estimate a "diffusivity age" for the soil (Hsu and Pelletier, 2004; Pelletier et al., 2006).

Global climate patterns have shifted dramatically over the last 65 Myr (Zachos et al., 2001). The majority of soils observed in the compiled chronosequence database span the Quaternary, including both the Holocene and Pleistocene. The Pleistocene was marked by a number of major glacial-interglacial cycles at approximately 100 000-year intervals (Imbrie et al., 1992; Wallace and Hobbs, 2006), which corresponded with shifting climatic conditions; e.g., for large portions of the northern midlatitudes glacial periods were generally cooler and wetter and interglacial periods were warmer and drier (Connin et al., 1998; Petit et al., 1999). Further, the Pleistocene climate shifts likely influenced the rates of weathering and clay production (Hotchkiss et al., 2000). Taking into account the differences in past and modern climate would partially reduce prediction errors between observed and modeled soil physical properties. Reconstructed global paleo-EEMT values would improve model accuracy and limit uncertainty in the probabilistic ranges of soil properties for soils older than Holocene age.

5 Conclusions

The present approach effectively predicts soil physical properties across a diverse range of geomorphic surfaces, lithologies, ecosystems, and climates. Further, this approach is mathematically simple and only requires knowledge of the probable age of a geomorphic surface and the effective energy and mass transfer value associated with a given location, making this approach universally applicable. The simplicity of the probabilistic approach lies in the lack of the need to consider the initial conditions of the soil-forming state or the processes driving soil property change. A probabilistic approach does not exactly predict a soil physical property value at a given location but constrains the probable values based upon the state of the external environment to the soil. Using probabilistic approaches, we can model probable soil-landscape evolution scenarios, greatly informing our understanding of the evolution of critical zone structure.

Competing interests. The authors declare that they have no conflict of interest.

Acknowledgements. We thank Molly Holleran, Rebecca Lybrand, and Ashlee Dere for providing data for this study. Support for C. Shepard was provided by the University Fellows program at the University of Arizona and by the University of Arizona/NASA Space Grant Graduate Fellowship. This research was funded by the US National Science Foundation grant no. EAR-1331408 provided in support of the Catalina-Jemez Critical Zone Observatory. Lidar data acquisition was supported by US National Science Foundation grant no. EAR-0922307 (P. I. Qinghua Guo).

Edited by: P. Finke

References

Almond, P., Roering, J., and Hales, T. C.: Using soil residence time to delineate spatial and temporal patterns of transient landscape response, J. Geophys. Res., 112, F03S17, doi:10.1029/2006JF000568, 2007.

Amundson, R., Heimsath, A., Owen, J., Yoo, K., and Dietrich, W. E.: Hillslope soils and vegetation, Geomorphology, 234, 122–132, doi:10.1016/j.geomorph.2014.12.031, 2015.

Anderson, R. S., Repka, J. L., and Dick, G. S.: Explicit treatment of inheritance in dating depositional surfaces using in site 10Be and 26Al, Geology, 24, 47–51, 1996.

Andre, J. and Anderson, H.: Variation of Soil Erodibility with Geology, Geographic Zone, Elevation, and Vegetation Type in Northern California Wildlands, J. Geophys. Res., 66, 3351–3358, 1961.

Bacon, A. R., Richter, D. D., Bierman, P. R., and Rood, D. H.: Coupling meteoric 10Be with pedogenic losses of 9Be to improve soil residence time estimates on an ancient North American interfluve, Geology, 40, 847–850, doi:10.1130/G33449.1, 2012.

Bierman, P. R.: Using in situ produced cosmogenic isotopes to estimate rates of landscape evolution: A review from the geomorphic perspective, J. Geophys. Res., 99, 13885–13896, doi:10.1029/94JB00459, 1994.

Birkeland, P. W.: Holocene soil chronofunctions, Southern Alps, New Zealand, Geoderma, 34, 115–134, doi:10.1016/0016-7061(84)90017-X, 1984.

Birkeland, P. W.: Soils and Geomorphology, Third., Oxford University Press, New York, New York, 1999.

Burke, R. M. and Birkeland, P. W.: Reevaluation of multiparameter relative dating techniques and their application to the glacial sequence along the eastern escarpment of the Sierra Nevada, California, Quat. Res., 11, 21–51, doi:10.1016/0033-5894(79)90068-1, 1979.

Chadwick, O. A. and Chorover, J.: The chemistry of pedogenic thresholds, Geoderma, 100, 321–353, doi:10.1016/S0016-7061(01)00027-1, 2001.

Chadwick, O. A., Brimhall, G. H., and Hendricks, D. M.: From a black to a gray box – a mass balance interpretation of pedogenesis, Geomorphology, 3, 369–390, doi:10.1016/0169-555X(90)90012-F, 1990.

Connin, S., Betancourt, J., and Quade, J.: Late Pleistocene C4 plant dominance and summer rainfall in the southwestern United States from isotopic study of herbivore teeth, Quat. Res., 50, 179–193, 1998.

Dethier, D. P., Birkeland, P. W., and McCarthy, J. A.: Using the accumulation of CBD-extractable iron and clay content to estimate soil age on stable surfaces and nearby slopes, Front Range, Colorado, Geomorphology, 173–174, 17–29, doi:10.1016/j.geomorph.2012.05.022, 2012.

Dietrich, W. E., Bellugi, D. G., Heimsath, A. M., Roering, J. J., Sklar, L. S., and Stock, J. D.: Geomorphic Transport Laws for Predicting Landscape Form and Dynamics, Geophys. Monogr., 135, 1–30, doi:10.1029/135GM09, 2003.

Dixon, J. L., Heimsath, A. M., and Amundson, R.: The critical role of climate and saprolite weathering in landscape evolution, Earth Surf. Process. Landforms, 34, 1507–1521, doi:10.1002/esp.1836, 2009.

Dokuchaev, V. V.: Russian Chernozem, edited by S. Monson, Israel Program for Scientific Translations Ltd. (For USDA-NSF), 1967 (Translated from Russian to English by N. Kaner), Jerusalem, Israel, 1883.

Egli, M., Merkli, C., Sartori, G., Mirabella, A., and Plotze, M.: Weathering, mineralogical evolution and soil organic matter along a Holocene soil toposequence developed on carbonate-rich materials, Geomorphology, 97, 675–696, doi:10.1016/j.geomorph.2007.09.011, 2008.

Favilli, F., Egli, M., Brandova, D., Ivy-Ochs, S., Kubik, P., Cherubini, P., Mirabella, A., Sartori, G., Giaccai, D., and Haeberli, W.: Combined use of relative and absolute dating techniques for detecting signals of Alpine landscape evolution during the late Pleistocene and early Holocene, Geomorphology, 112, 48–66, doi:10.1016/j.geomorph.2009.05.003, 2009.

Finke, P. A.: Modeling the genesis of luvisols as a function of topographic position in loess parent material, Quat. Int., 265, 3–17, doi:10.1016/j.quaint.2011.10.016, 2012.

Foster, M. A., Anderson, R. S., Wyshnytzky, C. E., Ouimet, W. B., and Dethier, D. P.: Hillslope lowering rates and mobile-regolith residence times from in situ and meteoric 10 Be analysis, Boulder Creek Critical Zone Observatory, Colorado, Geol. Soc. Am. Bull., 127, 862–878, doi:10.1130/B31115.1, 2015.

Fullen, M. A.: Compaction, hydrological processes and soil erosion on loamy sands in east Shropshire, England, Soil Tillage Res., 6, 17–29, doi:10.1016/0167-1987(85)90003-0, 1985.

Gent, P. R., Danabasoglu, G., Donner, L. J., Holland, M. M., Hunke, E. C., Jayne, S. R., Lawrence, D. M., Neale, R. B., Rasch, P. J., Vertenstein, M., Worley, P. H., Yang, Z. L., and Zhang, M.: The community climate system model version 4, J. Clim., 24, 4973–4991, doi:10.1175/2011JCLI4083.1, 2011.

Gosse, J. C. and Phillips, F. M.: Terrestrial in situ cosmogenic nuclides: theory and application, Quat. Sci. Rev., 20, 1475–1560, doi:10.1016/S0277-3791(00)00171-2, 2001.

Granger, D. E. and Muzikar, P. F.: Dating sediment burial with in situ-produced cosmogenic nuclides: Theory, techniques, and limitations, Earth Planet. Sci. Lett., 188, 269–281, doi:10.1016/S0012-821X(01)00309-0, 2001.

Grieve, I. C.: Human impacts on soil properties and their implications for the sensitivity of soil systems in Scotland, Catena, 42, 361–374, doi:10.1016/S0341-8162(00)00147-8, 2001.

Hamza, M. A. and Anderson, W. K.: Soil compaction in cropping systems: A review of the nature, causes and possible solutions, Soil Tillage Res., 82, 121–145, doi:10.1016/j.still.2004.08.009, 2005.

Harden, J.: A quantitative index of soil development from field descriptions: Examples from a chronosequence in central California, Geoderma, 28, 1–28, 1982.

Harden, J.: Soils Developed in Granitic Alluvium near Merced, California, USGS Bulletin 1590-A, Washington, DC, 1987.

Harden, J. W. and Taylor, E. M.: A quantitative comparison of Soil Development in four climatic regimes, Quat. Res., 20, 342–359, doi:10.1016/0033-5894(83)90017-0, 1983.

Heckman, K. and Rasmussen, C.: Lithologic controls on regolith weathering and mass flux in forested ecosystems of the southwestern USA, Geoderma, 164, 99–111, doi:10.1016/j.geoderma.2011.05.003, 2011.

Heimsath, A. M., Dietrich, W. E., Nishiizumi, K., and Finkel, R. C.: The soil production function and landscape equilibrium, Nature, 388, 358–361, 1997.

Heimsath, A. M., Chappell, J., Spooner, N. A., and Questiaux, D. G.: Creeping soil, Geology, 30, 111, doi:10.1130/0091-7613(2002)030<0111:CS>2.0.CO;2, 2002.

Holleran, M., Levi, M., and Rasmussen, C.: Quantifying soil and critical zone variability in a forested catchment through digital soil mapping, SOIL, 1, 47–64, doi:10.5194/soil-1-47-2015, 2015.

Hotchkiss, S., Vitousek, P. M., Chadwick, O. A., and Price, J.: Climate Cycles, Geomorphological Change, and the Interpretation of Soil and Ecosystem Development, Ecosystems, 3, 522–533, doi:10.1007/s100210000046, 2000.

Howard, J., Amos, D., and Daniels, W.: Alluvial soil chronosequence in the Inner Coastal Plain, Virginia, Quat. Res., 39, 201–213, 1993.

Hsu, L. and Pelletier, J. D.: Correlation and dating of Quaternary alluvial-fan surfaces using scarp diffusion, Geomorphology, 60, 319–335, doi:10.1016/j.geomorph.2003.08.007, 2004.

Huang, W.-S., Tsai, H., Tsai, C.-C., Hseu, Z.-Y., and Chen, Z.-S.: Subtropical Soil Chronosequence on Holocene Marine Terraces in Eastern Taiwan, Soil Sci. Soc. Am. J., 74, 1271, doi:10.2136/sssaj2009.0276, 2010.

Huckle, D., Ma, L., McIntosh, J., Vazquez-Ortega, A., Rasmussen, C., and Chorover, J.: U-series isotopic signatures of soils and headwater streams in a semi-arid complex volcanic terrain, Chem. Geo., 445, 68–83, doi:10.1016/j.chemgeo.2016.04.003, 2016.

Huggett, R. J.: Soil chronosequences, soil development, and soil evolution: a critical review, Catena, 32, 155–172, doi:10.1016/S0341-8162(98)00053-8, 1998.

Imbrie, J., Boyle, I. E. A., Clemens, S. C., Duffy, A., Howard, I. W. R., Kukla, G., Kutzbach, J., Martinson, D. G., McIntyre, A., Mix, A. C., Molfino, B., Morley, J. J., Pisias, N. G., Prell, W. L., Peterson, L. C., and Toggweiler, J. R.: On the structure and origin of major glaciation cycles 1. Linear responses to Milankovith forcing, Paleoceanography, 7, 701–738, 1992.

Jackson, M., Tyler, S., Willis, A., Bourbeau, G., and Pennington, R.: Weathering sequence of clay-size minerals in soils and sediments. I. Fundamental Generalizations, J. Phys. Colloid Chem., 52, 1237–1260, 1948.

Jenny, H.: Factors of Soil Formation: A System of Quantitative Pedology, Dover Publications, Inc, New York, New York, available at: http://books.google.com/books?hl=en&lr=&id=orjZZS3H-hAC&oi=fnd&pg=PP1&dq=Factors+of+Soil+Formation:+A+System+of+Quantitative+Pedology&ots=fIfMb5fWkk&sig=e6Ev-CJjgsMYaO8DzFszbQK6Sss (last access: 6 November 2014), 1941.

Jenny, H.: Derivation of state factor equations of soils and ecosystems, Soil Sci. Soc. Am. J., 385–388, 1961.

Johnson, D. and Watson-Stegner, D.: Evolution model of pedogenesis, Soil Sci., 143, 349–366, 1987.

Lybrand, R. A. and Rasmussen, C.: Quantifying Climate and Landscape Position Controls on Soil Development in Semiarid Ecosystems, Soil Sci. Soc. Am. J., 79, 104–116, doi:10.2136/sssaj2014.06.0242, 2015.

Maejima, Y., Matsuzaki, H., and Higashi, T.: Application of cosmogenic 10Be to dating soils on the raised coral reef terraces of Kikai Island, southwest Japan, Geoderma, 126, 389–399, doi:10.1016/j.geoderma.2004.10.004, 2005.

Matthews, J. A. and Shakesby, R. A.: The status of the "Little Ice Age" in southern Norway: relative-age dating of Neoglacial moraines with Schmidt hammer and lichenometry, Boreas, 13, 333–346, doi:10.1111/j.1502-3885.1984.tb01128.x, 1984.

McFadden, L. and Weldon, R.: Rates and processes of soil development on Quaternary terraces in Cajon Pass, California, Geol. Soc. Am. Bull., 98, 280–293, 1987.

McFadden, L., Wells, S., and Jercinovich, M.: Influences of eolian and pedogenic processes on the origin and evolution of desert pavements, Geology, 15, 504–508, 1987.

McKenzie, N. J. and Ryan, P. J.: Spatial prediction of soil properties using environmental correlation, Geoderma, 89, 67–94, doi:10.1016/S0016-7061(98)00137-2, 1999.

Merritts, D., Chadwick, O., and Hendricks, D.: Rates and processes of soil evolution on uplifted marine terraces, northern California, Geoderma, 51, 241–275, 1991.

Minasny, B. and McBratney, A.: A rudimentary mechanistic model for soil production and landscape development, Geoderma, 90, 3–21, 1999.

Minasny, B. and McBratney, A.: A rudimentary mechanistic model for soil formation and landscape development II. A two-dimensional model incorporating chemical weathering, Geoderma, 103, 161–179, 2001.

Muhs, D. R.: Intrinsic thresholds in soil systems., Phys. Geogr., 5, 99–110, doi:10.1080/02723646.1984.10642246, 1984.

Muhs, D. R.: Evolution of Soils on Quaternary Reef Terraces of Barbados, West Indies, Quat. Res., 56, 66–78, doi:10.1006/qres.2001.2237, 2001.

Muhs, D. R., Crittenden, R. C., Rosholt, J. N., Bush, C. A., and Stewart, K.: Genesis of marine terrace soils, Barbados, West Indies: evidence from mineralogy and geochemistry, Earth Surf. Process. Landforms, 12, 605–618, 1987.

Muhs, D. R., Bettis, E. a., Been, J., and McGeehin, J. P.: Impact of Climate and Parent Material on Chemical Weathering in Loess-derived Soils of the Mississippi River Valley, Soil Sci. Soc. Am. J., 65, 1761, doi:10.2136/sssaj2001.1761, 2001.

Neff, J., Reynolds, R., Belnap, J., and Lamothe, P.: Multi-decadal impacts of grazing on soil physical and biogeochemical properties in southeast Utah, Ecol. Appl., 15, 87–95, 2005.

New, M., Hulme, M., and Jones, P.: Representing Twentieth-Century Space – Time Climate Variability. Part I: Development of a 1961–90 Mean Monthly Terrestrial Climatology, J. Clim., 12, 829–856, 1999.

Nicholas, J. W. and Butler, D. R.: Application of Relative-Age Dating Techniques on Rock Glaciers of the La Sal Mountains, Utah: An Interpretation of Holocene Paleoclimates, Geogr. Ann. Ser. A, Phys. Geogr., 78, 1–18, 1996.

Parsons, R. and Herriman, R.: A Lithosequence in the Mountains of Southwestern Oregon, Soil Sci. Soc. Am. J., 39, 943–948, 1975.

Pelletier, J. D. and Rasmussen, C.: Geomorphically based predictive mapping of soil thickness in upland watersheds, Water Resour. Res., 45, W09417, doi:10.1029/2008WR007319, 2009a.

Pelletier, J. D. and Rasmussen, C.: Quantifying the climatic and tectonic controls on hillslope steepness and erosion rate, Lithosphere, 1, 73–80, doi:10.1130/L3.1, 2009b.

Pelletier, J. D., DeLong, S. B., Al-Suwaidi, a. H., Cline, M., Lewis, Y., Psillas, J. L., and Yanites, B.: Evolution of the Bonneville shoreline scarp in west-central Utah: Comparison of scarp-analysis methods and implications for the diffusions model of hillslope evolution, Geomorphology, 74, 257–270, doi:10.1016/j.geomorph.2005.08.008, 2006.

Petit, J., Jouzel, J., Raynaud, D., and Barkov, N.: Climate and atmospheric history of the past 420,000 years from the Vostok ice core, Antarctica, Nature, 399, 429–436, 1999.

Phillips, J. D.: An evaluation of the state factor model of soil ecosystems, Ecol. Modell., 45, 165–177, 1989.

Phillips, J. D.: Progressive and Regressive Pedogenesis and Complex Soil Evolution, Quat. Res., 40, 169–176, 1993a.

Phillips, J. D.: Stability implications of the state factor model of soils as a nonlinear dynamical system, Geoderma, 58, 1–15, doi:10.1016/0016-7061(93)90082-V, 1993b.

Portenga, E. W. and Bierman, P. R.: Understanding earth's eroding surface with 10Be, GSA Today, 21, 4–10, doi:10.1130/G111A.1, 2011.

Pouyat, R. V, Yesilonis, I. D., Russell-Anelli, J., and Neerchal, N. K.: Soil chemical and physical properties that differentiate urban land-use and cover types, Soil Sci. Soc. Am. J., 71, 1010–1019, doi:10.2136/sssaj2006.0164, 2007.

Pye, K.: Formation of quartz silt during humid tropical weathering of dune sands, Sediment. Geol., 34, 267–282, 1983.

Pye, K. and Sperling, C. H. B.: Experimental investigation of silt formation by static breakage processes: the effect of temperature, moisture and salt on quartz dune sand and granitic regolith, Sedimentology, 30, 49–62, doi:10.1111/j.1365-3091.1983.tb00649.x, 1983.

Rasmussen, C.: Mass balance of carbon cycling and mineral weathering across a semiarid environmental gradient, Geochim. Cosmochim. Acta, 72, A778, 2008.

Rasmussen, C. and Tabor, N. J.: Applying a Quantitative Pedogenic Energy Model across a Range of Environmental Gradients, Soil Sci. Soc. Am. J., 71, 1719, doi:10.2136/sssaj2007.0051, 2007.

Rasmussen, C., Southard, R. J., and Horwath, W. R.: Modeling Energy Inputs to Predict Pedogenic Environments Using Regional Environmental Databases, Soil Sci. Soc. Am. J., 69, 1266–1274, doi:10.2136/sssaj2003.0283, 2005.

Rasmussen, C., Troch, P. A., Chorover, J., Brooks, P., Pelletier, J., and Huxman, T. E.: An open system framework for integrating critical zone structure and function, Biogeochemistry, 102, 15–29, doi:10.1007/s10533-010-9476-8, 2011.

Rasmussen, C., Pelletier, J. D., Troch, P. A., Swetnam, T. L., and Chorover, J.: Quantifying Topographic and Vegetation Effects on the Transfer of Energy and Mass to the Critical Zone, Vadose Zo. J., 14, doi:10.2136/vzj2014.07.0102, 2015.

Riebe, C. S., Kirchner, J. W., and Finkel, R. C.: Erosional and climatic effects on long-term chemical weathering rates in granitic landscapes spanning diverse climate regimes, Earth Planet. Sci. Lett., 224, 547–562, doi:10.1016/j.epsl.2004.05.019, 2004.

Runge, E. C. A.: Soil Development Sequences and Energy Models, Soil Sci., 115, 183–193, doi:10.1097/00010694-197303000-00003, 1973.

Salvador-Blanes, S., Minasny, B., and McBratney, A. B.: Modelling long-term in situ soil profile evolution: application to the genesis of soil profiles containing stone layers, Eur. J. Soil Sci., 58, 1535–1548, doi:10.1111/j.1365-2389.2007.00961.x, 2007.

Schaetzl, R. and Anderson, S.: Soils: Genesis and Geomorphology, First, Cambridge University Press, Cambridge, UK, 2005.

Sharmeen, S. and Willgoose, G.: The interaction between armouring and particle weathering for eroding landscapes, Earth Surf. Process. Landforms, 31, 1195–1210, doi:10.1002/esp.1397, 2006.

Shoeneberger, P., Wysocki, D., Benham, E., and Soil Survey Staff: Field book for describing and sampling soils, Version 3., Natural Resources Conservation Service, National Soil Survey Center, Lincoln, NE, available at: http://scholar.google.com/scholar?hl=en&btnG=Search&q=intitle:Field+Book+for+Describing+and+Sampling+Soils#2 (last access: 24 June 2015), 2012.

Smeck, N., Runge, E., and Mackintosh, E.: Dynamics and genetic modelling of soil systems, in: Pedogenesis and Soil Taxonomy I. Concepts and Interactions, edited by: Wilding, L., Smeck, N., and Hall, G., 51–81, Elsevier, Amsterdam, ND, 1983.

Soil Survey Staff: Keys to Soil Taxonomy, 11th ed., United States Department of Agriculture, National Resources Conservation Service, 2010.

Taylor, K. E., Stouffer, R. J., and Meehl, G. A.: An overview of CMIP5 and the experiment design, B. Am. Meteorol. Soc., 93, 485–498, doi:10.1175/BAMS-D-11-00094.1, 2012.

Temme, A. J. A. M. and Vanwalleghem, T.: LORICA – A new model for linking landscape and soil profile evolution: Development and sensitivity analysis, Comput. Geosci., 90, doi:10.1016/j.cageo.2015.08.004, 2015.

Ugarte, M., Militino, A., and Arnholt, A.: Probability and Statistics with R, CRC Press, Boca Raton, FL, 2008.

Vanwalleghem, T., Stockmann, U., Minasny, B., and McBratney, A. B.: A quantitative model for integrating landscape evolution and soil formation, J. Geophys. Res.-Earth Surf., 118, 331–347, doi:10.1029/2011JF002296, 2013.

Volobuyev, V.: Ecology of soils, Academy of Sciences of the Azerbaijan SSR. Institute of Soil Science and Agronomy, Israel Program for Scientific Translations, Jerusalem, Israel, 1964.

Wallace, J. M. and Hobbs, P. V: Atmospheric Science: An Introductory Survey, Second, Academic Press Inc., Amsterdam, ND, 2006.

West, N., Kirby, E., Bierman, P., Slingerland, R., Ma, L., Rood, D., and Brantley, S.: Regolith production and transport at the Susquehanna Shale Hills Critical Zone Observatory, part 2: Insights from meteoric10Be, J. Geophys. Res.-Earth Surf., 118, 1877–1896, doi:10.1002/jgrf.20121, 2013.

White, A. F., Bullen, T. D., Schulz, M. S., Blum, A. E., Huntington, T. G., and Peters, N. E.: Differential rates of feldspar weathering in granitic regoliths, Geochim. Cosmochim. Acta, 65, 847–869, doi:10.1016/S0016-7037(00)00577-9, 2001.

Wilson, M. J.: Weathering of the primary rock-forming minerals: processes, products and rates, Clay Miner., 39, 233–266, doi:10.1180/0009855043930133, 2004.

Yaalon, D.: Conceptual models in pedogenesis: Can soil-forming functions be solved?, Geoderma, 14, 189–205, 1975.

Yoo, K. and Mudd, S. M.: Discrepancy between mineral residence time and soil age: Implications for the interpretation of chemical weathering rates, Geology, 36, 35–38, doi:10.1130/G24285A.1, 2008a.

Yoo, K. and Mudd, S. M.: Toward process-based modeling of geochemical soil formation across diverse landforms: A new mathematical framework, Geoderma, 146, 248–260, doi:10.1016/j.geoderma.2008.05.029, 2008b.

Yoo, K., Amundson, R., Heimsath, A. M., Dietrich, W. E., and Brimhall, G. H.: Integration of geochemical mass balance with sediment transport to calculate rates of soil chemical weathering and transport on hillslopes, J. Geophys. Res.-F Earth Surf., 112, F02013, doi:10.1029/2005JF000402, 2007.

Zachos, J., Pagani, M., Sloan, L., Thomas, E., and Billups, K.: Trends, rhythms, and aberrations in global climate 65 Ma to present, Science, 292, 686–693, doi:10.1126/science.1059412, 2001.

5

Thermal alteration of soil organic matter properties: a systematic study to infer response of Sierra Nevada climosequence soils to forest fires

Samuel N. Araya[1], Marilyn L. Fogel[1,2], and Asmeret Asefaw Berhe[1,2]

[1]Environmental Systems Graduate Group, University of California, Merced, CA 95343, USA
[2]Life and Environmental Sciences Unit, University of California, Merced, CA 95343, USA

Correspondence to: Samuel N. Araya (saraya@ucmerced.edu)

Abstract. Fire is a major driver of soil organic matter (SOM) dynamics, and contemporary global climate change is changing global fire regimes. We conducted laboratory heating experiments on soils from five locations across the western Sierra Nevada climosequence to investigate thermal alteration of SOM properties and determine temperature thresholds for major shifts in SOM properties. Topsoils (0 to 5 cm depth) were exposed to a range of temperatures that are expected during prescribed and wild fires (150, 250, 350, 450, 550, and 650 °C). With increase in temperature, we found that the concentrations of carbon (C) and nitrogen (N) decreased in a similar pattern among all five soils that varied considerably in their original SOM concentrations and mineralogies. Soils were separated into discrete size classes by dry sieving. The C and N concentrations in the larger aggregate size fractions (2–0.25 mm) decreased with an increase in temperature, so that at 450 °C the remaining C and N were almost entirely associated with the smaller aggregate size fractions (< 0.25 mm). We observed a general trend of ^{13}C enrichment with temperature increase. There was also ^{15}N enrichment with temperature increase, followed by ^{15}N depletion when temperature increased beyond 350 °C. For all the measured variables, the largest physical, chemical, elemental, and isotopic changes occurred at the mid-intensity fire temperatures, i.e., 350 and 450 °C. The magnitude of the observed changes in SOM composition and distribution in three aggregate size classes, as well as the temperature thresholds for critical changes in physical and chemical properties of soils (such as specific surface area, pH, cation exchange capacity), suggest that transformation and loss of SOM are the principal responses in heated soils. Findings from this systematic investigation of soil and SOM response to heating are critical for predicting how soils are likely to be affected by future climate and fire regimes.

1 Introduction

Fire is a common, widespread global phenomenon (Bowman et al., 2009) that conditions the dynamics of soil and soil organic matter (SOM). Vegetation fires burn an estimated 300 to 400 million ha of land globally every year (FAO, 2005). In the US alone, over 80 000 fires were reported in 2014 – including about 63 000 wildland fires and 17 000 prescribed burns that burned over 1.5 million and 970 000 ha of land, respectively (National Interagency Fire Center, 2015). In the Sierra Nevada, vegetation fires have a major influence on the landscape. Ecological functions such as plant regeneration, habitat revitalization, biomass accumulation, and nutrient cycling are influenced by fires (McKelvey et al., 1996). Historically most fires were caused by lightning fires, and vegetation fires play an important role in maintaining the health of many ecosystems around the world (Harrison et al., 2010). In recent decades, anthropogenic activities have become major causes of vegetation fires (Caldararo, 2002). Moreover, climate and climatic variations exert a strong influence on the distribution, frequency, and severity of fires (Harrison et al., 2010). Significant changes in global fire regimes are anticipated because of climate change, including increased frequency of fires in the coming decades (Pechony and Shindell,

2010; Westerling et al., 2006). However, our understanding of how climate change and changes in fire regimes will interact to influence topsoils in fire-affected ecosystems is limited.

In addition to combustion of aboveground biomass and alteration of vegetation dynamics, fires also affect the physical, chemical, and biological properties of soils (Certini, 2005; González-Pérez et al., 2004; Mataix-Solera et al., 2011). The degree of alteration caused by fires depends on the fire intensity and duration, which in turn depend on factors such as the amount and type of fuels, properties of aboveground biomass, air temperature and humidity, wind, topography, and soil properties such as moisture content, texture, and SOM content (DeBano et al., 1998). The first-order effects of fire on soil are caused by the input of heat, causing extreme soil temperatures in topsoil (Badía and Martí, 2003b; Neary et al., 1999), which results in loss and transformation of SOM, changes in soil hydrophobicity, changes in soil aggregation, loss of soil mass, and addition of charred material and other combustion products (Albalasmeh et al., 2013; Araya et al., 2016; Mataix-Solera et al., 2011; Rein et al., 2008; Santos et al., 2016).

The duration of burning regulates the amount of energy transferred through the soil. Fires with longer residence time and lower temperature typically impact the soil and SOM more than fires with shorter residence time that burn at a higher temperature (Frandsen and Ryan, 1986; González-Pérez et al., 2004). Penetration of heat down a soil profile depends on intensity and duration of fire as well as the thermal conductivity of the soil (Steward et al., 1990). Soils have low thermal conductivity and only experience extreme temperatures in the top few centimeters of soil during fires. For example, in short-duration or low-severity fires temperatures typically reach only 100–150 °C at 5 cm depth, with no significant change of temperature at 30 cm depth (DeBano, 2000; Janzen and Tobin-Janzen, 2008).

Fire has multiple complex effects on carbon (C) dynamics in soil. Wildfires alone lead to the release of up to $4.1 \, \mathrm{Pg \, C \, year^{-1}}$ to the atmosphere in the form of carbon dioxide, with an additional 0.05 to $0.2 \, \mathrm{Pg \, C \, year^{-1}}$ added to the soil as black or pyrogenic carbon ash (Singh et al., 2012). The changes in SOM characteristics due to combustion include reduced solubility of organic matter (OM) due to loss of external oxygen containing functional groups; reduced chain length of fatty acids, alcohols, and other alkyl compounds; higher aromaticity due to transformation of carbohydrates and lipids; production of pyrogenic carbon; formation of heterocyclic nitrogen (N) compounds; and macromolecular condensation of humic substances (González-Pérez et al., 2004). In the long term, fires can affect soils by altering and removing vegetation and topsoil biomass and increasing soil erodibility (Carroll et al., 2007; DeBano, 1991), subsequently leading to a shift in plant and microbial populations (Janzen and Tobin-Janzen, 2008).

The aim of this study is to determine the effects of heating temperatures on important SOM properties. We used a laboratory heating experiment on five soils from a well-characterized climosequence in the western Sierra Nevada mountain range (Dahlgren et al., 1997). We analyzed changes in SOM quantity and quality following heating treatment with the aim to (1) determine magnitudes of change in SOM properties associated with different fire heating temperatures, (2) identify critical thresholds for these changes, and (3) infer the implications of changing climate on topsoil SOM properties that might experience changing fire regime. This study aims to contribute to the systematic evaluation and development of the ability to predict the effect of fires of different intensities on soil properties under changing climate and fire regimes.

2 Materials and methods

Following the laboratory heating of five soils from the western Sierra Nevada to temperatures ranging from 150 to 650 °C, we analyzed changes in SOM quality and quantity. We measured the changes in C and N concentration in the soil and changes in the distribution of C and N to different aggregate size classes. We also measured changes in isotopic composition of $^{13}\mathrm{C}$ and $^{15}\mathrm{N}$ in the soils and in the different aggregate size classes. Changes in SOM quality was analyzed using Fourier transform infrared (FTIR) spectroscopy of soils. Description of the study site is given in Sect. 2.1 and details of the methods used are given in Sect. 2.2 to 2.4.

2.1 Study site and soil description

For this study, we collected soils from five sites across an elevation transect along the western slope of the central Sierra Nevada, California (Fig. 1); the sites were previously characterized by Dahlgren et al. (1997). We selected four forested sites that are likely to experience forest fires and a fifth lower-elevation grassland site. The thermal alterations in bulk soil physical and chemical properties from the same study soils were previously reported in Araya et al. (2016).

All the sites have a Mediterranean climate characterized by warm-to-hot, dry summers and cool-to-cold, wet winters. Mean annual air temperature ranges from 16.7 °C at the lowest site located at 210 m a.s.l. to 3.9 °C at the highest elevation site at an elevation of 2865 m a.s.l. Annual precipitation ranges from 33 cm at the lowest site to 127 cm at the highest site (Dahlgren et al., 1997; Rasmussen et al., 2007) (Table 1).

The lower elevation woodlands of the Sierra Nevada experience less frequent fires than further upslope and the fires are often fast moving and lower severity (Skinner and Chang, 1996). At the middle-elevation zone of the Sierran forest, the mixed conifer zones, frequent fires are low to moderate severity at lower altitudes, but fire frequency generally increases with altitude towards the upper elevation of the mixed conifer forest (Caprio and Swetnam, 1993). Fires are infre-

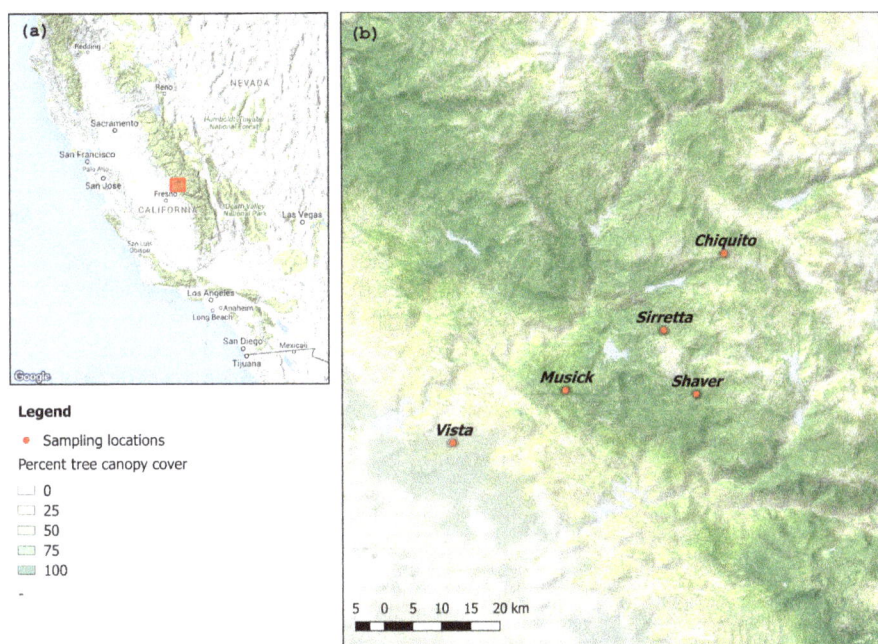

Figure 1. (a) Location of the sampling site on the western slopes of the Sierra Nevada, California, and **(b)** map of the five sampling locations and percent tree canopy cover (US Geological Survey, 2014).

Table 1. Soil classification and site description for the five sites along an elevational transect of the western slopes of the Sierra Nevada (adapted from Dahlgren et al., 1997).

Soil series	Elevation (m)	Ecosystem	MAT[a] (°C)	MAP[b] (cm)	Precip[c]	Dominant vegetation (listed in order of dominance)	Soil taxonomy (family)
Vista	210	Oak woodland	16.7	33	Rain	Annual grasses, *Quercus douglasii*, *Quercus wislizeni*	Coarse-loamy, mixed, superactive, thermic; Typic Haploxerepts
Musick	1384	Oak–mixed conifer forest	11.1	91	Rain	*Pinus ponderosa*, *Calocedrus decurrens*, *Quercus kelloggii*, *Chamaebatia foliolosa*	Fine-loamy, mixed, semiactive, mesic; Ultic Haploxeralf
Shaver	1737	Mixed conifer forest	9.1	101	Snow	*Abies concolor*, *Pinus lambertiana*, *Pinus ponderosa*, *Calocedrus decurrens*	Coarse-loamy, mixed, superactive, mesic; Humic Dystroxerepts
Sirretta	2317	Mixed conifer forest	7.2	108	Snow	*Pinus jeffreyi*, *Abies magnifica*, *Abies concolor*	Sandy-skeletal, mixed, frigid; Dystric Xerorthent
Chiquito[d]	2865	Subalpine mixed conifer forest	3.9	127	Snow	*Pinus contorta murrayana*, *Pinus monticola*, *Lupinus* species	Sandy-skeletal, mixed; Entic Cryumbrept

[a] Mean annual air temperature, calculated from regression equation of Harradine and Jenny (1958). [b] Mean annual precipitation. [c] Dominant form of precipitation. [d] Tentative soil series.

quent and low severity within the high altitude, subalpine zone of the Sierra Nevada (Skinner and Chang, 1996).

Soils from the lowest elevation site, Vista series soils (210 m a.s.l.), fall within the oak woodland zone (elevations < 1008 m a.s.l.). This is the only soil in our study that does not have an O-horizon. The soil has dense annual grass cover and its A-horizon SOM originates mainly from root turnover. The Musick series soils (1384 m a.s.l.) lie within oak–mixed

conifer forest (1008–1580 m a.s.l.) and mixed conifer forest (1580–2626 m a.s.l.). These soils receive the highest litter fall biomass. The Shaver and Sirretta series soils (1737 and 2317 m a.s.l., respectively) fall within the mixed conifer forest range zone, while the Chiquito series soils (2865 m a.s.l.) lie within the subalpine mixed conifer forest range (2626–3200 m a.s.l.). These soils have lower litter fall compared

Table 2. Bulk density, water content, pH, C concentration, cation exchange capacity (CEC), specific surface area (SSA), and particle size distribution for the five soils (mean ± standard error, $n = 3$).

Soil series and elevation (m)	Bulk density $(g\,cm^{-3})$	Gravimetric water content (%)	pH $(CaCl_2)$	Carbon (%)	CEC $(cmol_c\,kg^{-1})$	SSA $(m^2\,g^{-1})$	Particle size distribution[a] (%)		
							Sand	Silt	Clay
Vista (210)	1.26 ± 0.07	0.7 ± 0.0	5.53 ± 0.0	1.51 ± 0.2	8.40 ± 1.1	1.75 ± 0.2	79	11	10
Musick (1384)	0.90 ± 0.06	9.3 ± 1.6	4.67 ± 0.1	7.66 ± 0.8	25.20 ± 2.0	4.98 ± 0.3	60	27	15
Shaver (1737)	0.98 ± 0.06	8.3 ± 1.1	4.85 ± 0.3	2.84 ± 0.2	10.67 ± 2.1	3.08 ± 0.3	80	15	5
Sirretta (2317)	0.61 ± 0.09	9.9 ± 2.2	4.54 ± 0.1	4.74 ± 0.8	12.23 ± 2.6	6.63 ± 0.8	80	15	5
Chiquito (2865)	1.17 ± 0.03	6.1 ± 1.9	3.96 ± 0.1	4.10 ± 0.2	6.03 ± 1.8	1.00 ± 0.04	80	16	4

[a] Particle size distribution of topsoil profile from Dahlgren et al. (1997): Vista (0–14 cm), Musick (0–29 cm), Shaver (0–4 cm), Sirretta (0–6 cm), and Chiquito (0–6 cm).

to the lower elevation soils (van Wagtendonk and Fites-Kaufman, 2006).

The western slope of the central Sierra Nevada presents a remarkable climosequence of soils that developed under similar granitic parent material and are located in landscapes of similar age, relief, slope, and aspect (Trumbore et al., 1996), with significant developmental differences attributed to climate. The soils at mid-elevation range (1000 to 2000 m a.s.l.) tend to be highly weathered, while soils at high and low elevations are relatively less developed (Dahlgren et al., 1997; Harradine and Jenny, 1958; Huntington, 1954; Jenny et al., 1949). Among the most important changes in soil properties along the climosequence are changes in soil organic carbon (SOC) concentration, base saturation, mineral desilication, and hydroxyl-Al interlayering of 2 : 1 layer silicates. Soil pH generally decreases with elevation and the concentrations of clay and secondary iron oxides show a step change at the elevation of the present-day average effective winter snow line, i.e., 1600 m elevation (Tables 1 and 2) (California Department of Water Resources, 1952–1962; Dahlgren et al., 1997).

2.2   Experimental design and sample collection

Triplicate samples (0 to 5 cm depth) were collected at the five sites, approximately 10 m apart from each other. Any overlaying organic layer was removed prior to sampling so that only mineral soil was collected. The soils were air dried at room temperature and passed through a 2 mm sieve. Prior to furnace heating, the soils were oven dried at 60 °C overnight. Soil bulk density and field soil moisture were determined from separate undisturbed core samples collected from each site (Table 2).

Subsamples from each soil were heated in a muffle furnace to one of six selected maximum temperatures (150, 250, 350, 450, 550, and 650 °C). To ensure uniform soil heating and reduce formation of heating gradient inside, the soils were packed 1 cm high in 7 cm diameter porcelain flat capsule crucibles. Oxygen supply was not limited during the heat-

ing – the volume of soil sample to volume air in the furnace was approximately 1 : 50. Furnace temperature was ramped at a rate of 3 °C min^{-1} and soils were exposed to the maximum temperature for 30 min. Once cooled to touch, soils were stored in airtight polyethylene bags prior to analysis.

The six heating temperatures were selected to correspond with fire intensity categories that are based on maximum surface temperature (DeBano et al., 1977; Janzen and Tobin-Janzen, 2008; Neary et al., 1999), that is, low intensity (150 and 250 °C), medium intensity (350 and 450 °C), and high intensity (550 and 650 °C). These fire intensity classes generally correspond with thresholds for important thermal reactions in soils observed by differential thermal analyses (Giovannini et al., 1988; Soto et al., 1991; Varela et al., 2010). A heating rate of 3 °C min^{-1} is preferred in laboratory fire simulation experiments (Giovannini et al., 1988; Terefe et al., 2008; Varela et al., 2010); the slow heating rate prevents sudden combustion when soil ignition temperature is reached at about 220 °C (Fernández et al., 1997, 2001; Varela et al., 2010). The samples were exposed to the maximum set temperature for a period of 30 min. This length of time ensures that the entire sample is uniformly heated at the set temperature and is in keeping with the wide majority of similar laboratory soil heating experiments (for example Badía and Martí, 2003a; Fernández et al., 2001; Giovannini, 1994; Varela et al., 2010; Zavala et al., 2010). The duration of soil heating under vegetation fires is highly varied and not uniform across landscape (Parsons et al., 2010). The same heating procedure was used for all the soils so that it would be possible to compare how the soils from different climate regimes are likely to respond to the fires.

2.3   Laboratory analysis

Dry-aggregate size distribution was measured by sieving. Samples were dry sieved into three aggregate size classes: 2–0.25 mm (macroaggregates), 0.25–0.053 mm (microaggregates), and < 0.053 mm (silt- and clay-sized particles or composites). These aggregate size classes were selected to enable comparison with other studies that investigated the effect of different natural and anthropogenic properties on soil aggregate dynamics and aggregate-protected organic matter (Six et al., 2000).

C and N concentrations and stable isotope ratios were measured using an elemental combustion system (Costech ECS 4010 CHNSO Analyzer, Costech Analytical Technologies, Valencia, CA, USA) that was interfaced with a mass spectrometer (DELTA V Plus Isotope Ratio Mass Spectrometer, Thermo Fisher Scientific, Inc., Waltham, MA, USA). For the analyses, air-dried soil samples were ground to powder consistency on a ball mill (8000M Mixer/Mill, with a 55 mL tungsten carbide vial, SPEX SamplePrep, LLC, Metuchen, NJ, USA) and oven dried at 60 °C for over 36 h. This lower temperature and longer duration of oven drying was used to avoid possible heating-related C or N changes that might occur if drying was done at 105 °C (Kaiser et al., 2015). The C and N concentration results were corrected for moisture by oven drying subsamples at 105 °C overnight. The C and N concentration results were corrected by adjusting for moisture as $W_{adj} = W \times (100 - W_m)$, where W_{adj} is the adjusted percent concentration, W is the concentration before moisture adjustment, and W_m is the percent moisture content. All concentration changes resulting from moisture adjustment were a decrease of less than 1 % of the value. The stable isotope ratios are presented using the δ notation (per mill, ‰) as $\delta^{13}C$ and $\delta^{15}N$ calculated as $\delta = \left[\left(R_{sample} - R_{standard} \right) / R_{standard} \right] \times 1000$‰, where R is ratio of ^{13}C to ^{12}C for $\delta^{13}C$ and ^{15}N to ^{14}N for $\delta^{15}N$. The standards used for analyses are atmospheric $N_2 \delta^{15}N$ and Vienna Pee Dee Belemnite (VPDB) $\delta^{13}C$.

Bulk soil organic matter composition was analyzed using FTIR spectroscopy on a Bruker IFS 66v/S vacuum FTIR spectrometer (Bruker Biosciences Corporation, Billerica, MA, USA). We used the diffuse reflectance infrared Fourier transform (DRIFT) technique (Ellerbrock and Gerke, 2013; Parikh et al., 2014). Powder samples were dried overnight at 60 °C and scanned in mid-infrared from 4000 to 400 cm^{-1}. We used non-KBr-diluted samples after preliminary analyses showed that dilution was not necessary. KBr dilution is not required for soils with low (< 10 %) organic matter concentrations (Ellerbrock and Gerke, 2013; Reeves III, 2003). The FTIR spectrum was collected using KBr background and was baseline corrected using the rubber band correction method with the default 64 baseline points that is part of the OPUS software (Bruker Corporation, 2009).

2.4 Statistical analysis

All quantitative results are expressed as means of three replicates ± standard error, unless otherwise indicated. Differences in means were tested by analysis of variance (ANOVA) and pairwise comparison of treatments done using Tukey's honest significant difference (HSD) test at $p < 0.05$ significance level. The normality of the data and the homogeneity

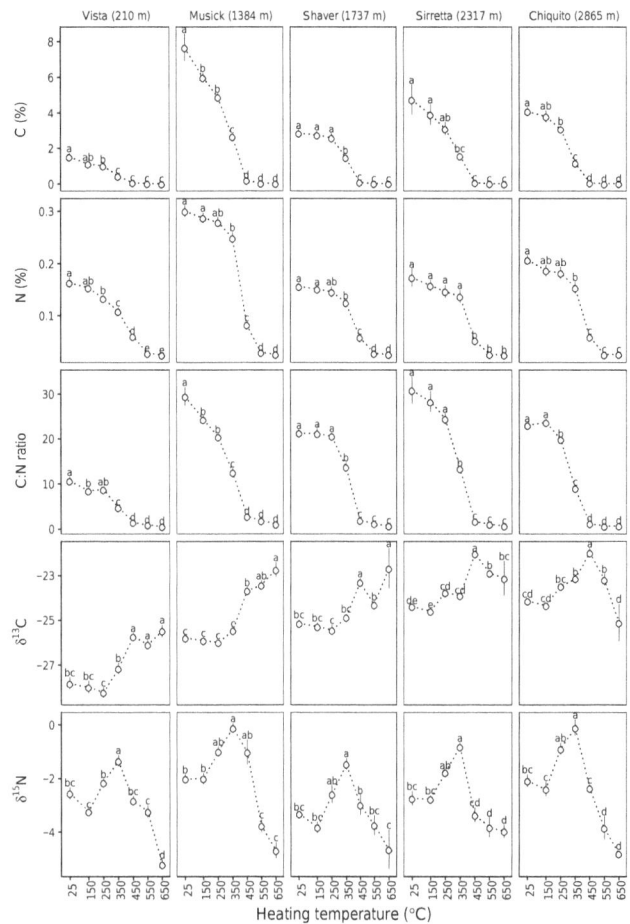

Figure 2. Bulk soil carbon and nitrogen concentrations, C : N atomic ratio, and $\delta^{13}C$ and $\delta^{15}N$ isotope (‰) changes with an increase in heating temperature. Error bars represent standard error, where $n = 3$. Different letters represent significantly different means ($p < 0.05$) at each temperature after Tukey's HSD testing.

of variances was checked using Shapiro–Wilk's and Levene's tests, respectively. All statistical analyses were performed using R statistical software (R Core Team, 2014). The Pearson's correlation coefficient was used to examine relationships between C concentration and changes in soil properties.

3 Results

3.1 Carbon and nitrogen concentration

The initial concentration of C ranged from 1.5 % (Vista soil, 210 m) to 7.7 % (Musick soils, 1384 m). Soil C concentration continuously decreased with increasing temperature. The largest decrease occurred between temperatures of 250 and 450 °C. At 450 °C, all soils lost more than 95 % of their original C. C concentration changes with heating above 450 °C were small and not statistically significant at $p < 0.05$. The C : N ratio ranged from 10 (Vista soils, 210 m) to 29 (Musick soils, 1384 m). Following a similar pattern to C concentration changes, the C : N ratio decreased with an increase in heating temperature (Fig. 2).

The loss of C and N from soils due to heating showed a similar response among all five soils (Fig. 2). After 250 °C, all the soils lost more than 25 % of their initial C (except Shaver soils that lost only about 10 %). At 350 °C all soils lost 50 to 70 % of C. Heating at 450 °C led to the loss of more than 95 % of their initial C for all soils in this study. However, the rate of loss of N was lower than that of C. At temperatures greater than 550 °C there was 5 to 15 % of soil N still remaining. Consequently, we observed a decrease in C : N ratio with increased heating temperature. All soils continued to lose about 15 % soil N for every 100 °C increase and maintained more than 60 % of their N at heating temperatures up to 350 °C. After heating at 450 °C, all soils lost more than 60 % of their original soil N and 85 % by 550 °C.

3.2   Carbon and nitrogen stable isotopes

The δ^{13}C composition of all soils was indicative of C3 vegetation. Soil δ^{13}C composition was most negative at about $-28‰$ for the lowest elevation Vista site (210 m), and the value became consistently less negative with an increase in elevation, reaching $-24‰$ for the highest two sites (i.e., > 2317 m elevation). For all soils, there was a general trend of δ^{13}C enrichment with temperature increase (Fig. 2). The largest change (2.5 to 3.0‰) occurred at heating temperatures between 250 and 450 °C for the lower elevation soils and between 150 and 450 °C for the two highest elevation soils. For the two highest elevation soils, there was a significant ($p < 0.05$) depletion above that temperature. For all soils, except Musick (1384 m) and Shaver (1737 m), the maximum enrichment occurred at 450 °C. All soils showed a similarly patterned δ^{15}N composition change with temperature. The soils were increasingly δ^{15}N enriched with temperature increase up to 350 °C. At temperatures above 350 °C, the soils got more δ^{15}N depleted, with the most negative δ^{15}N occurring at 650 °C (Fig. 2).

3.3   Carbon and nitrogen distribution in aggregate size fractions

C and N concentrations as well as ^{13}C- and ^{15}N-stable isotope ratios were measured for individual soil aggregate size class. The analysis was done on samples heated up to a temperature of 450 °C. The concentration of C and N in samples heated above 450 °C was too low to measure significant changes in C distribution in the different aggregate size classes.

The distribution of C in the three aggregate size fractions followed the same general pattern with increase in the heating temperatures. The macroaggregate size fraction (2–0.25 mm) had the least C concentration and silt–clay-sized particles (< 0.053 mm) had the largest concentration of C

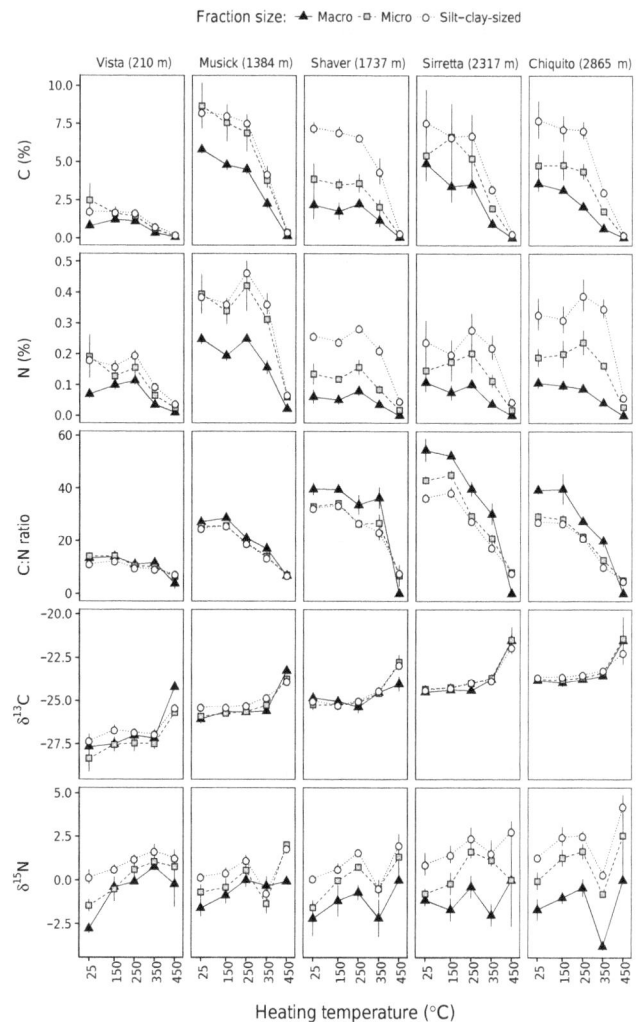

Figure 3. C and N concentrations, C : N atomic ratio, and δ^{13}C and δ^{15}N isotope (‰) changes in macroaggregates (2–0.25 mm), microaggregates (0.25–0.053 mm), and silt–clay-sized (< 0.053 mm) aggregates with increase in heating temperature. Error bars represent standard error, where $n = 3$.

(Fig. 3). N concentration for the macroaggregates was below the detection limit at 450 °C for Chiquito and Sirretta. The change in C and N concentration across heating temperature was similar for all soils.

The distribution of C and N in different size aggregates did not change noticeably except at 450 °C where concentration in all three fractions converged to zero. The distribution of N in the three aggregate size fractions was similar to that of C and followed a similar pattern across all the heating temperatures. Similarly, the macroaggregate size fraction (2–0.25 mm) had the least amount of N concentration, and silt–clay-sized particles (< 0.053 mm) had the largest concentration of N. For Shaver (1737 m), Sirretta (2317 m), and Chiquito (2865 m) soils, the macroaggregate N concentration was too low and could not be detected (Fig. 3). The atomic

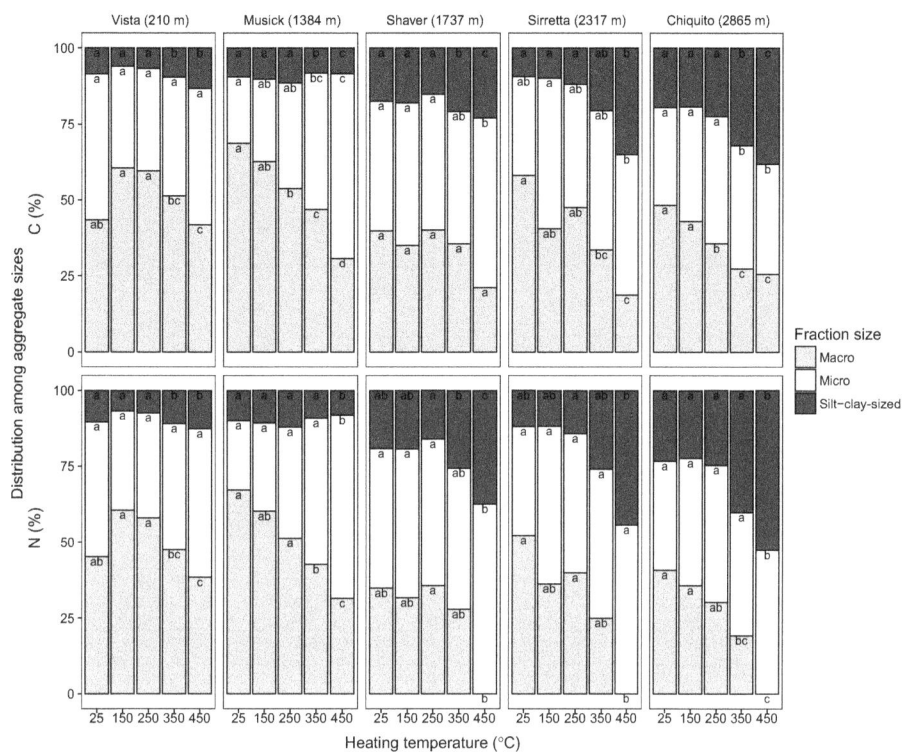

Figure 4. C and N distributions in macroaggregates (2–0.25 mm), microaggregates (0.25–0.053 mm), and silt–clay-sized (< 0.053 mm) aggregates.

C : N ratio generally stayed the same for all soils throughout the temperatures. C : N ratio was highest in macroaggregates, which had the lowest C and N concentrations, followed by microaggregates and silt–clay sizes for all soils.

The stable isotope composition of ^{13}C was very similar between aggregate sizes, with silt–clay-sized aggregates being slightly more enriched, except for Shaver (1737 m), which had slightly more enriched macroaggregates. Conversely, the δ^{15}N values showed clear differences among aggregate fractions even though the measured values of δ^{15}N did not change notably with combustion temperatures. δ^{15}N was highest in silt–clay-sized particles and lowest in macroaggregates, with the microaggregates showing intermediate values. The pattern of change in δ^{15}N across combustion temperatures did not affect this order of δ^{15}N values among aggregate fractions. Most of the C and N in the soils was associated with the larger macroaggregate and microaggregate fractions. With the exception of Vista (210 m) soils, the concentrations in macroaggregates continued to decrease with an increase in temperature, and the remaining C and N concentrations were distributed between the smaller aggregate fractions (Fig. 4). At 450 °C, most of the C and N of the higher altitude soils (Shaver, Sirretta, and Chiquito) was now associated with the silt–clay-sized fractions.

3.4   FTIR spectroscopy

Changes in chemical composition of SOM due to heating were analyzed by infrared spectroscopy using the DRIFT technique. The spectra and peaks after contrasting levels of thermal treatments exhibited qualitative similarities among the different soils. FTIR spectra for the soils are shown in Fig. 5. One notable change that occurred in the functional group composition of SOM with heating is the lowered absorbance intensity of aliphatic methylene groups (as represented by the aliphatic C–H stretching peak that appears at bands between 2950 and 2850 cm^{-1}) at > 250 °C in all soils. When comparing intensity of peaks at 2910–2930 and 2853 cm^{-1} wave numbers (from aliphatic methyl and methylene groups, band A) with those at 1653 and 1400 cm^{-1} (oxygen containing carboxyl and carbonyl groups, band B), the decrease in prominence in the aliphatic C–H peak occurs early in the heating sequence, while the C=O band shows little relative change. In addition, after heating at a temperature of 550 °C, all soils lost the O–H stretching peaks (between 3700 and 3200 cm^{-1}). In a pattern that is more prominent for the Musick soil that had the highest concentration of OM, the aromatic C=C stretch around 1600 cm^{-1} gets more resolved with increase in heating temperature. This pattern in the C=C is visible but not as well resolved in the rest of the soils, especially the Vista soil that showed the least-resolved aromatic C=C stretch peak in this region.

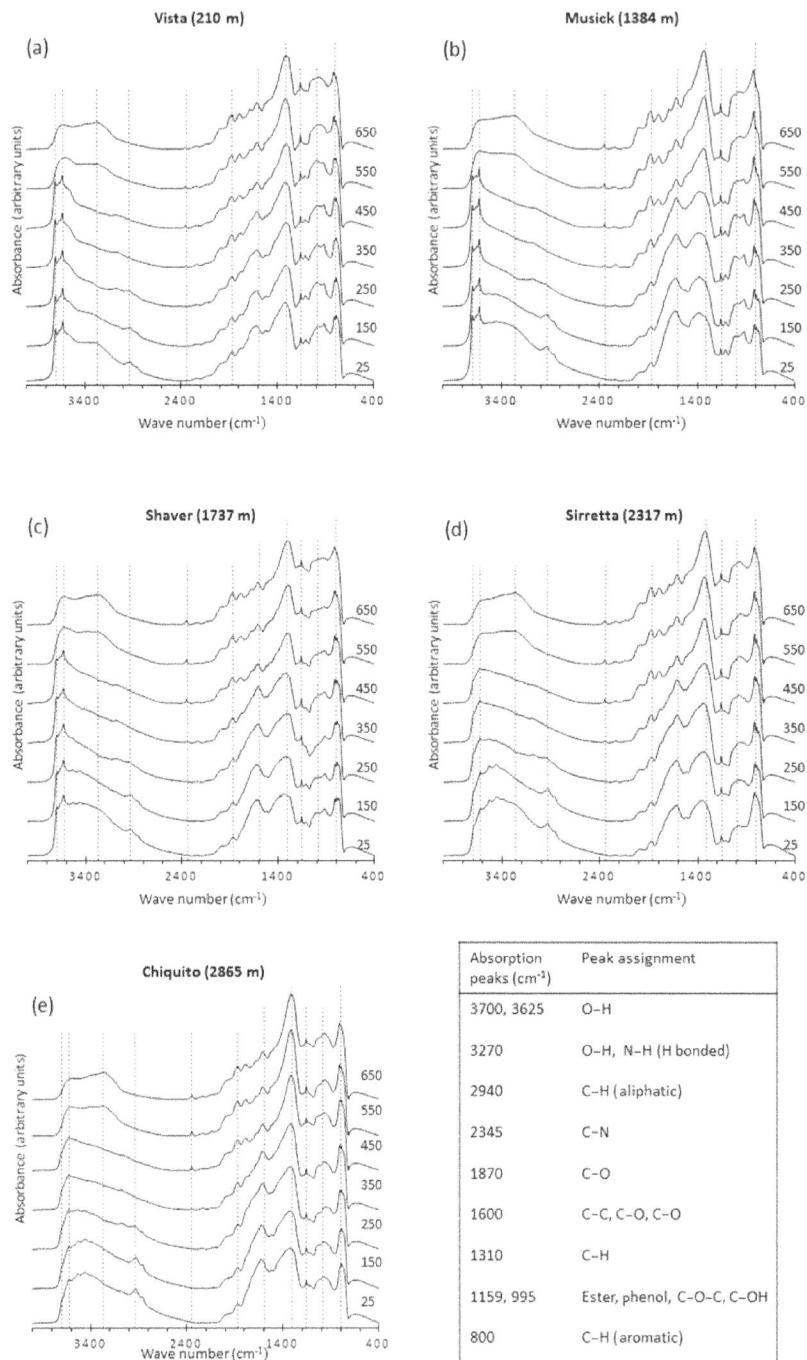

Vista (210 m), Musick (1384 m), Shaver (1737 m), Sirretta (2317 m), Chiquito (2865 m)

Absorption peaks (cm⁻¹)	Peak assignment
3700, 3625	O–H
3270	O–H, N–H (H bonded)
2940	C–H (aliphatic)
2345	C–N
1870	C–O
1600	C–C, C–O, C–O
1310	C–H
1159, 995	Ester, phenol, C–O–C, C–OH
800	C–H (aromatic)

Figure 5. FTIR spectra of the five soils at the different heating temperatures. Heating temperatures, in Celsius, are shown to the right of each spectrum.

4 Discussion

4.1 Changes in SOM concentration, distribution, and composition

Our results show significant effects of combustion temperature on concentration, distribution, and composition of SOM in topsoils that experience the most intense heating dur-

ing vegetation fires. Topsoils have relatively high OM and low clay content that render them more sensitive to heating since the SOM experiences significant changes during heating. In our study system, the effect of fire heating on SOM ranged from slight distillation (volatilization of minor constituents), typically at temperatures below 150 °C, to charring, which typically starts at temperatures above 350 °C, and complete combustion, consistent with findings of previous studies (Badía and Martí, 2003b; Certini, 2005). Our findings also confirmed that regardless of the differences our soils had in mineralogy and other soil physical and chemical properties, the heating treatments (as a proxy for wildfires) led to a consistent decrease in concentration of soil C. This was in agreement with previous studies that showed a decrease in soil C concentration in topsoil after fires (for example Badía et al., 2014; Certini, 2005). However, this loss of C is expected to be restricted to topsoil, while it is expected that the C concentration in subsoil is likely to remain unchanged or may even increase (for example Dennis et al., 2013; Kavdır et al., 2005) due to incorporation of necromass from surface biomass (Almendros et al., 1990; Knicker et al., 2005).

We observed significant changes in concentration, distribution, and composition of SOM with increasing heating temperature. The steep decline in concentration of C in soil that we observed in this study is consistent with a decrease of about 25 % C at 250 °C and an almost 99 % loss at 450 °C (Fig. 6). The magnitude of C loss with heating we observed is similar to the findings of Terefe et al. (2008); and Ulery and Graham (1993), who investigated changes in soil C using artificial heating experiments. Similarly, Giovannini et al. (1988) also found that OM decrease started at 220 °C with about 15 % loss of OM and about 90 % OM loss at 460 °C. Fernández et al. (1997) reported 37 % of SOM loss at 220 °C and 90 % at 350 °C. Furthermore, along with the change in C concentration, between 150 °C and before almost total loss of C above 450 °C, the SOM went through significant qualitative changes that included decrease in C : N ratio, enrichment in $\delta^{13}C$ isotope, changes in $\delta^{15}N$ isotope, and changes in FTIR spectra. Loss of N after fire heating is the result of combustion and volatilization (Fisher and Binkley, 2000). In this study, we observed that N is not as significantly reduced until 350 °C, with about 75 % N remaining as opposed to a loss of C concentration greater than 50 % at the same temperature (Fig. 6). Previous studies had shown that moderate-to high-intensity fires convert most organic-N into inorganic forms of N, such as ammonium (Certini, 2005; Huber et al., 2013). Ammonium is the immediate combustion product that contributes to formation of nitrate (NO_3^-) by nitrification reactions in the weeks or months after a fire. Other studies have shown that a considerable amount of N is transferred into pyrogenic OM products, to black N (de la Rosa and Knicker, 2011; Knicker, 2010), which would also explain the decrease in the C : N ratio. Decrease in the C : N ratio with fire heating has previously been observed in both laboratory and field fire studies (Badía and Martí, 2003a; Certini, 2005; Fernández et al., 1997; González-Pérez et al., 2004).

SOM has a C isotopic composition that reflects the $\delta^{13}C$ signature of native vegetation. Plants are depleted in $\delta^{13}C$ relative to atmosphere. The $\delta^{13}C$ composition for our soils indicated that the dominant source of OM in all soils was C3 plant biomass that had an average $\delta^{13}C$ of −27 ‰, with the higher-elevation soils having more positive $\delta^{13}C$ than the

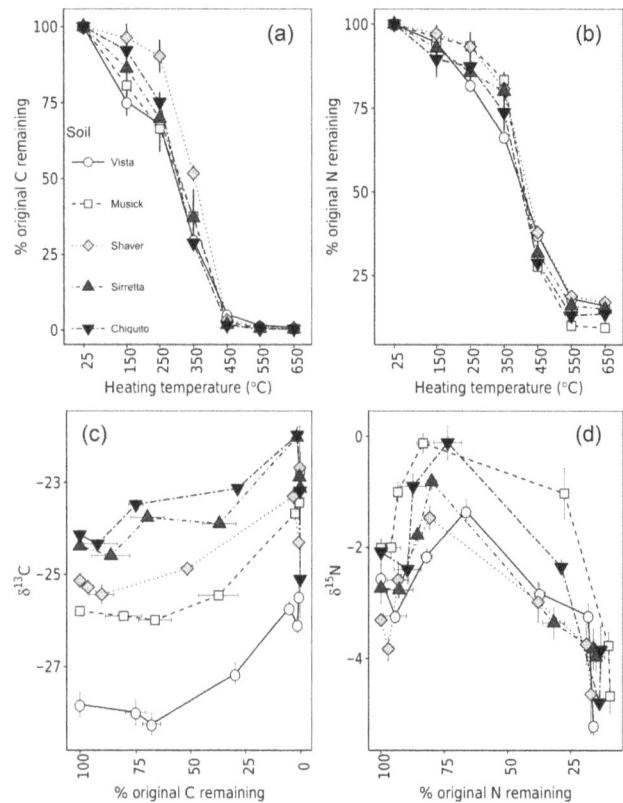

Figure 6. (a) Percentage of C and **(b)** N loss with heating, **(c)** change in $\delta^{13}C$, and **(d)** $\delta^{15}N$ versus percent of total C and N lost from soils (error bars represent standard error, where $n = 3$).

low-elevation soils. Enrichment of ^{13}C with heating is consistent with the loss of plant-derived C. In addition, the fact that lipids (that have relatively more $\delta^{13}C$ depleted than the woody materials) are combusted at lower temperatures than woody materials (such as cellulose and lignin) might contribute to the enrichment of $\delta^{13}C$ with heating (Czimczik et al., 2002). The stable C and N isotope composition of our soils showed significant fractionation with temperature. $\delta^{13}C$ values became more positive (enriched in $\delta^{13}C$) up to 450 °C, where up to 99 % of C was lost (Fig. 6). At higher temperatures there was a less uniform pattern among the soils. For the last < 1 % C, Sirretta and Chiquito soils continued to be more negative (depleted in $\delta^{13}C$) at higher temperatures, while for the rest of the soils there was a slight depletion at 550 °C followed by a slight enrichment at 650 °C (Fig. 2). The depletion of $\delta^{13}C$ at 550 and 650 °C we found in this study is likely a result of SOM charring since there was little or no decrease in C concentration between these temperatures. In a wood charring experiment (non-oxygen atmosphere) at 150, 340, and 480 °C, Czimczik et al. (2002) observed an enrichment of $\delta^{13}C$ at 150 °C where there was no C concentration change but a depletion of $\delta^{13}C$ at 340 and 480 °C, with charring where the C concentration increased over 50 % due to charring.

Table 3. Linear correlation coefficients of changes in soil properties with changes in C concentration. All correlation coefficients have p values < 0.01 unless otherwise indicated.

Soil	Correlation coefficient (r^2) values					
	Mass loss	SSA	Aggregate stability	pH (CaCl$_2$)	CEC	N concentration
Vista	0.74	0.73	0.21[a]	0.77	0.78	0.89
Musick	0.89	0.58	0.77	0.89	0.96	0.83
Shaver	0.82	0.58	0.68	0.74	0.78	0.93
Sirretta	0.60	0.34[b]	0.47	0.67	0.87	0.86
Chiquito	0.82	0.62	0.78	0.88	0.44	0.87

[a] $p = 0.078$. [b] $p = 0.035$.

Fires tend to lead to enrichment of ^{15}N. This is particularly observed in soils immediately in the aftermath of fires (Boeckx et al., 2005; Grogan et al., 2000; Herman and Rundel, 1989; Huber et al., 2013), but there is limited information available on the exact temperature ranges that cause specific levels of ^{15}N enrichment. In this study, we observed enrichment of ^{15}N up to 350 °C and depletion after 350 °C for all soils (Fig. 2). It is likely that the continued ^{15}N enrichment with heating is the result of fractionation due to combustion and volatilization of organic matter, which discriminate against ^{15}N. However, the exact mechanism behind continued depletion of ^{15}N when heated above 350 °C remains unclear. One potential explanation for the ^{15}N depletion at higher temperatures could be indiscriminate removal of N since higher temperatures cause the combustion and volatilization process to happen instantly, compared to charring of OM at lower temperatures. In a post-fire analysis of δ^{15}N on a subalpine ecosystem in Australia, Huber et al. (2013) found that the ^{15}N enrichment of bulk surface soil (from unburnt leaves) was higher than that of the charred OM, which was again higher than that of the ash. They attributed this difference in enrichment level to be the result of the lower heating intensity experienced by the bulk soil, which provided slower processes for greater fractionation. Conversely, higher heat intensity experienced by the ash results in full combustion of plant material, providing little opportunity for isotopic discrimination. The temperature range where we observed the depletion of ^{15}N in our experiment corresponds with the range where a steep decline in N concentration happened (Fig. 6), which would be consistent with the explanation.

Implication of SOM changes with heating

The alterations in and loss of SOM are likely more important causes of soil property changes rather than alterations in soil minerals. SOM is vulnerable to temperatures, while soil minerals are only affected at much higher temperatures (Araya et al., 2016). In addition, all of the soils in our study are characterized by low clay content and low concentration of reactive minerals, but they have a high concentration of SOM, especially in the topsoil, leading to strong relationships between SOM concentrations and soil physical properties.

Degradation of lignin and hemicellulose begins between 130 and 190 °C (Chandler et al., 1983), and carbohydrate signal is completely removed from ^{13}C nuclear magnetic resonance (NMR) spectra by 350 °C. Furthermore, Knicker (2007) observed loss of stable alkyl C and carboxyl C at 350 °C, leading to enrichment of aromatic functional groups in the remaining residue, consistent with what would be expected from incomplete combustion of OM during fires. This leads to transformation and production of charred products (Almendros et al., 2003; Knicker et al., 1996). FTIR analyses from our work showed that the aliphatic O−H stretch peak (bands 3700–3200 cm^{-1}) disappeared at temperatures above 550 °C for all soils accompanied by nitriles or methane nitrile C≡N stretch (2300–2200 cm^{-1}) at temperatures above 450 °C, suggesting condensation of aromatic functional groups.

Loss of OM from soil due to combustion has multiple implications for soil physical and chemical properties. Simple linear correlation between C concentration changes and other soil physical and chemical changes that we observed with heating (reported here and in Araya et al., 2016) show that more than 80 % of the variability in mass loss, aggregate strength, specific surface area (SSA), pH, cation exchange capacity (CEC), and N concentrations is associated with changes in C concentration at the different heating temperatures. Table 3 summarizes the correlation coefficients of soil property changes with change in C concentration. Analyses of associations between C concentration and several soil properties showed a linear association between C and N ($R^2 > 0.8$), mass loss ($R^2 > 0.8$, except for Vista and Sirretta soils), pH ($R^2 > 0.8$, except for Shaver and Sirretta), CEC ($R^2 > 0.7$, except for Chiquito). There was a linear association between C concentration and aggregate strength ($R^2 > 0.7$, except for Musick and Chiquito, which had $R^2 \sim 0.7$). Specific surface area showed relation with C ($R^2 > 0.7$, except for Vista and Musick).

In this study, the greatest changes in SOM occurred between the temperatures 250 and 450 °C, and we found that temperatures below 250 °C had little effect on the quality and quantity of SOM. This implies that lower intensity fires, such as typical prescribed fires, where soil surface temperatures do not exceed 250 °C (Janzen and Tobin-Janzen, 2008), have minimum impact on SOM.

4.2 Climate change implications

Investigation of the response of climosequence soils to different heating temperatures in this study enables us to infer how changes in climate are likely to alter the effect of fires on topsoil physical and chemical properties in the long term. Along our study climosequence, we observed critical differences in response of topsoils based mostly on concentration of OM in soil and soil development stages of each soil. Soil OM concentration and composition in particular have been shown to respond to changes in precipitation amount and distribution, as is expected in the Sierra Nevada (Berhe et al., 2012b). Consequently, changes in soil C storage associated with climate change are expected to lead to different amounts of C loss due to fires. This is evidenced by the observed highest total mass of C loss from the mid-elevation Musick soil that had the highest carbon stock, compared to soils on either side of that elevation range. Anticipated changes in climate in the Sierra Nevada mountain range are expected to include upward movement of the rain–snow transition line, exposing areas that now receive most of their precipitation as snow to rainfall and associated runoff (Arnold et al., 2015, 2014; Stacy et al., 2015). Upward movement of the rain–snow transition zone under anticipated climate change scenarios and associated more intense weathering at higher elevation zones can render more C to be lost during fires. More than 80 % of the variability in mass loss, aggregate strength, SSA, pH, CEC, and N concentrations is associated with changes in C concentration (Table 3). Hence, as the vulnerability of these ecosystems to increased fire frequency increases due to climate change (Westerling et al., 2006), we can expect more soil C loss with fires, along with associated changes in soil chemical and physical properties. In particular, our findings of important changes in soil physical and chemical properties occurring between 250 and 450 °C are important for recognizing that critical transformations of topsoil SOM are likely to occur when, as a result of climate change, systems that are adapted to low-severity fires experience medium- to high-severity fires.

The different responses of soil aggregation in our climosequence to the treatment temperatures also suggest potential loss and transformation of the physically protected C pool in topsoil. Degradation of aggregates during fire (Albalasmeh et al., 2013) is likely to render aggregate-protected C susceptible to potential losses through oxidative decomposition, leaching, and erosion. Moreover, in systems such as the Sierra Nevada, which are dominated by steep slopes, movement of the rain–snow transition zone upward is likely to increase proportion of precipitation that occurs as rain. The kinetic energy of raindrops and the observed increase in hydrophobicity of soils after fires (Johnson et al., 2007, 2004) can lead to higher rates of erosional redistribution, especially for the free light fraction or particulate C that is not associated with soil minerals (Berhe et al., 2012a; Berhe and Kleber, 2013; McCorkle et al., 2016; Stacy et al., 2015).

5 Conclusion

A considerable amount of work has been published to demonstrate how fires affect OM concentration and composition in biomass. This study fills critical gaps by determining how and to what extent OM in soil experiences changes due to heating. The findings of this study also showed that changes in soil properties during heating are closely related to changes in C concentrations in soil. The temperatures most critical to C loss and alteration were found to be 250 °C, where charring of organic matter starts, and 450 °C, where most of the SOM is combusted. Most soil properties exhibited a steep change in this temperature range. SOM exhibited largest change, i.e., soils became enriched in ^{13}C and ^{15}N isotopic composition until approximately 90 % of C and N was lost. At higher temperatures a slight depletion of ^{13}C and a steep depletion of ^{15}N was observed. FTIR spectroscopy showed the reduction and disappearance of aliphatic OH functional groups with temperature increase and accumulation of aromatic carbon groups.

This study presented the effects of heat input on topsoil properties. The study is necessary for understanding thermally induced changes in soil properties in isolation from other variables that accompany vegetation fires, such as the addition of pyrolysis products from plants and ash and the fire-induced soil moisture dynamics. Findings from this study will contribute towards estimating the amount and rate of change in carbon and nitrogen loss and other essential soil properties that can be expected from topsoil exposure to fires of different intensities under anticipated climate change scenarios.

6 Data availability

The data for this article are available online at doi:10.6084/m9.figshare.4614973 (Araya et al., 2017).

Competing interests. The authors declare that they have no conflict of interest.

Acknowledgements. The authors would like to thank Randy A. Dahlgren for providing georeferences for the study sites, background data, and for his comments on an earlier version of this paper. We thank Christina Bradley for her help and expertise in analysis of C and N and Samuel Traina for his comments on an earlier version of this paper. Funding for this work was provided by a UC Merced Graduate Research Council grant and the National Science Foundation (CAREER EAR – 1352627) award to A. A. Berhe.

Edited by: A. Jordán

References

Albalasmeh, A. A., Berli, M., Shafer, D. S., and Ghezzehei, T. A.: Degradation of moist soil aggregates by rapid temperature rise under low intensity fire, Plant Soil, 362, 335–344, doi:10.1007/s11104-012-1408-z, 2013.

Almendros, G., Gonzalez-Vila, F. J., and Martin, F.: Fire-induced transformation of soil organic matter from an oak forest: an experimental approach to the effects of fire on humic substances, Soil Sci., 149, 158–168, doi:10.1097/00010694-199003000-00005, 1990.

Almendros, G., Knicker, H., and González-Vila, F. J.: Rearrangement of carbon and nitrogen forms in peat after progressive thermal oxidation as determined by solid-state ^{13}C- and ^{15}N-NMR spectroscopy, Organic Geochem., 34, 1559–1568, 2003.

Araya, S. N., Meding, M., and Berhe, A. A.: Thermal alteration of soil physico-chemical properties: a systematic study to infer response of Sierra Nevada climosequence soils to forest fires, Soil, 2, 351–366, doi:10.5194/soil-2-351-2016, 2016.

Araya, S., Berhe, A. A., and Fogel, M. L.: Analysis data from soil heating experiment, figshare, doi:10.6084/m9.figshare.4614973.v1, 2017.

Arnold, C., Ghezzehei, T. A., and Berhe, A. A.: Early spring, severe frost events, and drought induce rapid carbon loss in high elevation meadows, PloS one, 9, e106058, doi:10.1371/journal.pone.0106058, 2014.

Arnold, C., Ghezzehei, T. A., and Berhe, A. A.: Decomposition of distinct organic matter pools is regulated by moisture status in structured wetland soils, Soil Biol. Biochem., 81, 28–37, doi:10.1016/j.soilbio.2014.10.029, 2015.

Badía, D. and Martí, C.: Effect of simulated fire on organic matter and selected microbiological properties of two contrasting soils, Arid Land Research and Management, 17, 55–69, doi:10.1080/15324980301594, 2003a.

Badía, D. and Martí, C.: Plant ash and heat intensity effects on chemical and physical properties of two contrasting soils, Arid Land Research and Management, 17, 23–41, doi:10.1080/15324980301595, 2003b.

Badía, D., Martí, C., Aguirre, A. J., Aznar, J. M., González-Pérez, J. A., De la Rosa, J. M., León, J., Ibarra, P., and Echeverría, T.: Wildfire effects on nutrients and organic carbon of a Rendzic Phaeozem in NE Spain: Changes at cm-scale topsoil, Catena, 113, 267–275, doi:10.1016/j.catena.2013.08.002, 2014.

Berhe, A. A. and Kleber, M.: Erosion, deposition, and the persistence of soil organic matter: mechanistic considerations and problems with terminology, Earth Surf. Proc. Land., 38, 908–912, doi:10.1002/esp.3408, 2013.

Berhe, A. A., Harden, J. W., Torn, M. S., Kleber, M., Burton, S. D., and Harte, J.: Persistence of soil organic matter in eroding versus depositional landform positions, J. Geophys. Res.-Biogeo., 117, G02019, doi:10.1029/2011jg001790, 2012a.

Berhe, A. A., Suttle, K. B., Burton, S. D., and Banfield, J. F.: Contingency in the Direction and Mechanics of Soil Organic Matter Responses to Increased Rainfall, Plant Soil, 358, 371–383, doi:10.1007/s11104-012-1156-0, 2012b.

Boeckx, P., Paulino, L., Oyarzun, C., van Cleemput, O., and Godoy, R.: Soil δ^{15}N patterns in old-growth forests of southern Chile as integrator for N-cycling, Isot. Environ. Healt. S., 41, 249–259, doi:10.1080/10256010500230171, 2005.

Bowman, D. M. J. S., Balch, J. K., Artaxo, P., Bond, W. J., Carlson, J. M., Cochrane, M. A., D'Antonio, C. M., DeFries, R. S., Doyle, J. C., Harrison, S. P., Johnston, F. H., Keeley, J. E., Krawchuk, M. A., Kull, C. A., Marston, J. B., Moritz, M. A., Prentice, I. C., Roos, C. I., Scott, A. C., Swetnam, T. W., van der Werf, G. R., and Pyne, S. J.: Fire in the Earth System, Science, 324, 481–484, doi:10.1126/science.1163886, 2009.

Bruker Corporation: OPUS Spectroscopy Software (Version 6.5), Ettlingen, Germany, 2009.

Caldararo, N.: Human ecological intervention and the role of forest fires in human ecology, Sci. Total Environ., 292, 141–165, doi:10.1016/S0048-9697(01)01067-1, 2002.

California Department of Water Resources: Water Conditions in California, Bull. 120, Sacramento, 1952–1962.

Caprio, A. C. and Swetnam, T. W.: Historic fire regimes along an elevational gradient on the west slope of the Sierra Nevada, California, Missoula, MT, 30 March–1 April 1993, 173–179, 1995.

Carroll, E. M., Miller, W. W., Johnson, D. W., Saito, L., Qualls, R. G., and Walker, R. F.: Spatial analysis of a large magnitude erosion event following a Sierran wildfire, J. Environ. Qual., 36, 1105–1111, doi:10.2134/jeq2006.0466, 2007.

Certini, G.: Effects of fire on properties of forest soils: a review, Oecologia, 143, 1–10, doi:10.1007/s00442-004-1788-8, 2005.

Chandler, C., Cheney, P., Thomas, P., Trabaud, L., and William, D.: Fire effects on soil, water and air, in: Fire in Forestry. Vol I: Forest fire behaviour and effects, John Wiley Sons, New York, 1983.

Czimczik, C. I., Preston, C. M., Schmidt, M. W. I., Werner, R. A., and Schulze, E.-D.: Effects of charring on mass, organic carbon, and stable carbon isotope composition of wood, Organic Geochem., 33, 1207–1223, doi:10.1016/S0146-6380(02)00137-7, 2002.

Dahlgren, R. A., Boettinger, J. L., Huntington, G. L., and Amundson, R. G.: Soil development along an elevational transect in the western Sierra Nevada, California, Geoderma, 78, 207–236, doi:10.1016/S0016-7061(97)00034-7, 1997.

DeBano, L. F.: The effect of fire on soil properties, Boise, ID, 10–12 April, 151–156, 1990–1991.

DeBano, L. F.: The role of fire and soil heating on water repellency in wildland environments: a review, J. Hydrol., 231, 195–206, doi:10.1016/S0022-1694(00)00194-3, 2000.

DeBano, L. F., Dunn, P. H., and Conrad, C. E.: Fire's effect on physical and chemical properties of Chaparral soils, USDA Forest Service General Technical Report WO-3, 1977.

DeBano, L. F., Neary, D. G., and Ffolliott, P. F.: Fire's effect on ecosystems John Wiley & Sons, Inc., New York, USA, 1998.

de la Rosa, J. M. and Knicker, H.: Bioavailability of N released from N-rich pyrogenic organic matter: An incubation study, Soil Biol. Biochem., 43, 2368–2373, doi:10.1016/j.soilbio.2011.08.008, 2011.

Dennis, E. I., Usoroh, A. D., and Ijah, C. J.: Soil properties dynamics induced by passage of fire during agricultural burning, International Journal of Advance Agricultural Research, 1, 43–52, doi:10.9734/IJPSS/2013/3121, 2013.

Ellerbrock, R. H. and Gerke, H. H.: Characterization of organic matter composition of soil and flow path surfaces based on physicochemical principles – a review, in: Advances in Agronomy, edited by: Sparks, D. L., Academic Press, Burlington, 2013.

FAO: State of the world's forest Food and Agricultural Organization of the United Nations, Rome, Italy, 2005.

Fernández, I., Cabaneiro, A., and Carballas, T.: Organic matter changes immediately after a wildfire in an Atlantic forest soil and comparison with laboratory soil heating, Soil Biol. Biochem., 29, 1–11, doi:10.1016/S0038-0717(96)00289-1, 1997.

Fernández, I., Cabaneiro, A., and Carballas, T.: Thermal resistance to high temperatures of different organic fractions from soils under pine forests, Geoderma, 104, 281–298, doi:10.1016/S0016-7061(01)00086-6, 2001.

Fisher, R. F. and Binkley, D.: Ecology and management of forest soils, 3, John Wiley and Sons, Inc., New York, USA, 2000.

Frandsen, W. H. and Ryan, K. C.: Soil moisture reduces belowground heat flux and soil temperatures under a burning fuel pile, Can. J. Forest Res., 16, 244–248, doi:10.1139/x86-043, 1986.

Giovannini, G.: The effect of fire on soil quality, in: Soil Erosion and Degradation as a Consequence of Forest Fires, edited by: Sala, M. and Rubio, J. L., Geoforma Ediciones, Logrono, 1994.

Giovannini, G., Lucchesi, S., and Giachetti, M.: Effect of heating on some physical and chemical parameters related to soil aggregation and erodibility, Soil. Sci., 146, 255–261, doi:10.1097/00010694-198810000-00006, 1988.

González-Pérez, J., González-Vila, F., Almendros, G., and Knicker, H.: The effect of fire on soil organic matter – a review, Environ. Int., 30, 855–870, doi:10.1016/j.envint.2004.02.003, 2004.

Grogan, P., Burns, T. D., and Chapin III, F. S.: Fire effects on ecosystem nitrogen cycling in a Californian bishop pine forest, Oecologia, 122, 537–544, doi:10.1007/s004420050977, 2000.

Harradine, F. and Jenny, H.: Influence of parent material and climate on texture and nitrogen and carbon contents of virgin California soils: texture and nitrogen contents of soils, Soil. Sci., 85, 235–243, 1958.

Harrison, S. P., Marlon, J. R., and Bartlein, P. J.: Fire in the Earth System, in: Changing Climates, Earth Systems and Society, edited by: Dodson, J., International Year of Planet Earth, Springer, New York, 2010.

Herman, D. J. and Rundel, P. W.: Nitrogen isotope fractionation in burned and unburned Chaparral soils, Soil Sci. Soc. Am. J., 53, 1229–1236, 1989.

Huber, E., Bell, T. L., and Adams, M. A.: Combustion influences on natural abundance nitrogen isotope ratio in soil and plants following a wildfire in a sub-alpine ecosystem, Oecologia, 173, 1063–1074, doi:10.1007/s00442-013-2665-0, 2013.

Huntington, G. L.: The effect of vertical zonality on clay content in residual granitic soils of the Sierra Nevada mountains, University of California, Berkeley, 1954.

Janzen, C. and Tobin-Janzen, T.: Microbial communities in fire-affected soils, in: Microbiology of Extreme Soils, edited by: Dion, P. and Nautiyal, C. S., Soil Biology, 13, Springer-Verlag Berlin Heidelberg, 2008.

Jenny, H., Gessel, S. P., and Bingham, F. T.: Comparative study of decomposition rates of organic matter in temperate and tropical regions, Soil Sci., 68, 419–432, 1949.

Johnson, D., Murphy, J., Walker, R., Glass, D., and Miller, W.: Wildfire effects on forest carbon and nutrient budgets, Ecol. Eng., 31, 183–192, 2007.

Johnson, D. W., Susfalk, R. B., Caldwell, T. G., Murphy, J. D., Miller, W. W., and Walker, R. F.: Fire Effects on Carbon and Nitrogen Budgets in Forests, Water Air Soil Poll., 4, 263–275, doi:10.1023/B:WAFO.0000028359.17442.d1, 2004.

Kaiser, M., Kleber, M., and Berhe, A. A.: How air-drying and rewetting modify soil organic matter characteristics: An assessment to improve data interpretation and inference, Soil Biol. Biochem., 80, 324–340, doi:10.1016/j.soilbio.2014.10.018, 2015.

Kavdır, Y., Ekinci, H., Yüksel, O., and Mermut, A. R.: Soil aggregate stability and ^{13}C CP/MAS-NMR assessment of organic matter in soils influenced by forest wildfires in Çanakkale, Turkey, Geoderma, 129, 219–229, doi:10.1016/j.geoderma.2005.01.013, 2005.

Knicker, H.: How does fire affect the nature and stability of soil organic nitrogen and carbon? A review, Biogeochemistry, 85, 91–118, doi:10.1007/s10533-007-9104-4, 2007.

Knicker, H.: "Black nitrogen" – an important fraction in determining the recalcitrance of charcoal, Org. Geochem., 41, 947–950, doi:10.1016/j.orggeochem.2010.04.007, 2010.

Knicker, H., Almendros, G., González-Vila, F. J., Martín, F., and Lüdemann, H.-D.: ^{13}C-and ^{15}N-NMR spectroscopic examination of the transformation of organic nitrogen in plant biomass during thermal treatment, Soil Biol. Biochem., 28, 1053–1060, 1996.

Knicker, H., Gonzalezvila, F., Polvillo, O., Gonzalez, J., and Almendros, G.: Fire-induced transformation of C- and N- forms in different organic soil fractions from a Dystric Cambisol under a Mediterranean pine forest, Soil Biol. Biochem., 37, 701–718, doi:10.1016/j.soilbio.2004.09.008, 2005.

Mataix-Solera, J., Cerda, A., Arcenegui, V., Jordan, A., and Zavala, L. M.: Fire effects on soil aggregation: A review, Earth-Sci. Rev., 109, 44–60, doi:10.1016/j.earscirev.2011.08.002, 2011.

McCorkle, E. P., Berhe, A. A., Hunsaker, C. T., McFarlane, K. J., Johnson, D., Fogel, M. L., and Hart, S. C.: Tracing the source of soil organic matter eroded from temperate forested catchments using carbon and nitrogen isotopes, Chem. Geol., 445, 172–184, doi:10.1016/j.chemgeo.2016.04.025, 2016.

McKelvey, K. S., Skinner, C. N., Chang, C.-R., Erman, D. C., Husari, S. J., Parsons, D. J., Wagtendonk, J. W. V., and Weatherspoon, C. P.: An overview of fire in the Sierra Nevada, Davis, California, 37, 1996.

National Interagency Fire Center: available at: https://www.nifc.gov/fireInfo/fireInfo_statistics.html, last access: 5 June 2015.

Neary, D. G., Klopatek, C. C., DeBano, L. F., and Ffolliott, P. F.: Fire effects on belowground sustainability: a review and synthesis, Forest Ecol. Manage., 122, 51–71, doi:10.1016/S0378-1127(99)00032-8, 1999.

Parikh, S. J., Goyne, K. W., Margenot, A. J., Mukome, F. N. D., and Calderón, F. J.: Soil chemical insights provided through vibrational spectroscopy, in: Advances in Agronomy, edited by: Donald, L. S., Academic Press, 2014.

Parsons, A., Robichaud, P. R., Lewis, S. A., Napper, C., and Clark, J.: Field guide for mapping post-fire soil burn severity, U.S. Department of Agriculture, Forest Service, Rocky Mountain Research Station, Fort Collins, CO, 2010.

Pechony, O. and Shindell, D. T.: Driving forces of global wildfires over the past millennium and the forthcoming century, P. Natl. Acad. Sci. USA, 107, 19167–19170, doi:10.1073/pnas.1003669107, 2010.

Rasmussen, C., Matsuyama, N., Dahlgren, R. A., Southard, R. J., and Brauer, N.: Soil Genesis and Mineral Transformation Acrossan Environmental Gradient on Andesitic Lahar, Soil Sci. Soc. Am. J., 71, 225–237, doi:10.2136/sssaj2006.0100, 2007.

R Core Team: R: A Language and Environment for Statistical Computing, R Foundation for Statistical Computing, Vienna, Austria, 2014.

Reeves III, J. B.: Mid-infrared diffuse reflectance spectroscopy: is sample dilution with KBr necessary, and if so, when?, Am. Lab., 35, 24–28, 2003.

Rein, G., Cleaver, N., Ashton, C., Pironi, P., and Torero, J. L.: The severity of smouldering peat fires and damage to the forest soil, Catena, 74, 304–309, doi:10.1016/j.catena.2008.05.008, 2008.

Santos, F., Russell, D., and Berhe, A. A.: Thermal alteration of water extractable organic matter in climosequence soils from the Sierra Nevada, California, J. Geophys. Res.-Biogeo., 121, 2877–2885, doi:10.1002/2016JG003597, 2016.

Singh, N., Abiven, S., Torn, M. S., and Schmidt, M. W. I.: Fire-derived organic carbon in soil turns over on a centennial scale, Biogeosciences, 9, 2847–2857, doi:10.5194/bg-9-2847-2012, 2012.

Six, J., Elliott, E. T., and Paustian, K.: Soil macroaggregate turnover and microaggregate formation: a mechanism for C sequestration under no-tillage agriculture, Soil Biol. Biochem., 32, 2099–2103, doi:10.1016/S0038-0717(00)00179-6, 2000.

Skinner, C. N. and Chang, C.-R.: Fire Regimes, Past and Present, in: Sierra Nevada Ecosystem Project, Final Report to Congress, Vol. II, Assessment and Scientific Basis for management Options, Davis: University of California, Center for Wate rand Wildland Resources, 1996.

Soto, B., Benito, E., Basanta, R., and Díaz-Fierros, F.: Influence of antecedent soil moisture on pedological effects of fire, in: Soil Erosion and Degradation as a Consequence of Forest Fires, edited by: Sala, M. and Rubio, J. L., Geoforma Ed., Logroño, Spain, 1991.

Stacy, E. M., Hart, S. C., Hunsaker, C. T., Johnson, D. W., and Berhe, A. A.: Soil carbon and nitrogen erosion in forested catchments: implications for erosion-induced terrestrial carbon se-

questration, Biogeosciences, 12, 4861–4874, doi:10.5194/bg-12-4861-2015, 2015.

Steward, F. R., Peter, S., and Richon, J. B.: A Method for Predicting the Depth of Lethal Heat Penetration into Mineral Soils Exposed to Fires of Various Intensities, Can. J. Forest Res., 20, 919–926, doi:10.1139/X90-124, 1990.

Terefe, T., Mariscal-Sancho, I., Peregrina, F., and Espejo, R.: Influence of heating on various properties of six Mediterranean soils: a laboratory study, Geoderma, 143, 273–280, doi:10.1016/j.geoderma.2007.11.018, 2008.

Trumbore, S. E., Chadwick, O. A., and Amundson, R.: Rapid exchange between soil carbon and atmospheric carbon dioxide driven by temperature change, Science, 272, 393–396, doi:10.1126/science.272.5260.393, 1996.

Ulery, A. L. and Graham, R. C.: Forest-fire effects on soil color and texture, Soil Sci. Soc. Am. J., 57, 135–140, doi:10.2136/sssaj1993.03615995005700010026x, 1993.

US Geological Survey: NLCD 2011 Land Cover (2011th Edn., ammended 2014), Sioux Falls, SD, USA, 2014.

van Wagtendonk, J. W. and Fites-Kaufman, J. A.: Sierra Nevada Bioregion, in: Fire in California's Ecosystems, edited by: Sugihara, N. G., van Wagtendonk, J. W., Shaffer, K. E., and Thode, A. E., University of California Press, Berkeley, CA, USA, 2006.

Varela, M. E., Benito, E., and Keizer, J. J.: Effects of wildfire and laboratory heating on soil aggregate stability of pine forests in Galicia: the role of lithology, soil organic matter content and water repellency, Catena, 83, 127–134, doi:10.1016/j.catena.2010.08.001, 2010.

Westerling, A. L., Hidalgo, H. G., Cayan, D. R., and Swetnam, T. W.: Warming and earlier spring increase western U.S. forest wildfire activity, Science, 313, 940–943, doi:10.1126/science.1128834, 2006.

Zavala, L. M., Granged, A. J. P., Jordán, A., and Bárcenas-Moreno, G.: Effect of burning temperature on water repellency and aggregate stability in forest soils under laboratory conditions, Geoderma, 158, 366–374, doi:10.1016/j.geoderma.2010.06.004, 2010.

6

Characterization of stony soils' hydraulic conductivity using laboratory and numerical experiments

Eléonore Beckers[1], Mathieu Pichault[1,2], Wanwisa Pansak[3], Aurore Degré[1], and Sarah Garré[2]

[1]Université de Liège, Gembloux Agro-Bio Tech, UR Biosystems Engineering, Passage des déportés 2, 5030 Gembloux, Belgium
[2]Université de Liège, Gembloux Agro-Bio Tech, UR TERRA, Passage des déportés 2, 5030 Gembloux, Belgium
[3]Naresuan University, Department of Agricultural Science, 65000 Phitsanulok, Thailand

Correspondence to: Sarah Garré (sarah.garre@ulg.ac.be)

Abstract. Determining soil hydraulic properties is of major concern in various fields of study. Although stony soils are widespread across the globe, most studies deal with gravel-free soils, so that the literature describing the impact of stones on the hydraulic conductivity of a soil is still rather scarce. Most frequently, models characterizing the saturated hydraulic conductivity of stony soils assume that the only effect of rock fragments is to reduce the volume available for water flow, and therefore they predict a decrease in hydraulic conductivity with an increasing stoniness. The objective of this study is to assess the effect of rock fragments on the saturated and unsaturated hydraulic conductivity. This was done by means of laboratory experiments and numerical simulations involving different amounts and types of coarse fragments. We compared our results with values predicted by the aforementioned predictive models. Our study suggests that it might be ill-founded to consider that stones only reduce the volume available for water flow. We pointed out several factors of the saturated hydraulic conductivity of stony soils that are not considered by these models. On the one hand, the shape and the size of inclusions may substantially affect the hydraulic conductivity. On the other hand, laboratory experiments show that an increasing stone content can counteract and even overcome the effect of a reduced volume in some cases: we observed an increase in saturated hydraulic conductivity with volume of inclusions. These differences are mainly important near to saturation. However, comparison of results from predictive models and our experiments in unsaturated conditions shows that models and data agree on a decrease in hydraulic conductivity with stone content, even though the experimental conditions did not allow testing for stone contents higher than 20 %.

1 Introduction

Determining soil hydraulic properties is of primary importance in various fields of study such as soil physics, hydrology, ecology, and agronomy. Information on hydraulic properties is essential to model infiltration and runoff, to quantify groundwater recharge, to simulate the movement of water and pollutants in the vadose zone, etc. (Bouwer and Rice, 1984). Most unsaturated flow studies characterize the hydraulic properties of the fine fraction (particles smaller than 2 mm in diameter) of supposedly uniform soils only (Bouwer and Rice, 1984; Buchter et al., 1994; Gusev and Novák, 2007). Nevertheless, in reality, soils are heterogeneous media and may contain coarse inclusions (stones) of various sizes and shapes.

Stony soils are widespread across the globe (Ma and Shao, 2008) and represent a significant part of the agricultural land (Miller and Guthrie, 1984). Furthermore, their usage tends to increase because of erosion and cultivation of marginal lands (García-Ruiz, 2010). Yet little attention has been paid to the effects of the coarser fraction on soil hydraulic characteristics, so that the relevant literature is still rather scarce (Ma and Shao, 2008; Novák and Šurda, 2010; Poesen and Lavee, 1994).

Many authors consider that the reduction in volume available for water flow is the only effect of stones on hydraulic conductivity. This hypothesis has led to models linking the hydraulic conductivity of the fine earth to that of the stony soils. They predict a decrease in saturated hydraulic conductivity of stony soil (K_{se}) with an increasing volumetric stoniness (R_v) (Bouwer and Rice, 1984; Brakensiek et al., 1986; Corring and Churchill, 1961; Hlaváčiková and Novák, 2014; Novák and Kňava, 2012; Peck and Watson, 1979; Ravina and Magier, 1984).

However, a number of studies do not observe this simple relationship between the hydraulic conductivity and the stoniness (Zhou et al., 2009; Ma et al., 2010; Russo, 1983; Sauer and Logsdon, 2002) and suggest that other factors, mainly changes in pore size distribution and structure, may play a substantial role in specific situations. Indeed, ambivalent phenomena can intervene simultaneously, which makes the understanding of the effective hydraulic properties of stony soils difficult. The reduced volume available for flow might be partially compensated for by other factors. One compensation factor might be, as pointed out by Ravina and Magier (1984), the creation of large pores in the rock fragments' vicinity. Indeed, the creation of new voids at the stone–fine earth interface could generate preferential flows and hence increase the saturated hydraulic conductivity (Zhou et al., 2009; Cousin et al., 2003; Ravina and Magier, 1984; Sauer and Logsdon, 2002).

These statements define the general context in which our study takes place. The main objectives are (i) to assess the effect of rock fragments on the saturated and unsaturated hydraulic conductivity of soil and (ii) to test the validity of the predictive models that have been proposed in the literature.

2   Material and methods

We studied the effect of R_v on saturated and unsaturated hydraulic conductivity by means of laboratory experiments (evaporation and permeability measurements) and numerical simulations involving different amounts and types of coarse fragments. The latter also serve to further investigate the effect of the stone size and shape on the K_{se}.

2.1   Models predicting soil hydraulic properties of stony soils

Multiple equations have been proposed to estimate the saturated hydraulic conductivity of stony soil (K_{se}) from the one of the fine earth (K_s) assuming that rock fragments only decrease the volume available for water flow. The relative saturated hydraulic conductivity (K_r) is defined as the ratio between the K_{se} and the K_s. Equations (1) and (2) were derived by Peck and Watson (1979) based on heat transfer theory for a homogeneous medium containing non-conductive

spherical and cylindrical inclusions, respectively. Assuming that stones are non-porous and do not alter the porosity of the fine earth, Ravina and Magier (1984) approximated the K_r to the volumetric percentage of fine earth (Eq. 3). Based on empirical relations, Brakensiek et al. (1986) proposed a similar equation, but involving the mass fraction of the rock fragments instead of the volumetric fraction (Eq. 4). On the basis of numerical simulations, Novák et al. (2011) proposed to describe the K_{se} of stony soils as a linear function of the R_v and a parameter that incorporates the hydraulic resistance of the stony fraction (Eq. 5).

$$K_r = \frac{2(1 - R_v)}{2 + R_v} \text{ (Peck and Watson , 1979, for spherical stones)} \quad (1)$$

$$K_r = \frac{(1 - R_v)}{1 + R_v} \text{ (Peck and Watson , 1979, for cylindrical stones)} \quad (2)$$

$$K_r = (1 - R_v) \text{ (Ravina and Magier, 1984)} \quad (3)$$

$$K_r = (1 - R_w) \text{ (Brakensiek et al., 1986)} \quad (4)$$

$$K_r = (1 - a R_v) \text{ (Novák et al., 2011)} \quad (5)$$

In the above, R_v is the volumetric stoniness [$L^3 L^{-3}$]; R_w is the mass fraction of the rock fragments (mass of stones divided by the total mass of the soil containing stones; the stone density is typically $2.5\,\mathrm{g\,cm^{-3}}$ in this case) [$M M^{-1}$]; and a is an empirical parameter that incorporates the hydraulic resistance of the stony fraction considering shape, size, and orientation of inclusions (the recommended value is 1.32 for clay soils according to Novák et al., 2011).

Two major characteristics are widely used to describe the hydraulic properties of unsaturated soil: the water retention curve $\theta(h)$ and the hydraulic conductivity curve $K(h)$. These are both non-linear functions of the pressure head h. One of the most commonly used analytical models was introduced by van Genuchten (1980), based on the pore-bundle model of Mualem (1976), and given by:

$$S_e(h) = \frac{\theta(h) - \theta_r}{\theta_s - \theta_r} = \begin{cases} (1 + |\alpha h|^n)^{-m} & \text{if } h < 0 \\ 1 & \text{if } h \geq 0, \end{cases} \quad (6)$$

$$K(S_e) = K_s S_e^l \left[1 - \left(1 - S_e^{1/m}\right)^m\right]^2 \text{ if } h < 0, \quad (7)$$

in which h is the pressure head [L]; $S_e(h)$ is the saturation state [$L^3 L^{-3}$]; $\theta(h)$ is the volumetric water content [$L^3 L^{-3}$]; θ_r and θ_s respectively represent the residual and saturated water content [$L^3 L^{-3}$]; K_s is the saturated hydraulic conductivity [$L T^{-1}$]; and n [–], l [–], and α [L^{-1}] are empirical shape parameters ($m = 1$-$1/n$, $n > 1$). To extend the hydraulic conductivity curves to stony soils, Hlaváčiková and Novák (2014) propose a simple method assuming that the shape parameters of the van Genuchten–Mualem (VGM) equations (α, n, and l) are independent of R_v. However, this model relies on assumptions that have not been verified. It might be worth mentioning that there are currently no extensive empirical studies available dealing with the influence of porous inclusions under unsaturated conditions. This gap in existing literature is probably due to experimental issues linked with this kind of study: while measuring the potential and the water content of fine earth has become a standard procedure, the opposite is true for soil with rock fragments, especially under transient infiltration processes.

2.2 Laboratory experiments

2.2.1 Sample preparation

We performed laboratory experiments on disturbed samples (height: 65 mm; diameter: 142 mm) containing a mixture of fine earth and coarse inclusions > 10 mm. Two types of inclusions were used: rock fragments (granite) with a diameter between 1 and 2 cm (1) and spherical glass beads with a diameter of 1 cm (2) (see Fig. 1). The fine earth is classified as a clay (sand: 26 %; silt: 19 %; clay: 55 %).

Before each measurement campaign, the fine earth was first oven-dried for 24 h at 105 °C and passed through a 2 mm sieve. To prepare a sample without any inclusion, the fine earth was compacted layer by layer to get an overall bulk density of 1.51 g cm^{-3} (equal to the mean bulk density of the fine earth measured in situ; Pichault, 2015). For samples containing rock fragments, stones were divided over four layers of soil application and laid on the fine earth bed on their flattest side. The samples were then compacted layer by layer in a way that maintains the same bulk density of fine earth as for samples without inclusions (as a result, the global bulk density of samples varies according to stoniness). Even though the filling and compaction procedure was conducted with precision, it is probably impossible to avoid local bulk density heterogeneity as stones can move and/or soil between stones can be less compacted due to difficult access of the area close to the stone during compaction. The same procedure was to prepare samples containing glass balls. Once the specimen was made, it was placed in a basket containing a thin layer of water for at least 24 h in order to saturate the soil from below.

2.2.2 Unsaturated hydraulic conductivity

Setup description

We used the evaporation method to determine the unsaturated hydraulic conductivity and the retention curve of a soil sample. The principle of this method is to simultaneously measure the matric head at different depths and the water content of an initially saturated soil sample submitted to evaporation.

The experiments were performed using cylindrical Plexiglas samples of 1 L (height: 65 mm; diameter: 142 mm), perforated at the bottom to allow saturation from below and open to the atmosphere on the upper side to allow evaporation of the soil moisture. Four 24.9 mm long and 6 mm diameter ceramic tensiometers (SDEC230) were introduced at 10, 25, 40, and 55 mm height, respectively denoted T1 to T4 (the reference level is located at the bottom of the sample). Tensiometers are introduced at saturation; a pin with similar di-

Figure 1. Preparation of disturbed samples containing glass balls (left panel) and gravels (right panel).

mensions is used to facilitate their insertion. In order to avoid preferential flow due to the introduction of the tensiometers on the same vertical axis, each tensiometer was introduced with a horizontal shift of 12° with respect to the centre of the column. The tensiometers are connected by a tube to a pressure transducer (DPT-100, Deltran). The setup was filled with degassed water. The variation in pressure of the drying soil was recorded every 15 min by a CR800 logger (Campbell Scientific). Tensions beyond the air entry point were not taken into account. The air entry point refers to the state from which the measured pressure head starts to decrease as bubbles appear and water vapour accumulates (typically 68 kPa in this case).

The total water loss as a function of time was monitored by a balance (OHAUS) with a sensitivity of 0.2 g, an accuracy of ±1 g, and a time resolution of 15 min. A 50 W infrared lamp was positioned 1 m above the sample surface to slightly speed up the evaporation process. The light was turned off for the first 24 h of every experiment, as the evaporation rate is already high in a saturated sample. A measuring campaign lasted until three of the four tensiometers ran dry (the tension sharply drops down to approximately a null value). At the end of the experiment, the sample was oven-dried for 24 h at 105 °C to estimate the θ.

Data processing

A simplified Wind's (1968) method was used to transform matric potential and total weight data over time into the hydraulic conductivity curve (Schindler, 1980, cited by Schindler and Müller, 2006; Schindler et al., 2010). The method is further adapted in order to take into account the data from four tensiometers. The method assumes that the distribution of water tension and water content is linear throughout the soil column. It further linearizes the water tension and the mass changes over time. The time step chosen to process the data is 1 h.

By calculating the hydraulic conductivity based on measurements of two tensiometers and linking it to the corresponding mean matric head, one can evaluate a point of the hydraulic conductivity curve. We used every possible combination of two tensiometers (six here) to obtain data points for the hydraulic conductivity curve.

Points of the hydraulic conductivity curve obtained at very small hydraulic gradients (defined here as $\nabla H = \frac{\Delta|h|}{\Delta z} - 1$) were rejected, because large errors occur in the near-saturation zone due to uncertainties in estimating small hydraulic gradients (Peters and Durner, 2008; Wendroth, 1993). This highlights in turn the necessity of reliable tensiometers to estimate the near-saturated hydraulic conductivity. In the current literature, acceptance limits of the hydraulic gradient vary between 5 and $0.2 \, \text{cm cm}^{-1}$ (Mohrath et al., 1997; Peters and Durner, 2008; Wendroth, 1993). Using the least restrictive filter criterion (hydraulic gradient > 0.2) requires fine calibration and outstanding performance of the tensiometers. Choosing a more restrictive criterion leads to a larger loss of conductivity points but provides more reliable and robust data. We decided to use a filter criterion that does not consider hydraulic conductivity points higher than the evaporation rate (from 0.1 to $0.2 \, \text{cm day}^{-1}$ in this case), resulting in a lower limit of $1 \, \text{cm cm}^{-1}$ for the hydraulic gradient.

As pointed out by Wendroth (1993) and Peters and Durner (2008), the main drawback associated with the evaporation experiment is that no estimates of conductivity in the wet range can be obtained due to the typically small hydraulic gradients, so that additional measurements of the K_{se} should be provided. To do so, we used constant-head permeability experiments (see below). Except for the K_{se} which is fixed using results from the constant-head permeability experiments, the parameters of the VGM model (1980) (Eq. 7) are obtained by fitting evaluation points from each combination of tensiometers using the so-called "integral method" (Peters and Durner, 2006).

2.2.3   Saturated hydraulic conductivity

Constant-head permeability experiments were used to determine the K_{se} of saturated cylindrical core samples. The flow through the sample is measured at a steady rate under a constant pressure difference. The K_{se} can thus be derived using the following equation:

$$K_{se} = \frac{VL}{A\Delta H \Delta t}, \qquad (8)$$

in which V is the volume of discharge [L^3], L is the length of the permeameter tube [L], A is the cross-sectional area of the permeameter [L^2], ΔH is the hydraulic head difference across the length L [L], and Δt is the time for discharge [T].

The soil sample used for permeability tests has the same size as the one from the evaporation experiment (height: 65 mm; diameter: 142 mm). A 2 cm thick layer of water was maintained on top of the sample thanks to a Mariotte bottle. Water was collected through a funnel in a burette and the volume of discharge V was deduced from measurements

Table 1. Parameters of the van Genuchten equations used in the numerical experiments.

θ_r [−]	θ_s [−]	α [cm^{-1}]	n [−]	l [−]	K_{se} [cm day^{-1}]
0.185	0.442	0.0064	2.11	−0.135	2.686

after 30 and 210 min after the beginning of the experiment ($\Delta t = 180 \, \text{min}$).

2.3   Numerical simulations

The HYDRUS-2D software was used to simulate water flow in variably saturated porous stony soils. HYDRUS-2D solves the two-dimensional Richards equation using the Galerkin finite-element method.

All the performed simulations assumed that rock fragments were non-porous so that "no-flux" boundaries conditions were specified along the stones' limits. Since we mimic the laboratory setup, rock fragments were modelled as circular inclusions. The soil domain over which simulations were performed had the same dimensions as the longitudinal section of the sampling ring used in the laboratory experiments (14×6.5 cm). We considered the 2-D fraction of stoniness equal to the volumetric fraction. The parameters of fine earth used in the simulations come from the fitting of the hydraulic conductivity and water retention curves obtained in our laboratory experiments on stone-free samples (Table 1).

As a general rule, the hydraulic conductivity of a heterogeneous medium tends to be higher for 3-D than for 2-D simulations (Dagan, 1993). Similarly, for a same level of heterogeneity, the flow will be more hampered using 1-D rather than 2-D simulations. In the present study, we performed 2-D simulations: the quantitative and qualitative conclusions from these experiments can be only extended to the third dimension for their corresponding 3-D form with an infinitely long axis.

2.3.1   Unsaturated hydraulic conductivity

We complemented our experimental evaporation results with an equivalent virtual evaporation experiment. The top boundary of the virtual sample was submitted to an evaporation rate q of $0.1 \, \text{cm day}^{-1}$ over 14 days. No fluxes were allowed across other boundaries. The calculation method applied to the output data was similar to the laboratory evaporation experiment, except that the conductivity and pressure head estimations resulted from two observation nodes placed at the top and the bottom of the profile instead of from four tensiometers. We are aware that these choices are debatable: on the one hand, numerical instabilities are more plausible at the limits of the sample and, on the other hand, the use of bigger samples than conventionally used (6.5 cm height) might reduce the accuracy of the evaporation method (see Peters et al., 2015). However, we did keep the observation nodes on the edges and the larger sample size for the following reasons. Firstly, we observed more changes in hydraulic gradient near stones. As small variations in the hydraulic gradient can lead to substantial changes in the hydraulic conductivity estimates, we chose to place observation nodes out of the influence of one specific inclusion. This difficulty, especially at high stone contents, is the reason why the nodes are not situated inside of the sample volume but rather at the edges. Secondly, we checked whether the pressure head was linearly distributed across the soil profile, which was the case. Finally, as we are studying clayey soils, and as we are considering a pressure head range between pF 1.5 and 2.5, these assumptions are likely to be fair (Peters et al., 2015).

As the relative mass balance error was large at the beginning of the simulations, we only started considering values from the moment when this relative error was lower than 5 %. This validation criterion was set arbitrarily, based on the comparison between evaluation points from the simulation of the evaporation experiment on stone-free samples and the expected values obtained from the inputs of the simulation. The hydraulic conductivity curve was obtained fitting the discrete conductivity data plus the simulated saturated hydraulic conductivity using the integral method (Peters and Durner, 2006), just like we did for the laboratory experiment.

2.3.2 Saturated hydraulic conductivity

The K_{se} was determined using a numerical constant-head permeability simulation. We simulated a steady-state water flow of a saturated soil profile, with a constant head of 10 cm applied on the upper boundary. The bottom boundary of the column was defined as a "seepage face", which means that water starts flowing out as soon as the soil at the boundary reaches saturation. The calculation method applied to the output data was identical to the permeability experiment.

2.4 Treatments

Table 2 presents a scheme of all the performed experiments. We duplicated each laboratory experiment with similar numerical simulations.

We first studied the effect of R_v on unsaturated hydraulic properties using laboratory experiments and numerical simulations. In the laboratory approach, we performed evaporation experiments on samples containing (i) fine earth only and (ii) on others with rock fragments (1) at a R_v of 20 %. Two replications per treatment were performed (four measurement campaigns in total). For the numerical approach, simulations of the evaporation experiment were done on homogeneous soil (without stones) and on soil with a R_v of 10, 20, and 30 %. Having fewer time and practical constraints in the numerical simulation, we added an increasing R_v to observe the evolution of the hydraulic conductivity curve. Simulations were performed on soil samples containing 12 reg-

Table 2. Schematic summary of the treatments.

	Effect of R_v on unsaturated hydraulic conductivity				Effect of R_v on saturated hydraulic conductivity			Effect of size and shape on saturated hydraulic conductivity	
Method	Evaporation experiment + permeameter				Permeameter ($R = 5$)[1]			Permeameter	Numerical
R_v^2 [%]	0 – 10 – 20 – 30	0 – 20			0 – 20 – 40 – 60			0 – 10 – 20 – 30	
Approach	Numerical	Laboratory			Numerical	Laboratory			
Inclusion type	•[3] (2-D) $n^4 = 12$	Rock fragments			•[3] (2-D) $n = 12$	Glass spheres	Rock fragments	•[3] (2-D) $n = 1, 12, 27$ ▲[3] (2-D) $n = 1, 12, 27$	▼[3] (2-D) $n = 1, 12, 27$ ■[3] (2-D) $n = 1, 12, 27$ ◆[3] (2-D) $n = 1, 12, 27$

[1] R = replications; [2] R_v is the volumetric stony fraction; [3] • ▲ ▼ ■ ◆ stand for shape, i.e. circular, triangular on its longest side, triangular on its shortest side, rectangular on its shortest side, and rectangular on its longest side; [4] n is the number of inclusions.

ularly distributed stones. One can notice that no investigations of the unsaturated properties with coarse fragments above 30 % of R_v were performed. Indeed, given that small variations in the hydraulic gradient can lead to substantial changes in the hydraulic conductivity estimates, the tensiometers should be ideally positioned out of the direct influence of one particular stone in order to obtain generalizable results. This implies the need for relatively low stone contents (< 30 % according to Zimmerman and Bodvarsson, 1995).

Then, to study the relationship between saturated hydraulic conductivity, K_{se}, and R_v, we performed five replications of four volumetric stone fractions (0, 20, 40, and 60 %) with rock fragments (1). We also tested a second type of inclusions, glass spheres (2), with a R_v of 20 % (1 replication). The first setup with rock fragments was concomitant with the one with glass spheres. Then, the four supplementary replications with rock fragments were processed for the different volumetric fractions altogether: between replications the soil was oven-dried for 24 h at 105 °C and passed through a 2 mm sieve. Numerical permeability simulations were also performed involving 12 circular regularly distributed inclusions for the same R_v (0, 20, 40, and 60 %).

Finally, we used supplementary numerical simulations to investigate the effect of the inclusion shape and size on K_{se}. To do so, simulations of the permeability test were performed on soil containing stones of five different shapes: circular, upward equilateral triangle, downward equilateral triangle, rectangle on its shortest side (L × 1.5 L), and rectangle on its longest side (1.5 L × L) with an R_v of 10, 20, and 30 %. We first performed simulations on soil containing only one centred inclusion. We also performed permeability simulations on soil containing 12 and 27 regularly distributed inclusions (for each R_v).

3 Results and discussion

In the following, results from laboratory experiments and numerical simulations will be compared to the predictions of the different models presented in Sect. 2.1. The K_{se} will be represented by the median value predicted by the five models linking the properties of fine earth to the ones of stony soil (Eqs. 1 to 5). This will be referred to as "results from the K_{se} predictive models" in the following and will be graphically represented by dotted lines. The same predictive models assume that the shape parameters of the VGM equations (n, l, and α) do not depend on the stoniness, as suggested by Hlaváčiková and Novák (2014). As mentioned above, unsaturated functions of stony soils have been barely studied. We will compare results from unsaturated experiments and numerical simulations to predictive models results following this assumption.

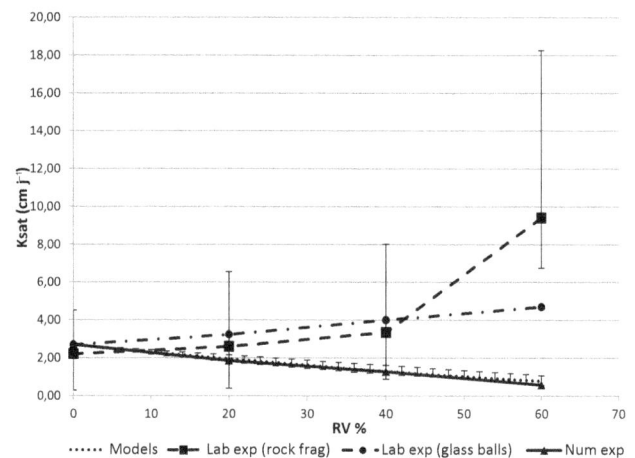

Figure 2. K_{se} depending on R_v obtained from laboratory experiments, numerical simulations with 12 circular inclusions, and the predictive models (the bars show the maximum and minimum intervals around the median predicted by these models).

3.1 Effect of stones on saturated hydraulic conductivity

Figure 2 shows the relationship between the saturated hydraulic conductivity (K_{se}) and the volumetric stone content (R_v) obtained from the constant-head permeability tests for laboratory experiments and numerical simulation (12 circular inclusions). The figure also depicts the median K_{se} of the predictive models (dashed line) and the bars show the 95 % intervals around the median predicted by these models.

The models predict a decreasing K_{se} for an increasing R_v. The numerical simulations show a decrease in K_{se} with an increasing R_v, similar to the predictive models. Looking at the average curve obtained with our five replications (Fig. 2), we observe an overall increase between a R_v of 0 and 60 %, with this global trend being observed for each replication individually (Fig. 3). Statistically speaking, there are significant differences between K_{se} at a R_v of 0 and 60 % and between K_{se} at a R_v of 20 and 60 %. However, at low stone content, we observe a local decrease in K_{se} for some replications. For example, for the first replication (Gravels 1, Fig. 3) K_{se} decreases until a R_v of 20 % and then K_{se} begins to increase. For the second replication (Gravels 2, Fig. 3), the K_{se} increases from a R_v of 0 to 20 % and then decreases at a R_v of 40 %. Analogous permeability tests conducted by Zhou et al. (2009) showed a similar behaviour: the K_{se} initially decreases at low rock content to a minimum value at $R_v = 22$ %, and then at higher R_v K_{se} tends to increase with R_v. Other laboratory tests carried out by Ma et al. (2010) displayed a larger K_{se} at $R_v = 8$ % than the one of the fine earth alone. While carrying out in situ infiltration tests, Sauer and Logsdon (2002) measured higher K_{se} with increasing R_v but decreasing K with increasing R_v under unsaturated conditions (and particularly at $h = -12$ cm). These considerations suggest that the relationship between K_{se} and R_v proposed by

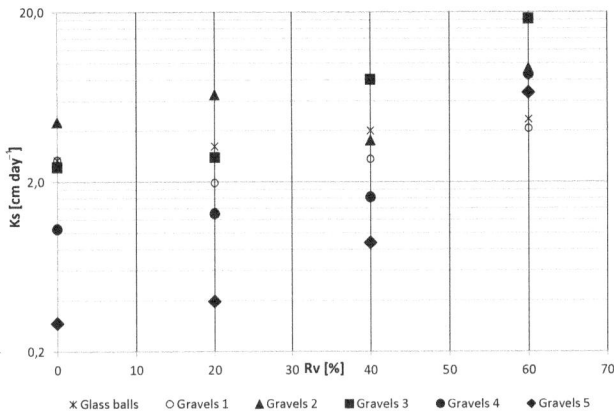

Figure 3. K_{se} depending on R_v obtained from laboratory experiments with gravels (five replications) and glass balls (one replication).

Table 3. Results from the investigation of the inclusion size and shape with regard to the saturated hydraulic conductivity by means of numerical simulations (n is the number of inclusions simulated in the profile for the corresponding R_v).

R_v	Shape	Relative saturated hydraulic conductivity		
		$n = 1$	$n = 12$	$n = 27$
10 %	■	0.88	0.88	0.88
	●	0.84	0.83	0.82
	▲	0.80	0.79	0.78
		0.80	0.79	0.78
	◆	0.84	0.83	0.82
20 %	■	0.76	0.71	0.68
	●	0.73	0.69	0.65
	▲	0.67	0.63	0.54
	▼	0.67	0.63	0.54
	◆	0.66	0.61	0.54
30 %	■	0.70	0.60	0.55
	●	0.64	0.58	0.48
	▲	0.59	0.50	0.46
	▼	0.59	0.50	0.47
	◆	0.56	0.48	0.31

the predictive models simplifies reality to a great extent. These contradictory results suggest that the variation in K_{se} depends on different factors that can counteract the reduction in the volume available for water flow. One possible explanation of our observations has been pointed out by Ravina and Magier (1984), who directly observed large voids by cutting across a stony clay soil sample after its compaction, presumably due to translational displacement of densely packed fragments. This compaction of a saturated sample creates voids near the stone surface and hence increases K_{se} with an increasing R_v. Our packing procedure, demanding the compaction of the sample layer by layer, could lead to the same kind of phenomena observed by Ravina and Magier (1984). Moreover, we have to keep in mind that these elements are very likely to have a different impact depending on soil texture, which was clay for both studies.

Glass beads were used to check the influence of rock characteristics on our conclusions about K_{se}. Since results with glass beads show a trend similar to the five replications with rock fragments, we infer that it is not the rock fragment itself that produces bigger K_{se} but rather the presence of a certain volume of inclusions. In addition, the variation observed between the trends of the curves with rock fragments and glass beads could be due to the inner variation in the hydraulic properties of samples, but it could suggest as well that K_{se} depends on the shape and the roughness of the inclusions. Nevertheless, we can only see the combined effect of these factors in this experiment. This leaves the understanding of the major drivers of the K_{se} and their relative importance unclear. These elements are further investigated through numerical simulations.

Besides the observed increase in K_{se} with rock content, we can also observe a decrease in K_{se} between replications (see Fig. 3). In fact, as mentioned above, the global trend of increasing K_{se} is observed for each replication individually, but packing procedure seems to have a large impact on

results too. There are significant differences ($p < 0.05$) between replication 2 and replication 5, the last one presenting lower K_{se}. The drying and wetting cycles and/or the sieving influence the hydrodynamic behaviour of soil fraction since the effect decreases when R_v increases. This underlines the effect of soil texture and is an important aspect to take into account in future studies.

3.2 Effect of the stone size and shape on the saturated hydraulic conductivity

To investigate the effect of the size of the inclusions and their shape on K_{se} separately from other factors of variation, we performed constant-head permeability simulations on samples containing 1, 12, and 27 inclusions of various shapes, for a R_v of 10, 20, and 30 %. Table 3 illustrates the tendency of the effects and their respective factors.

Table 3 presents the K_r for different sizes of circular inclusions and increasing overall stone content (R_v). When the size of the inclusions decreases (when the number of inclusions increases for a same R_v), the K_r tends to decrease. An interaction between the R_v and the size of inclusion can be observed: the effect of size is more marked with a higher R_v. For example, the decrease in K_r between 1 and 27 circular inclusions is limited to 2 % for a R_v of 10 % but rises up to 25 % for a R_v of 30 %. A similar behaviour is observed with simulations for different shapes of inclusions. One could think that this observation is directly related to change in the minimal cross section for water flow. Figure 4 plots K_r as a func-

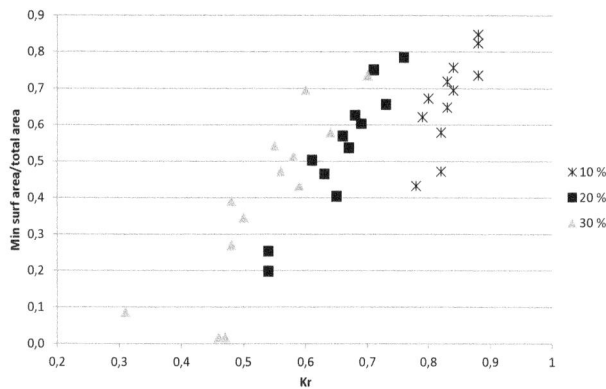

Figure 4. Relationship between minimum surface area and K_r for different R_v.

tion of the ratio between minimal surface area and total surface area. Minimal surface area was calculated as the sample width minus the maximal bulk of stones. Even if we observe a linear trend between these two variables, the relationship is not perfect as we could expect with numerical simulations, supporting the hypothesis that the reduction in the cross section is not the only factor for K_r variations. These statements support the findings of Novák et al. (2011): the smaller the stones, the higher the resistance to flow at a given stoniness. We suggest the decrease in K_{se} is due to a combination of the two following phenomena. The first one is the overlapping of the influence zone of each inclusion, causing further reduction in K_r. The concept of overlapping influence zones was first proposed by Peck and Watson (1979) to explain higher decrease in the hydraulic conductivity of stones very close to each other in comparison to isotropically distributed stones. The second phenomenon could be that, for a given R_v, the contact area between stones and fine earth is higher for small stones than for bigger ones. Hence, a higher tortuosity can be responsible for a lower flow rate.

The shape of the inclusions also has a visible impact on K_r. For a fixed number of inclusions, the K_r is higher with rectangular inclusions on their shortest side and smaller with rectangular inclusions on their longest side. Circular inclusions provoke a smaller reduction than triangular inclusions. The orientation of the triangles does not have a pronounced effect on K_r. Here again, we observe a stronger effect of the size for higher stoniness. As an illustration, the decrease in K_r between circular and triangular inclusions is limited to 5 % for a R_v of 10 % but rises up to 14 % for a R_v of 30 %. A similar behaviour is observed with simulations including either 1 or 27 fragments.

Considering a fixed R_v of 20 % (see Table 3), the effect of the shape of the inclusions depends on their size. For example, the decrease in K_r between rectangular inclusions positioned on their longest and shortest sides is limited to 13 % for samples containing one inclusion only, while it is as high as 21 % for samples containing 27 inclusions. Inversely, the

effect of the size of inclusions also depends on their shape. This effect is higher for triangular and rectangular inclusions positioned on their longest side, with a K_r decrease between 1 and 27 inclusions of 23 and 18 %, respectively. This effect is less significant for circular inclusions, as well as for rectangular inclusions positioned on their shortest sides. The associated K_r decrease between 1 and 27 inclusions is 11 and 10 %, respectively.

The median value of K_r predicted by the models for a R_v of 20 % (0.73) is similar to the simulated K_r for samples containing only one spherical inclusion (Table 3). The K_r predicted by the models is always higher than the K_r determined by the simulations, except for soils containing one inclusion on its shortest side. This can be a side effect of 2-D simulations versus 3-D measurements. Nevertheless, the numerical simulations show that the shape and the size of inclusions may have an effect on K_{se}, which is usually neglected by the current predictive models. In general there is a concordance between models and simulations, whatever the shape and orientation of stones. This strengthens our hypothesis that macropore creation or heterogeneity of bulk density close to the stones can occur and influence K_{se}. Indeed, numerical simulations cannot simulate the creation of voids, unless we create them manually and subjectively in the domain.

Eventually, we hypothesize that, from a certain R_v onwards (the exact R_v value depending on the sampling procedure and the shape and roughness of inclusions, as well as soil texture), stoniness is at the origin of a modification of pore size distributions and of a more continuous macropore system at the stone interface. This macropore system could overcome the other drivers reducing K_{se}.

3.3 Effect of stones on unsaturated hydraulic conductivity

Figure 5 represents the hydraulic conductivity curves obtained from the permeability and evaporation simulations for different stoniness ($R_v = 0$, 10, 20 and 30 %) as well as results predicted by the models for the corresponding R_v. The hydraulic conductivity curves from the predictive models and from the numerical simulations match hydraulic conductivity decreases for increasing R_v. According to these simulations, hydraulic conductivity in the unsaturated zone is well defined using a correct K_{se} and shape parameters do not depend on the stoniness. But this is not surprising since predictive models and numerical simulations rely on the same assumptions, i.e. imperviousness of stones and an identical porosity distribution of fine earth. As a result, these elements do not prove that shape parameters do not depend on the stoniness.

Figure 6 represents the hydraulic conductivity curves obtained from laboratory experiments on stone-free samples and on samples with a R_v of 20 % as well as the results predicted by the models for a R_v of 20 %. Even though the data points are dispersed, those coming from the evaporation

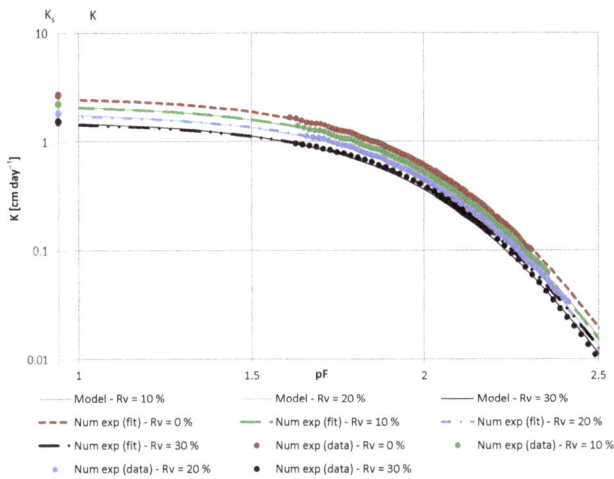

Figure 5. Hydraulic conductivity curves obtained from numerical experiments (data and fit for $R_v = 0$, 10, 20, 30 %) and results predicted by the models for the corresponding R_v.

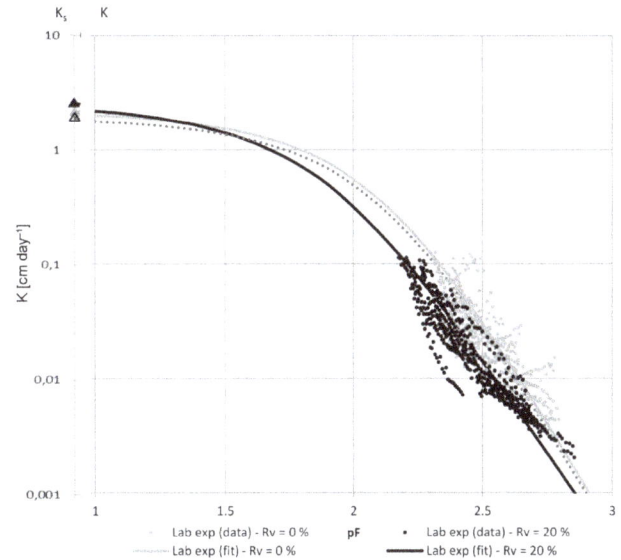

Figure 6. Hydraulic conductivity curves obtained from laboratory experiments (data and fit for $R_v = 0$ and 20 %) and results predicted by the models for a R_v of 20 % (dotted line). Triangles are saturated hydraulic conductivity: closed is measured with black for the stony and grey for the fine earth, and open is predicted by the model (median value of the models).

experiments measured on stony samples are globally lower and slightly more flattened than the ones measured on stone-free samples. This suggests that stones decrease unsaturated hydraulic conductivity. However, it must be noted that we do not have unsaturated K data for higher stone contents, whereas for K_{se} the effect of stoniness becomes more obvious for $R_v > 20$ %. In order to draw final conclusions, it might be necessary to find a way to conduct evaporation experiments for higher stone contents.

In the numerical simulations, the presence of stones reduces the hydraulic conductivity in the same way as predicted by the models, regardless of what the suction was. Similarly, the laboratory experiments suggest that stones reduce the unsaturated hydraulic conductivity, while laboratory experiments in saturated conditions indicated that stones content might increase the K_{se}. These elements support the hypothesis of the macropore creation: according to the well-known law of Jurin (1717), pores through which water will flow depend on both the pore size distribution and the effective saturation. Consequently, flow in the macropore system will only be "activated" in the near-saturation zone, while small pores will only be drained at high suction. Therefore, we could hypothesize that stones are always expected to decrease the hydraulic conductivity at low effective saturation states. However, under saturated conditions, the relationship between R_v and K_{se} seems to be less trivial and requires further investigations considering soil texture and stone characteristics.

4 Conclusion

Determining the effect of rock fragments on soil hydraulic properties is a major issue in soil physics and in the study of fluxes in soil–plant–atmosphere systems in general. Several

models aim at linking the hydraulic properties of fine earth to those of stony soil. Many of them assume that the only effect of stones is to reduce the volume available for water flow. We tested the validity of such models with various complementary experiments.

Our results suggest that it may be ill-founded to consider that stones only reduce the volume available for water flow. First, we observed that, contradictory to the predictive models, the saturated hydraulic conductivity of the clayey soil of this study increases with stone content. Furthermore, we pointed out several other potential drivers influencing K_{se} which are not considered by these K_{se} predictive models. We observed that, for a given stoniness, the resistance to flow is higher for smaller inclusions than for bigger ones. We explain this tendency by an overlap of the influence zones of each stone combined with a higher tortuosity of the flow path. We also pointed out the shape of stones as a factor affecting the hydraulic conductivity of the soil. We showed that the effect of the shape depends on the inclusion size and inversely that the effect of inclusion size depends on its shape. Finally, our results converge to the assumption that this contradictory variation in K_{se} could find its origin at the creation of voids at the stone–fine earth interface as pointed out by Ravina and Magier (1984). Even if the very mechanisms behind these observations remain unclear, they seem to strongly depend on R_v, shape, and roughness of inclusions. However, as we conducted these experiments on a specific clay soil only, and given the fact that structural modifications are textural dependent, our results cannot be extrapolated to other soil textures without similar experiments. Finally, as we worked with disturbed samples, our results do not include quantification of natural phenomena such as swelling and shrinking that occur naturally for clay soils.

These findings suggest that the aforementioned predictive models are not appropriate in all cases, particularly under saturated conditions. Models should take into account the counteracting factors, notably size and shape of stones. However, further investigations are required in order to explore the hydraulic properties of stony soils and to develop new models or adapt the existing ones. The direct observation of the porosity of undisturbed stony samples using X-ray computed tomography or magnetic resonance imaging could firstly confirm and then help to better understand the mechanism of supposed voids' creation at the stone–fine earth interface. However, under unsaturated conditions, these considerations should be more nuanced, as both numerical simulations and laboratory experiments corroborate the general trends from the predictive models. Finally, similar analyses should be conducted in view of determining the effect of the fine earth texture on the drivers of hydraulic properties as pointed out throughout our research.

5   Data availability

The laboratory measurements of this study are available at doi:10.5281/zenodo.32661.

Acknowledgements. We thank Stephane Becquevort of the soil physics lab for his support in setting up the experiments.

Edited by: J. Vanderborght

References

Bouwer, H. and Rice, R. C.: Hydraulic Properties of Stony Vadose Zones, Ground Water, 22, 696–705, 1984.

Brakensiek, D. L., Rawls, W. J., and Stephenson, G. R.: Determining the Saturated Hydraulic Conductivity of a Soil Containing Rock Fragments 1, Soil Sci. Soc. Am. J., 50, 834–835, 1986.

Buchter, B., Hinz, C., and Flühler, H.: Sample size for determination of coarse fragment content in a stony soil, Geoderma, 63, 265–275, 1994.

Corring, R. L. and Churchill, S. W.: Thermal conductivity of heterogeneous materials, Chem. Eng. Prog., 57, 53–59, 1961.

Cousin, I., Nicoullaud, B., and Coutadeur, C.: Influence of rock fragments on the water retention and water percolation in a calcareous soil, Catena, 53, 97–114, 2003. Dagan, G.: Higher-order correction of effective permeability of heterogeneous isotropic formations of lognormal conductivity distribution, Transp. Porous. Med., 12, 279–290, 1993.

García-Ruiz, J. M.: The effects of land uses on soil erosion in Spain: A review, Catena, 81, 1–11, 2010.

Gusev, Y. and Novák, V.: Soil water – main water resources for terrestrial ecosystems of the biosphere, J. Hydrol. Hydromech. Slovak Republ., 55, 3–15, 2007.

Hlaváčiková, H. and Novák, V.: A relatively simple scaling method for describing the unsaturated hydraulic functions of stony soils, J. Plant Nutr. Soil Sci., 177, 560–565, 2014.

Jurin, J.: An Account of Some Experiments Shown before the Royal Society; With an Enquiry into the Cause of the Ascent and Suspension of Water in Capillary Tubes, Philos. Trans., 30, 739–747, 1717.

Ma, D. H. and Shao, M. A.: Simulating infiltration into stony soils with a dual-porosity model, Eur. J. Soil Sci., 59, 950–959, 2008.

Ma, D. H., Zhang, J. B., Shao, M. A., and Wang, Q. J.: Validation of an analytical method for determining soil hydraulic properties of stony soils using experimental data, Geoderma, 159, 262–269, 2010.

Miller, F. T. and Guthrie, R. L.: Classification and distribution of soils containing rock fragments in the United States, Eros. Product. Soils Contain. Rock Fragm. SSSA Spec Publ., 13, 1–6, 1984.

Mohrath, D., Bruckler, L., Bertuzzi, P., Gaudu, J. C., and Bourlet, M.: Error Analysis of an Evaporation Method for Determining Hydrodynamic Properties in Unsaturated Soil, Soil Sci. Soc. Am. J., 61, 725–735, 1997.

Mualem, Y.: A new model for predicting the hydraulic conductivity of unsaturated porous media, Water Resour. Res., 12, 513–522, 1976.

Novák, V. and Kňava, K.: The influence of stoniness and canopy properties on soil water content distribution: simulation of water movement in forest stony soil, Eur. J. Forest Res., 131, 1727–1735, 2012.

Novák, V. and Šurda, P.: The water retention of a granite rock fragments in High Tatras stony soils, J. Hydrol. Hydromech., 58, 181–187, 2010.

Novák, V., Kňava, K., and Šimůnek, J.: Determining the influence of stones on hydraulic conductivity of saturated soils using numerical method, Geoderma, 161, 177–181, 2011.

Peck, A. J. and Watson, J. D.: Hydraulic conductivity and flow in non-uniform soil, Workshop on Soil Physics and Field Heterogeneity, CSIRO Div. Environ. Mech., Canberra, 1979.

Peters, A. and Durner, W.: Improved estimation of soil water retention characteristics from hydrostatic column experiments, Water Resour. Res., 42, W11401, doi:10.1029/2006WR004952, 2006.

Peters, A. and Durner, W.: Simplified evaporation method for determining soil hydraulic properties, J. Hydrol., 356, 147–162, 2008.

Peters, A., Iden, S. C., and Durner, W.: Revisiting the simplified evaporation method: Identification of hydraulic functions considering vapor, film and corner flow, J. Hydrol., 527, 531–542, 2015.

Pichault, M.: Characterization of stony soil's hydraulic properties and elementary volume using field, laboratory and numerical experiments, MsC thesis, ULg Gembloux Agro Bio Tech, p. 73, 2015.

Poesen, J. and Lavee, H.: Rock fragments in top soils: significance and processes, Catena, 23, 1–28, 1994.

Ravina, I. and Magier, J.: Hydraulic Conductivity and Water Retention of Clay Soils Containing Coarse Fragments, Soil Sci. Soc. Am. J., 48, 736–740, 1984.

Russo, D.: Leaching Characteristics of a Stony Desert Soil 1, Soil Sci. Soc. Am. J., 47, 431–438, 1983.

Sauer, T. J. and Longsdon, S. D.: Hydraulic and physical properties of stony soils in a small watershed, Soil Sci. Soc. Am. J., 66, 1947–1956, 2002.

Schindler, U.: Ein Schnellverfahren zur Messung der Wasserleitfähigkeit im teilgesättigten Boden an Stechzylinderproben, Arch. Acker- Pflanzenb. Bodenkd., 24, 1–7, 1980.

Schindler, U. and Müller, L.: Simplifying the evaporation method for quantifying soil hydraulic properties, J. Plant Nutr. Soil Sci., 169, 623–629, 2006.

Schindler, U., Durner, W., von Unold, G., and Müller, L.: Evaporation Method for Measuring Unsaturated Hydraulic Properties of Soils: Extending the Measurement Range, Soil Sci. Soc. Am. J., 74, 1071–1083, 2010.

van Genuchten, M. T.: A Closed-form Equation for Predicting the Hydraulic Conductivity of Unsaturated Soils, Soil Sci. Soc. Am. J., 44, 892–898, 1980.

Wendroth, E.: Reevaluation of the evaporation method for determining hydraulic functions in unsaturated soils, Soil Sci. Soc. Am. J., 57, 1436–1443, 1993.

Wind, G. P.: Capillary conductivity data estimated by a simple method, 181–191, in: Water in the unsaturated zone, edited by: Rijtema, P. E. and Wassink, H., Proc. Wageningen Symp., 1, 19–23 June 1968, Int. Assoc. Sci. Hydrol. Publ., Gentbrugge, the Netherlands, 1968.

Zhou, B. B., Shao, M. A. and Shao, H. B.: Effects of rock fragments on water movement and solute transport in a Loess Plateau soil, Comptes Rendus Geosci., 341, 462–472, 2009.

Zimmerman, R. W. and Bodvarsson, G. S.: The effect of rock fragments on the hydraulic properties of soils, Lawrence Berkeley Lab., CA, USA, USDOE, Washington, D.C., USA, 1995.

7

Process-oriented modelling to identify main drivers of erosion-induced carbon fluxes

Florian Wilken[1,2,3], **Michael Sommer**[3,4], **Kristof Van Oost**[5], **Oliver Bens**[6], and **Peter Fiener**[1]

[1]Institute for Geography, Universität Augsburg, Augsburg, Germany
[2]Chair of Soil Protection and Recultivation, Brandenburg University of Technology Cottbus-Senftenberg, Cottbus, Germany
[3]Institute of Soil Landscape Research, Leibniz Centre for Agricultural Landscape Research ZALF e.V., Müncheberg, Germany
[4]University of Potsdam, Institute of Earth and Environmental Sciences, Potsdam, Germany
[5]Earth & Life Institute, TECLIM, Université catholique de Louvain, Louvain-la-Neuve, Belgium
[6]Helmholtz Centre Potsdam GFZ German Research Centre for Geosciences, Potsdam, Germany

Correspondence to: Peter Fiener (peter.fiener@geo.uni-augsburg.de)

Abstract. Coupled modelling of soil erosion, carbon redistribution, and turnover has received great attention over the last decades due to large uncertainties regarding erosion-induced carbon fluxes. For a process-oriented representation of event dynamics, coupled soil–carbon erosion models have been developed. However, there are currently few models that represent tillage erosion, preferential water erosion, and transport of different carbon fractions (e.g. mineral bound carbon, carbon encapsulated by soil aggregates). We couple a process-oriented multi-class sediment transport model with a carbon turnover model (MCST-C) to identify relevant redistribution processes for carbon dynamics. The model is applied for two arable catchments (3.7 and 7.8 ha) located in the Tertiary Hills about 40 km north of Munich, Germany. Our findings indicate the following: (i) redistribution by tillage has a large effect on erosion-induced vertical carbon fluxes and has a large carbon sequestration potential; (ii) water erosion has a minor effect on vertical fluxes, but episodic soil organic carbon (SOC) delivery controls the long-term erosion-induced carbon balance; (iii) delivered sediments are highly enriched in SOC compared to the parent soil, and sediment delivery is driven by event size and catchment connectivity; and (iv) soil aggregation enhances SOC deposition due to the transformation of highly mobile carbon-rich fine primary particles into rather immobile soil aggregates.

1 Introduction

Soil organic carbon (SOC) is the largest terrestrial carbon (C) pool and has been identified as a cornerstone for the global C cycle. Globally, approx. 1400 Pg C is stored in the upper meter of soil, with approx. 700 Pg C in the upper 0.3 m (Hiederer and Köchy, 2011). As a result, exchange rates between soil and the atmosphere are a major concern with regards to climate change (Polyakov and Lal, 2004a). Earth system model-based estimates for terrestrial C storage in the year 2100 vary widely, ranging from a sink of approx. 8 Pg C yr^{-1} to a source of approx. 6 Pg C yr^{-1} (Friedlingstein

et al., 2014). This large uncertainty might even increase if process levels that are at this point not yet implemented in current models are taken into account (Doetterl et al., 2016). One such process is the lateral redistribution of SOC via erosion processes and the effect this has on vertical C fluxes. Global estimates of erosion-induced C fluxes show conflicting results, ranging from a source of 1 Pg C yr^{-1} to a sink of the same magnitude (for recent reviews see Doetterl et al., 2016; Kirkels et al., 2014). The main reasons for these large differences are a lack of appropriate data (Prechtel et al., 2009), oversimplified modelling approaches that ignore important processes, and differences in measuring approaches, e.g. extrapolating from arable plots (Hooke, 2000; Myers, 1993; Pimentel et al., 1995) vs. measuring continental delivery in river systems (Berhe et al., 2007; Wilkinson and McElroy, 2007).

Most challenging in developing and especially testing models that couple process-oriented SOC redistribution with SOC dynamics are the different spatial and temporal scales of the processes at play (Doetterl et al., 2016). Process-oriented erosion models need event-based data to be validated, while SOC dynamics can hardly be observed on timescales smaller than several decades. Consequently, most existing models that couple soil erosion and SOC turnover processes are based on long-term, USLE-type erosion models that ignore event dynamics. The most widespread of these is SPEROS-C, which was applied on scales ranging from micro- to mesoscale catchments (Fiener et al., 2015; Nadeu et al., 2015; Van Oost et al., 2005b).

The conventional approach to modelling coupled soil erosion and SOC turnover is to treat SOC as a stable part of the bulk parent soil and statistically model (long-term) erosion. However, this approach is likely to lead to biased estimates of both water-erosion-induced SOC redistribution and its effect on vertical C fluxes. Numerous studies have shown that the transport of SOC is selective (Schiettecatte et al., 2008), controlled by event characteristics (Sharpley, 1985; Van Hemelryck et al., 2010) and soil aggregation (Hu and Kuhn, 2014, 2016). The enrichment of SOC during transport has been explicitly addressed by a few modelling studies, using different approaches (Fiener et al., 2015; Lacoste et al., 2015). The effects of tillage erosion on vertical C fluxes have not yet been evaluated in detail, although a representation has been accounted for in some modelling studies (Lacoste et al., 2015; Van Oost et al., 2005a).

The aim of this study is to couple a spatially distributed, process-oriented and event-based water erosion model with a tillage erosion model and a SOC turnover model in order to analyse the importance of individual erosion processes in the erosion-induced C balance of agricultural catchments. The study intends to identify relevant processes that should be implemented in less data-demanding, more parsimonious models.

2 Materials and methods

2.1 Test site

The test site is located about 40 km north of Munich in the Tertiary Hills, an intensively used agricultural area in southern Germany. The site consists of two small arable catchments (48°29′ N, 11°26′ E; Fig. 1), catchments C1 and C2, covering an area of 3.7 and 7.8 ha, respectively. The rolling topography ranges from 454 to 496 m above sea level with a mean slope of 4.2° (±0.6°) for catchment C1 and 5.3° (±1.7°) for catchment C2. The soil landscape is characterized by Cambisols and Luvisols (partly redoximorphic), both developed from loess. Furthermore, Colluvic Regosols have developed in depressional areas due to long-term soil translocation processes. In both catchments, the dominant topsoil textures are loam and silty loam with a median grain size diameter between 12.5 and 16.0 µm (Sinowski and Auerswald, 1999). The average SOC content of the Ap horizons is 3.7 kg m^{-2}. The mean annual temperature and precipitation is 8.4 °C and 834 mm, respectively (measured 1994 to 2001). Agricultural management at the research farm is dedicated to soil conservation: the main cropping principle is to keep soil covered by vegetation or residues as long as possible (Auerswald et al., 2000). The crop rotation during the project was winter wheat (*Triticum aestivum* L.) – maize (*Zea mays* L.) – winter wheat – potato (*Solanum tuberosum* L.). This crop rotation allowed for the cultivation of mustard (*Sinapis alba* L.) cover crops before each row crop (i.e. potato and maize). For implementation, potato ridges were formed before mustard sowing and, later, potato was directly sown into the ridges covered by winterkilled mustard. Maize, on the other hand, was directly sown into the winterkilled mulched mustard (Auerswald et al., 2000). For the established mulch tillage system, the main soil tillage operation was performed with a chisel cultivator (tillage depths approx. 0.2 m). To avoid soil compaction and depressions, which could potentially induce concentrated runoff, wide and low-pressure tires were used on all farming machines (e.g. Fiener and Auerswald, 2007b). Catchment C1 drains one large field with an approx. 2–3 m wide grass filter strip along its downslope border, whereas catchment C2 consists of two fields draining into an approx. 300 m long and 30–40 m wide grassed waterway (Fig. 1).

2.2 Model description

For our study, we coupled three different models: (i) the process-oriented Multi-Class Sediment Transport Model (Fiener et al., 2008; Van Oost et al., 2004; Wilken et al., 2017), a spatially distributed and event-based water erosion model with a specific emphasis on grain size selectivity using the Hairsine and Rose equations (Hairsine et al., 1992; Hairsine and Rose, 1991); (ii) a tillage erosion model following a diffusion-type equation adopted from Govers et al. (1994), which derives tillage erosion from topography and tool-specific tillage erosion coefficients; and (iii) the Introductory Carbon Balance Model (ICBM; Andrén and Kätterer, 1997; Kätterer and Andrén, 2001), which models SOC turnover. The ICBM calculates yearly SOC dynamics using two SOC pools ("young" and "old") and four C fluxes (C input from plants, mineralization from the young and the old pool, and humification). Both the tillage erosion and ICBM model were adapted from SPEROS-C, which couples annual water erosion (based on the RUSLE; Renard et al., 1996), tillage erosion and SOC turnover (Fiener et al., 2015; Nadeu et al., 2015; Van Oost et al., 2005b). In the following, we describe only those features of the coupled MCST-C

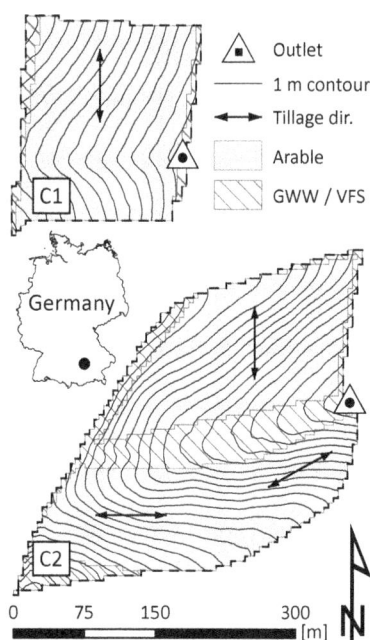

Figure 1. Land use, topography, and tillage direction for modelled catchments C1 and C2. In catchment C2, a grassed waterway (GWW) is located along the thalweg, while vegetated filter strips (VFS) are located along the upslope and downslope field borders.

model (Multi-Class Sediment Transport and Carbon dynamics model) that had to be adapted in order to couple the models or for the introduction of SOC-specific transport mechanisms. An overview of the main model concepts of MCST-C is given in Fig. 2. For more details regarding the three coupled models and processes modelled therein, we refer the reader to the original publications (see above).

2.3   Representation of grain-size-specific soil and associated SOC

The representation of soil texture and SOC in the model is three-dimensional. The horizontal distribution of grain-size-specific soil and SOC is grid-based, while the vertical distribution is represented by ten 10 cm layers. The two uppermost layers are assumed to be homogeneously mixed due to tillage. The grain size distribution is represented in 14 primary particle classes, described by class median particle diameter, particle density, and the class proportion relative to the bulk soil (kg kg^{-1}). The median class diameter is calculated by a logarithmic function that takes grain diameter class boundaries into account (Scheinost et al., 1997). The standard procedure (e.g. sieve–pipette method; Casagrande, 1934; DIN, 2002) to determine grain size distributions destroys soil aggregates in a pre-processing step and therefore only represents the primary particle distribution. However, soil aggregation has a large effect on the fall velocity distribution of soils and reduces the transport distance

of SOC-rich material (Hu and Kuhn, 2014, 2016). Therefore, to account for soil aggregation, two water-stable aggregate classes have been introduced following the hierarchy model of Oades (1984), which describes microaggregate formation inside macroaggregates: silt-sized small microaggregates (6.3–53 µm, median diameter (D50): 18 µm; Tisdall and Oades, 1982) and microaggregates (53–250 µm, D50: 115 µm; Six et al., 2002). In model parametrization, the small microaggregates are exclusively formed out of primary particles with diameters less than 6.3 µm, whereas microaggregates are formed from those with diameters less than 53 µm (i.e. the lower diameter boundary of the aggregate class). As a result, aggregation causes a certain number of primary particles to be moved into the aggregate classes. Hence, the absolute amount of soil aggregation is controlled by the availability of fine primary particles, i.e. sandy soils are less aggregated compared to clayey soils. Macroaggregates (250–2000 µm) are neglected since they are rather immobile during selective interrill erosion and are assumed to break into smaller aggregates during extreme events with high-precipitation kinetic energies (Legout et al., 2005; Oades and Waters, 1991; Tisdall and Oades, 1982). Furthermore, particulate organic matter (POM) is not treated as an individual class, as POM is assumed to be predominantly encapsulated within soil aggregates (Beuselinck et al., 2000; Wang et al., 2013; Wilken et al., 2017).

SOC transport is associated with various primary particle and aggregate classes. Based on the literature (Doetterl et al., 2012; Von Lützow et al., 2007), it is assumed that mineral bound SOC is primarily attached to fine particles (< 6.3 µm) or included in soil aggregates. To keep the mass balance, SOC in water-stable aggregates is allocated based on the SOC content of the primary particles that form these aggregates. This leads to a conservative estimate of SOC in aggregates, as measurements show that aggregates tend to encapsulate more C than found attached to mineral primary particles (Doetterl et al., 2012). As small microaggregates in the model consist solely of primary particles with diameters less than 6.3 µm, their C content equals that of the fine primary particles. Microaggregates show a somewhat smaller C content, since the larger primary particles from which they are also made tend to have less associated SOC.

2.4   Continuous tracking of catchment dynamics

In its original version, the MCST model treats events individually without considering changes caused by previous events. For a continuous application, the water erosion module of MCST-C simulates single events and keeps track of the following redistribution related changes in the catchment: spatial and vertical changes in (i) the grain size distribution and (ii) SOC content and (iii) the development of a rill network, which remains until the next tillage operation. A layer-specific mixture module continuously updates for spatial changes in the vertical grain size distribution and its asso-

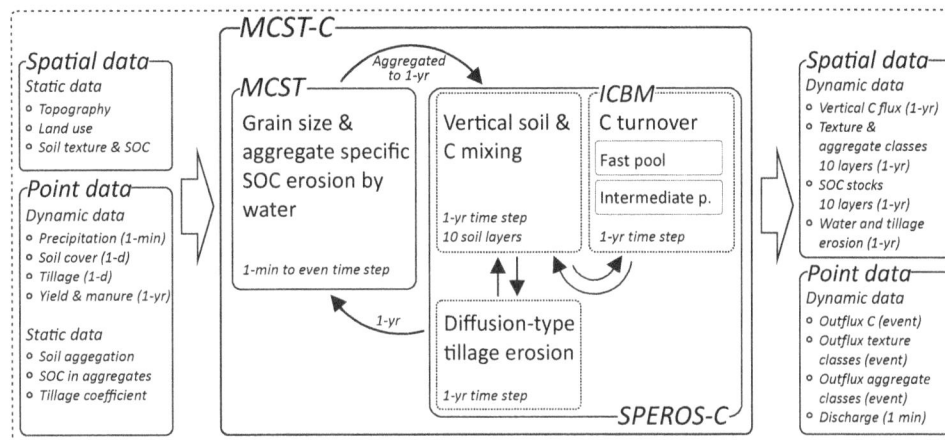

Figure 2. Modelling scheme of the Multi-Class Sediment Transport and Carbon dynamics model (MCST-C).

ciated SOC content, changes which are caused by selective redistribution of water and non-selective tillage erosion. In the case of net deposition, new material with a different grain size distribution is added to the top of the plough horizon (layer 1 and 2). Subsequently, the grain size distribution of the plough layer is mixed and assumed to be homogeneous. Furthermore, deposition leads to an upward movement of the layer borders such that soil material from the plough layer becomes incorporated into the subsoil layers. Any C content moving below 1 m depth is summarized and assumed to be stable in time. In contrast, erosion lifts new material from the subsoil horizons upwards. Assuming that the deepest horizon represents the original loess, the properties of uplifted subsoil remain constant, delivering infinite material of the same grain size distribution and C content.

2.5 Model validation

For a truly rigorous validation of MCST-C, there are numerous long-term data requirements: event-based data for surface runoff, sediment delivery and SOC delivery, long-term data regarding changes in spatially distributed SOC stocks, spatially distributed C loss and gain due to crop harvesting, and C input via plants and manure application. In addition to these validation data requirements, model input data would also be required over decades for a long-term validation. The research project (Auerswald et al., 2000) which is the basis of this study provided a very comprehensive database. However, continuous monitoring was "only" carried out for 8 years (1994 to 2001), and SOC inventories span roughly a decade (first inventory in 1990/91, second in 2001). Therefore, measured changes in SOC stocks are too small to be used for a long-term model validation (requires approx. 50 years; see implementation).

In consequence, we only use the measured continuous event-based surface runoff and sediment delivery from catchment C1 to validate the modelled erosion. The runoff was

collected at the lowest point of the catchment (Fig. 1), which was bordered by a small earthen dam. From the dam, the runoff was transmitted via an underground tile outlet (diameter 0.29 m) to a measuring system consisting of a Coshocton-type wheel runoff sampler (for details regarding the procedure and the precision of the measurements see Fiener and Auerswald, 2003). Corresponding precipitation was measured using a tipping bucket rain gauge of 0.2 mm volume resolution. To determine single erosion events, the precipitation data are filtered in two steps: first, all events with cumulative precipitation > 5 mm and without a 6 h gap in recorded precipitation are considered single erosion events. Second, we included all the largest events accounting for 90 % of total observed runoff. The model is not able to predict erosion under soil frost; hence, winter events, indicated by air temperatures below zero, are removed.

As the original MCST model was previously tested in catchment C1 (Fiener et al., 2008), we did not explicitly calibrate the surface runoff and erosion model. Instead, observed runoff and sediment delivery data was used to test whether our changes to the model still result in a reasonable model performance.

2.6 Model implementation

To run and test MCST-C, a variety of measured input data and parameters are required. This input data are partly calculated from measured data at the research farm and partly taken from literature (Table 1; Fig. 2). To model surface runoff and erosion, the most important input data requirements are (i) precipitation, measured at two meteorological stations about 100 to 300 m from the catchments using 0.2 mm tipping buckets, (ii) a lidar 5 m × 5 m digital elevation model, (iii) soil data taken from a 50 m × 50 m raster sampled during the soil survey in 1990/91 (Sinowski et al., 1997), and (iv) soil cover data, measured biweekly during the vegetation period, monthly in autumn and spring, and before

Table 1. Main input data and parameters used in the Multi-Class Sediment Transport and Carbon dynamics model (MCST-C).

Description	Unit	Temporal resolution	Range/value
Digital elevation model	m	static	(5 m × 5 m) 454–496
Land use	–	daily	–
Soil cover	%	biweekly	0–100
Curve number per crop to be modified by cover and soil crusting	–	daily	38–88
Tillage roughness and direction	m	vegetation period	0–0.25
Hydraulic roughness arable land	$s\,m^{-1/3}$	biweekly	0.016–0.101
Hydraulic roughness grass strip	$s\,m^{-1/3}$	static	0.20
Yield	$kg\,m^{-2}$	at harvest	0.6–4.3
Manure	$kg\,C\,m^{-2}$	at fertilization	0–0.13
Tillage operation	–	daily	–
Soil bulk density	$kg\,m^{-3}$	static	1350
Initial texture	μm	static	0.04–2000
Primary particle density	$kg\,m^{-3}$	static	2650
Small microaggregate density	$kg\,m^{-3}$	static	1300
Microaggregate density	$kg\,m^{-3}$	static	1300
Small microaggregate median diameter	μm	static	18
Microaggregate median diameter	μm	static	115

and after each soil management operation (1993–1997). A tillage transport coefficient (k_{til}) of 169 kg m^{-1} yr^{-1} was utilized for contour tillage by a chisel, following Van Muysen et al. (2000). For SOC redistribution and modelling of vertical C fluxes, the most important model inputs were yields and manure application, a topsoil SOC map (12.5 × 12.5 m^2; Sinowski et al., 1997), and assumptions regarding the allocation of C to different texture classes and in different aggregates. As texture and aggregate C allocation was not measured, we took measured data from Doetterl et al. (2012) and scaled these measurements according to the available bulk SOC (see Sect. 2.3: Representation of grain-size-specific soil and SOC). The parameters for the C turnover model are taken from Dlugoß et al. (2012), who worked under similar environmental conditions with loess-derived soils in a small catchment in western Germany. The C turnover decline with depth was determined by an inverse modelling approach and found a mean turnover rate of 0.268 yr^{-1} for the young pool and 0.002 yr^{-1} for the old pool over the 1 m soil profile. Further details regarding the monitoring data are given in Fiener and Auerswald (2003, 2007b) and Fiener et al. (2008).

As indicated above, it is difficult, if not impossible, to identify erosion-induced changes in SOC and vertical C fluxes if measurements or modelling efforts do not cover decadal time spans. Therefore, a 50-year synthetic input data set and parameter set was created for MCST-C in order to analyse C dynamics. This data set is based on the 8 years of measured data used to validate the erosion component of the model. First, a time series of precipitation was established by randomly choosing the data of one of the eight measured years (see Sect. 2.5: Model validation) and applying it for the first 42 years of the time series. This was followed by the original 8 measured years to reach the total of 50 years.

Next, this precipitation time series was combined with synthetic land use and soil management data representing two full crop rotations (1994 to 2001), which were repeatedly used for all 50 years. This combination leads to a wide variety of precipitation events (time step 1 min) occurring for different daily soil covers by vegetation as a major driver of soil erosion. In contrast to the erosion dynamics, C inputs via plants and manure are repeated every 8 years, which ignores any potential change in management and yields within the modelling period. The synthetic input data were applied for both catchments for the purpose of comparability.

2.7 Analysis of process-specific, erosion-induced C fluxes

Various model setups were chosen (Table 2) to analyse the effects of different erosion processes upon lateral SOC redistribution and the resulting vertical C fluxes. All of these model runs were compared to the 50-year reference run that was validated for the 8-year monitoring phase at the research farm (1994–2001). In general, we tested the effect of a number of water erosion processes and compared the relevance of water vs. tillage erosion. Firstly, the critical shear stress of rill initiation (τ_{crit}) was varied by ±50 % in comparison to its reference run value (0.9 Pa) in order to change the proportion of interrill vs. rill erosion, whereas interrill erosion is a selective SOC transport process and rill erosion is unselective. The reference run value for τ_{crit} was derived from flume experiments in loamy, loess-derived soils (Giménez and Govers, 2002) similar to those found at the test site. Next, the aggregation level was varied in an analogous way to modify the allocation of soil primary particles into the small microaggregate and microaggregate classes (Fig. 3). In another model

Table 2. Model parametrization to analyse the effects of different erosion processes upon C fluxes. Model runs are abbreviated as follows: reference run (Ref), without tillage erosion (Til_{off}), water erosion without grain size selectivity (GS_{off}), high threshold for rill initiation (Ril_{lo}), low threshold for rill initiation (Ril_{hi}), without soil aggregation (Agg_{off}), low soil aggregation (Agg_{lo}), high soil aggregation (Agg_{hi}), without water erosion (Wa_{off}), low tillage erosion (Til_{lo}), and high tillage erosion (Til_{hi}).

Processes	Parameter (unit)	Ref	Til_{off}	GS_{off}	Ril_{lo}	Ril_{hi}	Agg_{off}	Agg_{lo}	Agg_{hi}	Wa_{off}	Til_{lo}	Til_{hi}
Water erosion												
with vs. w/o tillage erosion	(−)	+[a]	−	+	+	+	+	+	+	+	+	+
with vs. w/o grain size selectivity	(−)	+	+	−	+	+	+	+	+	+	+	+
varying rill/interrill erosion	τ_{crit}[b] (Pa)	0.9	0.9	0.9	1.35	0.45	0.9	0.9	0.9	0.9	0.9	0.9
varying small micro & microaggregates	(%)	60	60	60	60	60	0	30	90	60	60	60
Tillage erosion												
with vs. w/o water erosion	(−)	+	+	+	+	+	+	+	+	−	+	+
varying tillage intensity	k_{til}[c] (kg m^{-1} yr^{-1})	169	0	169	169	169	169	169	169	169	85	254

[a] + and − indicate whether a process is modelled or not; [b] critical shear stress for rill initiation; [c] tillage erosion coefficient.

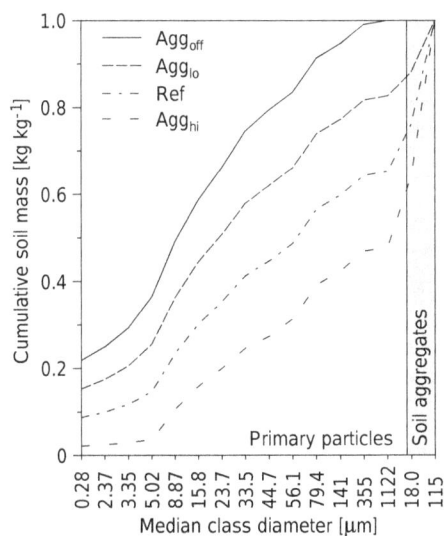

Figure 3. Median class diameter distribution (14 primary particle and 2 aggregate classes) in the plough layer assuming different aggregation levels, as described in Table 2.

run, grain size selectivity was switched off in order to produce a similar behaviour to more parsimonious models which only erode bulk soil (Table 2). To analyse the sensitivity of C fluxes to water and tillage erosion, we first compared model runs with pure water or pure tillage erosion. Secondly, we varied the reference run k_{til} coefficient of 169 kg m^{-1} yr^{-1} by ±50 %. All model runs altered only a single parameter, with all other parameters retaining their reference run values. Parameter variations and the abbreviations for each of the model runs are given in Table 2.

2.8 Analysis of erosion-induced C fluxes

To compare vertical C fluxes from erosional and depositional sites, the corresponding total and mean C flux was calculated on an annual basis. To isolate the C fluxes that result

Table 3. Model performance, as described by the Nash–Sutcliffe efficiency (NSE; Nash and Sutcliffe, 1970), root mean square error (RMSE), coefficient of determination (R^2), and Spearman's rank correlation coefficient (RHO).

	NSE	RMSE	R^2	RHO
Runoff	0.83	5.6 mm	0.94	0.89
Sediment delivery	0.92	165 kg ha^{-1}	0.95	0.71

solely from erosion processes, we first calculate all vertical C fluxes excluding erosion processes and then subtract these from the vertical C fluxes including erosion processes. In the following results section, positive C fluxes indicate an erosion-induced C gain for the catchment (input to the soil), while negative fluxes indicate an erosion-induced loss (from soil to the atmosphere or SOC delivery from the catchment by runoff). Subsequently, erosional and depositional sites were spatially subdivided and an average vertical C flux in kg C m^{-2} was calculated. Finally, the erosion-induced C balance of the catchment was calculated as the sum of the total vertical C flux and laterally delivered SOC.

3 Results

3.1 Validation

A number of goodness-of-fit parameters (Table 3) indicate a sufficient model performance to simulate event runoff and sediment delivery for the 8-year observation period. The Nash–Sutcliffe efficiency and coefficient of determination for runoff (NSE = 0.83; $R^2 = 0.94$) and sediment delivery (NSE = 0.92; $R^2 = 0.95$) are particularly satisfactory. However, a root mean square error of 165 kg ha^{-1} for sediment delivery indicates difficulties in predicting some small events.

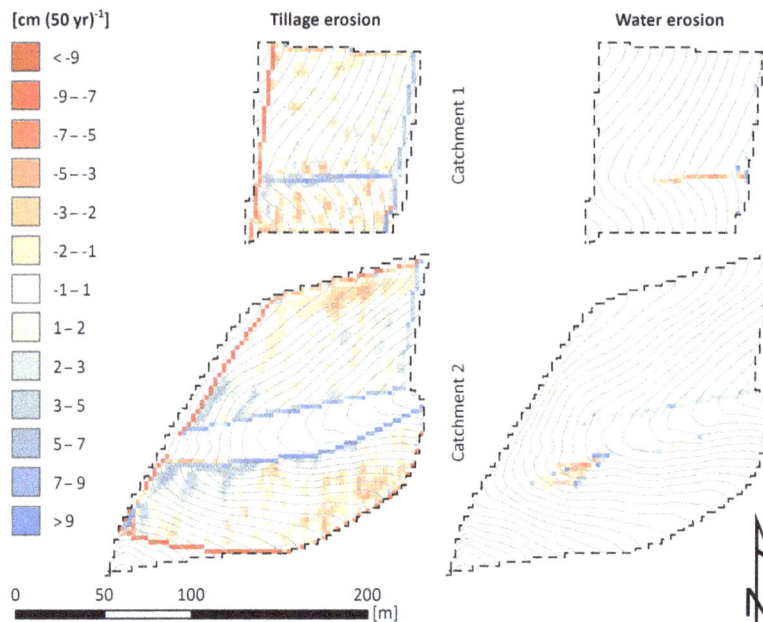

Figure 4. Spatial patterns of tillage and water erosion for the 50-year simulation period of the reference run.

3.2 Long-term erosion-induced C fluxes

The simulated tillage and water erosion shows distinct spatial patterns (Fig. 4). The highest rates of tillage erosion are found along the upslope boundaries of the arable field and on hilltops. The main areas for tillage-induced deposition are at the downslope arable field boundaries and in concavities (Fig. 4). Due to the well-established soil conservation system, water erosion takes place over a much smaller spatial extent and is limited to the main hydrological flow path, while deposition is dominantly found in the vegetated filter strips and grassed waterway (Fig. 4).

The reference run (validated against sediment delivery in catchment C1, 1994–2001) shows positive vertical C fluxes at erosional sites over the 50-year simulation period, with a cumulative flux of 40 g m^{-2} (50 yr)$^{-1}$ in C1 and 59 g m^{-2} (50 yr)$^{-1}$ in C2 (Fig. 5: Ero1, Ero2). The depositional C fluxes show a cumulative C loss of -27 g m^{-2} (50 yr)$^{-1}$ and -30 g m^{-2} (50 yr)$^{-1}$ for C1 and C2, respectively (Fig. 5: Dpo1, Dpo2). Lateral SOC delivery is mainly driven by three heavy erosion events causing 58 and 53 % of the total SOC delivery in C1 and C2, respectively. The total SOC delivery in C1 is -15.6 g m^{-2} (50 yr)$^{-1}$ and in C2 is -6.5 g m^{-2} (50 yr)$^{-1}$ (Fig. 5: Del1, Del2). In C1, the source function of lateral SOC delivery exceeds the sink function of vertical SOC sequestration and leads to a net C loss of -5.7 g m^{-2} (50 yr)$^{-1}$ (Fig. 5: Bal1, Bal2). In contrast, catchment C2 is a net C sink of 4.6 g m^{-2} (50 yr)$^{-1}$.

The event-based SOC enrichment in delivered sediments, compared to parent soil, ranges from 1.1 to 2.7 (2.4 mean) for C1 and from 2.5 to 2.7 (2.7 mean) for C2 over the 50-year time span (Fig. 6). Subdividing the events into tertiles (33 %

parts) according to sediment delivery, the mean enrichment in C1 is 2.5 ($n = 67$) for the low tertile (i.e. smallest 33 % of all event-specific sediment delivery masses), 1.4 ($n = 6$) for the middle tertile, and 1.2 ($n = 2$) for the high tertile (Fig. 6). In contrast, more or less no variation in SOC enrichment was modelled for C2 (Fig. 6).

3.3 Importance of individual erosion processes for long-term erosion-induced C fluxes

Vertical C fluxes show a large response to changes in the k_{til} coefficient but a negligible response to varying levels of water erosion (Fig. 5: Ero1, Ero2, Dpo1, Dpo2). Cumulative C flux at erosional and depositional sites is found to be lowest when no tillage (Til$_{off}$) is simulated and highest for strong tillage (Til$_{hi}$). When pure tillage erosion is simulated (Wa$_{off}$) in catchment C1, a C sequestration of 7 g m^{-2} (50 yr)$^{-1}$ is simulated (Fig. 5: Bal1). The majority of processes in catchment C2 lead to an erosion-induced C gain for the catchment. The highest C sequestration in catchment C2 is found for high tillage erosion (Til$_{hi}$: 10.3 g m^{-2} (50 yr)$^{-1}$). In contrast, catchment C2 acts as a source when there is no tillage (Til$_{off}$: -4.8 g m^{-2} (50 yr)$^{-1}$), as well as when tillage erosion is low (Til$_{lo}$: -0.4 g m^{-2} (50 yr)$^{-1}$; Fig. 5: Bal2).

Lateral SOC delivery is solely caused by water erosion. The model shows its smallest levels of lateral SOC delivery when grain size selectivity is ignored (GS$_{off}$), and delivered sediments therefore have the same SOC concentration as the parent soil (C1: -10 g m^{-2} (50 yr)$^{-1}$; C2: -2.4 g m^{-2} (50 yr)$^{-1}$). This effect is less pronounced for catchment C2 (Fig. 5: Del1, Del2). Catchment C1 shows the largest SOC delivery when the threshold for rill initiation is low (Ril$_{hi}$:

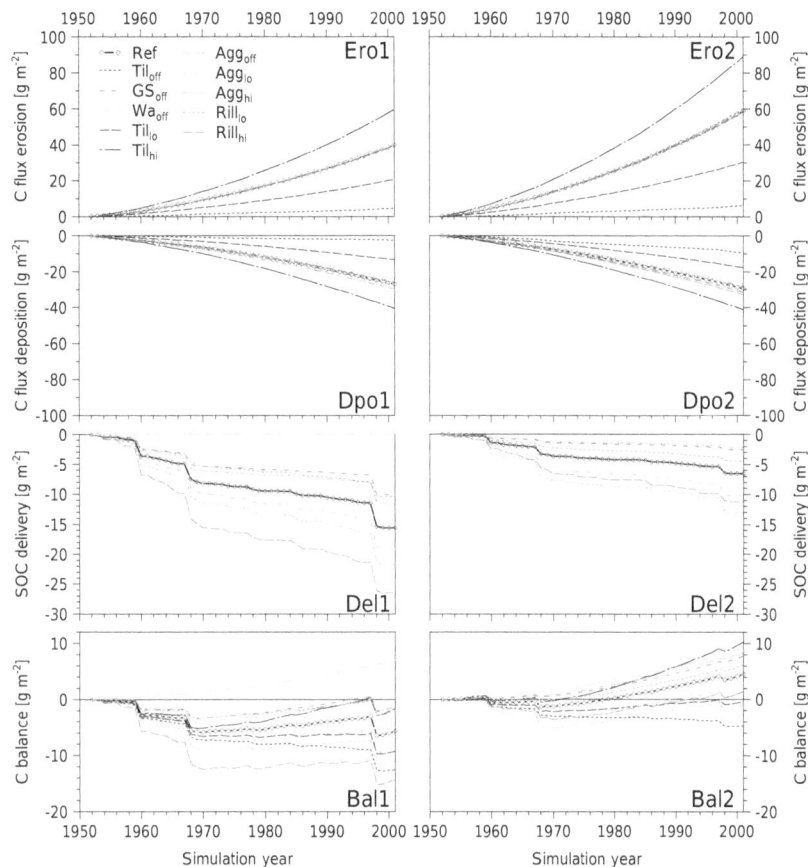

Figure 5. Simulated cumulative vertical C fluxes for erosional (Ero1, Ero2) and depositional (Dpo1, Dpo2) sites, lateral C delivery (Del1, Del2), and catchment C balance (Bal1, Bal2) for catchment C1 and C2. For details regarding the model runs and corresponding abbreviations see Table 2.

$-26.3 \, \mathrm{g \, m^{-2} \, (50 \, yr)^{-1}}$. In catchment C2, the highest lateral SOC delivery is achieved when there is assumed to be no soil aggregation (Agg$_\mathrm{off}$: $-13.0 \, \mathrm{g \, m^{-2} \, (50 \, yr)^{-1}}$). If water erosion is taken into account, catchment C1 is a net C source ranging from 1.3 (GS$_\mathrm{off}$) to 14.2 (Ril$_\mathrm{hi}$) $\mathrm{g \, m^{-2} \, (50 \, yr)^{-1}}$. In contrast, the tillage-induced sequestration potential of catchment C2 exceeds SOC delivery in most water erosion model runs, leading to a positive erosion-induced C balance (sink) as long as soil aggregation is included (Agg$_\mathrm{off}$: $-1 \, \mathrm{g \, m^2}$ $(50 \, \mathrm{yr})^{-1}$; Fig. 5: Bal1–Bal2).

Variations in SOC enrichment of delivered sediments is generally rather small for all model runs (Fig. 6). The most pronounced effect on SOC enrichment results from different aggregation levels (Agg$_\mathrm{off}$, Agg$_\mathrm{lo}$, Agg$_\mathrm{hi}$). However, differences in SOC enrichment were much more pronounced between the catchments. While C2 show high enrichment ratios (> 2.5) for all events, the enrichment ratios strongly decline with increasing event size in C1 (Fig. 6b–c).

4 Discussion

4.1 Vertical C fluxes

Tillage erosion dominates the erosion-induced vertical C fluxes in both catchments. Without water erosion (Wa$_\mathrm{off}$), total tillage-erosion-induced C sequestration potential was 7 and 9 $\mathrm{g \, m^2 \, (50 \, yr)^{-1}}$ in catchment C1 and C2, respectively. The higher sequestration potential in catchment C2 results from steeper slopes and more field boundaries, where tillage erosion is most pronounced (Fig. 4). This offsets its smaller relative proportion of arable land. However, this field boundary effect (Fig. 4) might be overestimated as we did not update the digital elevation model during the 50-year simulation period. The response of vertical C fluxes to changes in tillage erosion strength (Til$_\mathrm{lo}$; Til$_\mathrm{hi}$) further underlines the dominance of tillage redistribution in determining these fluxes (Fig. 5). This dominance results, in part, from the soil conservation system established at the research farm. Indeed, when compared to conventional soil management practices, water erosion was reduced by roughly a factor of 20 (Fiener and Auerswald, 2007a), while tillage erosion intensity (k_til) was only reduced by a factor of about 3 (Van Oost et al.,

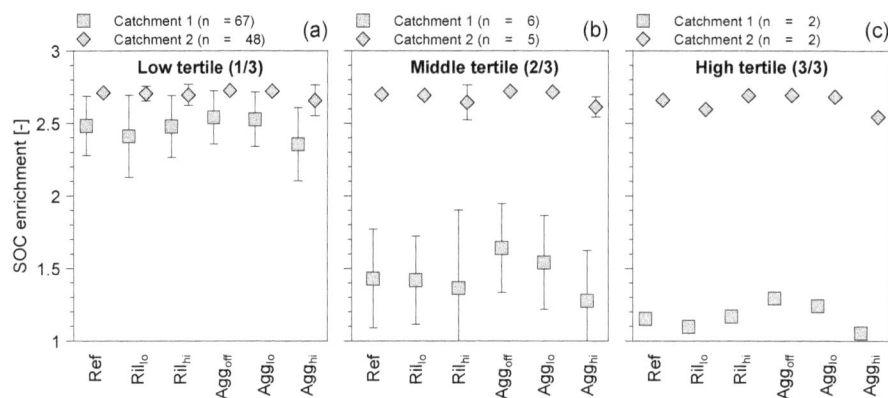

Figure 6. Event-size-specific simulated mean SOC enrichment in delivered sediments of catchment C1 and C2. Error bars indicate one standard deviation. Panels (a), (b), and (c) represent the smallest, middle, and largest 33.3 % of all event-specific sediment delivery masses. For details regarding different model runs and abbreviations see Table 2.

2006) as a result of the soil conservation system. However, independent from the soil tillage management, it is obvious that tillage erosion needs to be taken into account for reasonable estimates of vertical erosion-induced C fluxes on arable land (see also Van Oost et al., 2005a). Moreover, it should be noted that modelling tillage erosion is associated with large uncertainties since it is controlled by a large number of parameters (e.g. tool geometry and type, up-down or contour tillage, speed, depth, soil characteristics; Van Muysen et al., 2000; Van Oost and Govers, 2006). This uncertainty is illustrated by the large range of k_{til} coefficients which can be found in the literature (e.g. for chisel k_{til}: 70 to $657\,kg\,m^{-1}\,yr^{-1}$; Van Oost and Govers, 2006). Interestingly, different water erosion processes hardly affected the vertical erosion-induced C fluxes. This is even true for model parametrizations with very pronounced rill erosion (Ril_{hi}) and large sediment fluxes, because rills only affect a small area. Deposition is also restricted to a small number of raster cells (Fig. 4), particularly in the grassed waterway of catchment C2. The model does not account for changes in C mineralization at depositional sites that may occur as a result of aggregate breakdown shortly after deposition (Hu et al., 2016; Van Hemelryck et al., 2010, 2011). However, the potential underestimation of C mineralization at depositional sites is assumed to be small (< 2 % at a loess site in Belgium; Van Hemelryck et al., 2011). In addition, various drivers of additional C mineralization at depositional sites have been discussed in literature (soil moisture, crusting and crust recovery, deposition of large macroaggregates; Van Hemelryck et al., 2010, 2011) but there is still a substantial lack in process understanding. At this moment, this issue makes it difficult to transfer the specific experimental results into a modelling framework addressing other environmental conditions.

Overall, to achieve accurate estimates of vertical erosion-induced C fluxes, it seems to be more important to improve the representation of tillage erosion in the model, rather than

focusing on detailed process-oriented water erosion modelling, which is less important for vertical C fluxes.

4.2 Lateral C fluxes

In contrast to vertical C fluxes, lateral erosion-induced C fluxes are substantially affected by a number of event-specific processes. To assess these processes, a spatially distributed process-oriented modelling approach is needed.

Our synthetic 50-year data set (based on the 1994–2001 observations) produces three large SOC delivery events, representing nearly 60 % of the total SOC delivery in both catchments (Fig. 5: Del1–Del2). This underlines the importance of accounting for individual events, particularly for the enrichment of SOC in delivered sediment (Fig. 6). However, it should be noted that SOC enrichment is mostly affected by catchment characteristics (Fig. 6b–c). While catchment C1 follows the expected behaviour, i.e. decreasing SOC enrichment with increasing event size (Auerswald and Weigand, 1999; Menzel, 1980; Polyakov and Lal, 2004b; Sharpley, 1985), and is in good agreement with the results of Wang et al. (2010) for similar soils in the Belgian loam belt, event size had hardly any effect on the SOC enrichment in catchment C2, where any larger particles, including aggregates, are deposited in the grassed waterway due to consistently high hydraulic roughness throughout the year. Hence, a parsimonious approach solely relating annual erosion magnitude to SOC enrichment (e.g. Fiener et al., 2015, using the model SPEROS-C) might fail on the landscape scale due to varying inter-field connectivity characteristics of catchments. Underlining the results of recent studies (e.g. Hu and Kuhn, 2016), it seems to be essential to take detailed processes into account during erosion, transport, and deposition in order to accurately capture the SOC enrichment of delivered sediments. In our modelling example, neglecting enrichment would lead to a 36 % underestimation of the total SOC delivery in catchment C1 and an even more extreme 63 % underestimation in catchment C2. This large difference between catchment C1 and C2 suggests that the relevance of SOC enrichment in delivered sediments is controlled not only by event size but also by the catchment connectivity to the outlet.

SOC enrichment in delivered sediments is mainly controlled by the physical properties (e.g. soil texture) of the parent soil (Foster et al., 1985). Soil aggregation transforms unconsolidated fine primary particles, a highly mobile SOC fraction, into soil aggregates, a fraction in which SOC is far less mobile. Hu and Kuhn (2016) showed that soil aggregation reduces the transport distance and potentially enhances terrestrial SOC deposition up to 64 %. We found a similar trend: upon increasing the aggregation level of the model from non-aggregated (Agg$_{off}$) to heavily aggregated (Agg$_{hi}$) soil conditions, we found an increase in SOC deposition for both catchment C1 (47 %) and C2 (83 %). However, while soil texture clearly plays an important role, inter-field connectivity can be the dominant process driving lateral SOC delivery on the landscape scale. This is demonstrated by catchment C2, which shows its largest SOC delivery when it is assumed that there is no soil aggregation. Unfortunately, representing soil aggregation in models is challenging due to a pronounced seasonality (Angers and Mehuys, 1988; Coote et al., 1988; Six et al., 2004; Wang et al., 2010) and complex spatial patterns related to soil nutrients, moisture, grain size distribution, management practices, erosion, and soil biota (Denef et al., 2002). Especially for landscape-scale applications, this high degree of complexity needs to be substantially reduced in a conceptual way. In general, static soil parameters might underestimate dynamic feedbacks, but they are a necessary simplification for landscape-scale modelling approaches.

4.3 Erosion-induced C balance of different catchments

Under the same precipitation and field conditions, the simulated erosion-induced C balance of catchment C1 and C2 show opposing results (Fig. 5: Bal1–Bal2). While catchment C1 acts as a C source for the majority of simulated processes (controlled primarily by SOC delivery), the presence of the grassed waterway for catchment C2 substantially reduces lateral SOC delivery and leads the catchment to function as a C sink for most simulated processes. For both catchments, the majority of simulation years show a positive erosion-induced C balance (sink). However, three heavy erosion events in catchment C1 exceeded the positive cumulative vertical flux. Therefore, we underline that any analysis of landscape-scale erosion-induced C balances must consider inter-field connectivity.

5 Conclusions

In this study, the effect of individual SOC redistribution processes on SOC dynamics is assessed by utilizing a coupled process-oriented erosion and C turnover model. The erosion component of the model was successfully validated against a continuous 8-year data set of surface runoff and sediment delivery. The model was able to estimate the relevance of different processes in terms of their impact on vertical and lateral C fluxes for two catchments with distinct characteristics over an artificial time series of 50 years. We found that tillage erosion dominates on-field soil redistribution and vertical erosion-induced C fluxes on arable land, while water erosion processes have a much more limited effect. However, episodic lateral SOC delivery is critically important for the carbon balance. Ignoring SOC enrichment in delivered sediments leads to a pronounced underestimation of delivered SOC. Soil aggregates substantially reduce SOC delivery by turning highly mobile fine primary particles into less mobile soil aggregates. In general, the erosion-induced C balance is largely affected by inter-field deposition related to catchment connectivity.

Our results underline the importance of having an accurate and spatially distributed representation of tillage erosion. The episodic nature of water erosion calls for a sufficiently long simulation period and the inclusion of grain-size-selective transport in order to address the enrichment of delivered SOC. Furthermore, we stress the need for future investigations on seasonal and spatial variations in soil aggregation for a conceptual model implementation.

Competing interests. The authors declare that they have no conflict of interest.

Acknowledgements. The study was supported by the Terrestrial Environmental Observatory TERENO-Northeast of the Helmholtz Association. We would like to acknowledge the large number of scientists and technicians who collected the data used in this study, which was funded by the Bundesministerium für Bildung, Wissenschaft, Forschung und Technologie (BMBF No. 0339370) and the Bayerische Staatsministerium für Unterricht und Kultus, Wissenschaft und Kunst.

Edited by: N. J. Kuhn

References

Andrén, O. and Kätterer, T.: ICBM: The introductory carbon balance model for exploration of soil carbon balances, Ecol. Appl., 7, 1226–1236, 1997.

Angers, D. A. and Mehuys, G. R.: Effects of cropping on macroaggregation of a marine clay soil, Can. J. Soil Sci., 68, 723–732, 1988.

Auerswald, K. and Weigand, S.: Eintrag und Freisetzung von P durch Erosionsmaterial in Oberflächengewässern, VDLUFA-Schriftenreihe, 50, 37–54, 1999.

Auerswald, K., Albrecht, H., Kainz, M., and Pfadenhauer, J.: Principles of sustainable land-use systems developed and evaluated by the Munich Research Alliance on agro-ecosystems (FAM), Petermanns Geographische Mitteilungen, 144, 16–25, 2000.

Berhe, A. A., Harte, J., Harden, J. W., and Torn, M. S.: The significance of the erosion-induced terrestrial carbon sink, Bioscience, 57, 337–346, 2007.

Beuselinck, L., Steegen, A., Govers, G., Nachtergaele, J., Takken, I., and Poesen, J.: Characteristics of sediment deposits formed by intense rainfall events in small catchments in the Belgian Loam Belt, Geomorphology, 32, 69–82, 2000.

Casagrande, A.: Die Aräometermethode zur Bestimmung der Korngrößenverteilung von Böden, Arthur Casagrande, Berlin, 1934.

Coote, D. R., Malcolm-McGovern, C. A., Wall, G. J., Dickinson, W. T., and Rudra, R. P.: Seasonal variation of erodibility indices based on shear strength and aggregate stability in some Ontario soils, Can. J. Soil Sci., 68, 405–416, 1988.

Denef, K., Six, J., Merckx, R., and Paustian, K.: Short-term effects of biological and physical forces on aggregate formation in soils with different clay mineralogy, Plant Soil, 246, 185–200, 2002.

DIN: DIN ISO 11277: 2002-08 Bodenbeschaffenheit – Bestimmung der Partikelgrößenverteilung in Mineralböden – Verfahren mittels Siebung und Sedimentation, Beuth Verlag, Berlin, 2002.

Dlugoß, V., Fiener, P., Van Oost, K., and Schneider, K.: Model based analysis of lateral and vertical soil C fluxes induced by soil redistribution processes in a small agricultural watershed, Earth Surf. Proc. Land., 37, 193–208, 2012.

Doetterl, S., Six, J., Van Wesemael, B., and Van Oost, K.: Carbon cycling in eroding landscapes: geomorphic controls on soil organic C pool composition and C stabilization, Glob. Change Biol., 18, 2218–2232, 2012.

Doetterl, S., Berhe, A. A., Nadeu, E., Wang, Z., Sommer, M., and Fiener, P.: Erosion, deposition and soil carbon: A review on process-level controls, experimental tools and models to address C cycling in dynamic landscapes, Earth-Sci. Rev., 154, 102–122, 2016.

Fiener, P. and Auerswald, K.: Effectiveness of grassed waterways in reducing runoff and sediment delivery from agricultural watersheds, J. Environ. Qual., 32, 927–936, 2003.

Fiener, P. and Auerswald, K.: Möglichkeiten der Abfluss- und Stofftransportkontrolle durch landwirtschaftliche Maßnahmen und ihre Kombination im Landschaftsmaßstab, in: DWA Sonderdruck zum Tag der Hydrologie 2007, Rostock, 23–36, 2007a.

Fiener, P. and Auerswald, K.: Rotation effects of potato, maize and winter wheat on soil erosion by water, Soil Sci. Soc. Am. J., 71, 1919–1925, 2007b.

Fiener, P., Govers, G., and Van Oost, K.: Evaluation of a dynamic multi-class sediment transport model in a catchment under soil-conservation agriculture, Earth Surf. Processes Landforms, 33, 1639-1660, 2008.

Fiener, P., Dlugoß, V., and Van Oost, K.: Erosion-induced carbon redistribution, burial and mineralisation – Is the episodic nature of erosion processes important?, Catena, 133, 282–292, 2015.

Foster, G. R., Young, R. A., and Neibling, W. H.: Sediment composition for nonpoint source pollution analyses, Transactions of the American Society of Agricultural Engineers, 28, 133–139, 1985.

Friedlingstein, P., Andrew, R. M., Rogelj, J., Peters, G. P., Canadell, J. G., Knutti, R., Luderer, G., Raupach, M. R., Schaeffer, M., van Vuuren, D. P., and Le Quere, C.: Persistent growth of CO_2 emissions and implications for reaching climate targets, Nat. Geosci., 7, 709–715, 2014.

Giménez, R. and Govers, G.: Flow detachment by concentrated flow on smooth and irregular beds, Soil Sci. Soc. Am. J., 66, 1475–1483, 2002.

Govers, G., Vandaele, K., Desmet, P., Poesen, J., and Bunte, K.: The role of tillage in soil redistribution on hillslopes, Eur. J. Soil Sci., 45, 469–478, 1994.

Hairsine, P. B. and Rose, C. W.: Rainfall detachment and deposition: Sediment transport in the absence of flow-driven processes, Soil Sci. Soc. Am. J., 55, 320–324, 1991.

Hairsine, P. B., Moran, C. J., and Rose, C. W.: Recent developments regarding the influence of soil surface characteristics on overland flow and erosion, Aust. J. Soil Res., 30, 249–264, 1992.

Hiederer, R. and Köchy, M.: Global Soil Organic Carbon Estimates and the Harmonized World Soil Database, Publications Office of the EU, Luxenbourg, 2011.

Hooke, R. L.: On the history of humans as geomorphic agents, Geology, 28, 843–846, 2000.

Hu, Y. and Kuhn, N. J.: Aggregates reduce transport distance of soil organic carbon: are our balances correct?, Biogeosciences, 11, 6209–6219, doi:10.5194/bg-11-6209-2014, 2014.

Hu, Y. X. and Kuhn, N. J.: Erosion-induced exposure of SOC to mineralization in aggregated sediment, Catena, 137, 517–525, 2016.

Hu, Y. X., Berhe, A. A., Fogel, M. L., Heckrath, G. J., and Kuhn, N. J.: Transport-distance specific SOC distribution: Does it skew erosion induced C fluxes?, Biogeochemistry, 128, 339–351, 2016.

Kätterer, T. and Andrén, O.: The ICBM family of analytically solved models of soil carbon, nitrogen and microbial biomass dynamics descriptions and application examples, Ecol. Model., 136, 191–207, 2001.

Kirkels, F. M. S. A., Cammeraat, L. H., and Kuhn, N. J.: The fate of soil organic carbon upon erosion, transport and deposition in agricultural landscapes – A review of different concepts, Geomorphology, 226, 94–105, 2014.

Lacoste, M., Viaud, V., Michot, D., and Walter, C.: Landscape-scale modelling of erosion processes and soil carbon dynamics under land-use and climate change in agroecosystems, Eur. J. Soil Sci., 66, 780–791, 2015.

Legout, C., Leguédois, S., and Le Bissonnais, Y.: Aggregate breakdown dynamics under rainfall compared with aggregate stability measurements, Eur. J. Soil Sci., 56, 225–237, 2005.

Menzel, R. G.: Enrichment ratios for water quality modeling, in: CREAMS, edited by: Knisel, W. G., USDA Cons. Res. Rep., 1980.

Myers, N.: Gaia: An atlas of planet management, Anchor Press, Garden City, 1993.

Nadeu, E., Gobin, A., Fiener, P., Van Wesemael, B., and Van Oost, K.: Modelling the impact of agricultural management on soil carbon stocks at the regional scale: the role of lateral fluxes, Glob. Change Biol., 21, 3181–3192, 2015.

Nash, J. E. and Sutcliffe, J. V.: River flow forecasting through conceptual models: Part I. A discussion of principles, J. Hydrol., 10, 282-290, 1970.

Oades, J. M.: Soil organic matter and structural stability: mechanisms and implications for management, Plant Soil, 76, 319–337, 1984.

Oades, J. M. and Waters, A. G.: Aggregate hierachy in soils, Aust. J. Soil Res., 29, 815–828, 1991.

Pimentel, D., Harvey, C., Resosudarmo, P., Sinclair, K., Kurz, D., McNair, M., Crist, S., Shpritz, L., Fitton, L., Saffouri, R., and Blair, R.: Environmental and economic costs of soil erosion and conservation benefits, Science, 267, 1117–1123, 1995.

Polyakov, V. and Lal, R.: Modeling soil organic matter dynamics as affected by soil water erosion, Environ. Int., 30, 547–556, 2004a.

Polyakov, V. O. and Lal, R.: Soil erosion and carbon dynamics under simulated rainfall, Soil Sci., 169, 590–599, 2004b.

Prechtel, A., von Lützow, M., Schneider, B. U., Bens, O., Bannick, C. G., Kögel-Knabner, I., and Hüttl, R. F.: Organic carbon in soils of Germany: Status quo and the need for new data to evaluate potentials and trends of soil carbon sequestration, J. Plant Nutr. Soil Sci., 172, 601–614, 2009.

Renard, K. G., Foster, G. R., Weesies, G. A., McCool, D. K., and Yoder, D. C.: Predicting soil erosion by water: A guide to conservation planning with the Revised Universal Soil Loss Equation (RUSLE), USDA-ARS, Washington DC, 1996.

Scheinost, A. C., Sinowski, W., and Auerswald, K.: Regionalization of soil water retention curves in a highly variable soilscape, I. Developing a new pedotransfer function, Geoderma, 78, 129–143, 1997.

Schiettecatte, W., Gabriels, D., Cornelis, W. M., and Hofman, G.: Enrichment of organic carbon in sediment transport by interrill and rill erosion processes, Soil Sci. Soc. Am. J., 72, 50–55, 2008.

Sharpley, A. N.: The selective erosion of plant nutrients in runoff, Soil Sci. Soc. Am. J., 49, 1527–1534, 1985.

Sinowski, W. and Auerswald, K.: Using relief parameters in a discriminant analysis to stratify geological areas with different spatial variability of soil properties, Geoderma, 89, 113–128, 1999.

Sinowski, W., Scheinost, A. C., and Auerswald, K.: Regionalization of soil water retention curves in a highly variable soilscape, II. Comparison of regionalization procedures using a pedotransfer function, Geoderma, 78, 145–159, 1997.

Six, J., Conant, R. T., Paul, E. A., and Paustian, K.: Stabilization mechanisms of soil organic matter: implications for C-saturation of soils, Plant Soil, 241, 155–176, 2002.

Six, J., Bossuyt, H., Degryze, S., and Denef, K.: A history of research on the link between (micro)aggregates, soil biota, and soil organic matter dynamics, Soil Till. Res., 79, 7–31, 2004.

Tisdall, J. M. and Oades, J. M.: Organic matter and water-stable aggregates in soils, J. Soil Sci., 33, 141–163, 1982.

Van Hemelryck, H., Fiener, P., Van Oost, K., Govers, G., and Merckx, R.: The effect of soil redistribution on soil organic carbon: an experimental study, Biogeosciences, 7, 3971–3986, doi:10.5194/bg-7-3971-2010, 2010.

Van Hemelryck, H., Govers, G., Van Oost, K., and Merckx, R.: Evaluating the impact of soil redistribution on the in situ mineralization of soil organic carbon, Earth Surf. Proc. Land., 36, 427–438, 2011.

Van Muysen, W., Govers, G., Van Oost, K., and Van Rompaey, A.: The effect of tillage depth, tillage speed, and soil condition on chisel tillage erosivity, J. Soil Water Conserv., 55, 2–11, 2000.

Van Oost, K. and Govers, G.: Tillage erosion, in: Soil erosion in Europe, edited by: Boardman, J. and Poesen, J., Wiley, Chichester, 2006.

Van Oost, K., Beuselinck, L., Hairsine, P. B., and Govers, G.: Spatial evaluation of a multi-class sediment transport and deposition model, Earth Surf. Proc. Land., 29, 1027–1044, 2004.

Van Oost, K., Govers, G., Quine, T., Heckarth, G., Olesen, J. E., De Gryze, S., and Merckx, R.: Landscape-scale modeling of carbon cycling under the impact of soil redistribution: The role of tillage erosion, Global Biogeochem. Cy., 19, GB4014, doi:10.1029/2005GB002471, 2005a.

Van Oost, K., Quine, T., Govers, G., and Heckrath, G.: Modeling soil erosion induced carbon fluxes between soil and atmosphere on agricultural land using SPEROS-C, in: Advances in soil science. Soil erosion and carbon dynamics, edited by: Roose, E. J., Lal, R., Feller, C., Barthes, B., and Stewart, B. A., CRC Press, Boca Raton, 2005b.

Van Oost, K., Govers, G., De Alba, S., and Quine, T. A.: Tillage erosion: a review of controlling factors and implications for soil quality, Prog. Phys. Geogr., 30, 443–466, 2006.

Von Lützow, M., Kögel-Knabner, I., Ekschmitt, K., Flessa, H., Guggenberger, G., Matzner, E., and Marschner, B.: SOM fractionation methods: Relevance to functional pools and to stabilization mechanisms, Soil Biol. Biochem., 39, 2183–2207, 2007.

Wang, X., Cammeraat, L. H., Wang, Z., Zhou, J., Govers, G., and Kalbitz, K.: Stability of organic matter in soils of the belgian loess belt upon erosion and deposition, Eur. J. Soil Sci., 64, 219–228, 2013.

Wang, Z., Govers, G., Steegen, A., Clymans, W., Van den Putte, A., Langhans, C., Merckx, R., and Van Oost, K.: Catchment-scale carbon redistribution and delivery by water erosion in an intensively cultivated area, Geomorphology, 124, 65–74, 2010.

Wilken, F., Fiener, P., and Van Oost, K.: Modelling a century of soil redistribution processes and carbon delivery from small watersheds using a multi-class sediment transport model, Earth Surf. Dynam., 5, 113–124, 2017.

Wilkinson, B. H. and McElroy, B. J.: The impact of humans on continental erosion and sedimentation, Geol. Soc. Am. Bull., 119, 140–156, 2007.

8

Sensitivity analysis of point and parametric pedotransfer functions for estimating water retention of soils in Algeria

Sami Touil[1,2,3]**, Aurore Degre**[2]**, and Mohamed Nacer Chabaca**[1]

[1]Superior National School of Agronomy, El Harrach, Algiers, Algeria
[2]Gembloux Agro-Bio Tech, Biosystem Engineering, Soil–Water–Plant Exchanges, University of Liege,
Passage des Déportés, Gembloux, Belgium
[3]Laboratory of Crop Production and Sustainable Valorization of Natural Resources, University of Djilali
Bounaama Khemis Miliana, Ain Defla, Algeria

Correspondence to: Sami Touil (touil_sy@hotmail.fr)

Abstract. Improving the accuracy of pedotransfer functions (PTFs) requires studying how prediction uncertainty can be apportioned to different sources of uncertainty in inputs. In this study, the question addressed was as follows: which variable input is the main or best complementary predictor of water retention, and at which water potential? Two approaches were adopted to generate PTFs: multiple linear regressions (MLRs) for point PTFs and multiple nonlinear regressions (MNLRs) for parametric PTFs. Reliability tests showed that point PTFs provided better estimates than parametric PTFs (root mean square error, RMSE: 0.0414 and 0.0444 cm^3 cm^{-3}, and 0.0613 and 0.0605 cm^3 cm^{-3} at -33 and -1500 kPa, respectively). The local parametric PTFs provided better estimates than Rosetta PTFs at -33 kPa. No significant difference in accuracy, however, was found between the parametric PTFs and Rosetta H2 at -1500 kPa with RMSE values of 0.0605 cm^3 cm^{-3} and 0.0636 cm^3 cm^{-3}, respectively. The results of global sensitivity analyses (GSAs) showed that the mathematical formalism of PTFs and their input variables reacted differently in terms of point pressure and texture. The point and parametric PTFs were sensitive mainly to the sand fraction in the fine- and medium-textural classes. The use of clay percentage (C%) and bulk density (BD) as inputs in the medium-textural class improved the estimation of PTFs at -33 kPa.

1 Introduction

Predictive information on the spatial distribution of soil water and its availability for plants enables producers to take effective decisions (e.g. on nutrient management and plant cover) to maximise profitability. The soil-water balance is central to many processes that influence plant growth and the degradation of soil and water resources.

Hydrologists face the situation where soil hydraulic data such as water retention or hydraulic conductivity are often missing. Therefore, pedotransfer functions (PTFs) are used as an alternative to estimate these properties. The extrapolation of PTFs in different agropedoclimatic context limits their performance (Touil et al., 2016). The development of local PTFs could be useful in meeting the agricultural requirements for modelling with reasonable accuracy.

Soil water retention (SWR) curves can usually be estimated using two approaches: point PTFs and parameter PTFs. With point PTFs, SWR is estimated at defined pressure points (Pachepsky et al., 1996; Minasny et al., 1999). One of the most commonly used SWR curves is the van Genuchten (1980) model. With parameter PTFs, the parameters of SWR models, such as θ_s, θ_r, α and n, are estimated by fitting them to the data and then relating them by empirical correlation to basic soil properties (Vereecken et al., 1992; Wösten et al., 1995; Schaap and

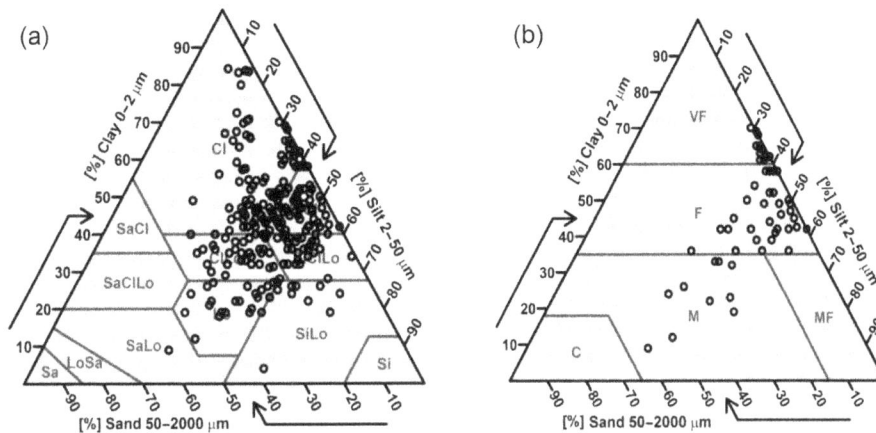

Figure 1. (a) Texture fractions of the dataset (242 samples), based on the USDA system. **(b)** Particle size distribution of 53 soil samples from Algeria according to the FAO textural triangle (FAO, 1990).

Leij, 1998; Minasny and McBratney, 2002; Rawls and Brakensiek, 1985; van Genuchten et al., 1992; Wösten et al., 2001; Vereecken et al., 2010). Schaap et al. (2001) developed the Rosetta package based on the artificial neural network (ANN) method, which uses five hierarchical models to predict the van Genuchten (VG) parameters (θ_s, θ_r, α and n) with only soil texture classes and the input data (texture, bulk density, BD, and one or two water content values at -33 and -1500 kPa).

PTFs for point and parametric estimation of SWR from basic soil properties can be developed using multiple regression methods (Lin et al., 1999; Mayr and Jarvis, 1999; Tomasella et al., 2000).

Some 97 % of water retention PTFs for soils in the tropics are based on multiple linear and polynomial regressions of nth-order techniques (Botula et al., 2014).

Using PTFs in environments that differ from those from which they were derived can lead to an under- or overestimation of SWR. Several studies have shown that SWR is a complex function of soil structure and composition (Rawls et al., 1991, 2003; Wösten et al., 2001; Mirus, 2015). Applying PTFs to different textural or structural classes could also be a source of uncertainty (Bruand et al., 2002; Pachepsky and Rawls, 2003). SWR and hydraulic conductivity vary widely and nonlinearly with soil water potential. Soil texture is the main determinant of the water-holding characteristics of most agricultural soils (Saxton et al., 1986). The relationship between the SWR curve and particle size distribution (PSD) has been investigated in many studies (Jonasson, 1992; Minasny et al., 2007; Ghanbarian-Alavijeh et al., 2009; Yang and You, 2013; Lee and Ro, 2014). SWR depends mainly on texture, with other factors such as BD, structure, organic matter (OM), clay type and hysteresis having a secondary impact (Williams et al., 1983; Saxton et al., 1986; Vereecken et al., 1989; Winfield et al., 2006).

The variability in PTF response depends on the variability and uncertainty of one or more of the input variables. Uncertainty analysis in the variety of available PTF approaches is necessary to minimise error in estimation and identify its source. Recently, sensitivity analysis techniques and uncertainty analysis have begun to receive considerable attention in PTF studies (Nemes et al., 2006; Kay et al., 1997; Grunwald et al., 2001; Deng et al., 2009; Moeys et al., 2012; Loosvelt et al., 2013). The question is as follows: which variable input is the main or best complementary predictor of SWR, and at which potential? Global sensitivity analysis (GSA) enables us to study how uncertainty in the output of a model can be apportioned to different sources of uncertainty in the model inputs (Saltelli et al., 2000). Generally, GSA is useful for identifying which variables make the main contribution to output variables (Jacques et al., 2006).

The objectives of this study were to

– develop and validate two PTF approaches using regression methods: point PTFs for estimating SWR in Algerian soils at -33 and -1500 kPa and parametric PTFs for estimating the VG parameters

– study the impact of each input on the PTF responses.

2   Materials and methods

2.1   The database

The soil dataset used for this study was collected from various regions in Algeria, mainly in the north, which has a Mediterranean climate. It contained 242 samples, with basic soil properties: texture fractions (based on the USDA system; clay and silty-clayey for most of the soils; Fig. 1a), BD, OM percentage and water content at -33 and -1500 kPa. Descriptive statistics of the development and validation datasets are presented in Table 1. The available database

Table 1. Soil characteristics of the developed and validated datasets.

	PSD					VWC ($cm^3\,cm^{-3}$)	
	S (%)	Si (%)	C (%)	BD ($g\,cm^3$)	OM (%)	-33 kPa	-1500 kPa
Samples used for deriving PTF ($n = 189$)							
Average	17.81	39.23	42.97	1.71	0.95	0.44	0.27
Standard deviation	10.32	10.76	13.90	0.20	0.93	0.09	0.08
Minimum	1.00	9.20	4.00	0.60	0.08	0.13	0.03
Maximum	50.00	67.00	84.30	2.10	8.40	0.73	0.56
Coefficient of variation	0.58	0.27	0.32	0.12	0.98	0.21	0.31
Samples used for testing PTF ($n = 53$)							
Average	12.50	41.58	45.92	1.49	0.87	0.40	0.21
Standard deviation	14.84	7.62	14.94	0.13	0.50	0.10	0.07
Minimum	–	29.00	9.00	1.15	0.20	0.14	0.07
Maximum	59.00	58.00	70.00	1.73	2.74	0.57	0.45
Coefficient of variation	1.19	0.18	0.33	0.09	0.57	0.24	0.35

PSD: particle size distribution, S: sand, Si: silt, C: clay, BD: bulk density, OM: organic matter, VWC: volumetric water content.

was split into two datasets. Subset 1, which was used to develop the PTFs, contained 78.1 % of the samples. Used as the calibration set, they were collected from the coastal plain of Annaba in north-eastern Algeria (13 samples), the Beni Slimane plain of Media (42 samples), the Kherba El Abadia plain of Ain Defla (54 samples) and the Lower Cheliff plain in north-western Algeria (80 samples). Subset 2 contained the remaining 21.9 % of the samples. Used to verify the PTFs, they were collected from Benziane valley in the lower south-western Cheliff plain. The depth of the two upper horizons varied from site to site, with a maximum of 30 cm for surface horizons and more than 30 cm for subsurface horizons.

Particle size distribution (PSD) analysis was conducted using the international Robinson's pipette method (Robinson, 1922). Undisturbed soil samples obtained with 500–1000 cm^3 cylinders were used to determine BD. The SWR values at -33 and -1500 kPa were obtained using Richards' apparatus (Richards and Fireman, 1943). Undisturbed soil samples were collected near field capacity with 100 cm^3 cylinders. Water content was measured using the gravimetric method at 105 °C (24 h). Organic carbon content was determined using the wet oxidation method (Walkley and Black, 1934). Variation in soil texture in the dataset is displayed using the textural triangle proposed by FAO (1990) in Fig. 1b.

The SWR model devised by van Genuchten (1980) is defined as

$$\theta(h) = \theta_r + \frac{\theta_s - \theta_r}{(1 + |\alpha h|^n)^m},\qquad(1)$$

where θ_r and θ_s are residual and saturated soil-water content ($cm^3\,cm^{-3}$), respectively, and α (cm^{-1}) and n are the shape factors of the SWR function. The VG parameters were in-

directly estimated for each soil sample from four levels of measured data inputs: sand, silt and clay percentages, and BD using the Rosetta model H3 (Schaap et al., 2001). The m parameter was calculated as follows:

$$m = 1 - 1/n.$$

2.2 PTF development

Two approaches were used in this study to develop the PTFs: point PTFs for estimating SWR for particular points of pressure (h) and parametric PTFs for predicting the VG parameters. Each water content level at selected water potentials of -33 and -1500 kPa and estimated VG parameters were related to basic soil properties (i.e. sand, silt, clay content, OM content and BD) using multiple regression techniques (Table 2). The most significant input variables were determined using the Pearson correlation ($\alpha = 5\%$). For the multiple linear regression (MLR) models, the general form of the resulting equations was thus expressed as

$$Y = a_0 + b_1 X_1 + b_2 X_2 + b_3 X_3 + b_4 X_4.\qquad(2)$$

For the multiple nonlinear regression (MNLR) models, it was thus expressed as

$$Y = a_0 + b_1 X_1 + b_2 X_2 + b_3 X_1^2 + b_4 X_2^2 + b_5 X_1^3$$
$$+ b_6 X_2^3 + b_7 X_1 \cdot X_2 + b_8 X_1^2 \cdot X_2 + b_9 X_1 \cdot X_2^2,\qquad(3)$$

where Y represents the dependent variable, a_0 is the intercept, $b_1 \ldots, b_n$ are the regression coefficients, and X_1 to X_4 refer to the independent variables representing the basic soil properties.

Table 2. Developed pedotransfer functions (PTFs).

Point PTFs	
at -33 kPa:	$\theta = 0.0246 - 0.0040\,S + 0.0012\,C + 0.2554\,BD + 0.0067\,OM$
at -1500 kPa:	$\theta = -0.0627 - 0.0029\,S + 0.00165\,C + 0.1837\,BD + 0.0017\,OM$

Parametric PTFs
$\theta_s = 0.44 - 0.0013369\,S + 0.0002\,C + 0.01771343\,BD - 0.0018272\,OM$
$\theta_r = 0.09 + 0.000777943\,S - 0.000319883\,C + 0.000063602\,S^2 + 0.000012\,C^2 + 0.00000093\,S^3 - 0.0000001\,C^3$
$\alpha = 0.003 - 0.0001\,S + 0.000089\,Si + 0.0000054\,S^2 - 0.0000045\,Si^2 - 0.000000073\,S^3 + 0.000000045\,Si^3$
$+ 0.0000077\,S\,Si - 0.000000031\,S^2\,Si - 0.000000062\,S\,Si^2$
$n = 2.9 - 0.00277395\,C - 0.09478943\,Si - 0.00036644\,C^2 + 0.00202592\,Si^2 + 0.00000249\,C^3$
$- 0.000015\,Si^3 + 0.00028374\,C\,Si + 0.00000491\,C^2\,Si - 0.00000532\,C\,Si^2$

S: sand (%), C: clay (%), Si: silt (%), BD: bulk density (g cm^{-3}), OM: organic matter (%), θ_r and θ_s are residual and saturated soil-water content (cm^3 cm^{-3}), respectively, and α (cm^{-1}) and n are the shape factors of the of van Genuchten model.

The prediction quality of the point and parametric PTFs developed from Algerian soils were then compared with three Rosetta PTFs (H1, H2 and H3). We chose the Rosetta model because it gives the user flexibility in inputting the data required (Stumpp et al., 2009), with the option of five levels based on input data (Schaap et al., 2001):

– H1 is textural classes (USDA system).

– H2 is clay + silt + sand.

– H3 is clay + silt + sand + BD.

– H4 is clay + silt + sand + BD + volumetric water at -33 kPa.

– H5 is clay + silt + sand + BD + volumetric water at -33 kPa + volumetric water at -1500 kPa.

The artificial neural network models were also chosen because they have given reasonable predictions in several evaluation studies (e.g. Nemes et al., 2003). In our study, the three Rosetta model levels (H1, H2 and H3) were selected to compare their performance in the Algerian soils because they require only texture data and BD as inputs, like locally developed PTFs do.

2.3 Evaluation criteria

PTFs are regularly assessed by comparing the values that they predict with the measured values (Pachepsky et al., 1999; FAO, 1990). In order to assess the validity of the PTFs developed, we used the following criteria: mean prediction error (ME) to indicate the bias of the estimate, root mean square error (RMSE) to assess the quality of the prediction (it is frequently used in studies on PTFs), and the index of agreement (d) developed by Willmott and Wicks (1980) and Willmott (1981) as a standardised measure of the degree of

model prediction error. They were calculated using the following equations, respectively:

$$\mathrm{ME} = \frac{1}{N}\sum_{i=1}^{n}\left(\theta_p - \theta_m\right), \tag{4}$$

where N is the number of horizons, and θ_p and θ_m predicted and measured volumetric water content, respectively. The estimate was better when ME was close to $0'$. Negative ME values indicated an average underestimation of θ_m, whereas positive values indicated overestimation.

$$\mathrm{RMSE} = \left\{\frac{1}{n}\sum_{i=1}^{n}\left(\theta_p - \theta_m\right)^2\right\}^{\frac{1}{2}} \tag{5}$$

Thus, the lower the RMSE the better the estimate.

$$d = 1 - \frac{\sum_{i=1}^{n}\left(\theta_p - \theta_m\right)^2}{\sum_{i=1}^{n}\left[\left|\left(\theta_p - \bar{\theta}_m\right)\right| + \left|\left(\theta_m - \bar{\theta}_m\right)\right|\right]^2} \tag{6}$$

The index of agreement varied from 0 to 1, with higher index values indicating that the modelled values θ_p were in better agreement with the observations θ_m.

2.4 Global sensitivity analysis

Global sensitivity analysis (GSA) involves determining which part of the variance in model response is due to variance in which input variables or groups of inputs. The impact of the parameters is quantified by calculating the global sensitivity indices.

The Sobol method (Sobol, 1990) is an independent GSA method based on decomposition of the variance. When the model is nonlinear and non-monotonic, the decomposition of the output variance is still defined and can be used. The Sobol model is represented by the following function:

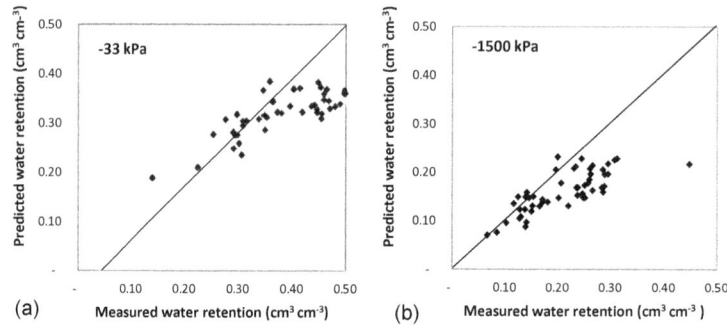

Figure 2. Scatter plots of measured versus predicted soil water retention by Rosetta H2.

$$Y = f\left(X_1, X_2, X_3, \ldots, X_p\right), \qquad (7)$$

where Y is the model output (or objective function) and $X = (X_1, \ldots, X_p)$ is the input variable set.

$$V(Y) = V(E(Y|X)) + E(\text{Var}(Y|X)), \qquad (8)$$

where $V(Y)$ is the total variance in the model, $V(E(Y|X))$ and $E(\text{Var}(Y|X))$ signify variance in the conditional expected value and expected value of the conditional variance, respectively. When the input variables X_i are independent, the variance decomposition of the model is

$$V(Y) = \sum_{i=1}^{p} V_i + \sum_i \sum_j V_{ij} + \sum_i \sum_j \sum_p V_{ijp} + \ldots + V_{1,2,3,\ldots p} \qquad (9)$$

$$V_i = V\left[E\left(Y|X_i\right)\right]$$

$$V_{ij} = V\left[E\left(Y|X_i X_j\right)\right] - V_i - V_j$$

$$V_{ijp} = V\left[E\left(Y|X_i, X_j, X_p\right)\right] - V_{ij} - V_{ip} - V_{jp} - V_i - V_j - V_p,$$

where V_i is the proportion of variance due to variable X_i. Dividing V_i by $V(Y)$ produces the expression of the first-order sensitivity index (S_i), such that

$$S_i = \frac{V_i}{V(Y)} = \frac{V\left[E\left(Y/X_i\right)\right]}{V(Y)}. \qquad (10)$$

The term S_i is the measure that guarantees an informed choice in cases where the factors are correlated and interact (Saltelli and Tarantola, 2002). This index is always between 0 and 1, and represents a proper measurement of the sensitivity used to classify the input variables in order of importance (Saltelli and Tarantola, 2002).

In order to quantify variation in the sensitivity index (V_{Si}) of an input factor X_i, we fixed it at $X_i = X_i^*$ (X_i^*: the average when the variable follows the normal distribution and the median when the variable follows the lognormal distribution). In order to calculate how much this assumption changed the variance of Y, we used the following formula:

$$V_{Si} = \left(\frac{V[E(Y/X)]}{V(Y)} - \frac{V\left[E\left(Y/X_i = X_i^*\right)\right]}{V(Y)}\right) \cdot 100. \qquad (11)$$

Table 3. Evaluation criteria of water retention pedotransfer functions (PTFs) at -33 and -1500 kPa.

			-33 kPa	-1500 kPa
ME ($cm^3 cm^{-3}$)	Point PTF	MLR	0.0188	0.0261
	Parametric PTF	MNLR	-0.0016	-0.0020
	Rosetta	H1	-0.0902	-0.0458
		H2	-0.0728	-0.0436
		H3	-0.0991	-0.0552
RMSE ($cm^3 m^{-3}$)	Point PTF	MLR	0.0414	0.0444
	Parametric PTF	MNLR	0.0613	0.0605
	Rosetta	H1	0.1170	0.0738
		H2	0.0970	0.0636
		H3	0.1280	0.0749
d ($cm^3 cm^{-3}$)	Point PTF	MLR	0.9975	0.9911
	Parametric PTF	MNLR	0.9938	0.9775
	Rosetta	H1	0.9623	0.9427
		H2	0.9775	0.9597
		H3	0.9519	0.9331

In addition, combining the RMSE and S_i enabled us to detect the contribution of each variable to improvement in the quality of prediction of the PTFs.

3   Results and discussion

In Table 3, most of the PTFs underestimated SWR except for the point PTF at the two pressure points (-33 and -1500 kPa). The Rosetta H2 model, which considers only texture as an input, gave ME values closer to zero than the H1 and H3 models (-0.0728 and -0.0436 $cm^3 cm^{-3}$ at -33 and -1500 kPa, respectively).

The poor ME values indicated better estimates of PTFs. They were produced after the application of point PTFs followed by parametric PTFs (Fig. 2).

Among the five tested models in the Lower Cheliff soils, the point PTFs (MLR) derived from a database taken from some Algerian soils had the lowest RMSE values (0.041 and 0.044 $cm^3 cm^{-3}$ at -33 and -1500 kPa, respectively). Performances equivalent or superior to PTFs derived by multiple regression methods have been reported in some studies

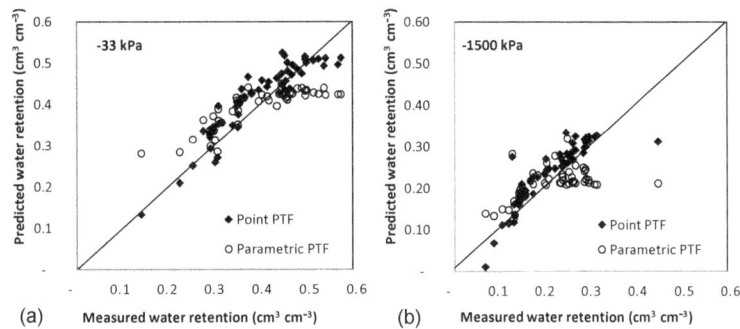

Figure 3. Scatter plots of measured soil water retention versus predicted soil water retention.

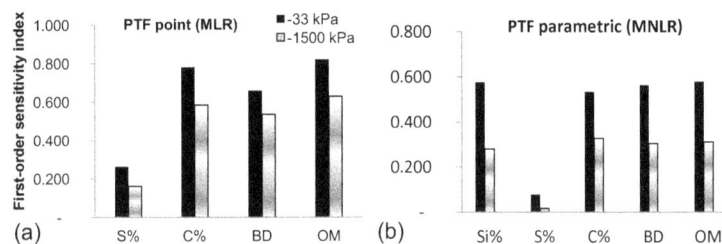

Figure 4. First-order sensitivity index.

(Minasny et al., 1999; Nemes et al., 2003). The nonlinear models (parametric PTFs), however, gave a better estimation than the Rosetta models based on ANN (RMSE: 0.0613 and 0.0605 cm^3 cm^{-3} at -33 and -1500 kPa, respectively). The RMSE and ME values of the three Rosetta models also showed that H2 was better than H1 or H3 (Table 3, Fig. 3).

The index of agreement results showed that point PTFs were more suitable for Lower Cheliff soils than parametric PTFs (Table 3) with values of 0.9975 and 0.9911 cm^3 cm^{-3}, respectively. Similar comparisons in different regions were undertaken by Minasny et al. (1999), Tomasella et al. (2003) and Ghorbani Dashtaki et al. (2010), who all reported similar differences between these two PTF approaches. As Table 3 shows, there was no significant difference in RMSE values between the parametric PTFs and Rosetta H2 at -1500 kPa (RMSE: 0.0605 cm^3 cm^{-3} and 0.0636 cm^3 cm^{-3}, respectively).

3.1 Sensitivity index before textural grouping

In the development of PTFs, using PSD as an input is the usual approach (texture as an overall expression of PSD, clay, silt and sand content) and its contribution is fundamental to understanding the process of retaining water at different pressure points, although various physical and chemical characteristics are used to describe the SWR curve, such as BD and OM.

The importance of each input variable was assessed by the first-order S_i. It was clear for the PTFs developed that OM % and clay percentages (C %) were the variables with the great-

est impact (Fig. 4). For the point PTFs (MLR), the most sensitive estimations were at two pressure points (S_i: 0.821 and 0.782 at -33 kPa, and 0.630 and 0.585 at -1500 kPa for OM % and C %, respectively). After OM, the percentage of silt (S_i %) was second in importance in parametric PTFs (0.576 at -33 kPa) followed by BD and C (Fig. 2). The S_i values placed sand content in third place in the MLR (0.262; 0.162), indicating that its impact on the parametric model was almost insignificant, with very low values (S_i: 0.077; 0.017) at -33 and -1500 kPa, respectively.

The prediction quality of point PTFs (MLR) can be explained, first, by taking into account the basic characteristics of soil as an input from the textural and structural information given by the BD. Second, point PTFs (MLR) are based mainly on these input variables, unlike parameter PTFs (MNLR), which have inputs other than texture and BD, as well as other parameters (VG parameters: θ_r, θ_s, α, n).

3.2 Sensitivity and uncertainty analysis after the textural grouping

The sensitivity of the multiple regression methods (linear and nonlinear) used to develop PTFs from basic soil characteristics for estimating SWR for different textural classes was analysed. We grouped the samples into three classes of particles (Fig. 1b) in line with FAO (1990) guidelines: very fine (12 samples), fine (31 samples) and medium (10 samples).

The results showed that after the textural grouping, there was an improvement in the quality estimation of PTFs in only the medium class. A better prediction at -1500 kPa

Table 4. Variation of first-order sensitivity index (S_i) in the different textural (Tex.) classes.

		Tex. class	Si (%)		S (%)		C (%)		BD (g cm^{-3})		OM (%)	
			V_{Si}	A.E.	V_{Si}	A.E.	V_{Si}	A.E.	V_{Si}	A.E.	V_{Si}	A.E.
RML	at −33 kPa	VF	Abs		−1.2		−0.4		−50.5	−	4.6	
		F	Abs		−43.2	−	−10.7	−	−39.9	−	0.2	
		M	Abs		−103.3	−	−27.5	+	−44.4	+	−5.7	
	at −1500 kPa	VF	Abs		−0.3		0.9		−27.3	−	1.1	
		F	Abs		−46.2	−	−20.7	−	−41.6	−	0.1	
		M	Abs		−86.4	−	−52.9	−	−22.9	−	−2.3	
MNLR	at −33 kPa	VF	0.4		−0.2		0.1		−00.1		−0.05	
		F	−1.6		−40.9	−	−1.1		−2.5		−0.1	
		M	15.0		−5.2		15.1	+	21.6	+	22.3	+
	at −1500 kPa	VF	−4.6		−0.3		−1.8		−1.4		−0.5	
		F	28.6	+	18.9	−	4.6		0.4		0.1	
		M	−36.7	−	−16.7	−	−22.6	−	8.9		−8.4	

Abs: absent in the model, V_{Si}: variation first sensitivity index, A.E.: improving estimation.

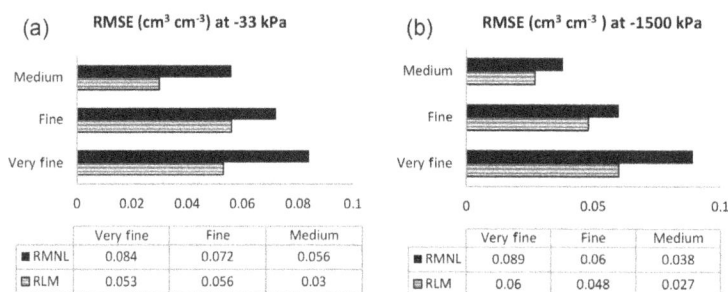

Figure 5. Root mean square error (RMSE) values calculated for the different textural classes.

was provided by point PTFs (RMSE $= 0.027$ cm^3 cm^{-3}) and parametric PTFs (RMSE $= 0.038$ cm^3 cm^{-3}) at −1500 kPa (Fig. 5).

3.2.1 Texture

After textural grouping, the MLR and MNLR PTFs developed were always sensitive, mainly to the sand fraction in the fine and medium classes (Table 4). The variation in the first S_i in the point PTFs was significantly greater in the medium-texture class at the two pressure points (−33 and −1500 kPa). In the MNLR, sand had the most influence, particularly with regard to the fine class (−40.9, 18.9 % at −33 and 1500 kPa) and the medium class (−16.7 % at −1500 kPa).

The S_i of a variable quantifies the influence of its uncertainty on the output. This is the part of the variability output explained by the variability input. What was confirmed after calculating the variation in the first-order S_i was that the PTFs developed were still more influenced by the variability in sand at −33 kPa than at −1500 kPa. This impact could be explained by the irregularity of the dispersion of sand content in the validation database, with a coefficient of

variation (CV) of about 119 % compared with the other input variables (33, 18, 9 and 57 % for clay, silt, BD and OM, respectively). This heterogeneity in the sand data series clearly influenced the uncertainty of the PTF response.

Looking at the matrix correlation (Table 5), the clay and silt fractions were significantly correlated with sand content. Saltelli and Tarantola (2002) observed that when X_1 and X_2 were correlated with a third factor, X_3, the S_i calculated depended on the force of this correlation as well as the distribution of X_3. In this case, the index power could be influenced by this statistical association, as it explains the higher value difference of index variation in the sand percentage compared with the other variables.

We observed that point PTF (MLR) produced a lower error of estimation when the variation of the first-order S_i for sand was the most important (MLR in the medium class: RMSE 0.030 and 0.027 cm^3 cm^{-3} with V_{Si} −103 and 86.4 % at −33 and −1500 kPa, respectively). A negative S_i variation in sand content when the latter was fixed was apparent in all texture classes (Table 4). This could be explained by the proportional relationship between sand and clay content, particularly in the validation dataset with a dominant clay texture. Insignificant sensitivity of sand was recorded for the very fine

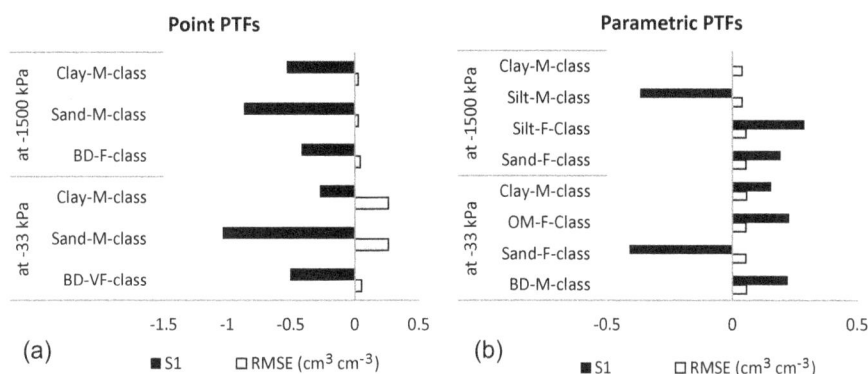

Figure 6. Variation in first sensitivity index with RMSE after textural grouping.

Table 5. Pearson correlation matrix between basic soil characteristics in the validation dataset of 53 soil samples.

Variables	S_i	C (%)	S (%)	BD ($g\,cm^{-3}$)	OM (%)
Si %	1				
S %	−0.334	1			
C %	−0.159	−0.878	1		
BD ($g\,cm^{-3}$)	0.164	−0.185	0.11	1	
OM ($g/100\,g$)	−0.174	−0.166	0.263	−0.19	1

The values in bold differ from 0 to a level of significance $\alpha = 0.05$, Si: silt, S: sand, C: clay, BD: bulk density, OM: organic matter.

texture. Rawls et al. (2003) observed that 10 % of sand provides an increase in SWR at low clay content and a decrease in SWR at high clay content of more than 50 %.

The relationship between the SWR curve parameters of VG (especially n and α) and PSD has been examined in many studies (e.g. Minasny and McBratney, 2007; Benson et al., 2014) in order to explain why the sand impact increases in the fine-texture class in parametric PTFs. It could be explained by the predominant presence of sand and clay content as inputs in parametric PTFs. For soils with clay content between 35 and 70 %, water content is greatly influenced by the percentage of sand in the soil (Loosvelt et al., 2013).

In addition, when the sand content of a sample increased to 60 %, the drying rate was faster and water absorbing ability was weaker than with the low sand content. When sand content falls to 20 %, the small pores occupy a large part of the pore structure, making the soil compact (Hao et al., 2015).

In the medium-texture class, there was increasing accuracy in PTFs at −33 kPa after fixing the clay content. This could be explained by the reduced clay percentage in the medium class (mean of clay = 23 %), which produced fewer errors at −33 kPa.

The accuracy of the PTFs decreased when they were applied to some soil samples with a clay content > 60 % (Fig. 5). In the very fine class, insignificant sensitivity was

recorded at all pressures defined in this study. In this class, the variation in clay was much lower because it is only the dominant solid fraction, which could explain the smaller variation in S_i after fixing the clay percentage. The greatest impact of clay (%) was observed at −1500 kPa in the point and parametric PTFs in different textural classes (Fig. 6). The clay content of soils is a major predictor for modelling the permanent wilting point of soils (Minasny et al., 1999).

The silt percentage was introduced as an explanatory variable only in parametric PTFs (MNLR). This fraction is known for its ability to retain water at high and medium soil water potentials. The GSA showed that the silt percentage had a stronger impact on the estimation of parametric PTFs at −1500 kPa than at −33 kPa with the MNLR model. After textural grouping, an important variation in the first-order S_i was observed in the medium class (−36.7 % to −1500 kPa). The lowest values were recorded in the very fine class. It was clear that the silt percentage has an important role in estimating parameters of VG (α, n), and that its use as an input influences the estimate in the medium and fine classes. There was an increasing accuracy, however, in the PTFs recorded in the fine class at −1500 kPa. With silt and clay as inputs, there was a better estimation. Plant-available water content variation is more related to sand and silt than to clay content (Reichert et al., 2009).

3.2.2 Bulk density

This is the second most influential variable on the point PTF (MLR) response on all textural class. The important variation of sensitivity index is noted mainly in the very fine textural class at −33 kPa ($V_{Si} = -50$, 5 %). In parametric PTFs, BD influenced the medium class at −33 kPa. The accuracy of quality estimation at −33 kPa in the medium class when fixing the BD for the two PTF approaches (Table 4). The very-fine-textural class represented 16 surface samples (0–30 cm) with a dominance of clay texture. In a similar study on clay soils, volumetric water content (VWC) was highly related to the inverse of BD at field capacity (Bruand et al., 1996). The inclusion of BD as an input provides information on pore volume, which can influence the performance of PTFs when applied to soil with high clay content. In addition, the soil structural information characterised by BD measurements is an indirect measurement of pore space and is affected mainly by texture and structure. For structureless soils, primarily coarse- and medium-textured soils, the pore-size distribution can be satisfactorily described by PSD. The medium texture is related in general to pore-size distribution, as large particles give rise to large pores between them, and therefore have a major influence on the SWR curve (Arya and Paris, 1981; Nimmo, 2004). With BD and texture as inputs in point PTF (MLR), predicted values very close to the experimental results are obtained.

3.2.3 Organic matter content

The less insignificant variation in the S_i after textural grouping is related to OM content. This could be explained, first, by the poor OM content in the Algerian soil samples. Lal (1979) did not find any effect of OM content on SWR. Danalatos et al. (1994) attributed this to the generally low OM content in their samples. Second, homogeneity of the data for OM content in every textural class reduced the variation in PTF response. The increasing accuracy of parametric PTFs, however, was apparent for medium-textured soils at -33 kPa, where OM was used as an input to predict θ_s. SWR at -33 kPa is affected more strongly by organic carbon than at -1500 kPa (Rawls et al., 2003). The sensitivity analysis conducted by Rawls et al. (2003) to study the role of OM content as a predictor showed that the SWR of coarse-textured soils is much more sensitive to changes in organic carbon than is the case with fine-textured soils. Bauer and Black (1981) found that the effect of organic carbon on SWR in disturbed samples was substantial in sandy soil and marginal in medium- and fine-textured soils.

4 Conclusions

The objective of this study was to analyse the sensitivity of estimating the SWR properties of Algerian soils using PTFs. We developed and validated point and parametric PTFs from basic soil properties using regression techniques and compared their predictive capabilities with the Rosetta models (H1, H2 and H3). The reliability tests showed that point PTFs produce more accurate estimations than parametric PTFs. The derived parametric PTFs, however, provided better estimates than the Rosetta models originally developed from a large intercontinental database.

The GSA showed that the mathematical formalism of the PTF models and their input variables reacted differently in terms of point pressure and textural class as follows:

– After textural grouping, the two PTF approaches developed (MLR and MNLR) were always sensitive primar-

ily to the sand fraction in the fine and medium classes at -33 kPa, rather than at -1500 kPa.

– The results illustrated the accuracy of estimation at -33 kPa in the medium class for the two PTF approaches when fixing the clay percentage (C%) and BD.

– The accuracy of PTFs decreased when they were applied to soil samples with a clay content > 60%.

– The most insignificant variation in the S_i after textural grouping was related to the OM content in Algerian soils.

5 Data availability

The data are available via the following database: http://hdl. handle.net/2268/204146. Otherwise, interested parties may email the corresponding author for datasets.

Acknowledgements. The authors want to gratefully acknowledge the topical editor and reviewers.

Edited by: D. Dunkerley

References

Arya, L. M. and Paris, J. F.: A physic empirical model to predict the soil moisture characteristic from particle-size distribution and bulk density data, Soil Sci. Soc. Am. J., 45, 1023–1030, 1981.

Bauer, A. and Black, A. L.: Soil carbon, nitrogen, and bulk density comparisons in two cropland tillage systems after 25 years and in virgin grassland, Soil Sci. Soc, Am. J., 45, 1166–1170, 1981.

Benson, C., Chiang, I., Chalermyanont, T., and Sawangsuriya, A.: Estimating van Genuchten parameters α and n for clean sands from particle size distribution data, Soil Behavior Fundamentals to Innovations in Geotechnical Engineering, GeoCongress 2014, ASCE, Reston, VA, 410–427, 2014.

Botula, Y. D., Van Ranst, E., and Cornelis, W. M.: Pedotransfer functions to predict water retention of soils from the humid tropics: a review, Revista Brasileira de Ciencia do Solo, 38, 679–698, 2014.

Bruand, A., Duval, O., Gaillard, H., Darthout, R., and Jamagne, M.: Variabilité des propriétés de rétention en eau des sols: importance de la densité apparente, Etude et Gestion des Sols, 3, 27–40, 1996.

Bruand, A., Perez-Fernandez, P., Duval, O., Quetin, P., Nicoullaud, B., Gaillard, H., Raison, L., Pessaud, J. F., and Prud'homme, L.: Estimation des propriétés de rétention en eau des sols: utilisation de classe de pédotransfert après stratifications texturale et texturo-structurale, Etud. Gest. Sols, 9, 105–125, 2002.

Danalatos, N. G., Kosmas, C. S., Driessen, P. M., and Yassoglou, N.: Estimation of the draining soil moisture characteristics from standard data as recorded in soil surveys, Geoderma, 64, 155–165, 1994.

Deng, H. L., Ye, M., Schaap, M. G., and Khaleel, R.: Quantification of uncertainty in pedotransfer function-based parameter estimation for unsaturated flow modeling, Water Resour. Res., 45, W04409, doi:10.1029/2008WR007477, 2009.

FAO – Food and Agriculture Organisation: Guidelines for soil description, 3rd Edn., FAO/ISRIC, Rome, 1990.

Ghanbarian-Alavijeh, B. and Liaghat, A. M.: Evaluation of soil texture data for estimating soil water retention curve, Can. J. Soil Sci., 89, 461–471, 2009.

Ghorbani Dashtaki, S., Homaee, M., and Khodaverdiloo, H.: Derivation and validation of pedotransfer functions for estimating soil water retention curve using a variety of soil data, Soil Use Manage, 26, 68–74, 2010.

Grunwald, S., McSweeney, K., Rooney, D. J., and Lowery, B.: Soil layer models created with profile cone penetrometer data, Geoderma, 103, 181–201, 2001.

Hao, D. R., Liao, H. J., Ning, C. M., and Shan, X. P.: The microstructure and soil water characteristic of unsaturated loess. Unsaturated Soil Mechanics – from Theory to Practice, Proceedings of the 6th Asia Pacific Conference on Unsaturated Soils, 23–26 October 2015, Guilin, China, 163–167, doi:10.1201/b19248-22, 2015.

Jacques, J., Lavergne, C., and Devictor, N.: Sensitivity analysis in presence of model uncertainty and correlated inputs, Reliabil. Eng. Syst. Safe., 91, 1126–1134, 2006.

Jonasson, S. A.: Estimation of the Van Genuchten parameters from grain-size distribution, Proc. of the Int. Workshop Indirect Methods for Estimating the Hydraulic Properties of Unsaturated Soils, 11–13 October 1989, Riverside, CA, Univ. of California, Riverside, 443–451, 1992.

Kay, B. D., da Silva, A. P., and Baldock, J. A.: Sensitivity of soil structure to changes in organic carbon content: predictions using pedotransfer functions, Can. J. Soil Sci., 77, 655–667, 1997.

Lal, R.: Physical properties and moisture retention characteristics of some Nigerian soils, Geoderma 21, 209–223, 1979.

Lee, T.-K. and Ro, H.-M.: Estimating soil water retention function from its particle-size distribution, Geosciences J., 18, 219–230, 2014.

Lin, H. S., McInnes, K. J., Wilding, L. P., and Hallmark, C. T.: Effects of soil morphology on hydraulic properties: II. Hydraulic pedotransfer functions, Soil Sci. Soc. Am. J., 63, 955–961, 1999.

Loosvelt, L., Vernieuwe, H., Pauwels, V. R. N., De Baets, B., and Verhoest, N. E. C.: Local sensitivity analysis for compositional data with application to soil texture in hydrologic modelling, Hydrol. Earth Syst. Sci., 17, 461–478, doi:10.5194/hess-17-461-2013, 2013.

Mayr, T. and Jarvis, N. J.: Pedotransfer functions to estimate soil water retention parameters for a modified Brooks-Corey type model, Geoderma, 91, 1–9, 1999.

Minasny, B. and McBratney, A. B.: Uncertainty analysis for pedotransfer functions, Eur. J. Soil Sci., 53, 417–429, 2002.

Minasny, B. and McBratney, A. B.: Estimating the water retention shape parameter from sand and clay content, Soil Sci. Soc. Am. J., 71, 1105–1110, 2007.

Minasny, B., McBratney, A. B., and Bristow, K.: Comparison of different approaches to the development of pedotransfer function for water-retention curves, Geoderma, 93, 225–253, 1999.

Mirus, B. B.: Evaluating the importance of characterizing soil structure and horizons in parameterizing a hydrologic process model, Hydrol. Process., 29, 4611–4623, 2015.

Moeys, J., Larsbo, M., Bergström, L., Brown, C. D., Coquet, Y., and Jarvis, N. J.: Functional test of pedotransfer functions to predict water flow and solute transport with the dual-permeability model MACRO, Hydrol. Earth Syst. Sci., 16, 2069–2083, doi:10.5194/hess-16-2069-2012, 2012.

Nemes, A., Schaap, M. G., and Wösten, J. H. M.: Functional evaluation of pedotransfer functions derived from different scales of data collection, Soil Sci. Soc. Am. J., 67, 1093–1102, 2003.

Nemes, A., Rawls, W. J., Pachepsky, Y. A., and van Genuchten, M. T.: Sensitivity analysis of the nonparametric nearest neighbor technique to estimate soil water retention, Vadose Zone J., 5, 1222–1235, 2006.

Nimmo, J. R.: Porosity and Pore Size Distribution, in: Encyclopedia of Soils in the Environment, edited by: Hillel, D., Elsevier, London, 295–303, 2004.

Pachepsky, Y. A. and Rawls, W. J.: Soil structure and pedotransfer function, Eur. J. Soil Sci., 54, 443–452, 2003.

Pachepsky, Y. A., Timlin, D., and Varallyay, G.: 1996, Artificial neural networks to estimate soil water retention from easily measurable data, Soil Sci. Soc. Am. J., 60, 727–733, 1996.

Pachepsky, Y. A., Rawls, W. J., and Timlin, D. J.: The current status of pedotransfer functions: Their accuracy, reliability, and utility in field and regional-scale modeling, in: Assessment of Non-point Source Pollution in the Vadose Zone, Geophys. Monogr. Ser., vol. 108, edited by: Corwin, D. L., Loague, K., and Ellsworth, T. R., AGU, Washington, D.C. 223–234, 1999.

Rawls, W. J. and Brakensiek, D. L.: Prediction of soil water properties for hydrologic modeling, Proc. Symp. Water shed Management in the Eighties, 30 April–1 May 1985, Denver, CO, Am. Soc. Civil Eng., New York, 293–299, 1985.

Rawls, W. J., Gish, T. J., and Brakensiek, D. L.: Estimating soil water retention from soil physical properties and characteristics, Adv. Soil Sci., 16, 213–234, 1991.

Rawls, W. J., Pachepsky, Y., and Ritchie, J.: Effect of soil organic carbon on soil water retention, Geoderma, 116, 61–76, 2003.

Reichert, J. M., Suzuki, L. E. A. S., Reinert, D. J., Horn, R., and Håkansson, I.: Reference bulk density and critical degree-of-compactness for no-till crop production in subtropical highly weathered soils, Soil Till. Res., 102, 242–254, 2009.

Richards, L. A. and Fireman, M.: Pressure-plate apparatus for measuring moisture sorption and transmission by soils, Soil Sci., 56, 395–404, 1943.

Robinson, G. W.: A new method for the mechanical analysis of soils and other dispersions, J. Agric. Sci., 12, 306–321, doi:10.1017/S0021859600005360, 1922.

Saltelli, A. and Tarantola, S.: On the relative importance of input factors in mathematical models, J. Am. Stat. Assoc., 97, 702–709, 2002.

Saltelli, A., Chan, K., and Scott, M. (Eds.): Sensitivity Analysis, in: Probability and Statistics Series, John Wiley and Sons, Ltd, Now York, 2000.

Saxton, K. E., Rawls, W. L., Rosenberger, J. S., and Papendick, R. I.: Estimating generalized soil-water characteristics from texture, Soil Sci. Soc. Am. J., 50, 1031–1036, 1986.

Schaap, M. G. and Leij, F. J.: Using neural networks to predict soil water retention and soil hydraulic conductivity, Soil Till. Res., 47, 37–42, 1998.

Schaap, M. G., Leij, F. J., and van Genuchten, M. T.: Rosetta: A computer program for estimating soil hydraulic parameters with hierarchical pedotransfer functions, J. Hydrol., 251, 163–176, 2001.

Sobol, I. M.: On sensitivity estimation for nonlinear mathematical models, Matematicheskoe Modelirovanie (in Russian), translated in: Math. Model. Comput. Exp., 2, 112–118, 1990.

Stumpp, C., Engelhardt, S., Hofmann, M., and Huwe, B.: Evaluation of Pedotransfer Functions for Estimating Soil Hydraulic Properties of Prevalent Soils in a Catchment of the Bavarian Alps, Eur. J. Forest Res., 128, 609–620, 2009.

Tomasella, J., Hodnett, M. G., and Rossato, L.: Pedotransfer functions for the estimation of soil water retention in Brazilian soils, Soil Sci. Soc. Am. J., 64, 327–338, 2000.

Tomasella, J., Pachepsky, Y. A., Crestana, S., and Rawls, W. J.: Comparison of two techniques to develop pedotransfer functions for water retention, Soil Sci. Soc. Am. J., 67, 1085–1092, 2003.

Touil, S., Degré, A., and Chabaca, M. N.: Transposability of pedotransfer functions for estimating water retention of Algerian soils, Desalin. Water Treat., 57, 5232–5240, 2016.

Van Genuchten, M. T.: A closed-form equation for predicting the hydraulic conductivity of unsaturated soils, Soil Sci. Soc. Am. J., 44, 892–898, 1980.

Van Genuchten, M. T., Leij, F. J., and Lund, L. J.: Indirect methods for estimating the hydraulic proprieties of unsaturated soils, US Salinity Lab., Riverside, CA, 1992.

Vereecken, H., Feyen, J., Maes, J., and Darius, P.: Estimating the soil moisture retention characteristic from texture, bulk density, and carbon content, Soil Sci., 148, 389–403, 1989.

Vereecken, H., Diels, J., Vanorshoven, J., Feyen, J., and Bouma, J.: Functional evaluation of pedotransfer functions for the estimation on of soil hydraulic properties, Soil Sci. Soc. Am. J., 56, 1371–1378, 1992.

Vereecken, H., Weynants, M., Javaux, M., Pachepsky, Y., Schaap, M. G., and van Genuchten, M. T.: Using pedotransfer functions to estimate the van Genuchten–Mualem soil hydraulic properties: A review, Vadose Zone J., 9, 795–820, 2010.

Walkley, A. and Black, I. A.: An examination of the Degtjareff method for determining organic carbon in soils: Effect of variations in digestion conditions and of inorganic soil constituent, Soil Sci., 63, 251–263, 1934.

Williams, J., Prebble, R. E., Williams, W. T., and Hignett, C. T.: The influence of texture, structure and clay mineralogy on the soil moisture characteristic, Aust. J. Soil Res., 21, 15–32, 1983.

Willmott, C. J.: On the validation of models, Phys. Geogr., 2, 184–194, 1981.

Willmott, C. J. and Wicks, D. E.: An empirical method for the spatial interpolation of monthly precipitation within California, Phys. Geogr., 1, 59–73, 1980.

Winfield, K. A., Nimmo, J. R., Izbicki, J. A., and MartinResolving, P. M.: Structural Influences on Water-Retention Properties of Alluvial Deposits, Vadose Zone J., 5, 706–719, 2006.

Wösten, J. H. M., Finke, P. A., and Jansen, M. J. W.: Comparison of class and continuous pedotransfer functions to generate soil hydraulic characteristics, Geoderma, 66, 227–237, 1995.

Wösten, J. H. M., Pachepsky, Y. A., and Rawls, W. J.: Pedotransfer functions: Bridging the gap between available basic soil data and missing soil hydraulic characteristics, J. Hydrol., 251, 123–150, 2001.

Yang, X. and You, X.: Estimating Parameters of Van Genuchten Model for Soil Water Retention Curve by Intelligent Algorithms, Appl. Math. Inf. Sci., 7, 1977–1983, 2013.

9

How Alexander von Humboldt's life story can inspire innovative soil research in developing countries

Johan Bouma[1],[*],[**]

[1]Wageningen University, Wageningen, the Netherlands
[*]retired
[**] *Invited contribution by Johan Bouma, recipient of the EGU Alexander von Humboldt Medal 2017.*

Correspondence to: Johan Bouma (johan.bouma@planet.nl)

Abstract. The pioneering vision of Alexander von Humboldt of science and society of the early 1800s is still highly relevant today. His open mind and urge to make many measurements characterizing the "interconnected web of life" are crucial ingredients as we now face the worldwide challenge of the UN Sustainable Development Goals. Case studies in the Philippines, Vietnam, Kenya, Niger, and Costa Rica demonstrate, in Alexander's spirit, interaction with stakeholders and attention to unique local conditions, applying modern measurement and modeling methods and allowing inter- and transdisciplinary research approaches. But relations between science and society are increasingly problematic, partly as a result of the information revolution and "post-truth", "fact-free" thinking. Overly regulated and financially restricted scientific communities in so-called developed countries may stifle intellectual creativity. Researchers in developing countries are urged to "leapfrog" these problems in the spirit of Alexander von Humboldt as they further develop their scientific communities. Six suggestions to the science community are made with particular attention to soil science. (The Humboldt lecture, presented by the 2017 recipient of the Alexander von Humboldt lecture, Johan Bouma, can be accessed at http://client.cntv.at/egu2017/ml1.)

1 Introduction

The scientific career of Alexander von Humboldt, a name linked with the medal for research in developing countries by the European Geosciences Union, is highly inspiring for scientists operating in the current scientific arena as one realizes after reading the impressive biography by Andrea Wulf (2015). He was the first in the early 1800s to emphasize the importance of the "interconnected web of life" rather than isolated disciplinary and taxonomic issues as was the custom at the time and still is in some quarters. He saw man as part of nature rather than as its justified and exclusive consumer, the dominant view at the time and still prevalent today. It is now generally accepted that the geosciences are not only closely linked with other environmental sciences but with society itself. Modern measuring, sensing, and modeling facilities now offer the possibility to express ecosystem dynamics in quantitative terms rather than in terms of the flowery illustrated books, reports, letters, and drawings by von Humboldt, but the basic message is the same. But perhaps his greatest contribution has been his enthusiastic and uncompromising dedication to be receptive to new ideas ("keep learning") and to maintain an open, inquisitive mind when observing phenomena in nature or when interacting with land users in Latin America and the United States. He always encouraged young colleagues and shared his data freely. As a scientist he carried his instruments everywhere, meticulously documenting his many observations to be systematically analyzed later, often deep into the night. At the same time he was in dialogue with poets like Goethe, quite aware that facts are experienced differently by different people, as the experience involves personal emotions and values. Two centuries before terms like inter- and transdisciplinarity were coined, they were acted out in real life by von Humboldt as, for example, he observed the misery of farmers in the Aragua valley in the Andes following erosion and soil degradation as a side effect of cutting upland forests. In general he warned against developments where science may feed the brain with abstract data while ignoring imagination in the process: a message with high relevance for the current scientific arena.

The relation between science and society has dramatically changed in the early 21st century and has become problematic not only in the so-called developed world but also globally as the internet reaches all corners of the world, mobile phones are used everywhere, and social media are prominent in daily life (e.g., Kahan, 2015; Bouma, 2015). "Citizen science" is promoted while many see science as providing only yet another opinion. The terms "post-truth", "fact free", and "alternative facts" have become prominent in public debate. Effects differ among scientific disciplines but issues develop, in particular, in land-related environmental and food science addressing concerns in the everyday lives of citizens. On the other side of the fence, governmental funding of environmental research is decreasing in many countries and researchers are increasingly forced to generate research contracts with industrial or commercial partners. Ever larger numbers of students follow tight curricula that leave limited time for side activities and scientific workers are being judged and squeezed by tight performance indicators where publication requirements figure prominently. The emphasis on publishing in, preferably, internationally refereed journals has resulted in an explosion of the number of new journals and a structural lack of competent referees (e.g., Munafo et al., 2017). As disciplinary papers, developing yet another new technique or introducing yet another new model, are relatively rapid to generate, very much needed time-consuming inter- and transdisciplinary approaches suffer. There are, in short, reasons for concern.

All these developments are of particular concern to researchers in so-called developing countries. They often do not have the facilities or equipment to generate papers that are acceptable to the big journals, restricting their professional development, while they perform good research in many cases. How to proceed? Are we not all, in fact, developing countries though development may move in opposite directions in different countries?

Considering these observations, the objectives of this paper are to briefly review and analyze work done by our group in developing countries and discuss how what appear to be unfavorable developments in science–society relations in so-called developed countries can be avoided in developing countries. How can they possibly "leapfrog" to a more stimulating, productive, less stressful, and sustainable condition in the spirit of Alexander von Humboldt? Attention will be confined to the soil science discipline, which is an essential part of environmental sciences.

2 Examples of eco-regional research in developing countries

Work by the former chair group Soil Inventarisation and Land Evaluation of Wageningen University in the Philippines, Vietnam, Kenya, Niger, and Costa Rica and projects executed in the context of the Ecoregional Methodology Fund will be briefly reviewed with reference to source publications, with the intention of reflecting and documenting the background of the 2017 von Humboldt medal. Certainly, many other reports and papers on research in developing countries have been written in the von Humboldt spirit.

2.1 Growing rice in the Philippines

Rice is the main food source in Southeast Asia and in approximately 75 % of the area, irrigation is used to submerge the growing rice plants with water on top of a slowly permeable puddled layer of topsoil. To produce 1 kg of rice, 5000 L of water are needed, and because fresh water is scarce in many areas, understanding soil water regimes is crucial to define optimal irrigation regimes intended to save water. Wopereis et al. (1994) developed a field test to measure infiltration rates through the puddled surface layer of soil, and this value varied significantly in different soils, following different puddling practices that could be refined based on such measurements. Statistical techniques were used to estimate the minimum number of samples needed (Wopereis et al., 1992, 1993). Rice is also grown without irrigation, and then natural soil moisture regimes determine development of the rice plants. Here, bypass flow (which is rapid downward movement of water and solutes beyond the root zone along air-filled cracks in the soil) is an important process that cannot be characterized with existing physical flow theory that implicitly assumes soils to be homogeneous. The application of a new technique to measure bypass flow allowed the development of innovative soil management procedures, restricting the potential for bypass flow by modifying surface structure and crack continuity (Wopereis et al., 1994). Data obtained were extended to Tarlac province (300 000 ha) for a regional analysis, predicting rice yields with the newly developed simulation model ORYZA as a function of soil differences and management practices. Stein et al. (1988) applied geostatistics to interpolate from points to areas using units of the soil map as a basis for sample stratification. It allows one to optimally use uniformity in soil units, in particular in terms of spatial variability of the soil variables. In this way the study contributes to collect information more efficiently and run simulation models in a more parsimonious way with quantitative uncertainty.

2.2 Managing acid sulfate soils in Vietnam

A major program in Vietnam by van Mensvoort, le Quang Tri, le Quang Minh, and Husson, in close cooperation with

Can Tho University, focused on the agricultural use of acid sulfate soils (Minh et al., 1997a, b; Husson et al., 2000a, b). These soils occur in marine deposits near the sea, containing pyrite that upon aeration and oxidation can result in strong acidification making soils unfit for plant growth. Chemical processes have been well documented in literature, but local implications for land and water management remained undefined. In Vietnam alone, there are 2 million ha of these soils occur, 12 million ha occur in the world overall. As long as the pyrite-containing soil layers are submerged nothing happens and soils can be highly productive. Properly managing water regimes is therefore of crucial importance to avoid aeration and oxidation of pyrite-containing layers and as local soil and hydrological conditions vary significantly over short distances, it is impossible to devise generalized modeling procedures. Farmers' experience, assisted by local measurements and observations, therefore played a key role when defining the appropriate management by digging ditches and heightening soil surfaces in between. Depending on soil conditions, different system dimensions were developed in three key areas. When vertical cracks form as soils dry out, the effects of bypass flow on acidification can be significant as was demonstrated by bypass measurements. Deeply penetrating cracks cause rapid acidification at much greater depth than in homogeneous soil. Next, bypass flow results in rapid leaching of highly acidic water (Minh et al., 1997c).

2.3 Integrated nutrient management in Africa

A major problem of African agriculture is the negative soil nutrient balance: more nutrients are extracted than supplied. Smaling et al. (1992) developed a framework for integrated nutrient management in the tropics, presenting Kenya as a case study and showing significant differences among regions and soils. He successfully continued the development of the QUEFTS model relating natural fertility and fertilization rates to nutrient uptake and crop yields, allowing the assessment of local potentials and limitations (Smaling and Janssen, 1993). Bypass flow was prominent in well-structured clay soils (Vertisols), leading to a loss of surface-applied nutrients of up to 60 %. Also, the evaporation of nitrogen compounds in these high-pH soils contributed to losses. Reductions could be achieved by modifying surface structures (Smaling and Bouma, 1992). The interdisciplinary approach of this work was emphasized by investigating the effects of parasitic weeds on crop yields (Smaling et al., 1991).

2.4 Agricultural development in the Sahelian region in Africa (Niger)

The arid Sahelian region in west Africa is particularly vulnerable in terms of agricultural productivity, not only because of the low natural fertility of sandy soils but also because of low and erratic rainfall that is expected to become even more problematic in future due to climate change. Brouwer and Bouma (1997), Gandah et al. (2003), and Brouwer (2008) studied farming systems in Niger in a comparable manner to the work in Vietnam by focusing on farmers' management practices when trying to cope with the severe constraints on farming. Theoretical studies on potential crop yields, assuming optimal water and nutrient availability, are irrelevant in an extremely poor country like Niger with limited agricultural policies, where farmers cannot afford fertilizers, and with no water for irrigation in agricultural land. Impressive ways of coping that have been developed by farmers include corraling cattle to gather manure that is spread over the land, leaving crop residues on the land, and growing crops near certain tree species or abandoned termite mounds where fertility is higher and competition for water limited. A scoring technique was introduced to allow reliable estimates of crop yields that had a very high spatial variability even within fields (with a factor up to 30), and this was quantified with statistical techniques, showing strong negative relations of yields with distances to settlements (Stein et al., 1997) or with a lack of shrubs (Van Groenigen et al., 2000). The research showed that corraling the cattle at night was effective for collecting manure, but application rates to the land, varying in practice from 3 to $17\,t\,ha^{-1}$, were often considerably higher than a threshold value of $3\,t\,ha^{-1}$ that was developed in the program (Brouwer and Powell, 1998; Gandah et al., 2003). Higher values result in the leaching of nutrients during rain in these highly permeable sandy soils. The locally developed *tesse* system, where plants are grown in individual small pits filled with fertilized soil, is effective but highly labor intensive. Studies also showed that surface sealing and runoff was a dominant process in areas farther away from settlements, explaining the results of a statistical analysis showing a higher effect of water shortages as compared with nutrient shortages (Gaze et al., 1997). Surface sealing and low infiltration rates were particularly prominent because of the addition of windblown silt particles, filling the pore spaces between the sand grains.

2.5 Soil-related land use patterns in Costa Rica

After studying land use in Costa Rica with remote sensing combined with extensive field work, Huising et al. (1994) introduced the concept of land use zones (LUZs) in the Guacimo region of $900\,km^2$ defining large geographic units with particular land use patterns and dynamic behavior. Applying this concept, the study indicated that soils in 18 % of the area were overused, leading to degradation processes, while in 50 % of the soils, more intensive management practices would be feasible, providing a guide for regional land use policies. Studies in banana fincas showed that soil differences could explain 67 % of the yield variation (Veldkamp et al., 1990). Soil moisture regimes could be related to nematode development (Stoorvogel et al., 1999). Stoorvogel et al. (1999) also developed innovative soil databases, based on soil surveys incorporated in Geographical Information Systems (GISs), based on a functional approach, in this case in terms of the risk of pesticide leaching. This way, simulation models of crop growth and nutrient regimes could logically be linked with the GIS system, strongly increasing its applicability, as was demonstrated when developing alternative land use systems for the Neguev area in Costa Rica and banana fincas (e.g., Stoorvogel et al., 2004). Comparing the behavior of young volcanic soils (Andosols) with old ones (Ultisols) showed that the former had a higher resilience as shown by a better recovery of soil structure after compaction following deforestation practices (Spaans et al., 1989, 1990). Recovery was a result of a higher biological activity.

2.6   The eco-regional methodology program

In addition to the above activities, considerable attention has been paid to coordinate and actively participate in an Ecoregional Methodology Program funded by the Dutch Government, as summarized in Bouma et al. (2007). Taking the policy cycle as a guiding principle, studies were made on (i) land use change in the Kenyan Highlands; (ii) the effects of trade liberalization in Peru; (iii) water resources on the Tibetan Plateau; (iv) land use problems and conflicts in the Philippines; (v) effects of environmental conditions on human health in Ecuador; (vi) soil erosion in the Kenyan Highlands, and (vii) reestablishing bank credits to be received by farmers in the Highveld region of South Africa.

3   Learning from developed countries when relating soil science to society: possibilities for leapfrogging?

As soil scientists we should not forget how our profession was established by people like Dukochaev in Russia and Marbut in the USA, who traveled widely in the late 19th century observing different soils in different landscapes, in a manner that resembles the endeavors of Alexander von Humboldt. Like Alexander, they saw something that others before them had not seen: then as well as now the very key to progress in science. Soil turned out to be more than "dirt obscuring rocks". Researchers in developing countries would be well advised to take note of developments in the so-called developed world and work actively to define ways and means to avoid problems encountered elsewhere, taking advantage of many available studies (e.g., World Bank, 2008; Rockström et al., 2009, 2010; IFAD, 2011; Schwilch et al., 2012; Lal, 2013; Bouma et al., 2014; FAO & ITPS, 2015; van Ittersum et al., 2016). The following six points are intended to stimulate discussions on future soil research that may also be relevant for other scientific disciplines:

3.1   Define clear goals

The amount of basic soil data in information systems, the number of available methods and models and continuous streams of monitoring data present an overwhelming challenge to the 21st-century soil scientist. Without clear goals, only trees will be seen, while the forest is obscured. Sustainable development is an excellent overall goal and the 17 Sustainable Development Goals (SDGs), defined by the UN and approved by all its members at the General Assembly in September 2015, are excellent aims (e.g., Bouma, 2014, 2016; Keesstra et al., 2016; Bouma and Montanarella, 2016) as is the 4 per 1000 proposal, accepted at the Paris Climate conference also in 2015 and focused on increasing the percentage of carbon in soils as a climate change mitigation measure.

3.2   Adhere strictly to scientific principles

The scientific method, when followed, does not produce just another opinion. Define the problem; reframe it in terms of an objective that can be researched; formulate a hypothesis to be tested, choose methods and procedures that have proven their reliability and reproducibility in previous research or develop new methods if needed. Make an adequate number of measurements in adequate number to allow the statistical expression of results in terms of reliability and accuracy and pay attention to the way results are communicated. Also, show a base level of hypothetical results when no or highly simplified soil data are applied. Only then can the effect and importance of using appropriate soil data be illustrated.

3.3   Engage stakeholders but stay in charge

The need to engage stakeholders and policy makers in transdisciplinary research has been acknowledged widely but the question still lingers of how to best do it. My suggestion: try the step-by-step approach (Hoosbeek and Bryant, 1992; Bouma, 1997, Bouma et al., 2008). In summary: stakeholder knowledge is empirical and qualitative in nature. As scientists get involved, presented procedures become more quantitative and underlying mechanisms can be better explained. From discussions with stakeholders, the conclusion can be reached that their knowledge is inadequate to solve the problem that has been identified or, better, the discussions may lead to the intended goal being reached. Next, relatively simple and available techniques are tested. If they yield satisfactory results, then the project can be terminated. If not, more elaborate techniques are needed and new methods may need to be developed. Costs will increase step by step and the ultimate project design will be the result of a cost–benefit analysis. This approach has a number of advantages.

   i. Stakeholders are taken along on a "scientific journey" involving joint learning and increasingly shared ownership. Stakeholders are shown that they are taken seriously by not being "talked-down" to (by the "elite"), the latter a major reason for the science–society divide and "fact-free" and "post-truth" attitudes.

ii. Many problems can be solved by applying existing data and methods, not needing new research. Bouma et al. (2015) showed six examples of research programs, three of which could be completed without developing new methods. But in three cases new basic research was needed and this was documented providing a rational argument for new research. Not having fancy equipment and supercomputers does not necessarily imply that effective research cannot be realized.

iii. The scientific method applies to all knowledge levels, ranging from empirical to mechanistic and from qualitative to quantitative.

3.4 Avoid atomization of soil science and associate with physical geographers

When interacting with other disciplines, stakeholders, and policy makers, soil science is more effective when the various subdisciplines work together rather than separately. The following sequence may be adhered to: start with pedology, defining the physical constitution of a dynamic soil in a landscape context; next add hydrology, chemistry, and biology in that particular sequence. When this sequence is not followed, the potential of soil science contributions to interdisciplinary research may not be reached. Physical geographers are not only a good source of spatial landscape data, but they also are experts in defining soil–landscape relations.

3.5 Develop storylines to facilitate communication

Soil research as presented in scientific papers or reports often makes an isolated, clinical impression. Putting research in a storyline context is quite useful for improving communication. Addressing the SDGs is one way; another is to express the work in the DPSIR scheme, defining drivers, pressures, state, impacts, and responses of external effects on land use (Van Camp et al., 2004; Bouma et al., 2008). Realizing that different stakeholder or policy groups have different goals, based on their particular perceptions of the truth, a range of scenarios can be formulated that do justice to each of these perceptions. The key premise in all of this should be that anything can be done anywhere. The scientific analysis will show what the likely economic, social, and environmental consequences are of each scenario. The stakeholders and the politicians decide by choosing the ultimate scenario, not the scientist.

3.6 Preserve intellectual vigor

Make sure that regulations, guidelines, indicators, and judgement criteria for research and researchers do not become too restricting leading to a routine, all too pragmatic, and risk-averse attitude of researchers. Initiate long-duration programs, allowing continued interaction with stakeholders when preparing, executing, and implementing research. Actively engage agronomists, hydrologists, ecologists, and climatologists in a proactive manner in joint programs focusing on SDGs. Judge researchers on their main papers not only on the number of papers or citations. Create conditions where research is fun rather than a burden. The education of young scientists is important, and innovative educational approaches are important and deserve to be followed widely (e.g., Field et al., 2011, 2013; Jarvis et al., 2012; Hartemink et al., 2014). And, finally, all soil science students should read the biography of Alexander von Humboldt as a source of inspiration.

As their scientific network is being established, researchers and policy makers in developing countries would be well advised to take these signals seriously and apply them to establishing vital scientific regimes. And, above all, look to Alexander von Humboldt as an inspiring example as to what science in the real world can be all about.

Competing interests. The author declares that he has no conflict of interest.

Acknowledgements. Thanks to our PhD students: Marco Wopereis, Tini van Mensvoort, le Quang Tri, le Qung Minh, Olivier Husson, Eric Smaling, Mohamadou Gandah, Jeroen Huising, and Jetse Stoorvogel. Thanks to the PhD (co-)supervisors: Martin Kropff, Vo Tong Xuan, To PhucTuong, Louise Fresco, Eric Smaling, Joost Brouwer, Niek van Duivenbooden, Alfred Stein, and Martin Moolenaar. Contributions by the Ecoregional Development Fund are also gratefully acknowledged: Rudy Rabbinge, Roberto Quiroz, Mario Herrera, Stephen Staal, Water Immerzeel, Raymund Roetter, Geert Sterk, Jetse Stoorvogel, Rick van den Bosch, Patrick Gichera, and Fred Muchena. Contributions by Damien Field are gratefully acknowledged.

Edited by: Boris Jansen

References

Bouma, J.: Role of quantitative approaches in soil science when interacting with stakeholders, Geoderma, 78, 1–12, 1997.

Bouma, J.: Soil science contributions towards Sustainable Development Goals and their implementation: linking soil functions with ecosystem services, J. Plant Nutr. Soil Sc., 177, 111–120, 2014.

Bouma, J.: Engaging soil science in transdisciplinary research facing wicked problems in the information society, Soil Sci. Soc.

Am. J., 79, 454–458, https://doi.org/10.2136/sssaj2014.11.0470, 2015.

Bouma, J.: Hydropedology and the societal challenge of realizing the 2015 United Nations Sustainable Development Goals, Vadose Zone J., 15, 36–48, https://doi.org/10.2136/vzj2016.09.0080, 2016.

Bouma, J. and Montanarella, L.: Facing policy challenges with inter- and transdisciplinary soil research focused on the UN Sustainable Development Goals, SOIL, 2, 135–145, https://doi.org/10.5194/soil-2-135-2016, 2016.

Bouma, J., Stoorvogel, J. J., Quiroz, R., Staal, S., Herrero, M., Immerzeel, W., Roetter, R. P., van den Bosch, H., Sterk, G., Rabbinge, R., and Chater, S.: Ecoregional Research for Development, Adv. Agron., 93, 257–311, 2007.

Bouma, J., de Vos, J. A., Sonneveld, M. P. W., Heuvelink, G. B. M., and Stoorvogel, J. J.: The role of scientists in multiscale land use analysis: lessons learned from Dutch communities of practice, Adv. Agron., 97, 177–239, 2008.

Bouma, J., Batjes, N., Sonneveld, M. P. W., and Bindraban, P.: Enhancing soil security for smallholder agriculture, in: Soil management of smallholder agriculture, Advances in Soil Science, edited by: Lal, R. and Stewart, B. A., CRC Press, Baco Raton (FL), 17–37, 2014.

Bouma, J., Kwakernaak, C., Bonfante, A., Stoorvogel, J. J., and Dekker, L. W.: Soil science input in Transdisciplinary projects in the Netherlands and Italy, Geoderma Regional, 5, 96–105, https://doi.org/10.1016/j.geodrs.2015.04.002, 2015.

Brouwer, J.: The importance of within-field soil and crop growth variability to improving food production in a changing Sahel. A summary in images based on five years of research at ICRISAT Sahelian Center, Niamey, Niger. IUCN Commission on Ecosystem Management, Gland, Switzerland, 12 pp., available at: http://cmsdata.iucn.org/downloads/cem_csd_16_brochure_sahel_hq.pdf (last access: 12 May 2017), 2008.

Brouwer, J. and Bouma, J.: Soil and crop growth variability in the Sahel, Infor. Bull. 49, ICRISAT-Sahelian Center and Agric. Univ. Wageningen Neth., Patencheru 502324, Andhra Pradesh, India, 1997.

Brouwer, J. and Powell, J. M.: Microtopography and leaching: possibilities for making more efficient use of nutrients in African agriculture, in: Nutrient Balances as Indicators of Productivity and Sustainability in sub-Saharan African Agriculture, edited by: Smaling, E. M. A., Agr. Ecosyst. Environ., 71, 229–239, 1998.

FAO & ITPS: Status of the World's Soil Resources. Main Report, FAO and Intergovernmental Technical Panel (ITPS), Rome, Italy, 2015.

Field, D. J., Koppi, A. J., Jarrett, L. A., Abbott, L. K., Cattle, S. R., Grant, C. D., McBratney, A. B., Menzies, N. W., and Weatherley, A. J.: Soil Science teaching principles, Geoderma, 167/168, 9–14, 2011.

Field, D. J., Koppi, A. J., Jarrett, L., and McBratney, A. B.: Engaging employers, graduates and students to inform the future curriculum needs of soil science, 19th Australian Conference on Science and Mathematics Education, 19–21 September 2013, Canberra, Australia, 2013.

Gandah, M., Bouma, J., Brouwer, J., Hiernaux, P., and van Duivenbooden, N.: Strategies to optimize allocation of limited nutrients to sandy soils of the Sahel: a case study from Niger, West Africa, Agr. Ecosyst. Environ., 94, 311–319, 2003.

Gaze, S. R., Simmonds, L. P., Brouwer, J., and Bouma, J.: Measurement of surface redistribution of rainfall and modeling its effect on water balance calculations for a millet field on sandy loam soil in Niger, J. Hydrol., 188/189, 267–284, 1997.

Hartemink, A. E., Balks, M. B., Chen, Z.-S., Drohan, P., Field, D. J., Krasilnikov, P., Lowe, D. J., Rabenhorst, M., van Rees, K., Schad, P., Schipper, L. A., Sonneveld, M., and Walter, C.: The joy of teaching soil science, Geoderma, 217–18, 1–9, 2014.

Hoosbeek, M. R. and Bryant, R. B.: Towards the quantitative modelling of pedogenesis – a review, Geoderma, 55, 183–210, 1992.

Huising, E. J., Wielemaker, W. G., and Bouma, J.: Evaluating land use at the sub-regional level in the Atlantic zone of Costa Rica, considering biophysical land potentials, Soil Use and Management, 10, 152–158, 1994.

Husson, O., Hanhart, K., Phung, M. T., and Bouma, J.: Water management for rice cultivation on acid sulphate soils in the Plain of Reeds, Vietnam, Agr. Water Manage., 46, 91–109, 2000a.

Husson, O., Phung, M. T., and Van Mensvoort, M. E. F.: Soil and water indicators for optimal practices when reclaiming acid sulphate soils in the Plain of Reeds, Viet Nam, Agr. Water Manage., 45, 127–143, 2000b.

IFAD: Rural Poverty Report: New Realities, New Challenges: New Opportunities for tomorrow's generation, Rome, Italy, 2011.

Jarvis, H. D., Collett, R., Wingenbach, G., Heilman, J. L., and Fowler, D.: Developing a Foundation for Constructing New Curricula in Soil, Crop and Turfgrass Sciences, Journal of Natural Resources & Life Sciences, 41, 7–14, 2012.

Kahan, D. M.: What is the Science of:"Science Communication"?, Journal of Science Communication, 14, 1–10, 2015.

Keesstra, S. D., Bouma, J., Wallinga, J., Tittonell, P., Smith, P., Cerdà, A., Montanarella, L., Quinton, J. N., Pachepsky, Y., van der Putten, W. H., Bardgett, R. D., Moolenaar, S., Mol, G., Jansen, B., and Fresco, L. O.: The significance of soils and soil science towards realization of the United Nations Sustainable Development Goals, SOIL, 2, 111–128, https://doi.org/10.5194/soil-2-111-2016, 2016.

Lal, R.: Food security in a changing climate, Ecohydrology & Hydrobiology, 13, 8–21, 2013.

Minh, L. Q., Tuong, T. P., van Mensvoort, M. E. F., and Bouma, J.: Contamination of surface water as affected by land use in acid sulphate soils in the Mekong River Delta, Vietnam, Agr. Ecosyst. Environ., 61, 19–27, 1997a.

Minh, L. Q., Tuong, T. P., van Mensvoort, M. E. F., and Bouma, J.: Tillage and water management for riceland productivity in acid sulphate soils of the Mekong Delta, Vietnam, Soil Till. Res., 42, 1–14, 1997b.

Minh, L. Q., Tuong, T. P., Booltink, H. W. G., van Mensvoort, M. E. F., and Bouma, J.: Bypass flow and its role in leaching of raised beds under different land use types on an acid sulphate soil, Agr. Water Manage., 32, 131–147, 1997c.

Munafo, M., Nosek, B. A., Bishop, D. V. M., Button, K. S., Chambers, C. O., Percie du Sert, N., Simonsohn, U., Wagenmakers, E. J., de Ware, J. J., and Ionnidis, J. P. A.: A manofesto for reproducible science, Nature Human Behaviour, 1, 0021, https://doi.org/10.1038/s41562-016-0021, 2017.

Rockström, J., Steffen, W., Noone, K., Persson, A., Chapin III, F. S., Lambin, E., Lenton, T. M., Scheffer, M., Folke, C., Schellnhuber, H. J., Nykvist, B., de Wit, C. A., Hughes, T., van der Leeuw, S., Rodhe, H., Sorlin, S., Snyder, P. K., Constanza, R., Svedin, U., Falkenmark, M., Karlberg, L., Correll, R. W., Fabbry, V. J., Hansen, J., Walker, B., Liverman, D., Richardson, K., Crutzen, P., and Foley, J.: Planetary boundaries: exploring the safe operating space for humanity, Ecol. Soc., 14, 32, http://www.ecologyandsociety.org/vol14/iss2/art32, 2009.

Rockström, J., Karlberg, L., Wani, S. P., Barron, J., Hatibie, N., Owas, Th., Bruggeman, A., Farahani, J., and Qiang, Z.: Managing water in rainfed agriculture. The need for a paradigm shift, Agr. Water Manage., 97, 543–550, 2010.

Schwilch, G., Hessel, R., and Verzandvoort, S. (Eds.): Desire for Greener Land. Options for Sustainable Land Management in Drylands, Bern, Switserland and Wageningen, the Netherlands, Univ. of Bern-CDE, Alterra-Wageningen-UR, ISRIC-World Soil Information and CTA-techn. Center for AGR and Rural Cooperation, 2012.

Smaling, E. M. A. and Bouma, J.: Bypass flow and leaching of nitrogen in a Kenyan Vertisol at the onset of the growing season, Soil Use Manage., 8, 44–48, 1992.

Smaling, E. M. A. and Janssen, B. H.: Calibration of QUEFTS, a model predicting nutrient uptake and yields from chemical soil fertility indices, Geoderma, 59, 21–44, 1993.

Smaling, E. M. A., Stein, A., and Sloot, P. H. M.: A statistical analysis of the influence of *Striga hermonthica* on maize yields in fertilizer trials in Southwestern Kenya, Plant Soil, 138, 1–8, 1991.

Smaling, E. M. A., Nandwa, S. M., Prestele, H., Roetter, R., and Muchena, F. N.: Yield response of maize to fertilizers and manure under different agro-ecological conditions in Kenya, Agr. Ecosyst. Environ., 41, 241–252, 1992.

Spaans, E., Bouma, J., Lansu, A., and Wielemaker, W. G.: Measuring soil hydraulic properties after clearing of tropical rain forest in a Costa Rican soil, Trop. Agr., 67, 61–65, 1990.

Spaans, E. J. A., Baltissen, G. A. M., Bouma, J., Miedema, R., Lansu, A.L.E., Schoonderbeek, D., and Wielemaker, W. G.: Changes in physical properties of young and old vulcanic surface soils in Costa Rica after clearing of tropical rain forest, Hydrol. Process., 3, 383–392, 1989.

Stein, A., Hoogerwerf, M., and Bouma, J.: Use of soil-map delineations to improve (co) kriging of point data on moisture deficits, Geoderma, 43, 163–177, 1988.

Stein, A., Brouwer, J., and Bouma, J.: Methods for comparing spatial variability patterns of millet, yield and soil data, Soil Sci. Soc. Am. J., 61, 861–870, 1997.

Stoorvogel, J. J., Kooistra, L., and Bouma, J.: Spatial and temporal variation in nematocide leaching, management implications for a Costa Rica banana plantation, in: Assessment of non-point source pollution in the vadose zone, Geophysical Monograph 108, Am. Geophys. Un., 281–289, 1999.

Stoorvogel, J. J., Bouma, J., and Orlich, R. A.: Participatory research for systems analysis: prototyping for a Costa Rican Banana Plantation, Agron. J., 96, 323–336, 2004.

Van Camp, L., Bujarrabal, B., Gentile, A. R., Jones, R. J. A., Montanarella, L., Olazabal, C., and S-Kumar Selvaradjou (Eds.): Reports of the TechnicalWorking Groups established under the Thematic Strategy for Soil Protection, EUR 2131'9EN/6, Office for the official publications of the European Commuinities, Luxembourg, 2004.

van Groenigen, J. W., Gandah, M., and Bouma, J.: Soil sampling strategies for precision agriculture research under Sahalian conditions, Soil Sci. Soc. Am. J., 64, 1674–1680, 2000.

van Ittersum, M. K., van Bussel, L. G. J., Wolf, J., Guilport, N., Claessens, L., de Groot, H., Wiebe, K., Mason-D'Croz, D., Yang, H., Boogaard, H., van Oort, P. A. J., van Loon, M. P., Saito, K., Adimo, O., Adjei-Nsiah, S., Agali, A., Bala, A., Chicowo, R., Kaizzi, K., Kouressy, M., Makoi, J. H. R., Ouatarra, K., Tesfaye, K., and Cassman, K. G.: Can Sub-Saheran Africa feed itself?, P. Natl. Acad. Sci. USA, 113, 14964–14969, https://doi.org/10.1073/pnas.1610359113, 2016.

Veldkamp, E., Huising, J. E., Stein, A., and Bouma, J.: Variability of measured banana yields in a Costarican plantation as expressed by soil survey and thematic mapper data, Geoderma, 47, 337–349, 1990.

Wopereis, M. S. C., Stein, A., Bouma, J., and Woodhead, T.: Sampling number and design for measurements of infiltration rates into puddled rice fields, Agr. Water Manage., 22, 281–295, 1992.

Wopereis, M. C. S., Kropff, M. J., Wösten, J. H. M., and Bouma, J.: Sampling strategies for measurement of soil hydraulic properties to predict rice yield using simulation models, Geoderma, 59, 1–20, 1993.

Wopereis, M. C. S., Bouma, J., Kropff, M. J., and Sanidad, W.: Reducing bypass flow through a cracked, previously puddled clay soil, Soil Till. Res., 29, 1–11, 1994.

World Bank: Agriculture for Development, The World Bank, Washington D.C., USA, 2008.

Wulf, A.: The invention of nature. The life of Alexander von Humboldt, John Murray Publ., London, UK, 2015.

10

Deriving pedotransfer functions for soil quartz fraction in southern France from reverse modeling

Jean-Christophe Calvet, Noureddine Fritz, Christine Berne, Bruno Piguet, William Maurel, and Catherine Meurey

CNRM, UMR 3589 (Météo-France, CNRS), Toulouse, France

Correspondence to: Jean-Christophe Calvet (jean-christophe.calvet@meteo.fr)

Abstract. The quartz fraction in soils is a key parameter of soil thermal conductivity models. Because it is difficult to measure the quartz fraction in soils, this information is usually unavailable. This source of uncertainty impacts the simulation of sensible heat flux, evapotranspiration and land surface temperature in numerical simulations of the Earth system. Improving the estimation of soil quartz fraction is needed for practical applications in meteorology, hydrology and climate modeling. This paper investigates the use of long time series of routine ground observations made in weather stations to retrieve the soil quartz fraction. Profile soil temperature and water content were monitored at 21 weather stations in southern France. Soil thermal diffusivity was derived from the temperature profiles. Using observations of bulk density, soil texture, and fractions of gravel and soil organic matter, soil heat capacity and thermal conductivity were estimated. The quartz fraction was inversely estimated using an empirical geometric mean thermal conductivity model. Several pedotransfer functions for estimating quartz content from gravimetric or volumetric fractions of soil particles (e.g., sand) were analyzed. The soil volumetric fraction of quartz (f_q) was systematically better correlated with soil characteristics than the gravimetric fraction of quartz. More than 60 % of the variance of f_q could be explained using indicators based on the sand fraction. It was shown that soil organic matter and/or gravels may have a marked impact on thermal conductivity values depending on which predictor of f_q is used. For the grassland soils examined in this study, the ratio of sand-to-soil organic matter fractions was the best predictor of f_q, followed by the gravimetric fraction of sand. An error propagation analysis and a comparison with independent data from other tested models showed that the gravimetric fraction of sand is the best predictor of f_q when a larger variety of soil types is considered.

1 Introduction

Soil moisture is the main driver of temporal changes in values of the soil thermal conductivity (Sourbeer and Loheide II, 2015). The latter is a key variable in land surface models (LSMs) used in hydrometeorology or in climate models for the simulation of the vertical profile of soil temperature in relation to soil moisture (Subin et al., 2013). Shortcomings in soil thermal conductivity models tend to limit the impact of improving the simulation of soil moisture and snowpack in LSMs (Lawrence and Slater, 2008; Decharme et al., 2016). Models of the thermal conductivity of soils are affected by uncertainties, especially in the representation of the impact of soil properties such as the volumetric fraction of quartz

(f_q), soil organic matter and gravels (Farouki, 1986; Chen et al., 2012). As soil organic matter (SOM) and gravels are often neglected in LSMs, the soil thermal conductivity models used in most LSMs represent the mineral fine earth, only. Nowadays, f_q estimates are not given in global digital soil maps, and it is often assumed that this quantity is equal to the fraction of sand (Peters-Lidard et al., 1998).

Soil thermal properties are characterized by two key variables: the soil volumetric heat capacity (C_h) and the soil thermal conductivity (λ), in $J\,m^{-3}\,K^{-1}$ and $W\,m^{-1}\,K^{-1}$, respectively. Provided the volumetric fractions of moisture, minerals and organic matter are known, C_h can be calculated easily. The estimation of λ relies on empirical models and is affected by uncertainties (Peters-Lidard et al., 1998; Tarnawski et al., 2012). The construction and the verification of the λ models is not easy. The λ values of undisturbed soils are difficult to observe directly. They are often measured in the lab on perturbed soil samples (Abu-Hamdeh and Reeder, 2000; Lu et al., 2007). Although recent advances in line-source probe and heat pulse methods have made it easier to monitor soil thermal conductivity in the field (Bristow et al., 1994; Zhang et al., 2014), such measurements are currently not made in operational meteorological networks. Moreover, for given soil moisture conditions, λ depends to a large extent on the fraction of soil minerals presenting high thermal conductivities such as quartz, hematite, dolomite or pyrite (Côté and Konrad, 2005). In midlatitude regions of the world, quartz is the main driver of λ. The information on quartz fraction in a soil is usually unavailable as it can only be measured using X-ray diffraction (XRD) or X-ray fluorescence (XRF) techniques. These techniques are difficult to implement because the sensitivity to quartz is low. In practise, using XRD and XRF together is necessary to improve the accuracy of the measurements (Schönenberger et al., 2012). This lack of observations has a major effect on the accuracy of thermal conductivity models and their applications (Bristow, 1998).

Most of the land surface models (LSMs) currently used in meteorology and hydrometeorology simulate λ following the approach proposed by Peters-Lidard et al. (1998). This approach consists of an updated version of the Johansen (1975) model and assumes that the gravimetric fraction of quartz (Q) is equal to the gravimetric fraction of sand within mineral fine earth. This is a strong assumption, as some sandy soils (e.g., calcareous sands) may contain little quartz and as quartz may be found in the silt and clay fractions of the soil minerals (Schönenberger et al., 2012). Moreover, the λ models used in most LSMs represent only the mineral fine earth. Yang et al. (2005) and Chen et al. (2012) have shown the importance of accounting for SOM and gravels in λ models for organic top soil layers of grasslands of the Tibetan plateau.

The main goals of this study are to (1) assess the feasibility of using routine automatic soil temperature profile subhourly measurements (one observation every 12 min) to retrieve instantaneous soil thermal diffusivity values at a depth of 0.10 m; (2) retrieve instantaneous λ values from the soil thermal diffusivity estimates, accounting for the impact of soil vertical heterogeneities; (3) obtain, from reverse modeling, the quartz fraction together with soil thermal conductivity at saturation ($λ_{sat}$); (4) assess the impact of gravels and SOM on $λ_{sat}$; (5) derive pedotransfer functions for the soil quartz fraction.

For this purpose, we use the data from 21 weather stations of the Soil Moisture Observing System – Meteorological Automatic Network Integrated Application (SMOSMA-NIA) network (Calvet et al., 2007) in southern France. The soil temperature and the soil moisture probes are buried in the enclosure around each weather station. Most of these stations are located in agricultural areas. However, the vegetation

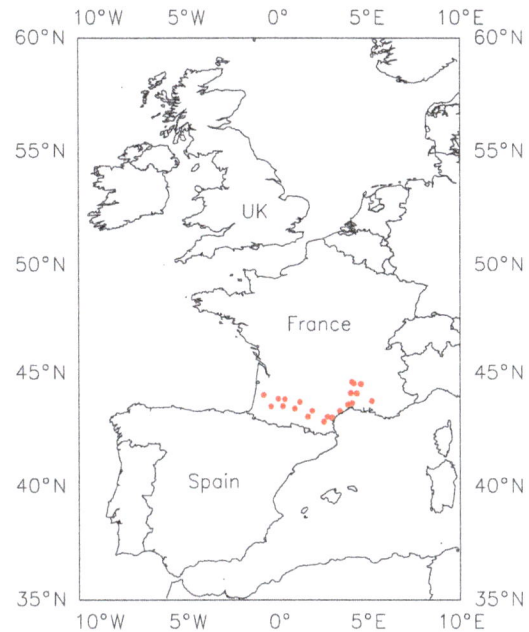

Figure 1. Location of the 21 SMOSMANIA stations in southern France (see station names in Supplement 1).

cover in the enclosure around the stations consists of grass. Along the Atlantic–Mediterranean transect formed by the SMOSMANIA network (Fig. 1), the grassland cover fraction ranges between 10 and 40 % (Zakharova et al., 2012). Various mineral soil types can be found along this transect, ranging from sand to clay and silt loam (see Supplement 1). During the installation of the probes, we collected soil samples which were used to determine soil characteristics: soil texture, soil gravel content, soil organic matter and bulk density.

Using this information together with soil moisture, λ values are derived from soil thermal diffusivity and heat capacity. The response of λ to soil moisture is investigated. The feasibility of modeling the λ value at saturation ($λ_{sat}$) with or without using SOM and gravel fraction observations is assessed using a geometric mean empirical thermal conductivity model based on Lu et al. (2007). The volumetric fraction of quartz, f_q, is retrieved by reverse modeling together with Q. Pedotransfer functions are further proposed for estimating quartz content from soil texture information.

The field data and the method to retrieve λ values are presented in Sect. 2. The λ and f_q retrievals are presented in Sect. 3 together with a sensitivity analysis of $λ_{sat}$ to SOM and gravel fractions. Finally, the results are discussed in Sect. 4, and the main conclusions are summarized in Sect. 5. Technical details are given in Supplement.

2 Data and methods

2.1 The SMOSMANIA data

The SMOSMANIA network was developed by Calvet et al. (2007) in southern France. The main purposes of SMOS-MANIA are to (1) validate satellite-derived soil moisture products (Parrens et al., 2012); (2) assess land surface models used in hydrological models (Draper et al., 2011) and in meteorological models (Albergel et al., 2010); and (3) monitor the impact of climate change on water resources and droughts (Laanaia et al., 2016). The station network forms a transect between the Atlantic coast and the Mediterranean sea (Fig. 1). It consists of preexisting automatic weather stations operated by Météo-France, upgraded with four soil moisture probes at four depths: 0.05, 0.10, 0.20 and 0.30 m. Twelve SMOSMANIA stations were activated in 2006 in southwestern France. In 2008, nine more stations were installed along the Mediterranean coast, and the whole network (21 stations) was gradually equipped with temperature sensors at the same depths as soil moisture probes. The soil moisture and soil temperature probes consisted of ThetaProbe ML2X and PT100 sensors, respectively. Soil moisture and soil temperature observations were made every 12 min at four depths. The soil temperature observations were recorded with a resolution of 0.1 °C.

In this study, the sub-hourly measurements of soil temperature and soil moisture at a depth of 0.10 m were used, together with soil temperature measurements at 0.05 and 0.20 m from 1 January 2008 to 30 September 2015.

The ThetaProbe soil moisture sensors provide a voltage signal (V). In order to convert the voltage signal into volumetric soil moisture content ($m^3 m^{-3}$), site-specific calibration curves were developed using in situ gravimetric soil samples for all stations and for all depths (Albergel et al., 2008). We revised the calibration in order to avoid spurious high soil moisture values during intense precipitation events. Logistics curves were used (see Supplement 1) instead of exponential curves in the previous version of the data set.

The observations from the soil moisture (48) and from the temperature (48) probes are automatically recorded every 12 min. The data are available to the research community through the International Soil Moisture Network web site (https://ismn.geo.tuwien.ac.at/).

Figure 2 shows soil temperature time series in wet conditions at various soil depths for a station presenting an intermediate value of λ_{sat} (Table 2) and of soil texture (see Fig. S1.1 in Supplement 1). The impact of recording temperature with a resolution of 0.1 °C is clearly visible at all depths as this causes a leveling of the curves.

2.2 Soil characteristics

In general, the stations are located on formerly cultivated fields and the soil in the enclosure around the stations is cov-

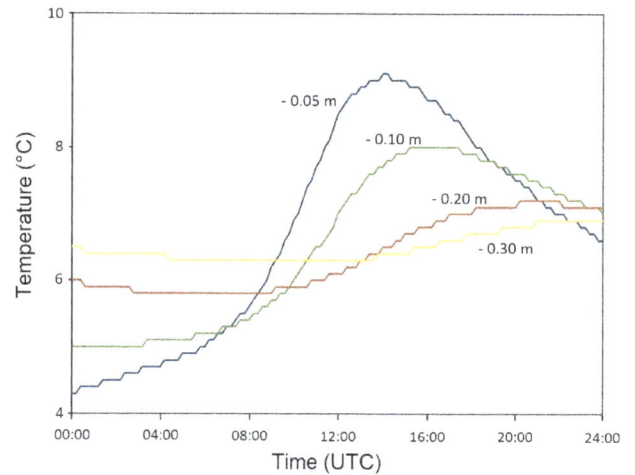

Figure 2. Soil temperature measured in wet conditions at the St-Félix-Lauragais (SFL) station on 23 February 2015 at depths of 0.05, 0.10, 0.20 and 0.30 m. Leveling is due to the low resolution of the temperature records (0.1 °C).

ered with grass. Soil properties were measured at each station by an independent laboratory we contracted (INRA-Arras) from soil samples we collected during the installation of the probes. The 21 stations cover a very large range of soil texture characteristics. For example, SBR is located on a sandy soil, PRD on a clay loam, and MNT on a silt loam (Table 1 and Supplement 1). Other properties such as the gravimetric fraction of SOM and of gravels were determined from the soil samples. Table 1 shows that 12 soils present a volumetric gravel content (f_{gravel}) larger than 15 %. Among these, three soils (at PRD, BRN and MJN) have f_{gravel} values larger than 30 %.

In addition, we measured bulk density (ρ_d) using undisturbed oven-dried soil samples we collected using metal cylinders of known volume (about $7 \times 10^{-4} m^3$; see Fig. S1.10 in the Supplement).

The porosity values at a depth of 0.10 m are listed in Table 1 together with gravimetric and volumetric fractions of soil particle-size ranges (sand, clay, silt, gravel) and SOM. The porosity, or soil volumetric moisture at saturation (θ_{sat}), is derived from the bulk dry density ρ_d, with soil texture and soil organic matter observations as

$$\theta_{sat} = 1 - \rho_d \left[\frac{m_{sand} + m_{clay} + m_{silt} + m_{gravel}}{\rho_{min}} + \frac{m_{SOM}}{\rho_{SOM}} \right]$$

or

$$\theta_{sat} = 1 - f_{sand} - f_{clay} - f_{silt} - f_{gravel} - f_{SOM}, \tag{1}$$

where m_x (f_x) represents the gravimetric (volumetric) fraction of the soil component x. The f_x values are derived from the measured gravimetric fractions, multiplied by the ratio of ρ_d observations to ρ_x, the density of each soil component x. Values of $\rho_{SOM} = 1300 kg m^{-3}$ and $\rho_{min} = 2660 kg m^{-3}$ are used for soil organic matter and soil minerals, respectively.

Table 1. Soil characteristics at 10 cm for the 21 stations of the SMOSMANIA network. Porosity values are derived from Eq. (1). Solid fraction values higher than 0.3 are in bold. The stations are listed from west to east (from top to bottom). ρ_d, θ_{sat}, f and m stand for soil bulk density, porosity, volumetric fractions and gravimetric fractions, respectively. Soil particle fractions larger than 0.3 are in bold. Full station names are given in Supplement 1 (Table S1.1).

Station	ρ_d (kg m^{-3})	θ_{sat} (m^3 m^{-3})	f_{sand} (m^3 m^{-3})	f_{clay} (m^3 m^{-3})	f_{silt} (m^3 m^{-3})	f_{gravel} (m^3 m^{-3})	f_{SOM} (m^3 m^{-3})	m_{sand} (kg kg^{-1})	m_{clay} (kg kg^{-1})	m_{silt} (kg kg^{-1})	m_{gravel} (kg kg^{-1})	m_{SOM} (kg kg^{-1})
SBR	1680	0.352	**0.576**	0.026	0.013	0.002	0.032	**0.911**	0.041	0.020	0.003	0.024
URG	1365	0.474	0.076	0.078	**0.341**	0.005	0.025	0.149	0.153	**0.665**	0.009	0.024
CRD	1435	0.438	**0.457**	0.027	0.033	0.000	0.045	**0.848**	0.051	0.060	0.000	0.041
PRG	1476	0.431	0.051	0.138	0.138	0.214	0.028	0.092	0.250	0.248	**0.385**	0.025
CDM	1522	0.413	0.073	0.241	0.231	0.012	0.030	0.128	**0.422**	**0.404**	0.020	0.026
LHS	1500	0.416	0.102	0.202	0.189	0.051	0.039	0.181	**0.359**	**0.335**	0.091	0.034
SVN	1453	0.445	0.127	0.073	0.176	0.162	0.017	0.233	0.133	**0.322**	0.296	0.015
MNT	1444	0.447	0.135	0.066	0.230	0.102	0.020	0.248	0.121	**0.424**	0.188	0.018
SFL	1533	0.413	0.127	0.071	0.118	0.250	0.021	0.221	0.123	0.205	**0.434**	0.018
MTM	1540	0.405	0.110	0.081	0.076	0.297	0.032	0.189	0.140	0.131	**0.512**	0.027
LZC	1498	0.429	0.129	0.066	0.068	0.292	0.015	0.229	0.117	0.121	**0.519**	0.013
NBN	1545	0.401	0.063	0.135	0.075	0.290	0.035	0.109	0.232	0.130	**0.499**	0.030
PZN	1311	0.495	0.222	0.074	0.131	0.054	0.023	**0.450**	0.151	0.266	0.111	0.023
PRD	1317	0.494	0.038	0.052	0.069	**0.326**	0.021	0.076	0.105	0.139	**0.659**	0.021
LGC	1496	0.428	0.253	0.044	0.042	0.214	0.019	**0.451**	0.078	0.074	**0.380**	0.017
MZN	1104	0.560	0.212	0.037	0.045	0.097	0.049	**0.510**	0.089	0.109	0.234	0.057
VLV	1274	0.506	0.294	0.054	0.086	0.031	0.029	**0.614**	0.112	0.179	0.064	0.030
BRN	1630	0.379	0.105	0.009	0.016	**0.474**	0.016	0.171	0.015	0.027	**0.774**	0.013
MJN	1276	0.506	0.064	0.029	0.056	**0.317**	0.028	0.133	0.060	0.118	**0.661**	0.029
BRZ	1280	0.508	0.097	0.074	0.109	0.190	0.020	0.202	0.154	0.228	**0.396**	0.021
CBR	1310	0.501	0.120	0.057	0.068	0.241	0.013	0.243	0.116	0.139	**0.489**	0.013

2.3 Retrieval of soil thermal diffusivity

The soil thermal diffusivity (D_h) is expressed in m^2 s^{-1} and is defined as

$$D_h = \frac{\lambda}{C_h}. \tag{2}$$

We used a numerical method to retrieve instantaneous values of D_h at a depth of 0.10 m using three soil temperature observations at 0.05 m, 0.10 and 0.20 m, performed every 12 min, by solving the Fourier thermal diffusion equation. The latter can be written as

$$C_h \frac{\partial T}{\partial t} = \frac{\partial}{\partial z}\left(\lambda \frac{\partial T}{\partial z}\right). \tag{3}$$

Given that soil properties are relatively homogeneous in the vertical (Sect. 2.1), values of D_h can be derived from the Fourier one-dimensional law:

$$\frac{\partial T}{\partial t} = D_h \frac{\partial^2 T}{\partial z^2}. \tag{4}$$

However, large differences in soil bulk density, from the top soil layer to deeper soil layers, were observed for some soils (see Supplement 1). In order to limit this effect as much as possible, we only used the soil temperature data presenting a relatively low vertical gradient close to the soil surface, where most differences with deeper layers are found. This data sorting procedure is described in Supplement 2.

Given that three soil temperatures T_i (i ranging from 1 to 3) are measured at depths $z_1 = -0.05$ m, $z_2 = -0.10$ m and

$z_3 = -0.20$ m, the soil diffusivity D_{hi} at $z_i = z_2 = -0.10$ m can be obtained by solving the one-dimensional heat equation, using a finite-difference method based on the implicit Crank–Nicolson scheme (Crank and Nicolson, 1996). When three soil depths are considered (z_{i-1}, z_i, z_{i+1}), the change in soil temperature T_i at depth z_i, from time t_{n-1} to time t_n, within the time interval $\Delta t = t_n - t_{n-1}$, can be written as

$$\frac{T_i^n - T_i^{n-1}}{\Delta t} = D_{hi}\left[\frac{1}{2}\left(\frac{\gamma_{i+1}^n - \gamma_i^n}{\Delta z_m}\right)\right.$$
$$\left. + \frac{1}{2}\left(\frac{\gamma_{i+1}^{n-1} - \gamma_i^{n-1}}{\Delta z_m}\right)\right], \text{ with}$$
$$\gamma_i^n = \frac{T_i^n - T_{i-1}^n}{\Delta z_i}, \quad \Delta z_m = \frac{\Delta z_i + \Delta z_{i+1}}{2}$$
$$\text{and } \Delta z_i = z_i - z_{i-1}. \tag{5}$$

In this study, $\Delta z_i = -0.05$ m, $\Delta z_{i+1} = -0.10$ m and a value of $\Delta t = 2880$ s (48 min) are used.

It is important to ensure that D_h retrievals are related to diffusion processes only and not to the transport of heat by water infiltration or evaporation (Parlange et al., 1998; Schelde et al., 1998). Therefore, only situations for which changes in soil moisture at all depths do not exceed 0.001 m^3 m^{-3} within the Δt time interval are considered.

2.4 From soil diffusivity to soil thermal conductivity

The observed soil properties and volumetric soil moisture are used to calculate the soil volumetric heat capacity C_h at a depth of 0.10 m, using the de Vries (1963) mixing model. The C_h values, in units of $\mathrm{J\,m^{-3}\,K^{-1}}$, are calculated as

$$C_h = \theta\,C_{h\,\mathrm{water}} + f_{\mathrm{min}}C_{h\,\mathrm{min}} + f_{\mathrm{SOM}}\,C_{h\,\mathrm{SOM}}, \tag{6}$$

where θ and f_{min} represent the volumetric soil moisture and the volumetric fraction of soil minerals, respectively. Values of 4.2×10^6, 2.0×10^6 and $2.5 \times 10^6 \mathrm{\,J\,m^{-3}\,K^{-1}}$ are used for $C_{h\,\mathrm{water}}$, $C_{h\,\mathrm{min}}$ and $C_{h\,\mathrm{SOM}}$, respectively.

The λ values at 0.10 m are then derived from the D_h and C_h estimates (Eq. 2).

2.5 Soil thermal conductivity model

Various approaches can be used to simulate thermal conductivity of unsaturated soils (Dong et al., 2015). We used an empirical approach based on thermal conductivity values in dry conditions and at saturation.

In dry conditions, soils present low thermal conductivity values (λ_{dry}). Experimental evidence shows that λ_{dry} is negatively correlated with porosity. For example, Lu et al. (2007) give

$$\lambda_{\mathrm{dry}} = 0.51 - 0.56 \times \theta_{\mathrm{sat}} \ (\mathrm{in\ W\,m^{-1}\,K^{-1}}). \tag{7}$$

When soil pores are gradually filled with water, λ tends to increase towards a maximum value at saturation (λ_{sat}). Between dry and saturation conditions, λ is expressed as

$$\lambda = \lambda_{\mathrm{dry}} + Ke\left(\lambda_{\mathrm{sat}} - \lambda_{\mathrm{dry}}\right), \tag{8}$$

where Ke is the Kersten number (Kersten, 1949). The latter is related to the volumetric soil moisture, θ, i.e., to the degree of saturation (S_d). We used the formula recommended by Lu et al. (2007):

$$Ke = \exp\left\{\alpha\left(1 - S_d^{(\alpha-1.33)}\right)\right\},$$

with $\alpha = 0.96$ for $\mathrm{Mn_{sand}} \geq 0.4\,\mathrm{kg\,kg^{-1}}$, $\alpha = 0.27$ for $\mathrm{Mn_{sand}} < 0.4\,\mathrm{kg\,kg^{-1}}$ and

$$S_d = \theta / \theta_{\mathrm{sat}}. \tag{9}$$

$\mathrm{Mn_{sand}}$ represents the sand mass fraction of mineral fine earth (values are given in Supplement 1).

The geometric mean equation for λ_{sat} proposed by Johansen (1975) for the mineral components of the soil can be generalized to include the SOM thermal conductivity (Chen et al., 2012) as

$$\ln(\lambda_{\mathrm{sat}}) = f_q \ln\left(\lambda_q\right) + f_{\mathrm{other}} \ln\left(\lambda_{\mathrm{other}}\right) + \theta_{\mathrm{sat}} \ln\left(\lambda_{\mathrm{water}}\right)$$
$$+ f_{\mathrm{SOM}} \ln\left(\lambda_{\mathrm{SOM}}\right), \tag{10}$$

where f_q is the volumetric fraction of quartz, and $\lambda_q = 7.7\,\mathrm{W\,m^{-1}\,K^{-1}}$, $\lambda_{\mathrm{water}} = 0.594\,\mathrm{W\,m^{-1}\,K^{-1}}$ and $\lambda_{\mathrm{SOM}} = 0.25\,\mathrm{W\,m^{-1}\,K^{-1}}$ are the thermal conductivities of quartz, water and SOM, respectively. The λ_{other}

term corresponds to the thermal conductivity of soil minerals other than quartz. Following Peters-Lidard et al. (1998), λ_{other} is taken as $2.0\,\mathrm{W\,m^{-1}\,K^{-1}}$ for soils with $\mathrm{Mn_{sand}} > 0.2\,\mathrm{kg\,kg^{-1}}$ and as $3.0\,\mathrm{W\,m^{-1}\,K^{-1}}$ otherwise. In this study, $\mathrm{Mn_{sand}} > 0.2\,\mathrm{kg\,kg^{-1}}$ for all soils, except for URG, PRG and CDM. The volumetric fraction of soil minerals other than quartz is defined as

$$f_{\mathrm{other}} = 1 - f_q - \theta_{\mathrm{sat}} - f_{\mathrm{SOM}}, \ \text{with}$$
$$f_q = Q \times (1 - \theta_{\mathrm{sat}}). \tag{11}$$

2.6 Reverse modeling

The λ_{sat} values are retrieved through reverse modeling using the λ model described above (Eqs. 7–11). This model is used to produce simulations of λ at the same soil moisture conditions as those encountered for the λ values derived from observations in Sect. 2.4. For a given station, a set of 401 simulations is produced for λ_{sat} ranging from 0 to $4\,\mathrm{W\,m^{-1}\,K^{-1}}$, with a resolution of $0.01\,\mathrm{W\,m^{-1}\,K^{-1}}$. The λ_{sat} retrieval corresponds to the λ simulation presenting the lowest root mean square difference (RMSD) value with respect to the λ observations. Only λ observations for S_d values higher than 0.4 are used because in dry conditions: (1) conduction is not the only mechanism for heat exchange in soils, as the convective water vapor flux may become significant (Schelde et al., 1998; Parlange et al., 1998); (2) the Ke functions found in the literature display more variability; and (3) the λ_{sat} retrievals are more sensitive to uncertainties in λ observations. The threshold value of $S_d = 0.4$ results from a compromise between the need of limiting the influence of convection, of the shape of the Ke function on the retrieved values of λ_{sat}, and of using as many observations as possible in the retrieval process. Moreover, the data filtering technique to limit the impact of soil heterogeneities, described in Supplement 2, is used to select valid λ observations.

Finally, the f_q value is derived from the retrieved λ_{sat} solving Eq. (10).

2.7 Scores

Pedotransfer functions for quartz and λ_{sat} are evaluated using the following scores:

- the Pearson correlation coefficient (r) and the squared correlation coefficient (r^2) are used to assess the fraction of explained variance

- the RMSD

- the mean absolute error (MAE), i.e., the mean of absolute differences

- the mean bias, i.e., the mean of differences.

Table 2. Thermal properties of 14 grassland soils in southern France: λ_{sat}, f_{q} and Q retrievals using the λ model (Eqs. 7–9 and 10) for degree of saturation values higher than 0.4, together with the minimized RMSD between the simulated and observed λ values and the number of used λ observations (n). The soils are sorted from the largest to the smallest ratio of m_{sand} to m_{SOM}. Full station names are given in Supplement 1 (Table S1.1).

Station	λ_{sat} $(\mathrm{W\,m^{-1}\,K^{-1}})$	RMSD $(\mathrm{W\,m^{-1}\,K^{-1}})$	n	f_{q} $(\mathrm{m^3\,m^{-3}})$	Q $(\mathrm{kg\,kg^{-1}})$	$\frac{m_{\mathrm{sand}}}{m_{\mathrm{SOM}}}$
SBR	2.80	0.255	6	0.62	0.96	37.2
LGC	2.07	0.311	20	0.44	0.77	26.6
CBR	1.92	0.156	20	0.44	0.88	18.4
LZC	1.71	0.107	20	0.29	0.51	17.3
SVN	1.78	0.163	20	0.34	0.61	15.4
MNT	1.96	0.058	20	0.42	0.76	13.8
BRN	1.71	0.131	20	0.25	0.40	13.5
SFL	1.57	0.134	20	0.22	0.37	12.5
MTM	1.52	0.095	20	0.21	0.35	7.0
URG	1.37	0.066	20	0.05	0.10	6.2
LHS	1.57	0.136	20	0.26	0.45	5.3
CDM	1.82	0.086	20	0.26	0.44	5.0
PRG	1.65	0.086	20	0.18	0.32	3.7
PRD	1.26	0.176	20	0.14	0.28	3.7

In order to test the predictive and generalization power of the pedotransfer regression equations, a simple bootstrapping resampling technique is used. It consists of calculating a new estimate of f_{q} for each soil using the pedotransfer function obtained without using this specific soil. Gathering these new f_{q} estimates, one can calculate new scores with respect to the retrieved f_{q} values. Also, this method provides a range of possible values of the coefficients of the pedotransfer function and permits assessing the influence of a given f_{q} retrieval on the final result.

3 Results

3.1 λ_{sat} and f_{q} retrievals

Retrievals of λ_{sat} and f_{q} could be obtained for 14 soils. Figure 3 shows retrieved and modeled λ values against the observed degree of saturation of the soil, at a depth of $0.10\,\mathrm{m}$ for contrasting retrieved values of λ_{sat}, from high to low values (2.80, 1.96, 1.52 and $1.26\,\mathrm{W\,m^{-1}\,K^{-1}}$) at the SBR, MNT, MTM and PRD stations, respectively.

All the obtained λ_{sat} and f_{q} retrievals are listed in Table 2, together with the λ RMSD values and the number of selected λ observations. For three soils (CRD, MZN and VLV), the reverse modeling technique described in Sect. 2.6 could not be applied as not enough λ observations could be obtained for S_{d} values higher than 0.4. For four soils (NBN, PZN, BRZ and MJN), all the λ retrievals were filtered out as the obtained values were influenced by heterogeneities in soil density (see Supplement 2). For the other 14 soils, λ_{sat} and f_{q} retrievals were obtained using a subset of 20 λ retrievals per soil, at most, corresponding to the soil temperature data pre-

senting the lowest vertical gradient close to the soil surface (Supplement 2).

3.2 Pedotransfer functions for quartz

The f_{q} retrievals can be used to assess the possibility of estimating f_{q} using other soil characteristics, which can be easily measured. Another issue is whether volumetric or gravimetric fraction of quartz should be used. Figure 4 presents the fraction of variance (r^2) of Q and f_{q} explained by various indicators. A key result is that f_{q} is systematically better correlated with soil characteristics than Q. More than 60 % of the variance of f_{q} can be explained using indicators based on the sand fraction (either f_{sand} or m_{sand}). The use of other soil mineral fractions does not give good correlations, even when they are associated to the sand fraction as shown by Fig. 4. For example, the f_{gravel} and $f_{\mathrm{gravel}} + f_{\mathrm{sand}}$ indicators present low r^2 values of 0.04 and 0.24, respectively.

The f_{q} values cannot be derived directly from the indicators as illustrated by Fig. 5: assuming $f_{\mathrm{q}} = f_{\mathrm{sand}}$ tends to markedly underestimate λ_{sat}. Therefore, more elaborate pedotransfer equations are needed. They can be derived from the best indicators, using them as predictors of f_{q}. The modeled f_{q} is written as

$$f_{\mathrm{qMOD}} = a_0 + a_1 \times P \quad \text{and}$$
$$f_{\mathrm{qMOD}} \leq 1 - \theta_{\mathrm{sat}} - f_{\mathrm{SOM}}, \tag{12}$$

where P represents the predictor of f_{q}.

The a_0 and a_1 coefficients are given in Table 3 for four pedotransfer functions based on the best predictors of f_{q}. The pedotransfer functions are illustrated in Fig. 6. The scores are displayed in Table 4. The bootstrapping indicates that the

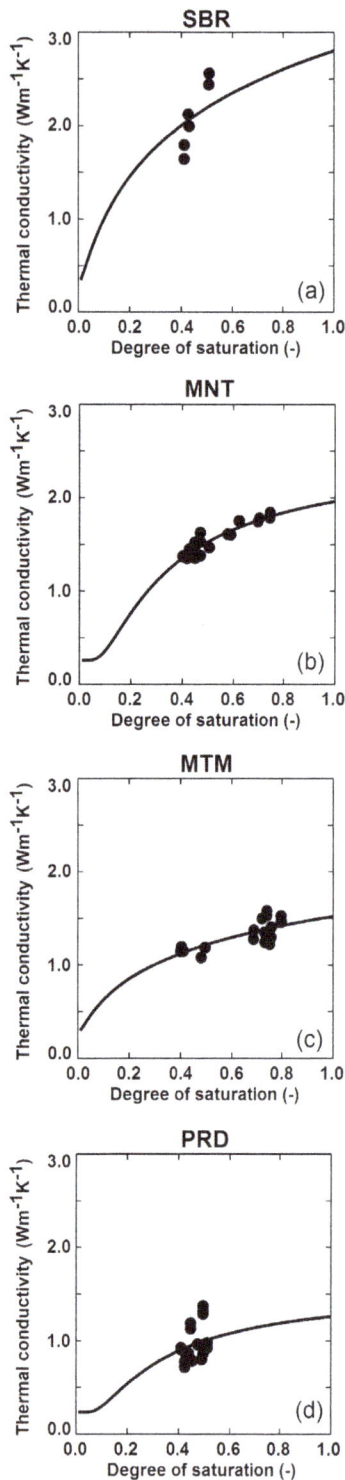

Figure 3. Retrieved λ values (dark dots) vs. the observed degree of saturation of the soil at a depth of $0.10\,\text{m}$ for (from top to bottom) Sabres (SBR), Montaut (MNT), Mouthoumet (MTM) and Prades-le-Lez (PRD), together with simulated λ values from dry to wet conditions (dark lines).

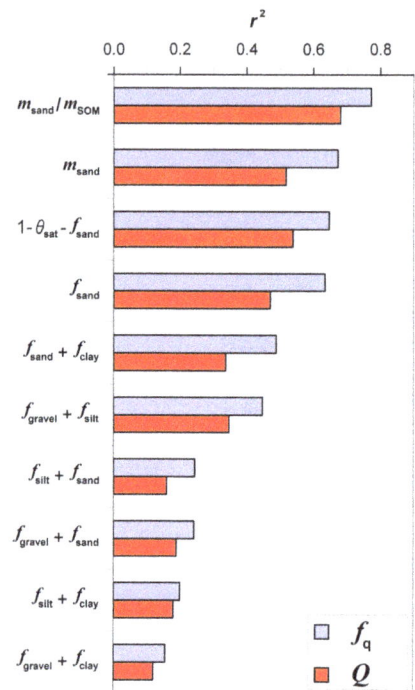

Figure 4. Fraction of variance (r^2) of gravimetric and volumetric fraction of quartz (Q and f_q, red and blue bars, respectively) explained by various predictors.

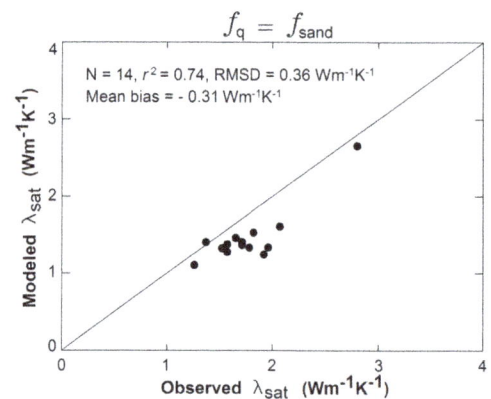

Figure 5. $\lambda_{\text{sat MOD}}$ values derived from volumetric quartz fractions f_q assumed equal to f_sand, using observed θ_sat values, vs. λ_sat retrievals.

SBR sandy soil has the largest individual impact on the obtained regression coefficients. This is why the scores without SBR are also presented in Table 4.

For the m_sand predictor, an r^2 value of 0.56 is obtained without SBR against a value of 0.67 when all the 14 soils are considered. An alternative to this m_sand pedotransfer function consists of considering only m_sand values smaller than $0.6\,\text{kg}\,\text{kg}^{-1}$ in the regression, thus excluding the SBR soil. The corresponding predictor is called m_sand^*. In this configuration, the sensitivity of f_q to m_sand is much increased (with

Table 3. Coefficients of four pedotransfer functions of f_q (Eq. 12) for 14 soils of this study (all with $m_{sand}/m_{SOM} < 40$), together with indicators of the coefficient uncertainty, derived by bootstrapping and by perturbing the volumetric heat capacity of soil minerals ($C_{h\,min}$). The best predictor is in bold.

Predictor of f_q	Coefficients for 14 soils		Confidence interval from bootstrapping		Impact of a change of $\pm 0.08 \times 10^6\,\mathrm{J\,m^{-3}\,K^{-1}}$ in $C_{h\,min}$	
	a_0	a_1	a_0	a_1	a_0	a_1
$\boldsymbol{m_{sand}/m_{SOM}}$	0.12	0.0134	[0.10, 0.14]	[0.012, 0.014]	[0.11, 0.13]	[0.013, 0.013]
m_{sand}^*	0.08	0.944	[0.00, 0.11]	[0.85, 1.40]	[0.07, 0.09]	[0.919, 0.966]
m_{sand}	0.15	0.572	[0.08, 0.17]	[0.54, 0.94]	[0.14, 0.17]	[0.55, 0.56]
$1 - \theta_{sat} - f_{sand}$	0.73	-1.020	[0.71, 0.89]	[-1.38, -0.99]	[0.70, 0.73]	[-1.00, -0.99]

* Only m_{sand} values smaller than $0.6\,\mathrm{kg\,kg^{-1}}$ are used in the regression.

Table 4. Scores of four pedotransfer functions of f_q for 14 soils of this study, together with the scores obtained by bootstrapping, without the sandy SBR soil. The MAE score of these pedotransfer functions for three Chinese soils of Lu et al. (2007) for which $m_{sand}/m_{SOM} < 40$ is given (within brackets and in italics). The best predictor and the best scores are in bold.

Predictor of f_q	Regression scores			Bootstrap scores			Scores without SBR (and MAE for three Lu soils)		
	r^2	RMSD $(\mathrm{m^3\,m^{-3}})$	MAE $(\mathrm{m^3\,m^{-3}})$	r^2	RMSD $(\mathrm{m^3\,m^{-3}})$	MAE $(\mathrm{m^3\,m^{-3}})$	r^2	RMSD $(\mathrm{m^3\,m^{-3}})$	MAE $(\mathrm{m^3\,m^{-3}})$
$\boldsymbol{m_{sand}/m_{SOM}}$	**0.77**	**0.067**	0.053	**0.72**	**0.074**	**0.059**	**0.62**	**0.070**	0.057 *(0.135)*
m_{sand}^*	0.74	0.072	**0.052**	0.67	0.126	0.100	0.56	0.075	**0.056** *(0.071)*
m_{sand}	0.67	0.081	0.060	0.56	0.121	0.084	0.56	0.075	**0.056** *(0.086)*
$1 - \theta_{sat} - f_{sand}$	0.65	0.084	0.064	0.56	0.102	0.079	0.45	0.084	0.061 *(0.158)*

* Only m_{sand} values smaller than $0.6\,\mathrm{kg\,kg^{-1}}$ are used in the regression.

$a_1 = 0.944$, against $a_1 = 0.572$ with SBR). For SBR, f_q is overestimated by the m_{sand}^* equation, but this is corrected by the f_{qMOD} limitation of Eq. (12), and in the end a better r^2 score is obtained when the 14 soils are considered ($r^2 = 0.74$).

Values of r^2 larger than 0.7 are obtained for two predictors of f_q: m_{sand}/m_{SOM} and m_{sand}^*. A value of $r^2 = 0.65$ is obtained for $1 - \theta_{sat} - f_{sand}$ (the fraction of soil solids other than sand). The m_{sand}/m_{SOM} predictor presents the best r^2 and RMSD scores in all the configurations (regression, bootstrap and regression without SBR). Another characteristic of the m_{sand}/m_{SOM} pedotransfer function is that the confidence interval for the a_0 and a_1 coefficients derived from bootstrapping is narrower than for the other pedotransfer functions (Table 3), indicating a more robust relationship of f_q with m_{sand}/m_{SOM} than with other predictors.

An alternative way to evaluate the quartz pedotransfer functions is to compare the simulated λ_{sat} with the retrieved values presented in Table 2. Modeled values of λ_{sat} ($\lambda_{sat\,MOD}$) can be derived from f_{qMOD} using Eq. (10) together with θ_{sat}

observations. The $\lambda_{sat\,MOD} r^2$, RMSD and mean bias scores are given in Table 5. Again, the best scores are obtained using the m_{sand}/m_{SOM} predictor of f_q, with r^2, RMSD and mean bias values of 0.86, 0.14 $\mathrm{W\,m^{-1}\,K^{-1}}$ and +0.01 $\mathrm{W\,m^{-1}\,K^{-1}}$, respectively (Fig. 7).

Finally, we investigated the possibility of estimating θ_{sat} from the soil characteristics listed in Table 1 and of deriving a statistical model for θ_{sat} (θ_{satMOD}). We found the following statistical relationship between θ_{satMOD}, m_{clay}, m_{silt} and m_{SOM}:

$$\theta_{satMOD} = 0.456 - 0.0735 \frac{m_{clay}}{m_{silt}} + 2.238\, m_{SOM} \qquad (13)$$

($r^2 = 0.48$, F test p value $= 0.0027$, RMSD $= 0.036\,\mathrm{m^3\,m^{-3}}$).

Volumetric fractions of soil components need to be consistent with θ_{satMOD} and can be calculated using the modeled bulk density values derived from θ_{satMOD} using Eq. (1).

Equations (10) to (13) constitute an empirical end-to-end model of λ_{sat}. Table 5 shows that using θ_{satMOD} (Eq. 13) in-

Table 5. Ability of the Eqs. (10)–(13) empirical model to estimate λ_{sat} values for 14 soils and impact of changes in gravel and SOM volumetric content: $f_{gravel} = 0\,m^3\,m^{-3}$ and $f_{SOM} = 0.013\,m^3\,m^{-3}$ (the smallest f_{SOM} value, observed for CBR). r^2 values smaller than 0.60, RMSD values higher than $0.20\,W\,m^{-1}\,K^{-1}$ and mean bias values higher (smaller) than $+0.10$ (-0.10) are in bold.

Model configuration	Predictor of f_q	r^2	RMSD $(W\,m^{-1}\,K^{-1})$	Mean bias $(W\,m^{-1}\,K^{-1})$
Model using θ_{sat} observations	m_{sand}/m_{SOM}	0.86	0.14	+0.01
	m^*_{sand}	0.83	0.15	−0.01
	m_{sand}	0.81	0.16	−0.03
	$1 - \theta_{sat} - f_{sand}$	0.82	0.16	−0.03
Full model using θ_{satMOD} (Eq. 13)	m_{sand}/m_{SOM}	0.85	0.14	+0.03
	m^*_{sand}	0.85	0.14	−0.03
	m_{sand}	0.84	0.15	−0.03
	$1 - \theta_{sat} - f_{sand}$	0.82	0.16	−0.02
Same with $f_{SOM} = 0.013\,m^3\,m^{-3}$	m_{sand}/m_{SOM}	**0.57**	**0.35**	**+0.20**
	m^*_{sand}	0.83	0.15	+0.00
	m_{sand}	0.81	0.16	−0.02
	$1 - \theta_{sat} - f_{sand}$	0.83	0.15	−0.02
Same with $f_{gravel} = 0\,m^3\,m^{-3}$	m_{sand}/m_{SOM}	0.87	0.19	**−0.12**
	m^*_{sand}	0.70	**0.23**	**+0.11**
	m_{sand}	0.79	0.17	+0.04
	$1 - \theta_{sat} - f_{sand}$	0.81	0.17	+0.05
Same with $f_{SOM} = 0.013\,m^3\,m^{-3}$ and $f_{gravel} = 0\,m^3\,m^{-3}$	m_{sand}/m_{SOM}	0.63	**0.31**	**+0.16**
	m^*_{sand}	0.52	**0.36**	**+0.24**
	m_{sand}	0.59	**0.29**	**+0.16**
	$1 - \theta_{sat} - f_{sand}$	0.70	**0.25**	**+0.16**

* Only m_{sand} values smaller than $0.6\,kg\,kg^{-1}$ are used in the regression.

stead of the θ_{sat} observations has little impact on the $\lambda_{sat\,MOD}$ scores.

3.3   Impact of gravels and SOM on λ_{sat}

Gravels and SOM are often neglected in soil thermal conductivity models used in LSMs. The Eqs. (10)–(13) empirical model obtained in Sect. 3.2 permits the assessment of the impact of f_{gravel} and f_{SOM} on λ_{sat}. Table 5 shows the impact on $\lambda_{sat\,MOD}$ scores of imposing a null value of f_{gravel} and a small value of f_{SOM} to all the soils. The combination of these assumptions is evaluated, also.

Imposing $f_{SOM} = 0.013\,m^3\,m^{-3}$ (the smallest f_{SOM} value, observed for CBR) has a limited impact on the scores, except for the m_{sand}/m_{SOM} pedotransfer function. In this case, λ_{sat} is overestimated by $+0.20\,W\,m^{-1}\,K^{-1}$ and r^2 drops to 0.57.

Neglecting gravels ($f_{gravel} = 0\,m^3\,m^{-3}$) also has a limited impact but triggers the underestimation (overestimation) of λ_{sat} for the m_{sand}/m_{SOM} (m^*_{sand}) pedotransfer function by $-0.12\,W\,m^{-1}\,K^{-1}$ ($+0.11\,W\,m^{-1}\,K^{-1}$).

On the other hand, it appears that combining these assumptions has a marked impact on all the pedotransfer functions. Neglecting gravels and imposing $f_{SOM} = 0.013\,m^3\,m^{-3}$ has a major impact on λ_{sat}: the modeled λ_{sat} is overestimated by all the pedotransfer functions (with a mean bias rang-

ing from $+0.16$ to $+0.24\,W\,m^{-1}\,K^{-1}$) and r^2 is markedly smaller, especially for the m_{sand} and m^*_{sand} pedotransfer functions. These results are illustrated in Fig. 8 in the case of the m^*_{sand} pedotransfer function. Figure 8 also shows that using the θ_{sat} observations instead of θ_{satMOD} (Eq. 13) has little impact on $\lambda_{sat\,MOD}$ (Sect. 3.2) but tends to enhance the impact of neglecting gravels. A similar result is found with the m_{sand} pedotransfer function (not shown).

4   Discussion

4.1   Can uncertainties in heat capacity estimates impact retrievals ?

In this study, the de Vries (1963) mixing model is applied to estimate soil volumetric heat capacity (Eq. 6), and a fixed value of $2.0 \times 10^6\,J\,m^{-3}\,K^{-1}$ is used for soil minerals. Soil-specific values for $C_{h\,min}$ may be more appropriate than using a constant standard value. For example, Tarara and Ham (1997) used a value of $1.92 \times 10^6\,J\,m^{-3}\,K^{-1}$. However, we did not measure this quantity and we were not able to find such values in the literature.

We investigated the sensitivity of our results to these uncertainties, considering the following minimum and maximum $C_{h\,min}$ values: $C_{h\,min} = 1.92 \times 10^6\,J\,m^{-3}\,K^{-1}$ and $C_{h\,min} = 2.08 \times 10^6\,J\,m^{-3}\,K^{-1}$. The impact of changes in

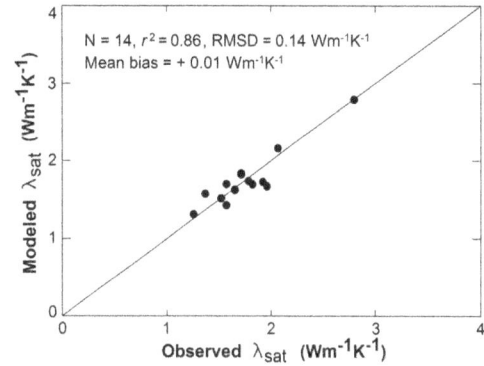

Figure 7. $\lambda_{sat\,MOD}$ values derived from the m_{sand}/m_{SOM} pedotransfer function for the volumetric quartz fractions, using observed θ_{sat} values, vs. λ_{sat} retrievals.

$C_{h\,min}$ on the retrieved values of λ_{sat} and f_q is presented in Supplement 3 (Fig. S3.1). On average, a change of $+ (-)$ $0.08 \times 10^6 \, J \, m^{-3} \, K^{-1}$ in $C_{h\,min}$ triggers a change in λ_{sat} and f_q of $+1.7\%$ (-1.8%) and $+4.8\%$ (-7.0%), respectively.

The impact of changes in $C_{h\,min}$ on the regression coefficients of the pedotransfer functions is presented in Table 3 (last column). The impact is very small, except for the a_1 coefficient of the m_{sand}^* pedotransfer function. However, even in this case, the impact of $C_{h\,min}$ on the a_1 coefficient is much lower than the confidence interval given by the bootstrapping, indicating that the relatively small number of soils we considered (as in other studies, e.g., Lu et al., 2007) is a larger source of uncertainty.

Moreover, uncertainties in the f_{clay}, f_{silt}, f_{gravel} or f_{SOM} fractions may be caused by (1) the natural heterogeneity of soil properties, (2) the living root biomass or (3) stones that may not be accounted for in the gravel fraction.

In particular, during the installation of the probes, it was observed that stones are present at some stations. Stones are not evenly distributed in the soil, and it is not possible to investigate whether the soil area where the temperature probes were inserted contains stones as it must be left undisturbed.

The grasslands considered in this study are not intensively managed. They consist of set-aside fields cut once or twice a year. Calvet et al. (1999) gave an estimate of $0.160 \, kg \, m^{-2}$ for the root dry matter content of such soils for a site in southwestern France, with most roots contained in the 0.25 m top soil layer. This represents a gravimetric fraction of organic matter smaller than $0.0005 \, kg \, kg^{-1}$, i.e., less than 4 % of the lowest m_{SOM} values observed in this study ($0.013 \, kg \, kg^{-1}$) or less than 5 % of f_{SOM} values. We checked that increasing f_{SOM} values by 5 % has negligible impact on heat capacity and on the λ retrievals.

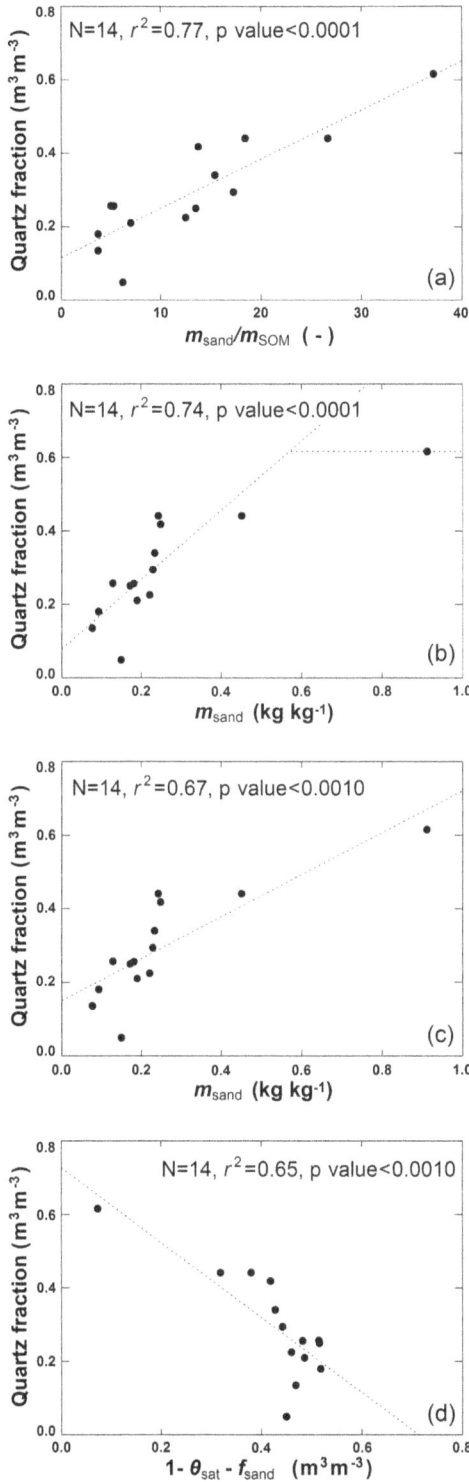

Figure 6. Pedotransfer functions for quartz: f_q retrievals (dark dots) vs. the four predictors of f_q given in Table 3. The modeled f_q values are represented by the dashed lines.

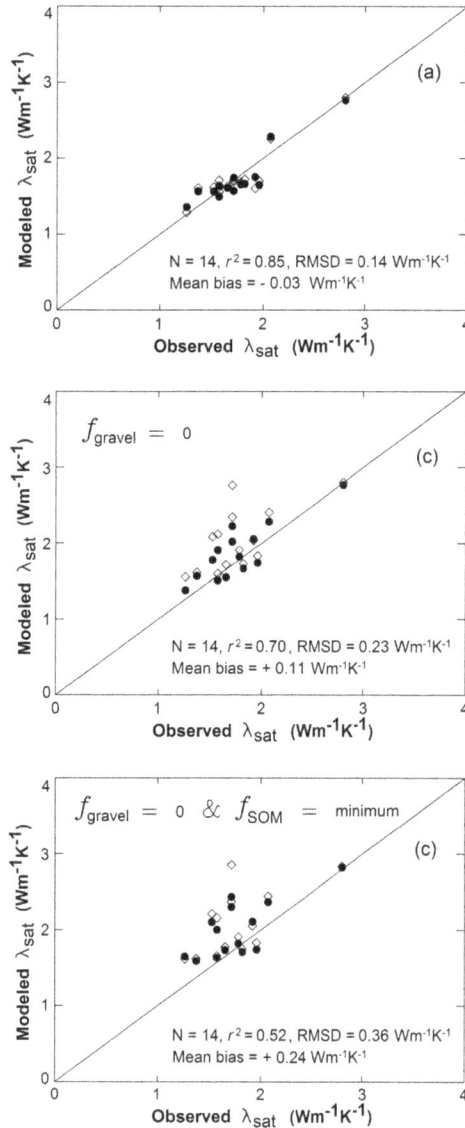

Figure 8. λ_{satMOD} values derived from the m^*_{sand} pedotransfer function for the volumetric quartz fractions, using θ_{satMOD} (Eq. 13) or the observed θ_{sat} (dark dots and open diamonds, respectively), vs. λ_{sat} retrievals: (top) full model, (middle) $f_{SOM} = 0.013\,\mathrm{m^3\,m^{-3}}$, and (bottom) $f_{SOM} = 0.013\,\mathrm{m^3\,m^{-3}}$ and $f_{gravel} = 0\,\mathrm{m^3\,m^{-3}}$. Scores are given for the θ_{satMOD} configuration.

4.2 Can the new λ_{sat} model be applied to other soil types?

The λ_{sat} values we obtained are consistent with values reported by other authors. In this study, λ_{sat} values ranging between 1.26 and 2.80 $\mathrm{W\,m^{-1}\,K^{-1}}$ are found (Table 2). Tarnawski et al. (2011) gave λ_{sat} values ranging between 2.5 and 3.5 $\mathrm{W\,m^{-1}\,K^{-1}}$ for standard sands. Lu et al. (2007) gave λ_{sat} values ranging between 1.33 and 2.2 $\mathrm{W\,m^{-1}\,K^{-1}}$.

A key component of the λ_{sat} model is the pedotransfer function for quartz (Eq. 12). The f_q pedotransfer functions

we propose are based on available soil characteristics. The current global soil digital maps provide information about SOM, gravels and bulk density (Nachtergaele et al., 2012). Therefore, using Eqs. (1) and (6)–(12) on a large scale is possible, and porosity can be derived from Eq. (1). On the other hand, the suggested f_q pedotransfer functions are obtained for temperate grassland soils containing a rather large amount of organic matter and are valid for m_{sand}/m_{SOM} ratio values lower than 40 (Table 2). These equations should be evaluated for other regions. In particular, hematite has to be considered together with quartz for tropical soils (Churchman and Lowe, 2012). Moreover, the pedotransfer function we get for θ_{sat} (Eq. 13) and we use to conduct the sensitivity study of Sect. 3.3 is valid for the specific sites we considered. Equation (13) cannot be used to predict porosity in other regions.

In order to assess the applicability of the pedotransfer function for quartz obtained in this study, we used the independent data from Lu et al. (2007) and Tarnawski et al. (2009) for 10 Chinese soils (see Supplement 4 and Table S4.1). These soils consist of reassembled sieved soil samples and contain no gravel, while our data concern undisturbed soils. Moreover, most of these soils contain very little organic matter and the m_{sand}/m_{SOM} ratio can be much larger that the m_{sand}/m_{SOM} values measured at our grassland sites. For the 14 French soils used to determine pedotransfer functions for quartz, the m_{sand}/m_{SOM} ratio ranges from 3.7 to 37.2 (Table 2). Only three soils of Lu et al. (2007) present such low values of m_{sand}/m_{SOM}. The other seven soils of Lu et al. (2007) present m_{sand}/m_{SOM} values ranging from 48 to 1328 (see Table S4.1).

We used λ_{sat} experimental values derived from Table 3 in Tarnawski et al. (2009) to calculate Q and f_q for the 10 Lu et al. (2007) soils. These data are presented in Supplement 4. Figure S4.1 shows the statistical relationship between these quantities and m_{sand}. Very good correlations of Q and f_q with m_{sand} are observed, with r^2 values of 0.72 and 0.83, respectively. This is consistent with our finding that f_q is systematically better correlated with soil characteristics than Q (Sect. 3.2).

The pedotransfer functions derived from French soils tend to overestimate f_q for the Lu et al. (2007) soils, especially for the seven soils presenting m_{sand}/m_{SOM} values larger than 40. Note that Lu et al. (2007) obtained a similar result for coarse-textured soils with their model, which assumed $Q = m_{sand}$. For the three other soils, presenting m_{sand}/m_{SOM} values smaller than 40, f_q MAE values are given in Table 4. The best MAE score (0.071 $\mathrm{m^3\,m^{-3}}$) is obtained for the m^*_{sand} predictor of f_q.

These results are illustrated by Fig. 9 for the m_{sand} predictor of f_q. Figure 9 also shows the f_q and λ_{sat} estimates obtained using specific coefficients in Eq. (12), based on the seven Lu et al. (2007) soils presenting m_{sand}/m_{SOM} values larger than 40. These coefficients are given together with the scores in Table 6. Table 6 also presents these values for other

Table 6. Pedotransfer functions of f_q (Eq. 12) for seven soils of Lu et al. (2007) with $m_{sand}/m_{SOM} > 40$. The best predictor and the best scores are in bold. The regression p values are within brackets and in italics.

Predictor of f_q	Regression scores for seven Lu soils with $m_{sand}/m_{SOM} > 40$			Coefficients	
	r^2 (p value)	RMSD ($m^3\,m^{-3}$)	MAE ($m^3\,m^{-3}$)	a_0	a_1
m_{sand}/m_{SOM}	0.40 (0.13)	0.089	0.075	0.20	0.000148
m^*_{sand}	0.82 (0.005)	0.073	0.054	0.07	0.425
m_{sand}	**0.82** (0.005)	**0.048**	**0.042**	0.04	0.386
$1 - \theta_{sat} - f_{sand}$	0.81 (0.006)	0.050	0.043	0.44	−0.814

* Only m_{sand} values smaller than $0.6\,kg\,kg^{-1}$ are used in the regression.

predictors of f_q. It appears that m_{sand} gives the best scores. The contrasting coefficient values between Tables 6 and 3 (Chinese and French soils, respectively) illustrate the variability of the coefficients of pedotransfer functions from one soil category to another, and the m_{sand}/m_{SOM} ratio seems to be a good indicator of the validity of a given pedotransfer function.

On the other hand, the m_{sand}/m_{SOM} ratio is not a good predictor of f_q for the Lu et al. (2007) soils presenting m_{sand}/m_{SOM} values larger than 40, and r^2 presents a small value of 0.40 (Table 6). This can be explained by the very large range of m_{sand}/m_{SOM} values for these soils (see Table S4.1). Using $\ln(m_{sand}/m_{SOM})$ instead of m_{sand}/m_{SOM} is a way to obtain a predictor linearly correlated with f_q. This is shown by Fig. S4.2 for the 10 Lu et al. (2007) soils: the correlation is increased to a large extent ($r^2 = 0.60$).

4.3 Can m_{sand}-based f_q pedotransfer functions be used across soil types?

Given the results presented in Tables 3, 4 and 6, it can be concluded that m_{sand} is the best predictor of f_q across mineral soil types. The m_{sand}/m_{SOM} predictor is relevant for the mineral soils containing the largest amount of organic matter.

Although the m_{sand}/m_{SOM} predictor gives the best r^2 scores for the 14 grassland soils considered in this study, it seems more difficult to apply this predictor to other soils, as shown by the high MAE score ($MAE = 0.135\,m^3\,m^{-3}$) for the corresponding Lu et al. (2007) soils in Table 4. Moreover, the scores are very sensitive to errors in the estimation of m_{SOM} as shown by Table 5. Although the m^*_{sand} predictor gives slightly better scores than m_{sand} (Table 4), the a_1 coefficient in more sensitive to errors in $C_{h\,min}$ (Table 3), and the bootstrapping reveals large uncertainties in a_0 and a_1 values.

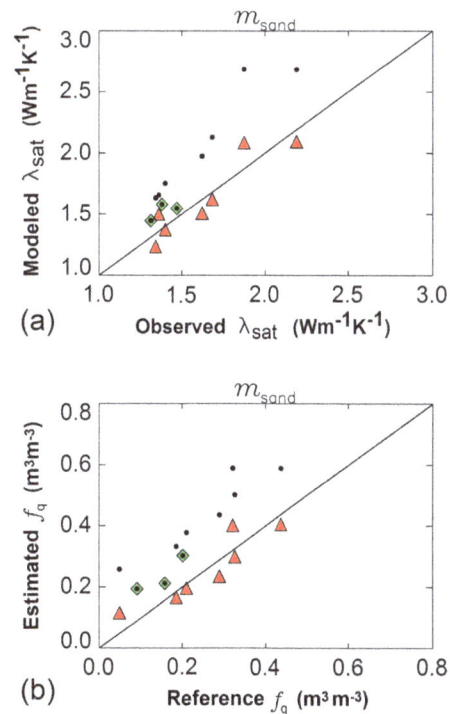

Figure 9. Estimated λ_{sat} and volumetric fraction of quartz f_q (top and bottom, respectively) vs. values derived from the λ_{sat} observations of Lu et al. (2007) given by Tarnawski et al. (2009) for 10 Chinese soils, using the gravimetric fraction of sand m_{sand} as a predictor of f_q. Dark dots correspond to the estimations obtained using the m_{sand} pedotransfer function for southern France, and the three soils for which $m_{sand}/m_{SOM} < 40$ are indicated by green diamonds. Red triangles correspond to the estimations obtained using the m_{sand} pedotransfer function for the seven soils for which $m_{sand}/m_{SOM} > 40$ (see Table 6).

The results presented in this study suggest that the m_{sand}/m_{SOM} ratio can be used to differentiate between temperate grassland soils containing a rather large amount of organic matter ($3.7 < m_{sand}/m_{SOM} < 40$) and soils containing less organic matter ($m_{sand}/m_{SOM} > 40$). The m_{sand} predictor can be used in both cases to estimate the volumetric fraction of quartz, with the following a_0 and a_1 coefficient values in Eq. (12): 0.15 and 0.572 for m_{sand}/m_{SOM} ranging between 3.7 and 40 (Table 3) and 0.04 and 0.386 for $m_{sand}/m_{SOM} > 40$ (Table 6).

4.4 Prospects for using soil temperature profiles

Using standard soil moisture and soil temperature observations is a way to investigate soil thermal properties over a large variety of soils, as the access to such data is facilitated by online databases (Dorigo et al., 2011).

A limitation of the data set we used, however, is that soil temperature observations (T_i) are recorded with a resolution of $\Delta T_i = 0.1\,°C$ only (see Sect. 2.1). This low resolution affects the accuracy of the soil thermal diffusivity estimates. In order to limit the impact of this effect, a data filtering technique is used (see Supplement 5) and D_h is retrieved with a precision of 18 %.

It can be noticed that if T_i data were recorded with a resolution of 0.03 °C (which corresponds to the typical uncertainty of PT100 probes), D_h could be retrieved with a precision of about 5 % in the conditions of Eq. (S5.3). Therefore, one may recommend revising the current practise of most observation networks consisting in recording soil temperature with a resolution of 0.1 °C only. More precision in the λ estimates would permit investigating other processes of heat transfer in the soil such as those related to water transport (Rutten, 2015).

5 Conclusions

An attempt was made to use routine soil temperature and soil moisture observations of a network of automatic weather stations to retrieve instantaneous values of the soil thermal conductivity at a depth of 0.10 m. The data from the SMOSMANIA network, in southern France, are used. First, the thermal diffusivity is derived from consecutive measurements of the soil temperature. The λ values are then derived from the thermal diffusivity retrievals and from the volumetric heat capacity calculated using measured soil properties. The relationship between the λ estimates and the measured soil moisture at a depth of 0.10 m permits the retrieval of λ_{sat} for 14 stations. The Lu et al. (2007) empirical λ model is then used to retrieve the quartz volumetric content by reverse modeling. A number of pedotransfer functions is proposed for volumetric fraction of quartz for the considered region in France. For the grassland soils examined in this study, the ratio of sand to SOM fractions is the best predictor of f_q. A sensitivity study shows that omitting gravels and the SOM information has

a major impact on λ_{sat}. Finally, an error propagation analysis and a comparison with independent λ_{sat} data from Lu et al. (2007) show that the gravimetric fraction of sand within soil solids, including gravels and SOM, is a good predictor of the volumetric fraction of quartz when a larger variety of soil types is considered.

6 Data availability

The SMOSMANIA data are available to the research community through the International Soil Moisture Network web site (https://ismn.geo.tuwien.ac.at/).

Acknowledgements. We thank Xinhua Xiao (NC State University Soil Physics, Raleigh, USA), Tusheng Ren (China Agricultural University, Beijing, China) and a third anonymous referee for their review of the manuscript and for their fruitful comments. We thank Aaron Boone (CNRM, Toulouse, France) for his helpful comments. We thank our Météo-France colleagues for their support in collecting and archiving the SMOSMANIA data: Catherine Bienaimé, Marc Bailleul, Laurent Brunier, Anna Chaumont, Jacques Couzinier, Mathieu Créau, Philippe Gillodes, Sandrine Girres, Michel Gouverneur, Maryvonne Kerdoncuff, Matthieu Lacan, Pierre Lantuejoul, Dominique Paulais, Fabienne Simon, Dominique Simonpietri, Marie-Hélène Théron and Marie Yardin.

Edited by: A. Cerdà

References

Abu-Hamdeh, N. H. and Reeder, R. C.: Soil thermal conductivity: effects of density, moisture, salt concentration, and organic matter, Soil Sci. Soc. Am. J., 64, 1285–1290, 2000.

Albergel, C., Rüdiger, C., Pellarin, T., Calvet, J.-C., Fritz, N., Froissard, F., Suquia, D., Petitpa, A., Piguet, B., and Martin, E.: From near-surface to root-zone soil moisture using an exponential filter: an assessment of the method based on in-situ observations and model simulations, Hydrol. Earth Syst. Sci., 12, 1323–1337, doi:10.5194/hess-12-1323-2008, 2008.

Albergel, C., Calvet, J.-C., de Rosnay, P., Balsamo, G., Wagner, W., Hasenauer, S., Naeimi, V., Martin, E., Bazile, E., Bouyssel, F., and Mahfouf, J.-F.: Cross-evaluation of modelled and remotely sensed surface soil moisture with in situ data in southwestern France, Hydrol. Earth Syst. Sci., 14, 2177–2191, doi:10.5194/hess-14-2177-2010, 2010.

Bristow, K. L.: Measurement of thermal properties and water content of unsaturated sandy soil using dual-probe heat-pulse probes, Agr. Forest Meteorol., 89, 75–84, 1998.

Bristow, K. L., Kluitenberg, G. J., and Horton R.: Measurement of soil thermal properties with a dual-probe heat-pulse technique, Soil Sci. Soc. Am. J., 58, 1288–1294, doi:10.2136/sssaj1994.03615995005800050002x, 1994.

Calvet, J.-C., Bessemoulin, P., Noilhan, J., Berne, C., Braud, I., Courault, D., Fritz, N., Gonzalez-Sosa, E., Goutorbe, J.-P., Haverkamp, R., Jaubert, G., Kergoat, L., Lachaud, G., Laurent, J.-P., Mordelet, P., Olioso, A., Péris, P., Roujean, J.-L.,

Thony, J.-L., Tosca, C., Vauclin, M., and Vignes, D.: MUREX: a land-surface field experiment to study the annual cycle of the energy and water budgets, Ann. Geophys., 17, 838–854, doi:10.1007/s00585-999-0838-2, 1999.

Calvet, J.-C., Fritz, N., Froissard, F., Suquia, D., Petitpa, A., and Piguet, B.: In situ soil moisture observations for the CAL/VAL of SMOS: the SMOSMANIA network, International Geoscience and Remote Sensing Symposium, IGARSS, Barcelona, Spain, 23–28 July 2007, 1196–1199, doi:10.1109/IGARSS.2007.4423019, 2007.

Chen, Y. Y., Yang, K., Tang, W., Qin, J., and Zhao, L.: Parameterizing soil organic carbon's impacts on soil porosity and thermal parameters for Eastern Tibet grasslands, Sci. China Earth Sci., 55, 1001–1011, doi:10.1007/s11430-012-4433-0, 2012.

Churchman, G. J. and Lowe, D. J.: Alteration, formation, and occurrence of minerals in soils, in: Handbook of soil sciences: properties and processes, edited by: Huang, P. M., Li, C., and Summer, M. E., Chapter 20, 40–42, isbn:978-1-4398-0306-6, CRC Press, Boca Raton (FL), 2012.

Côté, J. and Konrad, J.-M.: A generalized thermal conductivity model for soils and construction materials, Can. Geotech. J., 42, 443–458, doi:10.1139/T04-106, 2005.

Crank, J. and Nicolson, P.: A practical method for numerical evaluation of solutions of partial differential equations of the heat-conduction type, Adv. Comput. Math., 6, 207–226, doi:10.1007/BF02127704, 1996.

Decharme, B., Brun, E., Boone, A., Delire, C., Le Moigne, P., and Morin, S.: Impacts of snow and organic soils parameterization on northern Eurasian soil temperature profiles simulated by the ISBA land surface model, The Cryosphere, 10, 853–877, doi:10.5194/tc-10-853-2016, 2016.

de Vries, D. A.: Thermal properties of soils, in: Physics of plant environment, edited by: Van Wijk, W. R., 210–235, North-Holland Publ. Co., Amsterdam, 1963.

Dong, Y., McCartney, J. S., and Lu, N.: Critical review of thermal conductivity models for unsaturated soils, Geotech. Geol. Eng., 33, 207–221, doi:10.1007/s10706-015-9843-2, 2015.

Dorigo, W. A., Wagner, W., Hohensinn, R., Hahn, S., Paulik, C., Xaver, A., Gruber, A., Drusch, M., Mecklenburg, S., van Oevelen, P., Robock, A., and Jackson, T.: The International Soil Moisture Network: a data hosting facility for global in situ soil moisture measurements, Hydrol. Earth Syst. Sci., 15, 1675–1698, doi:10.5194/hess-15-1675-2011, 2011.

Draper, C., Mahfouf, J.-F., Calvet, J.-C., Martin, E., and Wagner, W.: Assimilation of ASCAT near-surface soil moisture into the SIM hydrological model over France, Hydrol. Earth Syst. Sci., 15, 3829–3841, doi:10.5194/hess-15-3829-2011, 2011.

Farouki, O. T.: Thermal properties of soils, Series on Rock and Soil Mechanics, 11, Trans. Tech. Pub., Rockport, MA, USA, 136 pp., 1986.

Johansen, O.: Thermal conductivity of soils, PhD thesis, University of Trondheim, 236 pp., Universitetsbiblioteket i Trondheim, Høgskoleringen 1, 7034 Trondheim, Norway, available at: http://www.dtic.mil/dtic/tr/fulltext/u2/a044002.pdf (last access: January 2016), 1975.

Kersten, M. S.: Thermal properties of soils, University of Minnesota Engineering Experiment Station Bulletin, 28, 227 pp., University of Minnesota Agricultural Experiment Station, St. Paul, MN 55108, 1949.

Laanaia, N., Carrer, D., Calvet, J.-C., and Pagé, C.: How will climate change affect the vegetation cycle over France? A generic modeling approach, Climate Risk Management, 13, 31–42, doi:10.1016/j.crm.2016.06.001, 2016.

Lawrence, D. M. and Slater, A. G.: Incorporating organic soil into a global climate model, Clim. Dynam., 30, 145–160, doi:10.1007/s00382-007-0278-1, 2008.

Lu, S., Ren, T., Gong, Y., and Horton, R.: An improved model for predicting soil thermal conductivity from water content at room temperature, Soil Sci. Soc. Am. J., 71, 8–14, doi:10.2136/sssaj2006.0041, 2007.

Nachtergaele, F., van Velthuize, H., Verelst, L., Wiberg, D., Batjes, N., Dijkshoorn, K., van Engelen, V., Fischer, G., Jones, A., Montanarella, L., Petri, M., Prieler, S., Teixeira, E., and Shi, X.: Harmonized World Soil Database, Version 1.2, FAO/IIASA/ISRIC/ISS-CAS/JRC, FAO, Rome, Italy and IIASA, Laxenburg, Austria, available at: http://webarchive.iiasa.ac.at/Research/LUC/External-World-soil-database/HWSD_Documentation.pdf (last access: January 2016), 2012.

Parlange, M. B., Cahill, A. T., Nielsen, D. R., Hopmans, J. W., and Wendroth, O.: Review of heat and water movement in field soils, Soil Till. Res., 47, 5–10, 1998.

Parrens, M., Zakharova, E., Lafont, S., Calvet, J.-C., Kerr, Y., Wagner, W., and Wigneron, J.-P.: Comparing soil moisture retrievals from SMOS and ASCAT over France, Hydrol. Earth Syst. Sci., 16, 423–440, doi:10.5194/hess-16-423-2012, 2012.

Peters-Lidard, C. D., Blackburn, E., Liang, X., and Wood, E. F.: The effect of soil thermal conductivity parameterization on surface energy fluxes and temperatures, J. Atmos. Sci., 55, 1209–1224, 1998.

Rutten, M. M.: Moisture in the topsoil: From large-scale observations to small-scale process understanding, PhD Thesis, Delft university of Technology, doi:10.4233/uuid:89e13a16-b456-4692-92f0-7a40ada82451, 2015.

Schelde, K., Thomsen, A., Heidmann, T., Schjonning, P., and Jansson, P.-E.: Diurnal fluctuations of water and heat flows in a bare soil, Water Resour. Res., 34, 2919–2929, 1998.

Schönenberger, J., Momose, T., Wagner, B., Leong, W. H., and Tarnawski, V. R.: Canadian field soils I. Mineral composition by XRD/XRF measurements, Int. J. Thermophys., 33, 342–362, doi:10.1007/s10765-011-1142-4, 2012.

Sourbeer, J. J. and Loheide II, S. P.: Obstacles to long-term soil moisture monitoring with heated distributed temperature sensing, Hydrol. Process., 30, 1017–1035, 2015.

Subin, Z. M., Koven, C. D., Riley, W. J., Torn, M. S., Lawrence, D. M., and Swenson, S. C.: Effects of soil moisture on the responses of soil temperatures to climate change in cold regions, J. Climate, 26, 3139–3158, doi:10.1175/JCLI-D-12-00305.1, 2013.

Tarara, J. M. and Ham, J. M.: Measuring soil water content in the laboratory and field with dual-probe heat-capacity sensors, Agron. J., 89, 535–542, 1997.

Tarnawski, V. R., Momose, T., and Leong, W. H.: Assessing the impact of quartz content on the prediction of soil thermal conductivity, Géotechnique, 59, 331–338, doi:10.1680/geot.2009.59.4.331, 2009.

Tarnawski, V. R., Momose, T., and Leong, W. H.: Thermal conductivity of standard sands II. Saturated conditions, Int. J. Thermophys., 32, 984–1005, doi:10.1007/s10765-011-0975-1, 2011.

Tarnawski, V. R., McCombie, M. L., Leong, W. H., Wagner, B., Momose, T., and Schönenberger J.: Canadian field soils II. Modeling of quartz occurrence, Int. J. Thermophys., 33, 843–863, doi:10.1007/s10765-012-1184-2, 2012.

Yang, K., Koike, T., Ye, B., and Bastidas, L.: Inverse analysis of the role of soil vertical heterogeneity in controlling surface soil state and energy partition, J. Geophys. Res., 110, D08101, doi:10.1029/2004JD005500, 2005.

Zakharova, E., Calvet, J.-C., Lafont, S., Albergel, C., Wigneron, J.-P., Pardé, M., Kerr, Y., and Zribi, M.: Spatial and temporal variability of biophysical variables in southwestern France from airborne L-band radiometry, Hydrol. Earth Syst. Sci., 16, 1725–1743, doi:10.5194/hess-16-1725-2012, 2012.

Zhang, X., Heitman, J., Horton, R., and Ren, T.: Measuring near-surface soil thermal properties with the heat-pulse method: correction of ambient temperature and soil–air interface effects, Soil Sci. Soc. Am. J., 78, 1575–1583, doi:10.2136/sssaj2014.01.0014, 2014.

Soil CO$_2$ efflux in an old-growth southern conifer forest (*Agathis australis*) – magnitude, components and controls

Luitgard Schwendenmann[1] and Cate Macinnis-Ng[2]

[1]School of Environment, University of Auckland, Private Bag 92019, 1142 Auckland, New Zealand
[2]School of Biological Sciences, University of Auckland, Private Bag 92019, 1142 Auckland, New Zealand

Correspondence to: Luitgard Schwendenmann (l.schwendenmann@auckland.ac.nz)

Abstract. Total soil CO$_2$ efflux and its component fluxes, autotrophic and heterotrophic respiration, were measured in a native forest in northern Aotearoa–New Zealand. The forest is dominated by *Agathis australis* (kauri) and is on an acidic, clay rich soil. Soil CO$_2$ efflux, volumetric soil water content and soil temperature were measured bi-weekly to monthly at 72 sampling points over 18 months. Trenching and regression analysis was used to partition total soil CO$_2$ efflux into heterotrophic and autotrophic respiration. The effect of tree structure was investigated by calculating an index of local contribution (I_c, based on tree size and distance to the measurement location) followed by correlation analysis between I_c and total soil CO$_2$ efflux, root biomass, litterfall and soil characteristics. The measured mean total soil CO$_2$ efflux was 3.47 µmol m^{-2} s^{-1}. Autotrophic respiration accounted for 25 % (trenching) or 28 % (regression analysis) of total soil CO$_2$ efflux. Using uni- and bivariate models showed that soil temperature was a poor predictor of the temporal variation in total soil CO$_2$ efflux (< 20 %). In contrast, a stronger temperature sensitivity was found for heterotrophic respiration (around 47 %). We found significant positive relationships between kauri tree size (I_c) and total soil CO$_2$ efflux, root biomass and mineral soil CN ratio within 5–6 m of the sampling points. Using multiple regression analysis revealed that 97 % of the spatial variability in total soil CO$_2$ efflux in this kauri-dominated stand was explained by root biomass and soil temperature. Our findings suggest that biotic factors such as tree structure should be investigated in soil carbon related studies.

1 Introduction

Soil CO$_2$ efflux (soil respiration) is the largest CO$_2$ flux from terrestrial ecosystems into the atmosphere (Raich and Potter, 1995; Janssens et al., 2001a; Bond-Lamberty and Thomson, 2010a). Quantifying the magnitude of soil CO$_2$ efflux and examining the spatial and temporal heterogeneity is critical in characterising the carbon (C) dynamics in terrestrial ecosystems (Schlesinger and Andrews, 2000; Trumbore, 2006; Smith and Fang, 2010) as even a small change in soil CO$_2$ efflux could have a strong impact on atmospheric CO$_2$ concentration (Andrews et al., 1999; Rustad et al., 2000). Advancing the understanding of soil CO$_2$ efflux and its driving factors is also important to predict the effects of land-use

conversion and climate change on the net C sink of the terrestrial biosphere (Giardina et al., 2014).

Soil CO$_2$ efflux varies widely in space and time according to changes in various abiotic and biotic factors. Soil temperature is often the main abiotic factor explaining the temporal variability of soil CO$_2$ efflux (Raich and Schlesinger, 1992; Jassal et al., 2005; Bond-Lamberty and Thomson, 2010b). Many studies show a positive correlation between soil temperature and soil CO$_2$ efflux (Reich and Schlesinger, 1992; Lloyd and Taylor, 1994; Rustad et al., 2000). This relationship is often expressed as a Q_{10} function (relative increase in soil CO$_2$ efflux rate per 10 °C difference) (van't Hoff, 1898) or modified Arrhenius function (Lloyd and Taylor, 1994). However, other abiotic factors have been found to influence the temporal and spatial variation in soil CO_2 efflux. For example, several studies have shown a parabolic relationship between soil water content and soil CO_2 efflux with the highest soil CO_2 efflux occurring at an intermediate soil water content (Davidson et al., 1998, 2000; Schwendenmann et al., 2003). Other soil factors driving the spatial variability in soil CO_2 efflux in forest ecosystems include the quality and quantity of soil organic matter (Rayment and Jarvis, 2000; Epron et al., 2004) and microbial biomass (Xu and Qi, 2001).

Biotic factors that influence rates of soil CO_2 efflux include plant and microbial components. Vegetation type and structure are important determinants of soil CO_2 efflux because they influence the quantity and quality of litter and root biomass supplied to the soil and they also mediate the soil microclimate (Fang et al., 1998; Raich and Tufekcioglu, 2000; Metcalfe et al., 2007, 2011). For example, litter addition experiments have shown that increasing litterfall enhances soil CO_2 efflux (Sulzman et al., 2005; Sayer et al., 2011). A few studies have investigated the effect of stand structure and tree size on soil CO_2 efflux in temperate (Longdoz et al., 2000; Søe and Buchmann, 2005; Ngao et al., 2012) and tropical forests (Ohashi et al., 2008; Katayama et al., 2009; Brechet et al., 2011). Those findings demonstrate that the spatial distribution of emergent trees strongly affects the root distribution and litterfall, partly explaining the spatial variation of soil CO_2 efflux (Katayama et al., 2009; Bréchet et al., 2011). Some studies show that soil CO_2 efflux at the base of emergent trees is significantly higher than soil CO_2 efflux at greater distances from the trees (Katayama et al., 2009; Ohashi et al., 2008).

Soil CO_2 efflux is the result of CO_2 production by heterotrophic and autotrophic respiration and gas transport (Fang and Moncrieff, 1999; Maier et al., 2011; Maier and Schack-Kirchner, 2014). Heterotrophic respiration mainly originates from microbes decomposing plant detritus and soil organic matter while autotrophic (= root/rhizosphere) respiration comes from plant roots, mycorrhizal fungi and the rhizosphere (Hanson et al., 2000; Bond-Lamberty et al., 2011). The relative contribution of autotrophic respiration to total soil CO_2 efflux varies widely (10–90 %) depending on the type of ecosystem studied (Hanson et al., 2000; Subke et al., 2006; Bond-Lamberty et al., 2011). Various methods (i.e. trenching, regression analysis, isotopic methods) have been developed to separate heterotrophic and autotrophic respiration under both laboratory and field conditions and are described in the review papers by Hanson et al. (2000), Kuzyakov (2006) and Bond-Lamberty et al. (2011). Separating total soil CO_2 efflux into autotrophic and heterotrophic sources is important to more accurately predict C fluxes under changing environmental conditions as heterotrophic and autotrophic respiration respond differently to abiotic and biotic factors (Boone et al., 1998; Davidson et al., 2006; Brüggemann et al., 2011). For example, heterotrophic respiration was found to be more susceptible to seasonal drought in a *Pinus contorta* forest (Scott-Denton et

al., 2006). Other studies showed that autotrophic respiration is more temperature-sensitive than heterotrophic respiration and total soil CO_2 efflux (Boone et al., 1998; Högberg, 2010).

Soil CO_2 efflux has been measured in a wide range of mature and old-growth forests across the globe (Schwendenmann et al., 2003; Epron et al., 2004; Sulzman et al., 2005; Adachi et al., 2006; Bahn et al., 2010; Bond-Lamberty and Thompson, 2014). Southern conifer forests are an exception to this (but see Urrutia-Jalabert, 2015). They include kauri (*Agathis australis* D. Don Lindl. ex Loudon, Araucariaceae) forests in Aotearoa–New Zealand. Old-growth kauri forests are considered to be one of the most C-dense forests worldwide (Keith et al., 2009) with up to $670\,\mathrm{Mg\,C\,ha^{-1}}$ in living woody biomass (Silvester and Orchard, 1999). Kauri is endemic to northern New Zealand (north of latitude $38°$ S) (Ecroyd, 1982) and is the largest and longest living tree species in the country. Kauri has significant effects on the soil environment (Whitlock, 1985; Verkaik et al., 2007) and plant community composition (Wyse et al., 2014). Phenolic compounds in kauri leaf litter (Verkaik et al., 2006) and low pH values (around 4) (Silvester, 2000; Wyse and Burns, 2013) partly explain the slow decomposition rates of kauri litter (Enright and Ogden, 1987) which result in thick organic layers in undisturbed kauri stands (Silvester and Orchard, 1999).

Organic layers (= forest floor composed of leaves, twigs and bark in various stages of decomposition above the soil surface) are important C reservoirs (Gaudinski et al., 2000) and can be a considerable source of CO_2. Organic layers can also contain a large amount of roots which may result in increased soil CO_2 efflux (Cavagnaro et al., 2012). Mature kauri trees have an extensive network of fine roots which extends from the lateral roots into the interface between organic layer and the mineral soil (Bergin and Steward, 2004; Steward and Beveridge, 2010). A recent study also showed that roots and root nodules of kauri harbour arbuscular mycorrhizal fungi (Padamsee et al., 2016). Roots colonized by mycorrhizal fungi have been found to release more CO_2 than non-mycorrhizal roots (Valentine and Kleinert, 2007; Nottingham et al., 2010).

It remains unknown, however, how much soil CO_2 is released from these C-rich southern conifer forests and which factors are driving the temporal and spatial variability in soil CO_2 efflux. It has been shown that kauri has a significant influence on soil properties but the influence of kauri tree distribution on soil carbon related ecosystem processes is untested. Quantifying the magnitude of soil C loss and identifying the controls of this significant C flux are essential for the assessment of the C balance of these C-rich and long-lived forest stands.

The aim of this study was to determine the magnitude, components and the driving factors of soil CO_2 efflux in an old-growth southern conifer forest. The specific objectives of our study were (i) to quantify total soil CO_2 efflux, (ii) to test the effect of collar insertion depth on soil CO_2

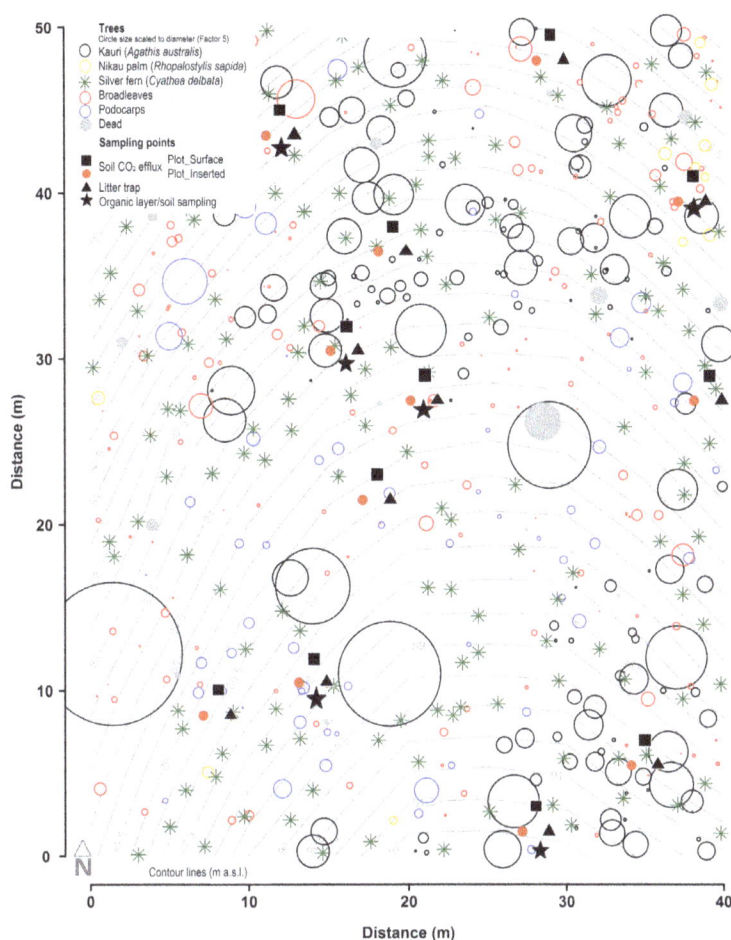

Figure 1. Overview of the research plot showing the position of all trees ≥ 2.5 cm diameter (circle size scaled to diameter – Factor 5) and sampling points for soil CO_2 efflux measurements (Plot_Surface, Plot_Inserted), litter and soil sampling. The trench plots are located adjacent (upslope) of the research plot at around 92 m a.s.l.

efflux, (iii) partition total soil CO_2 efflux into autotrophic and heterotrophic respiration, (iv) to identify the factors controlling the temporal variation of total soil CO_2 efflux and its component fluxes, and (v) to test the effect of kauri tree size and distribution on total soil CO_2 efflux and soil properties. We used direct (trenching) and indirect (regression technique) approaches to partition total soil CO_2 efflux into the autotrophic and heterotrophic components. Given that old-growth kauri forests are often characterised by thick organic layers, deep collars were deployed to assess the effect of insertion depth on total soil CO_2 efflux and to quantify the proportion of autotrophic and heterotrophic respiration in this layer.

2 Material and methods

2.1 Study site

The study was conducted in the University of Auckland Huapai reserve. The reserve is a 15 ha forest remnant surrounded

by farmland (Thomas and Ogden, 1983) and is located approximately 25 km west of central Auckland on the northern fringe of the Waitakere Ranges (36°47.7′ S, 174°29.5′ E). Within the long-term research plot (50 × 40 m), the diameter at breast height (DBH) of all trees ≥ 2.5 cm was measured, the species were identified and their location mapped (Wunder et al., 2010) (Fig. 1). The plot is dominated by kauri (770 stems ha^{-1}) with a basal area of 75 m^2 ha^{-1}, equating to approximately 80 % of the stand basal area (Wunder et al., 2010). Kauri tree size distribution differs within the plot. Four emergent kauri trees (up to 180 cm in DBH, ~ 300 year old) are found on the upper slope of the plot. At the lower slope tree fall and removal of five large kauri trees in the 1950s created gaps which are now dominated by a cohort of younger kauri trees. Silver ferns (*Cyathea dealbata*) are also highly abundant (785 stems ha^{-1}) (Wunder et al., 2010). Less-numerous species are a mixture of podocarps and broadleaved species, including *Phyllocladus trichomanoides*, *Myrsine australis*, *Coprosma arborea* and *Geniostoma ligustrifolium*.

Total annual rainfall, measured from 2011 to 2013 at a weather station located in the vicinity of the reserve, is approximately 1200 mm with 70 % occurring during austral winter (June–August). Annual mean temperature is 14 °C (Macinnis-Ng and Schwendenmann, 2015). The soils are derived from andesitic tuffs and are classified as Orthic Granular Soils (New Zealand soil classification; Hewitt, 1992) or Humults (US soil classification; Soil Survey Staff, 2014). The clayey soil is fairly sticky when wet, and hard and fragile when dry (Thomas and Ogden, 1983). The thickness of the organic layer varies between 5 and 15 cm and consists mainly of partly decomposed kauri leaves and twigs.

2.2   Experimental setup

The long-term research plot was subdivided into six equal quadrats. Within each quadrant two soil CO_2 efflux sampling points (in total 12) were randomly located (Fig. 1). For each sampling point we measured the distance to the closest tree with a DBH ≥ 2.5 cm. At each of these 12 sampling points, a cluster of measurements was made. There was one surface measurement and three inserted measurements as described below.

Total soil CO_2 efflux was measured on the surface of the forest floor by gently pressing a polyvinyl chloride (PVC) ring attached to the soil respiration chamber (see below for details) down on the forest floor during measurements to avoid cutting fine roots. The sampling points were marked with flags and kept free of vegetation. Total soil CO_2 efflux was measured over 18 months from August 2012 to January 2014 at each location. These sampling points were named Plot_Surface.

To measure the effect of collar insertion depth and to quantify the proportion of autotrophic and heterotrophic respiration to total soil CO_2 efflux in the organic layer, a cluster of three "deep" PVC collars (10 cm in diameter, 20 cm in height) was inserted next to each sampling point for surface soil CO_2 efflux measurements. Three collars per cluster were spaced evenly around the circumference of a circle 2 m in diameter, with small adjustments in the spacing to avoid large roots. Each collar was driven right through the organic layer and 1–2 cm into the mineral soil layer to cut off the roots growing in the organic layer. In order to prevent CO_2 uptake, any vegetation inside the collars was regularly removed. The thickness of the organic layer at each grid point was measured using a ruler outside each collar. The deep collars were inserted in November 2011 and left in place over the measurement period. Efflux was measured from August 2012 (9 months after insertion) to January 2014. Here after, these sampling points are known as Plot_Inserted.

We used the trenching approach to separate heterotrophic and autotrophic respiration in the organic layer plus mineral soil to 30 cm depth. To avoid disturbing the long-term research plot the trenching experiment was set-up directly adjacent (upslope) to the research plot. In July 2012, six

2×2 m plots were trenched to 30 cm depth based on a preliminary study showing that the majority of fine roots (over 80 %) are located in the organic layer and top 30 cm of the mineral soil. The trenches were double-lined with a water permeable polypropylene fabric and backfilled. During trenching, trampling and disturbance inside the 2×2 m plots were avoided as far as possible. Two types of measurements were conducted. First, total soil CO_2 efflux was measured at two sampling points outside each trenched plot (Outside_Trench_Surface, $n = 12$) in the same way as the Plot_Surface samples were measured (see above). Second, two collars were randomly placed inside the trenched plots (Trench_Inserted). The collars were inserted 1–2 cm into the mineral soil layer (deep collars) as described above. Soil CO_2 efflux was measured bi-weekly to monthly from August 2012 until December 2013.

2.3   Soil CO_2 efflux measurements

Soil CO_2 efflux was measured with a portable infrared gas analyser (EGM-4, PP Systems, Amesbury, MA, USA) equipped with a soil respiration chamber (SRC-1, PP Systems, Amesbury, MA, USA). The CO_2 concentration was measured every 5 sec over 90-120 sec between 09:00 and 02:00 LT (local time) and the change in CO_2 concentration over time was recorded. Diurnal soil CO_2 efflux measurements conducted in January 2013 showed that soil CO_2 efflux rates between 09:00 and 02:00 LT were comparable as there was no significant diurnal trend (data not shown).

Soil CO_2 efflux (μmol m^{-2} s^{-1}) was calculated as follows (Eq. 1):

$$\text{Soil } CO_2 \text{ efflux} \left(\mu\text{mol m}^{-2} \text{ s}^{-1} \right) = (\Delta CO_2/\Delta t)$$
$$\times (P \times V)/(R \times T \times A), \qquad (1)$$

where $\Delta CO_2/\Delta t$ is the change in CO_2 concentration over time (t), calculated as the slope of the linear regression (μmol mol^{-1} s^{-1} = ppm s^{-1}); P is the atmospheric pressure (Pa), V is the volume of the chamber including collar (m^3), R is the universal gas constant, 8.314 m^3 Pa K^{-1} mol^{-1}), T is the temperature (K), and A is the surface area of ground covered by each chamber (0.007854 m^2).

Soil temperature (Soil temperature probe, 10 cm probe, Novel Ways Ltd, Hamilton, New Zealand) and volumetric soil water content (Hydrosense II, 12 cm probe, Campbell Scientific Inc., Logan, UT, USA) were measured at the same time next to each of the collars.

2.4   Litterfall, root and soil characteristics

Litterfall (including leaves, twigs, fruits, flowers, cone scales, etc.) was collected from 12 litter traps (pop-up planters, 63 cm in diameter) located next to each soil CO_2 efflux

Table 1. Descriptive statistics for litter, root, and soil characteristics. Samples were taken in the vicinity of the surface soil CO_2 efflux sampling points ($n = 12$, except for root biomass, $n = 10$).

Parameter	mean	STDEV	SE	median	min–max	CV %
Litterfall						
$\sum 2012$ (kg m^{-2})	0.9	0.2	0.1	0.9	0.6–1.1	21.7
$\sum 2013$ (kg m^{-2})	1.6	0.3	0.1	1.7	1.0–1.9	21.3
Organic layer (OL)						
Thickness (cm)	8.8	2.3	0.9	8.2	6.2–12.2	26.1
Root biomass (kg m^{-2})	0.8	0.9	0.3	0.3	0.02–2.7	115.6
pH	4.85	0.57	0.23	5.06	3.88–5.51	11.8
C/N ratio	43.9	10.4	4.2	43.2	31.4–58.7	23.7
Carbon stock (kg m^{-2})	18.7	7.7	3.1	18.4	7.9–28.9	41.2
Nitrogen stock (kg m^{-2})	0.45	0.18	0.07	0.45	0.22–0.77	40.0
Mineral soil						
Root biomass, 0–15 cm (kg m^{-2})	2.2	1.6	0.5	1.6	0.7–6.3	75.1
Root biomass, 15–30 cm (kg m^{-2})	0.7	1.2	0.4	0.4	0.2–3.9	97.7
Root biomass, $\sum$ OL + 0–30 cm (kg m^{-2})	3.8	2.2	0.7	3.8	0.9–8.0	57.9
pH, 0–10 cm	4.68	0.52	0.21	4.91	3.75–5.13	11.1
C/N ratio, 0–10 cm	16.1	1.9	0.8	16.2	13.7–19.0	12.1
Carbon stock, 0–10 cm (kg m^{-2})	8.4	1.9	0.8	8.6	6.0–10.7	22.7
Nitrogen stock, 0–10 cm (kg m^{-2})	0.53	0.13	0.05	0.52	0.40–0.75	24.1
Soil temperature, 0–10 cm (°C)	14.2	0.2	0.1	14.2	14.0–14.5	1.4
Volumetric soil water content, 0–12 cm (%)	43.9	2.1	0.9	44.3	41.2–46.1	4.9

STDEV = standard deviation; SE = standard error; min = minimum; max = maximum; CV = Coefficient of Variation.

cluster within the long-term research plot (Fig. 1). Litterfall was collected bi-weekly from January 2012 to January 2014, dried at 80 °C to constant mass, sorted and weighed (Macinnis-Ng and Schwendenmann, 2015).

Organic layer and mineral soil samples (0–10 cm depths) were taken next to each collar with a core sampler in November 2011 (research plot) and July 2012 (trenched locations). Samples were ground and analysed for total C and N concentration using an elemental analyser (TruSpec, LECO Corporation, St. Joseph, Michigan, USA). Soil (LECO Lot 1016, 1007) and leaf (NIST SRM 1515 – Apple Leaves) standards were used for calibration. The coefficient of variation was 0.5 % for C and 1 % N for plant material (45 % C, 2.3 % N) and 1 % for C and N for soil (2–12 % C, 0.2–1 % N). 10 % of samples were replicated and results were within the range of variation given for the standards.

Organic layer and mineral soil samples (0–15, 15–30 cm) were collected for soil analysis and root biomass estimation adjacent to six clusters within the plot (Fig. 1) and the trenched plots. Organic layer samples were collected from 20 cm × 20 cm quadrats. Mineral soil samples were taken using a 15 cm diameter steel cylinder. Samples were dried at 60 °C (forest floor) and 40 °C (mineral soil). Mineral soil samples were sieved at 2 mm. pH was measured in a 1 : 2.5 soil-water suspension (SensION 3 pH meter, HACH, Love-

land, CO, USA). The organic layer samples were wetted and fine roots were manually picked with tweezers. The flotation method was used to separate roots from the clay rich mineral soil. Roots were dried at 60 °C to constant mass and weighed by size class (fine roots: < 2 mm, and small (coarse) roots: 2–20 mm). Litterfall, root and soil data are summarized in Table 1.

2.5 Data analysis

Efflux, soil temperature and volumetric soil water content measured at the three individual deep collars per cluster (Plot_Inserted) and the two samples of Outside_Trench_Surface and Trench_Inserted were averaged before statistical analysis. Normality of the data distribution was examined using a Kolmogorov–Smirnov test.

Descriptive statistics (minimum, maximum, mean and median values, standard deviation, standard error, coefficient of variation) were used to describe soil CO_2 efflux, soil temperature and volumetric soil water content. Differences between total soil CO_2 efflux between treatments (Plot_Surface vs. Plot_Inserted; Outside_Trench_Surface vs. Trench_Inserted) and seasons were tested using a mixed model where deep collar insertion and trenching were considered as a fixed effect and sampling dates as a random effect.

To estimate annual total soil CO_2 efflux, a linear temperature function (see below) and continuous half-hourly soil temperature measurements (plot centre, 10 cm depth, 107 temperature probe, Campbell Scientific, Logan, UT, USA) were used.

Two methods (trenching and regression analysis) were used for partitioning of total soil CO_2 efflux. In the trenching approach, the Plot_Inserted and Trench_Inserted sampling points represents heterotrophic respiration in the organic layer and organic layer plus mineral soil to 30 cm depth, respectively. Measurements from the soil surface (Plot_Surface and Outside_Trench_Surface) represent total soil CO_2 efflux. Autotrophic respiration in the organic layer and organic layer plus mineral soil to 30 cm depth was calculated as the difference between total soil CO_2 efflux and the efflux measured from the Plot_Inserted and Trench_Inserted sampling points, respectively. Heterotrophic respiration from Plot_Inserted and Trench_Inserted sampling points was not corrected for decomposing root-derived CO_2 efflux. For the regression-analysis approach (organic layer plus mineral soil to 30 cm depth), the heterotrophic respiration was derived analytically as the y-intercept of the linear regression between root biomass (independent variable) and total soil CO_2 efflux (dependent variable) (Kucera and Kirkham, 1971; Kuzyakov, 2006). Autotrophic respiration was then estimated by subtracting the heterotrophic respiration from total soil CO_2 efflux.

Univariate and bivariate models were used to investigate the relationship between total soil CO_2 efflux, heterotrophic and autotrophic respiration and the abiotic factors soil temperature and volumetric soil water content. Data from within the research plot and trench sampling points were combined. The temperature response of soil CO_2 efflux was tested using a linear, exponential (Q_{10}, van't Hoff, 1898) and modified Arrhenius function (Lloyd and Taylor, 1994). Linear and quadratic functions were used to assess the soil water dependence of soil CO_2 efflux. The combined effect of soil temperature and soil water content on soil CO_2 efflux was tested using a polynomial function. Coefficient of determination (R^2), standard error of estimate (SEE), and Akaike Information Criterion (AIC) were used to evaluate model performance. The analysis was conducted using Sigma Plot (Version 13, Systat Software Inc., Chicago, IL, USA).

The influence of kauri tree size and distribution on total soil CO_2 efflux, litterfall, root biomass and soil properties was tested using an index of local contribution (I_c). As described in Bréchet et al. (2011), the I_c index is a function of (1) the trunk cross section area (S, m^2) and (2) the distance (d, m) of kauri trees from the sampling points. The following functions were tested: uniform, $I_c = S$; linear, $I_c = S \times (1 - (d/r))$; parabolic, $I_c = S \times (1 - (d/r)^2))$; exponential, $I_c = S \times (\exp - (d/r - d))$ and power, $I_c = S \times (1 - (d/r)^a))$ where a is a coefficient of form and r is a fitted radius of influence (r, in m) (Bréchet et al., 2011). It was assumed that all kauri trees had the same

radius of influence (r, i.e. the distance above which their contribution would become negligible). The coefficient of determination was used to assess the strength of the relationships between litterfall, root biomass or soil CO_2 efflux and the sum of the I_c.

The spatial variability in soil CO_2 efflux was quantified at the plot scale using the coefficient of variation. Multiple regression analysis was used to assess the spatial controls (soil temperature, soil moisture, organic layer thickness, soil C and N, root biomass) of total soil CO_2 efflux.

Statistical analyses were performed using SPSS v. 22 (IBM SPSS Statistics, IBM Corporation, Chicago, IL, USA). The local contribution analysis (I_c) was conducted using R (v3.1.0, R Core Team, 2014). Significance for all statistical analyses was accepted at $p < 0.05$. Data are available in the figshare data repository (Schwendenmann and Macinnis-Ng, 2016).

3 Results

3.1 Seasonal variations in soil CO_2 efflux, soil temperature and volumetric soil water content and the effect of deep collar insertion and trenching

During the study period, soil temperature and volumetric soil water content (SWC) varied with season (Fig. 2b and c). Soil temperatures peaked at about 17 °C during austral summer and early autumn (January–March) while minimum temperatures of around 11 °C (Fig. 2b) were measured in late winter–early spring (July–September). Annual mean soil temperature was 14.2 ± 0.1 °C (Table 1). The SWC was highest during late winter–early spring (July–September) with values of 55 % and soil was driest during late summer–early autumn (March–May) with around 25 % (Fig. 2c). Annual average was 43.9 ± 0.9 % (Table 1). Across the study period, an average of 1.9 ± 0.1 kg m^{-2} litter fell at the sampling points and the organic layer was 8.8 ± 0.9 cm thick (Table 1). Other information on soil and vegetation characteristics is summarised in Table 1.

Total soil CO_2 efflux rates (Plot_Surface) measured at 12 sampling points within the research plot varied from 0.7 to 9.9 μmol CO_2 m^{-2} s^{-1} during the 18-month study period. Total soil CO_2 efflux was positively skewed with the mean larger than the median (Table 2). The mean total soil CO_2 efflux ($\pm$ SE), averaged over the 12 plot sampling points and all sampling dates, was 3.6 ± 0.1 μmol CO_2 m^{-2} s^{-1}. Higher efflux rates were measured during austral summer and early autumn (December–March, 2.7–4.7 μmol CO_2 m^{-2} s^{-1}) compared to winter (June–August, 1.8–3.9 μmol CO_2 m^{-2} s^{-1}) (Fig. 2a). However, differences among seasons were not significant ($p > 0.05$). In contrast, soil temperature differed significantly between summer (16.5 °C) and winter (11.8 °C). We also detected significant seasonal differences in SWC with drier conditions during summer (mean SWC = 31 %) compared to winter (mean SWC = 47 %).

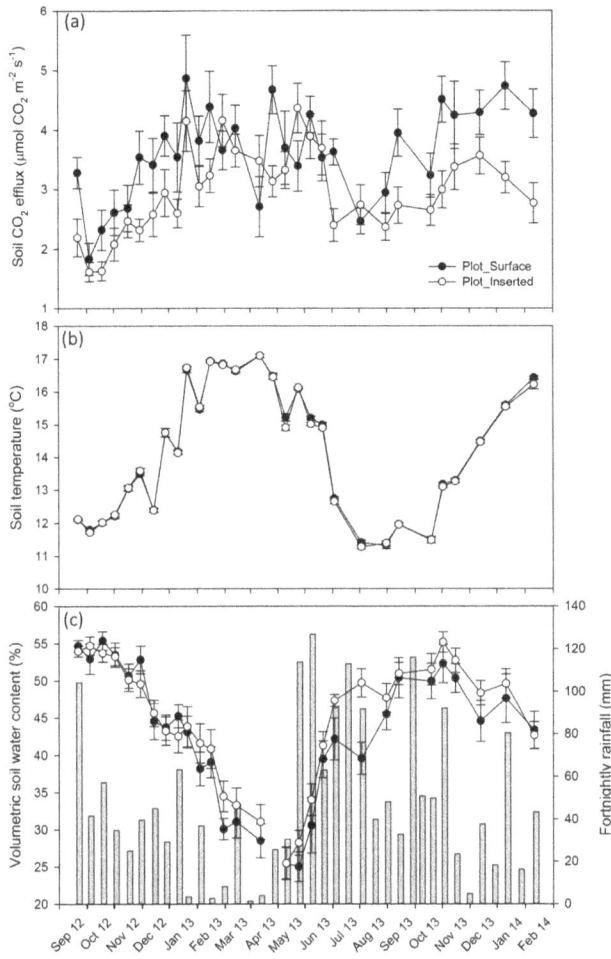

Figure 2. Soil CO_2 efflux (**a**), soil temperature (**b**) and volumetric soil water content (**c**) measured in the research plot from August 2012 to January 2014. Values show mean $\pm$ standard error of the Plot_Surface ($n = 12$) and Plot_Inserted (deep collar, $n = 12$) sampling points, respectively. Volumetric soil water content was not measured in March 2013 due to equipment failure. Fortnightly rainfall (**c**) was measured in a paddock in the vicinity of the research plot.

Table 2. Descriptive statistics of soil CO_2 efflux, soil temperature and volumetric soil water content measured within the research plot (Plot) and adjacent to it (Trench). Measurements were conducted between August 2012 and January 2014. Different letters (a, b for plot; c, d for trench) for a given variable indicate a significant difference between total soil CO_2 efflux (Plot_Inserted, Outside_Trench_Surface) versus deep collar insertion (Plot_Inserted) and trenching (Trench_Inserted).

	N	n	Soil CO_2 efflux ($\mu mol\ CO_2\ m^{-2}\ s^{-1}$)					Soil temperature (°C)					Vol. soil water content (%)				
			mean	SE	med	min / max	CV	mean	SE	med	min / max	CV	mean	SE	med	min / max	CV
												Plot					
Plot_Surface	12	30	3.61a	0.09	3.37	0.65 / 9.96	42.6	14.2a	0.11	14.4	10.9 / 17.5	13.5	43.1a	0.65	44.7	15.2 / 66.6	27.1
Plot_Inserted	36	30	2.98b	0.07	2.72	0.69 / 8.02	43.6	14.1a	0.10	14.1	10.9 / 17.4	13.8	44.7a	0.56	46.6	15.2 / 62.3	23.0
												Trench					
Outside_Trench_Surface	12	17	3.11c	0.14	2.92	0.55 / 6.92	43.0	13.1c	0.17	13.2	10.2 / 17.2	12.5	44.0c	1.27	44.2	17.4 / 72.5	25.2
Trench_Inserted	12	17	2.34d	0.08	2.14	0.67 / 5.30	41.0	12.9c	0.14	13.0	10.1 / 16.9	13.1	56.8d	0.74	56.4	20.2 / 76.5	14.8

N = number of sampling points; n = number of sampling dates between August 2012 and January 2014; SE = standard error; med = median; min = minimum; max = maximum.

Deep collar insertion (Plot_Inserted) had a significant effect on total soil CO_2 efflux but no effect on soil temperature and SWC was found (Table 2). Soil CO_2 efflux from inserted collars ($3.0 \pm 0.1\ \mu mol\ CO_2\ m^{-2}\ s^{-1}$) was 17 % lower compared to total soil CO_2 efflux ($3.6 \pm 0.1\ \mu mol\ CO_2\ m^{-2}\ s^{-1}$) (Table 2). The overall temporal pattern of soil CO_2 efflux was similar between inserted and surface collars (Fig. 2a). However, soil CO_2 efflux from surface collars varied considerably during the dry summer in 2013 (Fig. 2a and c). Higher soil CO_2 efflux from surface collars in April 2013 coincided with rain events after a long dry period with high litter input (see Macinnis-Ng and Schwendenmann (2015) for details).

Total soil CO_2 efflux measured outside the trenched plots ranged from 0.6 to 6.9 $\mu mol\ CO_2\ m^{-2}\ s^{-1}$ with a mean of

$3.1 \pm 0.1 \, \mu mol \, CO_2 \, m^{-2} \, s^{-1}$ (Outside_Trench_Surface, Table 2). Soil CO_2 efflux from Trench_Inserted collars was significantly lower (25 %) compared to total soil CO_2 efflux (Table 2). Volumetric soil water content in the trenched plots was significantly higher (56.8 %) compared to the untrenched sampling points (44 %). In contrast, soil temperature was not significantly affected by trenching (Table 2).

3.2 Contribution of autotrophic respiration to total soil CO_2 efflux

Mean autotrophic respiration derived from the trenching approach was $0.8 \pm 0.1 \, \mu mol \, CO_2 \, m^{-2} \, s^{-1}$. The contribution of autotrophic respiration to total soil CO_2 efflux (to 30 cm depth) was 25 %. Excluding the roots from the organic layer through deep collar insertion showed that roots in the organic layer contribute around 17 % to total soil CO_2 efflux. The proportion of autotrophic respiration to total soil CO_2 efflux tended to be lower during summer–early autumn (December–April) compared to winter (June–August). However, differences were not statistically significant due to high variability in autotrophic respiration, especially during summer (data not shown).

Mean total soil CO_2 efflux (Plot_Surface plus Outside_Trench_Surface; $n = 18$, mean $= 3.47 \, \mu mol \, CO_2 \, m^{-2} \, s^{-1}$; SE $= 0.20 \, \mu mol \, CO_2 \, m^{-2} \, s^{-1}$) was positively correlated with total (organic layer plus mineral soil to 30 cm depth) root biomass ($R^2 = 0.394$, $p = 0.042$, intercept $= 2.49 \, \mu mol \, CO_2 \, m^{-2} \, s^{-1}$) (Fig. 3). Using the regression approach produced an autotrophic respiration estimate of $0.98 \, \mu mol \, CO_2 \, m^{-2} \, s^{-1}$. The proportion of autotrophic respiration to total soil CO_2 efflux derived from the root biomass regression approach was 28 %.

3.3 Effect of soil temperature and volumetric soil water content on the temporal variability in total soil CO_2 efflux, heterotrophic and autotrophic respiration

Independent of the model used, soil temperature explained less than 20 % of the temporal variation in total soil CO_2 efflux (Fig. 4a, Table 3). The Q_{10} values for total soil CO_2 efflux was 1.6 (Table 3). A slightly stronger soil temperature response was found for heterotrophic respiration (Fig. 4b, Table 3) with a Q_{10} value of 2.2 (Table 3). However, all temperature models for heterotrophic respiration had higher AIC values compared to total soil CO_2 efflux (Table 3) which suggests a poorer performance. No significant relationship was found between soil temperature and autotrophic respiration (Fig. 4c, Table 3).

Neither a linear nor a quadratic function resulted in a significant relationship between SWC and total soil CO_2 efflux (Fig. 4d, Table 3). Heterotrophic respiration decreased significantly with increasing SWC (Fig. 4e, Table 3). In contrast a weak, but significant quadratic relationship was found between SWC and autotrophic respiration (Table 3).

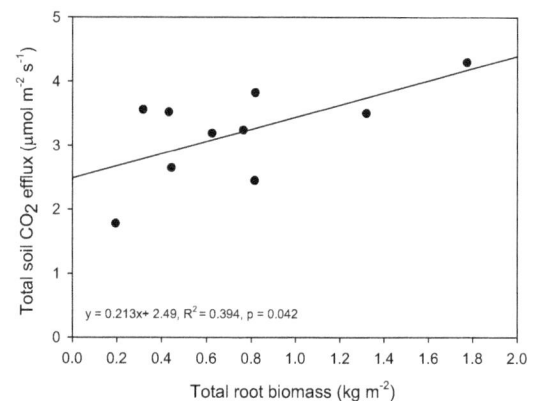

Figure 3. Regression of total root biomass (organic layer plus mineral soil to 30 cm depth) vs. total soil CO_2 efflux.

Bivariate polynomial functions did not result in higher R^2 or better AIC values compared to univariate models (Table 3).

3.4 Spatial variation in total soil CO_2 efflux and environmental factors

The spatial variability of total soil CO_2 efflux between the 12 sampling points in the research plot was relatively high, with a coefficient of variation (CV) of 43 % (Table 2).

We found a good relationship between the tree local contribution index (I_c) and total soil CO_2 efflux. The relationship was strongest (coefficient of determination, $R^2 = 0.342$, $p = 0.030$, linear model) within a radius of 5 m (Fig. 5a and b).

The spatial variation in total root biomass (organic layer plus mineral soil to 30 cm depth, 0.9 to 8 kg m^{-2}) was very high (CV > 95 %, Table 1). Similar to total soil CO_2 efflux, a radius of 5 m provided also the best correlation between root biomass and I_c. The coefficient of determination was $R^2 = 0.985$ ($p = 0.021$, univariate model, Fig. 5c and d).

Compared to root biomass and soil CO_2 efflux the spatial variation in litterfall (total amount over the 18-month period, 1.1–2.2 kg m^{-2}, Table 1) was small (CV = 20 %, Table 1). We did not find any significant correlations between litterfall and I_c (data not shown).

Between 8 and 29 kg C m^{-2} were stored in the 6–12 cm thick organic layer (Table 1). C : N ratio differed considerably between the organic layer (31–58) and mineral soil (13–19). Differences in pH were greater among sampling points compared to differences between organic layer and mineral soil (Table 1). Except for C : N ratio in the mineral soil ($R^2 = 0.655$, $p = 0.000$, linear model, Fig. 5d and e), no correlations were found between I_c and soil characteristics.

Using multiple regression analysis revealed that most of the spatial variability in total soil CO_2 efflux within the plot could be explained by soil temperature and root biomass ($R^2 = 0.977$, Adjusted $R^2 = 0.953$, $F = 41.972$, $p = 0.023$).

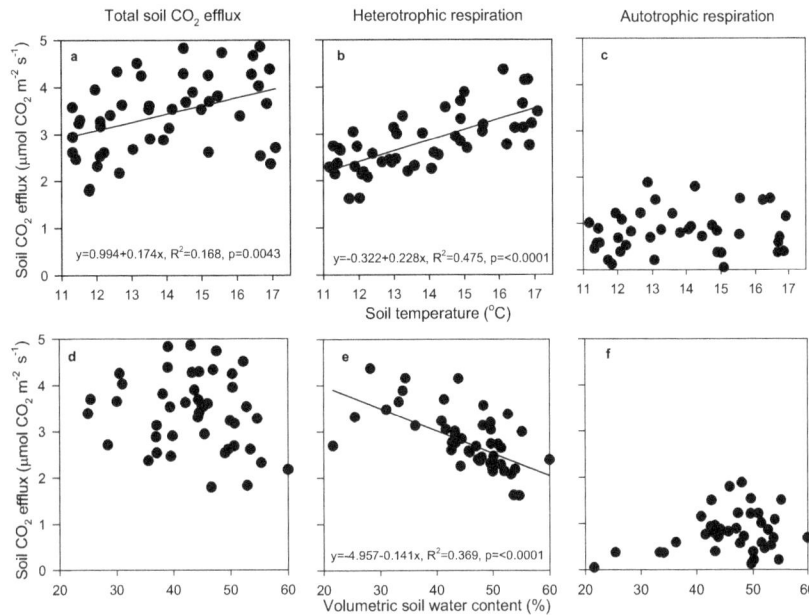

Figure 4. Upper panels: relationship between soil temperature and total soil CO_2 efflux (**a**), heterotrophic respiration (**b**) and autotrophic respiration (**c**). Lower panels: relationship between soil volumetric water content and total soil CO_2 efflux (**d**), heterotrophic respiration (**e**) and autotrophic respiration (**f**). Regression lines are only displayed for significant linear relationships. The results for other uni- and bivariate functions are shown in Table 3.

4 Discussion

4.1 Total soil CO_2 efflux: magnitude and temporal variation

Annual total soil CO_2 efflux (1324 ± 121 g C m^{-2} yr^{-1}; estimated using the linear temperature response function) in this kauri dominated forest was higher than mean values from mature conifer and mixed conifer-hardwood temperate rainforests along the Pacific coast of North America (500—2300 g C m^{-2} yr^{-1}; mean: 1100 ± 65 g C m^{-2} yr^{-1}; $n = 55$) (Campbell and Law, 2005; Hibbard et al., 2005; Bond-Lamberty and Tompson, 2014) and southern conifer (*Fitzroya cupressoides* forests in southern Chile (500–800 g C m^{-2} yr^{-1}; Urratia-Jalabert, 2015). Soil CO_2 emissions from the kauri stand were also higher than efflux rates measured in other New Zealand forests. For example, approximately 1000 g C m^{-2} yr^{-1} were measured in a rimu (*Dacryidium cupressinum*, conifer) dominated podocarp forest in South Westland (Hunt et al., 2008) and annual soil CO_2 efflux in *Leptospermum scoparium/Kunzea ericoides* var. *ericoides* shrublands range between 980 and 1030 g C m^{-2} yr^{-1} (Hedley et al., 2013). In contrast, our values are within the range of values reported for mature unmanaged tropical moist broadleaf forests (900–2000 g C m^{-2} yr^{-1}; mean: 1336 ± 70 g C m^{-2} yr^{-1}; $n = 27$) (Raich and Schlesinger, 1992; Schwendenmann et al., 2003; Bond-Lamberty and Tompson, 2014).

Our findings suggest that soil CO_2 efflux in a conifer dominated forest can be as high or even exceed the efflux rates from broadleaf forests. This is in contrast to previous studies which found that soil CO_2 efflux in conifer forests are lower than those in broadleaf forests (Raich and Tufekcioglu, 2000; Curiel Yuste et al., 2005a). However, these former studies were limited to temperate study sites and based on direct comparisons of sites where forest type was the principal variable differing among pairs. Mean annual soil temperature has been shown to be a good predictor of large-scale variation in total soil CO_2 efflux in non-water limited systems independent of vegetation types and biome (Bahn et al., 2010). With a mean annual temperature of $14\,^\circ$C, this study site was relatively warm compared to sites along the Pacific coast of North America partly explaining the high soil CO_2 efflux rates in this kauri dominated forest.

The amount of litterfall has also been associated with differences in soil CO_2 efflux at the scales of biomes (Davidson et al., 2002; Reichstein et al., 2003; Oishi et al., 2013). Annual C input via litterfall in this kauri dominated forest was 410 and 760 g C m^{-2} in 2012 and 2013, respectively (Macinnis-Ng and Schwendenmann, 2015). This litter C flux is substantially higher than those values from conifer and mixed conifer-hardwood forests in the Northern Hemisphere (50–400 g C m^{-2} yr^{-1}; mean: 164 ± 14 g C m^{-2} yr^{-1}; $n = 43$; Bond-Lamberty and Tompson, 2014; Holland et al., 2015). Kauri litterfall is within the range of values (110–700 g C m^{-2} yr^{-1}; mean: 345 ± 30 g C m^{-2} yr^{-1}; $n = 22$) reported for old-growth tropical forests (Chave et al., 2010;

Table 3. Relationships of total soil CO_2 efflux, heterotrophic and autotrophic respiration with soil temperature (T) and volumetric soil water content (W) using uni- and bivariate functions.

Model	Var	Total soil CO_2 efflux					Heterotrophic respiration					Autotrophic respiration				
		R^2	SEE	p	AIC	Q_{10}	R^2	SEE	p	AIC	Q_{10}	R^2	SEE	p	AIC	Q_{10}
Linear	T	0.168	0.747	0.0043	−22.9		0.475	0.461	<0.0001	−68.2		0.012	0.463	0.5061	−55.4	
Exponential (Q_{10})	T	0.160	0.750	0.0054	−22.5	1.6	0.475	0.461	<0.0001	−68.3	2.2	0.011	0.463	0.5229	−55.4	1.3
Mod. Arrhenius	T	0.181	0.741	0.0029	−23.7		0.471	0.463	<0.0001	−67.9		0.016	0.462	0.4493	−55.6	
Linear	W	0.015	0.799	0.4262	−16.1		0.369	0.510	<0.0001	−57.5		0.072	0.440	0.1043	−57.7	
Quadratic	W	0.123	0.763	0.0597	−19.1		0.408	0.499	<0.0001	−58.0		0.168	0.423	0.0399	−59.4	
Polynomial	T, W	0.177	0.739	0.0150	442		0.508	0.455	<0.0001	442		0.158	0.427	0.0490	345	

Var = variables; R^2 = coefficient of determination; SEE = standard error of estimate; AIC = Akaike Information Criterion; y = soil CO_2 efflux, heterotrophic or autotrophic respiration; x = soil temperature; z = volumetric soil water content; functions: linear: $y = a \times x/y + b$; exponential: $y = a \times \exp(b \times x)$; modified Arrhenius $y = a \times \exp(-b/(x - x_0))$; quadratic: $y = a \times y^2 + b \times y + c$; polynomial: $y = a + b \times x + c \times z$; $Q_{10} = \exp(10 \times b)$, b derived from exponential function.

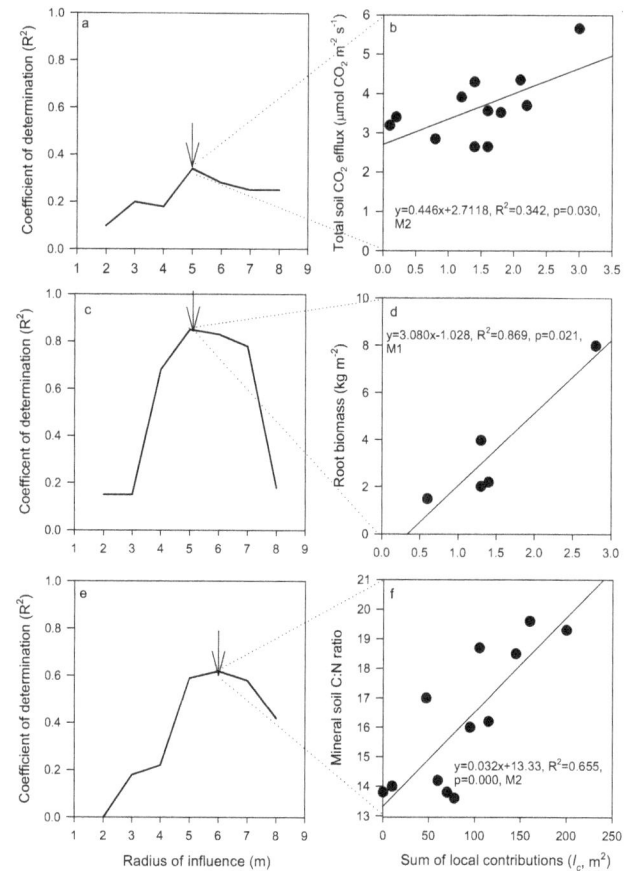

Figure 5. Relationships between the sum of local contribution indices of surrounding trees within the fitted radius of influence and total soil CO_2 efflux (**a, b**), root biomass (**c, d**) and mineral soil CN ratio (**e, f**). The arrows in (**a**), (**c**) and (**e**) indicate the best coefficients of variation (highest R^2 value) with models shown in (**b**), (**d**) and (**f**). M1 = univariate model ($I_c = S$); M2 = linear model ($I_c = S \times (1 - (d/r))$) where S = trunk cross section area (S, m²), d = distance between the trees and the measurement point (d, m), r = fitted radius of influence (r, m).

Holland et al., 2015; Bond-Lamberty and Tompson, 2014). High litter input, together with high annual temperature, can be another major factor explaining the comparatively high soil CO_2 efflux rate in this southern conifer forest. This is somewhat surprising as one would assume that organic matter mineralisation and thus soil CO_2 efflux is reduced given the slow decomposition rate of kauri litter. In four kauri forests ranging from pole to mature forests mean residence times between 9 and 78 years were estimated for 8 to 46 cm thick organic layers (Silvester and Orchard, 1999). According to Silvester and Orchard (1999), sites with higher litter-fall were accompanied by faster breakdown and no relationship was found between litterfall and the depth of the organic layer. The organic layer in our study sites was only 5 to 15 cm thick. Possible reasons for a lack of litter accumulation and build-up of a thick organic layer are the following: removal and disturbance of the organic layer as a consequence of tree fall and removal of five large kauri trees in the 1950s (Thomas and Ogden, 1983) and topography. The topography of the study site (moderate to steep slope) likely explains the negative correlation between organic layer thickness and elevation ($r = -0.539$, $p = 0.021$). Erosive removal of the organic layer and mineral soil on steep slopes and deposition downslope have been shown to affect soil characteristics and C cycling (Vitousek et al., 2003; Yoo et al., 2005; Stacy et al., 2015). For example, in a temperate forest in Japan (Nakane et al., 1984) and a tropical seasonal forest in Thailand (Takahashi et al., 2011) soil CO_2 efflux decreased with increasing slope. However, we did not find any significant correlation between elevation and total soil CO_2 efflux, root biomass, and soil water content suggesting that forest structure (see Sect. 4.2) may have had a stronger effect on soil characteristics than topography.

While mean annual soil temperature partly explains the overall high mean soil CO_2 efflux measured in this forest, soil temperature was not a very good predictor of the temporal variation in total soil CO_2 efflux. Independent of the regression model used, soil temperature explained a small proportion ($< 20\%$, Fig. 4a, Table 3) of the seasonal variation in total soil CO_2 efflux. In temperate forest ecosystems in the Northern Hemisphere (Ngao et al., 2012; Bond-Lamberty and Tompson, 2014) soil temperature often explains more than 50% of the temporal variability in total soil CO_2 efflux. It is important to note that the soil temperature range in this kauri forest was narrow (around $7\,°C$) compared to other temperate forests with a larger seasonal soil temperature amplitude ($> 10\,°C$, Paul et al., 2004). Thus, a seasonal temperature effect may not have been visible in this kauri forest. The Q_{10} value (1.6, Table 3) was at the lower end of the range reported for mixed and evergreen forests ($Q_{10\_10-20°C}$; 0.5-5.6; Bond-Lamberty and Tompson, 2014). However, low Q_{10} values have also been reported for other conifer forests, especially at sites characterized by mild winters (Borken et al., 2002; Curiel Yuste et al., 2005b; Sulzman et al., 2005). Low Q_{10} values in evergreen forests have been explained by the lack of a distinct seasonality in photosynthesis and substrate supply (Curiel Yuste et al., 2005b).

No significant relationship was found between SWC and total soil CO_2 efflux (Fig. 4d, Table 3). However, total soil CO_2 efflux tended to decline with increasing SWC. Excess SWC may negatively affect CO_2 efflux rates by reducing soil aeration and thus CO_2 diffusivity (Janssens and Pilegaard, 2003). Further, low levels of oxygen as a result of high SWC decreases activity of plant roots (Adachi et al., 2006) and the heterotrophic decomposition of soil organic matter (Linn and Doran, 1984). This may be particularly relevant in the clayey soils under study.

4.2 Forest structure and the spatial variation in soil CO_2 efflux

The spatial variability ($CV = 43\%$) of total soil CO_2 efflux in this study is slightly higher compared to other studies with similar numbers of measurements and/or plot size (32–39 %; Epron et al., 2006; Kosugi et al., 2007; Bréchet et al., 2011). The higher spatial variation might be related to differences in tree size and distribution across the plot. The stand is clearly dominated by kauri trees in all size classes (Fig. 1). However, kauri occurs in clusters around the four largest and emergent kauri individuals whose neighbourhood is generally characterised by relatively few trees (see lower centre of Fig. 1). The influence of forest structure (here: kauri tree distribution and tree size, I_c) on total soil CO_2 efflux is confirmed by the significant relationships between I_c and total soil CO_2 efflux, root biomass and mineral soil C : N ratio. Previous studies have shown that kauri has significant effects on soil processes such as pH and nitrogen cycling (Silvester, 2000; Jongkind et al., 2007; Verkaik et al., 2007; Wyse et al., 2014). This is the first study showing that kauri exerts a substantial influence on soil C related processes. Our results also corroborate a study by Katayama et al. (2009) suggesting that the spatial arrangement of emergent trees in a tropical forest is an important factor for generating spatial variation of soil CO_2 efflux. Studies in European beech forests also show that the combination of root, soil and stand structure help to understand the mechanisms underlying soil CO_2 efflux and that forest structure has some influence on the spatial variability of soil CO_2 efflux (Søe and Buchmann, 2005; Ngao et al., 2012).

The relationship between total soil CO_2 efflux and forest structure was strongest within a radius of 5 m (Fig. 5a and b). In a tropical forest, the strongest correlation between soil CO_2 efflux and forest structural parameters was within 6 m from the sampling points (Katayama et al., 2009). A radius of 5 m also provided the best correlation between root biomass and I_c. As measurements of the lateral root extension are not available for kauri, it remains unknown if this distance equals the maximum lateral extension of fine roots from the trunk or represents the distance where fine root density is highest. Based on observations, large lateral roots of mature kauri trees often extend beyond the width of the crown and an extensive network of fine roots extends from the lateral roots into the interface between organic layer and the mineral soil (Bergin and Steward, 2004). The radial fine root spread in mature Northern Hemisphere conifer stands varies considerably (6–20 m) depending on site characteristics and stand structure (Stone and Kalisz, 1991).

In contrast to other studies (e.g. Bréchet et al., 2011; Katayama et al., 2009), we did not find a significant correlation between litterfall and forest structure. Tree size and architecture have been reported to affect the pattern of litterfall distribution on the forest floor (Ferrari and Sugita 1996; Staelens et al., 2004; Zalamea et al., 2012). However, despite a 3-fold difference in tree size across the plot we did not see a significant effect of tree size on total litterfall. This is also reflected in a small within-plot variation in litterfall (CV = 21 %, Table 1). This is confirmed by a litterfall study in four remnant kauri forests where a small variation in litterfall (CV = 17–26 %) was found across a wide range of litter trap positions (Silvester and Orchard, 1999).

Spatial variability in total soil CO_2 efflux was largely attributed to soil temperature and the amount of fine root biomass and associated rhizosphere, with 97 % of the variation explained. This implies a relationship with tree productivity which is in agreement with findings from other conifer forests (Janssens et al., 2001b; Luo and Zhou, 2006). Although roots accounted for less than 30 % of total CO_2 efflux, recent research has shown that both recent photosynthate and fine root turnover can be important sources of C for forest soil CO_2 efflux (Epron et al., 2011; Warren et al., 2012) as discussed below.

4.3 Components of total soil CO_2 efflux

Collar insertion through the organic layer into the mineral soil resulted in a 17 % reduction in soil CO_2 efflux. Similar reductions were found in other ecosystems and demonstrates that collar insertion by only a few centimetres cuts off fine roots (Heinemeyer et al., 2011) and contributions by ectomycorrhizal fungal mats (Phillips et al., 2012) reducing total soil respiration. Thus, collar insertion can cause underestimation of total soil CO_2 efflux. This may be a particular problem in ecosystems where a large amount of roots and mycorrhiza are found in the organic layer and at the interface between the organic layer and an organic rich mineral soil as in this kauri forest.

The partitioning of total soil CO_2 efflux into its main components: heterotrophic respiration (oxidation of soil organic matter) and autotrophic respiration (root and associated mycorrhiza respiration) remains technically challenging. Differences in the proportion of autotrophic or heterotrophic respiration to total soil CO_2 efflux might vary not only among species and ecosystems but also with the method used for partitioning total soil CO_2 efflux (Kuzyakov, 2006; Subke et al., 2006; Millard et al., 2010). Cutting roots through inserting deep collars and trenching increases the dead root biomass (Heinemeyer et al., 2011). As we did not correct our estimates of soil CO_2 efflux for decomposing root-derived CO_2 fluxes the heterotrophic respiration may have been slightly overestimated (Hanson et al., 2000; Kuzyakov, 2006; Ngao et al., 2012). However, both techniques used in this study, trenching and regression-analysis, showed similar results. The proportion of autotrophic respiration in this kauri was between 25 % (trenching) and 28 % (regression analysis) of total soil surface CO_2 efflux. The contribution of autotrophic respiration to total soil CO_2 efflux can account for as little as 10 % to more than 90 % worldwide (Hanson et al., 2000) but values of 45–50 % are typical (Subke et al.,

2006). Our estimate is at the lower end of values observed for Northern Hemisphere conifer and tropical broadleaf forests (30–70 %, Epron et al., 2001; Högberg et al., 2001; Bond-Lamberty and Tompson, 2014; Taylor et al., 2015). This suggests that root and/or rhizosphere activity in this forest is comparatively low. However, a similar proportion of autotrophic respiration (23 %) was estimated for a New Zealand old-growth beech forest (Tate et al., 1993) and an old-growth Douglas-fir site in the Cascades, Oregon (23 %) (Sulzman et al., 2005). Another factor accounting for the differences in values is the depth of trenching (Hansen et al., 2000; Kuzyakov, 2006; Bond-Lamberty et al., 2011). The contribution of autotrophic respiration may have been underestimated as we only trenched to 30 cm depth. It is recommended to trench to a depth beyond the main rooting zone (Subke et al., 2006) and in some studies the trenched plots are dug down to the solid bedrock (Díaz-Pinés et al., 2010).

Total soil CO_2 efflux is not only directly affected by the amount of autotrophic respiration but also by the supply of C through root turnover and root exudates. The decomposition of root debris has been shown to increase microbial activity and thus heterotrophic respiration (Göttlicher et al., 2006). Despite a low root and/or rhizosphere activity the total soil CO_2 efflux in a mycorrhizally associated Douglas-fir forest was dominated by belowground contributions due to the large pool of rhizospheric litter with a relatively high turnover rate (Sulzman et al., 2005). In addition, root exudates containing carbohydrates, sugars and amino acids supply energy for the decomposition of soil C ("priming") (Högberg et al., 2001). Further, a recent study showed that a common root exudate, oxalic acid, promotes soil C loss by releasing organic compounds from mineral-protected aggregates. This indirect mechanism has been found to result in higher C losses compared to simply increasing the supply of energetically more favourable substrates (Keiluweit et al., 2015).

Root activity may also affect physical soil conditions. In some studies, SWC and fine root biomass were negatively correlated (Coomes and Grubb, 2000; Ammer and Wagner, 2002). High uptake of water by kauri fine roots concentrated in the organic layer may lead to lower SWC and slightly higher soil temperatures (Verkaik et al., 2007; Verkaik and Braakhekke, 2007). The drier conditions at the base of trees might be an indicator of good soil aeration that enhances the diffusivity of soil CO_2 into the air (De Jong and Schappert, 1972; Tang et al., 2003).

The soil temperature response was stronger for heterotrophic respiration (Fig. 4b, Table 3). This is in line with other studies conducted in temperate mixed forests (Kirschbaum, 1995; Boone et al., 1998) and suggests a higher sensitivity of heterotrophic respiration to temperature. Below 50 % SWC autotrophic respiration increased with increasing water content (Fig. 4f). A positive correlation between soil water content and autotrophic respiration have also been reported for temperate and tropical forests (Zang et al., 2014; Brunner et al., 2015; Doughty et al., 2015). This is in contrast to other studies which reported that dry conditions enhanced the growth of fine roots in the surface soil resulting in higher proportions of autotrophic respiration (Bhupinderpal-Singh et al., 2003; Noguchi et al., 2007).

5 Conclusion

Our study has two significant findings for southern conifer forests. Firstly, this is the only study quantifying the amount of soil CO_2 efflux in an old-growth kauri forest. Our findings suggest that the loss of soil CO_2 ($1324 \pm 121 \, \mathrm{g\,C\,m^{-2}\,yr^{-1}}$) from this forest type is considerable. Although the contribution of autotrophic respiration is comparatively low ($< 30\%$), root biomass explained a high proportion of the spatial variation in soil CO_2 efflux. This suggests that the total soil CO_2 efflux in this forest is not only directly affected by the amount of autotrophic respiration but also by the supply of C through roots and mycorrhiza. Any modification in root/rhizosphere will most likely result in long-term modifications of the soil CO_2 efflux. This is of relevance given that many kauri forests are threatened by *Phytophthora agathidicida* (Weir et al., 2015) which infects the roots and can lead to tree death (Than et al., 2013). Secondly, this study is the first to confirm that kauri not only exerts a strong control on soil pH and nitrogen cycling but also on soil carbon related processes. The effect of kauri tree size and distribution on total soil CO_2 efflux demonstrates the need to include biotic parameters for better prediction of the spatial variability in soil CO_2 efflux.

6 Data availability

The underlying research data set is available in the figshare data repository (doi:10.17608/k6.auckland.3505796).

Acknowledgements. We thank Andrew Wheeler for his assistance in installing the soil CO_2 efflux chambers, setting up the trenching experiment, measuring soil CO_2 efflux and developing R scripts for calculating soil CO_2 efflux; Roland Lafaele-Pereira and Chris Goodwin for assisting with root sampling and sorting; Tristan Webb for helping with the soil CO_2 efflux measurements; Hasinur Rahman for analysing the soil samples and reviewers for their constructive comments. This research was funded by a Faculty Research Development Fund grant (Project number: 3700359) from the Faculty of Science, University of Auckland to Luitgard Schwendenmann and Cate Macinnis-Ng.

Edited by: A. Don

References

Adachi, M., Bekku, Y. S., Rashidah, W., Okuda, T., and Koizumi, H.: Differences in soil respiration between different tropical ecosystems, Appl. Soil Ecol., 34, 258–265, 2006.

Ammer, C. and Wagner, S.: Problems and options in modelling fine-root biomass of single mature Norway spruce trees at given points from stand data, Can. J. Forest Res., 32, 581–590, 2002.

Andrews, J. A., Harrison, K. G., Matamala, R., and Schlesinger, W. H.: Separation of root respiration using carbon-13 labeling during free-air carbon enrichment (FACE), Soil Sci. Soc. Am. J., 63, 1429–1435, 1999.

Bahn, M., Reichstein, M., Davidson, E. A., Grünzweig, J., Jung, M., Carbone, M. S., Epron, D., Misson, L., Nouvellon, Y., Roupsard, O., Savage, K., Trumbore, S. E., Gimeno, C., Curiel Yuste, J., Tang, J., Vargas, R., and Janssens, I. A.: Soil respiration at mean annual temperature predicts annual total across vegetation types and biomes, Biogeosciences, 7, 2147–2157, doi:10.5194/bg-7-2147-2010, 2010.

Bergin, D. and Steward, G.: Kauri: Establishment, growth and management, New Zealand Indigenous Tree Bulletin No. 2, New Zealand Forest Research Institute, Rotorua, 2004.

Bhupinderpal-Singh, Nordgren, A., Löfvenius, M. O., Högberg, M. N., Mellander, P., and Högberg, P.: Tree root and soil heterotrophic respiration as revealed by girdling of boreal Scots pine forest: Extending observations beyond the first year, Plant Cell Environ., 26, 1287–1296, 2003.

Bond-Lamberty, B. and Thomson, A.: A global database of soil respiration data, Biogeosciences, 7, 1915–1926, doi:10.5194/bg-7-1915-2010, 2010a.

Bond-Lamberty, B. and Thomson, A.: Temperature-associated increases in the global soil respiration record, Nature, 464, 579–582, 2010b.

Bond-Lamberty, B. and Thomson, A.: A Global Database of Soil Respiration Data, Version 3.0, Data set, available at: http://daac.ornl.gov from Oak Ridge National Laboratory Distributed Active Archive Center, Oak Ridge, Tennessee, USA, doi:10.3334/ORNLDAAC/1235, 2014.

Bond-Lamberty, B., Bronson, D., Bladyka, E., and Gower, S. T.: A comparison of trenched plot techniques for partitioning soil respiration, Soil Biol. Biochem., 43, 2108–2114, 2011.

Boone, R. D., Nadelhoffer, K. J., Canary, J. D. and Kaye, J. P.: Roots exert a strong influence on the temperature sensitivity of soil respiration, Nature, 396, 570–572, 1998.

Borken, W., Xu, Y. J., Davidson, E. A., and Beese, F.: Site and temporal variation of soil respiration in European beech, Norway spruce, and Scots pine forests, Global Change Biol., 8, 1205–1216, 2002.

Bréchet, L., Ponton, S., Alméras, T., Bonal, D., and Epron, D.: Does spatial distribution of tree size account for spatial variation in soil respiration in a tropical forest?, Plant Soil, 347, 293–303, 2011.

Brüggemann, N., Gessler, A., Kayler, Z., Keel, S. G., Badeck, F., Barthel, M., Boeckx, P., Buchmann, N., Brugnoli, E., Esperschütz, J., Gavrichkova, O., Ghashghaie, J., Gomez-Casanovas, N., Keitel, C., Knohl, A., Kuptz, D., Palacio, S., Salmon, Y., Uchida, Y., and Bahn, M.: Carbon allocation and carbon isotope fluxes in the plant-soil-atmosphere continuum: a review, Biogeosciences, 8, 3457–3489, doi:10.5194/bg-8-3457-2011, 2011.

Brunner, I., Herzog, C., Dawes, M. A., Arend, M., and Sperisen, C.: How tree roots respond to drought, Front. Plant Sci., 6, 547, 2015.

Campbell, J. L. and Law, B. E.: Forest soil respiration across three climatically distinct chronosequences in Oregon, Biogeochemistry, 73, 109–125, 2005.

Cavagnaro, T. R., Barrios-Masias, F. H., and Jackson, L. E.: Arbuscular mycorrhizas and their role in plant growth, nitrogen interception and soil gas efflux in an organic production system, Plant Soil, 353, 181–194, 2012.

Chave, J., Navarrete, D., Almeida, S., Álvarez, E., Aragão, L. E. O. C., Bonal, D., Châtelet, P., Silva-Espejo, J. E., Goret, J.-Y., von Hildebrand, P., Jiménez, E., Patiño, S., Peñuela, M. C., Phillips, O. L., Stevenson, P., and Malhi, Y.: Regional and seasonal patterns of litterfall in tropical South America, Biogeosciences, 7, 43–55, doi:10.5194/bg-7-43-2010, 2010.

Coomes, D. A. and Grubb, P. J.: Impacts of root competition in forests and woodlands: A theoretical framework and review of experiments, Ecol. Monogr., 70, 171–207, 2000.

Curiel Yuste, J., Janssens, I. A., and Ceulemans, R.: Calibration and validation of an empirical approach to model soil CO_2 efflux in a deciduous forest, Biogeochemistry, 73, 209–230, 2005a.

Curiel Yuste, J., Janssens, I. A., Carrara, R., and Ceulemans, R.: Annual Q_{10} of soil respiration reflects plant phenological patterns as well as temperature sensitivity, Global Change Biol., 10, 161–169, 2005b.

Davidson, E. A., Belk, E., and Boone, R. D.: Soil water content and temperature as independent or confounded factors controlling soil respiration in a temperate mixed hardwood forest, Global Change Biol., 4, 217–227, 1998.

Davidson, E. A., Verchot, L. V., Henrique Cattânio, J., Ackerman, I. L., and Carvalho, J. E. M.: Effects of soil water content on soil respiration in forests and cattle pastures of eastern Amazonia, Biogeochemistry, 48, 53–69, 2000.

Davidson, E. A., Savage, K., Bolstad, P., Clark, D. A., Curtis, P. S., Ellsworth, D. S., Hanson, P. J., Law, B. E., Luo, Y., Pregitzer, K. S., Randolph, J. C., and Zak, D.: Belowground carbon allocation in forests estimated from litterfall and IRGA-based soil respiration measurements, Agr, Forest Meterol., 113, 39–51, 2002.

Davidson, E. A., Janssens, I. A., and Lou, Y.: On the variability of respiration in terrestrial ecosystems: Moving beyond Q_{10}, Global Change Biol., 12, 154–164, 2006.

De Jong, E. and Schappert, H. J.: Calculaton of soil respiration and activity from CO_2 profiles in the soil, Soil Sci., 113, 328–333, 1972.

Díaz-Pinés, E., Schindlbacher, A., Pfever, M., Jandl, R., Zechmeister-Boltenstern, S., and Rubio, A.: Root trenching: A useful tool to estimate autotrophic soil respiration? A case study in an austrian mountain forest, Eur. J. Forest Res., 129, 101–109, 2010.

Doughty, C. E., Metcalfe, D. B., Girardin, C. A. J., Amézquita, F. F., Cabrera, D. G., Huasco, W. H., Silva-Espejo, J. E., Araujo-Murakami, A., Da Costa, M. C., Rocha, W., Feldpausch, T. R., Mendoza, A. L. M., Da Costa, A. C. L., Meir, P., Phillips, O. L., and Malhi, Y.: Drought impact on forest carbon dynamics and fluxes in Amazonia, Nature, 519, 78–82, 2015.

Ecroyd, C. E.: Biological flora of New Zealand 8. *Agathis australis* (D. Don) Lindl. (Araucariaceae) kauri, New Zealand J. Bot., 20, 17–36, 1982.

Enright, N. J. and Ogden, J.: Decomposition of litter from common woody species of kauri (*Agathis australis* Salisb.) forest in northern New Zealand, Aust. J. Ecol., 12, 109–124, 1987.

Epron, D., Dantec, V. L., Dufrene, E., and Granier, A.: Seasonal dynamics of soil carbon dioxide efflux and simulated rhizosphere respiration in a beech forest, Tree Physiol., 21, 145–152, 2001.

Epron, D., Ngao, J., and Granier, A.: Interannual variation of soil respiration in a beech forest ecosystem over a six-year study, Ann. Forest Sci., 61, 499–505, 2004.

Epron, D., Bosc, A., Bonal, D., and Freycon, V.: Spatial variation of soil respiration across a topographic gradient in a tropical rain forest in French Guiana, J. Trop. Ecol., 22, 565–574, 2006.

Epron, D., Ngao, J., Dannoura, M., Bakker, M. R., Zeller, B., Bazot, S., Bosc, A., Plain, C., Lata, J. C., Priault, P., Barthes, L., and Loustau, D.: Seasonal variations of belowground carbon transfer assessed by in situ $^{13}CO_2$ pulse labelling of trees, Biogeosciences, 8, 1153–1168, doi:10.5194/bg-8-1153-2011, 2011.

Fang, C. and Moncrieff, J. B.: A model for soil CO_2 production and transport 1: Model development, Agr. Forest Meterol., 95, 225–236, 1999.

Fang, C., Moncrieff, J. B., Gholz, H. L., and Clark, K. L.: Soil CO_2 efflux and its spatial variation in a Florida slash pine plantation, Plant Soil, 205, 135–146, 1998.

Ferrari, J. B. and Sugita, S.: A spatially explicit model of leaf litter fall in hemlock-hardwood forests, Can. J. Forest Res., 26, 1905–1913, 1996.

Gaudinski, J. B., Trumbore, S. E., Davidson, E. A., and Zheng, S.: Soil carbon cycling in a temperate forest: Radiocarbon-based estimates of residence times, sequestration rates and partitioning of fluxes, Biogeochemistry, 51, 33–69, 2000.

Giardina, C. P., Litton, C. M., Crow, S. E., and Asner, G. P.: Warming-related increases in soil CO_2 efflux are explained by increased below-ground carbon flux, Nat. Clim. Change, 4, 822–827, 2014.

Göttlicher, S., Knohl, A., Wanek, W., Buchmann, N. and Richter, A.: Short-term changes in carbon isotope composition of soluble carbohydrates and starch: From canopy leaves to the root system, Rapid Commun. Mass Spectrom., 20, 653–660, 2006.

Hanson, P. J., Edwards, N. T., Garten, C. T., and Andrews, J. A.: Separating root and soil microbial contributions to soil respiration: A review of methods and observations, Biogeochemistry, 48, 115–146, 2000.

Hedley, C. B., Lambie, S. M., and Dando, J. L.: Edaphic and environmental controls of soil respiration and related soil processes under two contrasting manuka and kanuka shrubland stands in North Island, New Zealand, Soil Res., 51, 390–405, 2013.

Heinemeyer, A., Di Bene, C., Lloyd, A. R., Tortorella, D., Baxter, R., Huntley, B., Gelsomino, A. and Ineson, P.: Soil respiration: Implications of the plant-soil continuum and respiration chamber collar-insertion depth on measurement and modelling of soil CO_2 efflux rates in three ecosystems, Eur. J. Soil Sci., 62, 82–94, 2011.

Hewitt, A. E.: Soil classification in New Zealand: legacy and lessons, Aust. J. Soil Res., 30, 843–854, 1992.

Hibbard, K. A., Law, B. E., Reichstein, M., and Sulzman, J.: An analysis of soil respiration across northern hemisphere temperate ecosystems, Biogeochemistry, 73, 29–70, 2005.

Högberg, P.: Is tree root respiration more sensitive than heterotrophic respiration to changes in soil temperature?, New Phytol., 188, 9–10, 2010.

Högberg, P., Nordgren, A., Buchmann, N., Taylor, A. F. S., Ekblad, A., Högberg, M. N., Nyberg, G., Ottosson-Löfvenius, M., and Read, D. J.: Large-scale forest girdling shows that current photosynthesis drives soil respiration, Nature, 411, 789–792, 2001.

Holland, E. A., Post, W. M., Matthews, E., Sulzman, J., Staufer, R., and Krankina, O.: A Global Database of Litterfall Mass and Litter Pool Carbon and Nutrients, Oak Ridge National Laboratory Distributed Active Archive Center, doi:10.3334/ORNLDAAC/1244, 2015.

Hunt, J. E., Walcroft, A. S., McSeveny, T. M., Rogers, G. N. and Whitehead, D.: Ecosystem respiration in an undisturbed, old-growth, temperate rain forest, Abstract, American Geophysical Union, Fall Meeting, San Francisco, CA, USA, 2008.

Janssens, I. A., Kowalski, A. S., and Ceulemans, R.: Forest floor CO_2 fluxes estimated by eddy covariance and chamber-based model, Agr. Forest Meterol., 106, 61–69, 2001a.

Janssens, I. A., Lankreijer, H., Matteucci, G., Kowalski, A. S., Buchmann, N., Epron, D., Pilegaard, K., Kutsch, W., Longdoz, B., Grünwald, T., Montagnani, L., Dore, S., Rebmann, C., Moors, E. J., Grelle, A., Rannik, Ü., Morgenstern, K., Oltchev, S., Clement, R., Guomundsson, J., Minerbi, S., Berbigier, P., Ibrom, A., Moncrieff, J., Aubinet, M., Bernhofer, C., Jensen, N. O., Vesala, T., Granier, A., Schulze, E., Lindroth, A., Dolman, A. J., Jarvis, P. G., Ceulemans, R., and Valentini, R.: Productivity overshadows temperature in determining soil and ecosystem respiration across European forests, Global Change Biol., 7, 269–278, 2001b.

Janssens, I. A. and Pilegaard, K.: Large seasonal changes in Q_{10} of soil respiration in a beech forest, Global Change Biol., 9, 911–918, 2003.

Jassal, R., Black, A., Novak, M., Morgenstern, K., Nesic, Z., and Gaumont-Guay, D.: Relationship between soil CO_2 concentrations and forest-floor CO_2 effluxes, Agr. Forest Meterol., 130, 176–192, 2005.

Jongkind, A. G., Velthorst, E., and Buurman, P.: Soil chemical properties under kauri (Agathis australis) in The Waitakere Ranges, New Zealand, Geoderma, 141, 320–331, 2007.

Katayama, A., Kume, T., Komatsu, H., Ohashi, M., Nakagawa, M., Yamashita, M., Otsuki, K., Suzuki, M., and Kumagai, T.: Effect of forest structure on the spatial variation in soil respiration in a Bornean tropical rainforest, Agr. Forest Meterol., 149, 1666–1673, 2009.

Keiluweit, M., Bougoure, J. J., Nico, P. S., Pett-Ridge, J., Weber, P. K., and Kleber, M.: Mineral protection of soil carbon counteracted by root exudates, Nat. Clim. Change, 5, 588–595, 2015.

Keith, H., Mackey, B. G. and Lindenmayer, D. B.: Re-evaluation of forest biomass carbon stocks and lessons from the world's most carbon-dense forests, P. Natl. Acad. Sci. USA, 106, 11635–11640, 2009.

Kirschbaum, M. U. F.: The temperature dependence of soil organic matter decomposition, and the effect of global warming on soil organic C storage, Soil Biol. Biochem., 27, 753–760, 1995.

Kosugi, Y., Mitani, T., Itoh, M., Noguchi, S., Tani, M., Matsuo, N., Takanashi, S., Ohkubo, S. and Rahim Nik, A.: Spatial and temporal variation in soil respiration in a Southeast Asian tropical rainforest, Agr. Forest Meterol., 147, 35–47, 2007.

Kuzera, C. L. and Kirkham, D. R.: Soil respiration studies in tallgrass prarie in Missouri, Ecology, 52, 912–915, 1971.

Kuzyakov, Y.: Sources of CO_2 efflux from soil and review of partitioning methods, Soil Biol. Biochem., 38, 425–448, 2006.

Linn, D. M. and Doran, J. W.: Effect of water-filled pore space on carbon dioxide and nitrous oxide production in tilled and non-tilled soils, Soil Sci. Soc. Am. J., 48, 1267-1272, 1984.

Lloyd, J. and Taylor, J. A.: On the temperature dependence of soil respiration, Funct. Ecol., 8, 315–323, 1994.

Longdoz, B., Yernaux, M., and Aubinet, M.: Soil CO_2 efflux measurements in a mixed forest: Impact of chamber disturbances, spatial variability and seasonal evolution, Global Change Biol., 6, 907–917, 2000.

Luo, Y. and Zhou, X.: Temporal and spatial variations in soil respiration, in: Soil Respiration and the Environment, Academic Press, Elsevier, San Diego, CA, USA, 107–131, 2006.

Macinnis-Ng, C. and Schwendenmann, L.: Litterfall, carbon and nitrogen cycling in a southern hemisphere conifer forest dominated by kauri (Agathis australis) during drought, Plant Ecol., 216, 247–262, 2015.

Maier, M. and Schack-Kirchner, H.: Using the gradient method to determine soil gas flux: A review, Agr. Forest Meterol., 192–193, 78–95, 2014.

Maier, M., Schack-Kirchner, H., Hildebrand, E. E., and Schindler, D.: Soil CO_2 efflux vs. soil respiration: Implications for flux models, Agr. Forest Meterol., 151, 1723–1730, 2011.

Metcalfe, D. B., Meir, P., Aragão, L. E. O. C., Malhi, Y., da Costa, A. C. L., Braga, A., Gonçalves, P. H. L., de Athaydes, J., de Almeida, S. S., and Williams, M.: Factors controlling spatio-temporal variation in carbon dioxide efflux from surface litter, roots, and soil organic matter at four rain forest sites in the eastern Amazon, J. Geophys. Res.-Biogeo., 112, G04001, 2007.

Metcalfe, D. B., Fisher, R. A., and Wardle, D. A.: Plant communities as drivers of soil respiration: pathways, mechanisms, and significance for global change, Biogeosciences, 8, 2047–2061, doi:10.5194/bg-8-2047-2011, 2011.

Millard, P., Midwood, A. J., Hunt, J. E., Barbour, M. M., and Whitehead, D.: Quantifying the contribution of soil organic matter turnover to forest soil respiration, using natural abundance $\delta^{13}C$, Soil Biol. Biochem., 42, 935–943, 2010.

Nakane, K., Tsubota, H., and Yamamoto, M.: Cycling of soil carbon in a Japanese red pine forest I. Before a clear-felling, Bot. Mag. Tokyo, 97, 39–60, 1984.

Ngao, J., Epron, D., Delpierre, N., Bréda, N., Granier, A., and Longdoz, B.: Spatial variability of soil CO_2 efflux linked to soil parameters and ecosystem characteristics in a temperate beech forest, Agr. Forest Meterol., 154–155, 136–146, 2012.

Noguchi, K., Konôpka, B., Satomura, T., Kaneko, S., and Takahashi, M.: Biomass and production of fine roots in Japanese forests, J. Forest Res., 12, 83–95, 2007.

Nottingham, A. T., Turner, B. L., Winter, K., van der Heijden, M. G. A., and Tanner, E. V. J.: Arbuscular mycorrhizal mycelial respiration in a moist tropical forest, New Phytol., 186, 957–967, 2010.

Ohashi, M., Kumagai, T., Kume, T., Gyokusen, K., Saitoh, T. M., and Suzuki, M.: Characteristics of soil CO_2 efflux variability in an aseasonal tropical rainforest in Borneo Island, Biogeochemistry, 90, 275–289, 2008.

Oishi, A. C., Palmroth, S., Butnor, J. R., Johnsen, K. H., and Oren, R.: Spatial and temporal variability of soil CO_2 efflux in three proximate temperate forest ecosystems, Agr. Forest Meterol., 171–172, 256–269, 2013.

Padamsee, M., Johansen, R. B., Stuckey, S. A., Williams, S. E., Hooker, J. E., Burns, B. R., and Bellgard, S. E.: The arbuscular mycorrhizal fungi colonising roots and root nodules of New

Zealand kauri *Agathis australis*, Fungal Biol., 120, 807–817, 2016.

Paul, K. I., Polglase, P. J., Smethurst, P. J., Connell, A. C., Carlyle, C. L., and Khanna, P. A.: Soil temperature under forests: a simple model for predicting soil temperature under a range of forest types, Agr. Forest Meterol., 121, 167–182, 2004.

Phillips, C. L., Kluber, L. A., Martin, J. P., Caldwell, B. A., and Bond, B. J.: Contributions of ectomycorrhizal fungal mats to forest soil respiration, Biogeosciences, 9, 2099–2110, doi:10.5194/bg-9-2099-2012, 2012.

Raich, J. W. and Potter, C. S.: Global patterns of carbon dioxide emissions from soils, Global Biogeochem. Cy., 9, 23–36, 1995.

Raich, J. W. and Schlesinger, W. H.: The global carbon dioxide flux in soil respiration and its relationship to vegetation and climate, Tellus B, 44, 81–99, 1992.

Raich, J. W. and Tufekcioglu, A.: Vegetation and soil respiration: Correlations and controls, Biogeochemistry, 48, 71–90, 2000.

Rayment, M. B. and Jarvis, P. G.: Temporal and spatial variation of soil CO_2 efflux in a Canadian boreal forest, Soil Biol. Biochem., 32, 35–45, 2000.

R Core Team: R: A language and environment for statistical computing. R Foundation for Statistical Computing, Vienna, Austria, available at: http://www.R-project.org/ (last access: 14 January 2016), 2014

Reichstein, M., Rey, A., Freibauer, A., Tenhunen, J., Valentini, R., Banza, J., Casals, P., Cheng, Y., Grünzweig, J. M., Irvine, J., Joffre, R., Law, B. E., Loustau, D., Miglietta, F., Oechel, W., Ourcival, J., Pereira, J. S., Peressotti, A., Ponti, F., Qi, Y., Rambal, S., Rayment, M., Romanya, J., Rossi, F., Tedeschi, V., Tirone, G., Xu, M., and Yakir, D.: Modeling temporal and large-scale spatial variability of soil respiration from soil water availability, temperature and vegetation productivity indices, Global Biogeochem. Cy., 17, 1104, 2003.

Rustad, L. E., Huntington, T. G., and Boone, R. D.: Controls on soil respiration: Implications for climate change, Biogeochemistry, 48, 1–6, 2000.

Sayer, E. J., Heard, M. S., Grant, H. K., Marthews, T. R., and Tanner, E. V. J.: Soil carbon release enhanced by increased tropical forest litterfall, Nat. Clim. Change, 1, 304–307, 2011.

Schlesinger, W. H. and Andrews, J. A.: Soil respiration and the global carbon cycle, Biogeochemistry, 48, 7–20, 2000.

Schwendenmann, L. and Macinnis-Ng, C.: Soil CO_2 efflux in an old-growth southern conifer forest – data sets, doi:10.17608/k6.auckland.3505796, 2016.

Schwendenmann, L., Veldkamp, E., Brenes, T., O'Brien, J. J., and Mackensen, J.: Spatial and temporal variation in soil CO_2 efflux in an old-growth neotropical rain forest, La Selva, Costa Rica, Biogeochemistry, 64, 111–128, 2003.

Scott-Denton, L. E., Rosenstiel, T. N., and Monson, R. K.: Differential controls by climate and substrate over the heterotrophic and rhizospheric components of soil respiration, Global Change Biol., 12, 205–216, 2006.

Silvester, W. B.: The biology of kauri (*Agathis australis*) in New Zealand II. Nitrogen cycling in four kauri forest remnants, New Zealand J. Bot., 38, 205–220, 2000.

Silvester, W. B. and Orchard, T. A.: The biology of kauri (*Agathis australis*) in New Zealand. I. Production, biomass, carbon storage, and litter fall in four forest remnants, New Zealand J. Bot., 37, 553–571, 1999.

Smith, P. and Fang, C.: Carbon cycle: A warm response by soils, Nature, 464, 499–500, 2010.

Soil Survey Staff.: Keys to Soil Taxonomy, 12th Edn., USDA – Natural Resources Conservation Service, Washington, D.C., 2014.

Søe, A. R. B. and Buchmann, N.: Spatial and temporal variations in soil respiration in relation to stand structure and soil parameters in an unmanaged beech forest, Tree Physiol., 25, 1427–1436, 2005.

Stacy, E. M., Hart, S. C., Hunsaker, C. T., Johnson, D. W., and Berhe, A. A.: Soil carbon and nitrogen erosion in forested catchments: implications for erosion-induced terrestrial carbon sequestration, Biogeosciences, 12, 4861–4874, doi:10.5194/bg-12-4861-2015, 2015.

Staelens, J., Nachtergale, L., and Luyssaert, S.: Predicting the spatial distribution of leaf litterfall in a mixed deciduous forest, Forest Sci., 50, 836–847, 2004.

Steward, G. A. and Beveridge, A. E.: A review of New Zealand kauri (*Agathis australis* (D. Don) Lindl.): Its ecology, history, growth and potential for management for timber, New Zealand J. Forest Sci., 40, 33–59, 2010.

Stone, E. L. and Kalisz, P. J.: On the maximum extent of tree roots, Forest Ecol. Manage., 46, 59–102, 1991.

Subke, J., Inglima, I., and Cotrufo, M. F.: Trends and methodological impacts in soil CO_2 efflux partitioning: A metaanalytical review, Global Change Biol., 12, 921–943, 2006.

Sulzman, E. W., Brant, J. B., Bowden, R. D., and Lajtha, K.: Contribution of aboveground litter, belowground litter, and rhizosphere respiration to total soil CO_2 efflux in an old growth coniferous forest, Biogeochemistry, 73, 231–256, 2005.

Takahashi, M., Hirai, K., Limtong, P., Leaungvutivirog, C., Panuthai, S., Suksawang, S., Anusontpornperm, S., and Marod, D.: Topographic variation in heterotrophic and autotrophic soil respiration in a tropical seasonal forest in Thailand, Soil Sci. Plant Nutr., 57, 452–465, 2011.

Tang, J., Baldochi, D. D., Qi, Y., and Xu, L.: Assessing soil CO_2 efflux using continuous measurements of CO_2 within the soil profile with small solid-stat sensors, Agr. Forest Meteorol., 118, 207–220, 2003.

Tate, K. R., Ross, D. J., O'Brien, B. J., and Kelliher, F. M.: Carbon storage and turnover, and respiratory activity, in the litter and soil of an old-growth southern beech (*Nothofagus*) forest, Soil Biol. Biochem., 25, 1601–1612, 1993.

Taylor, A. J., Lai, C., Hopkins, F. M., Wharton, S., Bible, K., Xu, X., Phillips, C., Bush, S. and Ehleringer, J. R.: Radiocarbon-based partitioning of soil respiration in an old-growth coniferous forest, Ecosystems, 18, 459–470, 2015.

Than, D. J., Hughes, K. J. D., Boonhan, N., Tomlinson, J. A., Woodhall, J. W., and Bellgard, S. E.: A TaqMan real-time PCR assay for the detection of *Phytophthora* "taxon Agathis" in soil, pathogen of Kauri in New Zealand, Forest Pathol., 43, 324–330, 2013.

Thomas, G. M. and Ogden, J.: The scientific reserves of Auckland University. I. General introduction to their history, vegetation, climate and soils, Tane, 29, 143–162, 1983.

Trumbore, S.: Carbon respired by terrestrial ecosystems – Recent progress and challenges, Global Change Biol., 12, 141–153, 2006.

Urrutia-Jalabert, R.: Primary Productivity and Soil Respiration in Fitzroya Cupressoides Forests of Southern Chile and Their Environmental Controls, PhD Thesis, University of Oxford, Oxford, 2015.

Valentine, A. J. and Kleinert, A.: Respiratory responses of arbuscular mycorrhizal roots to short-term alleviation of P deficiency, Mycorrhiza, 17, 137–143, 2007.

van't Hoff, J. H.: Lectures on Theoretical and Physical Chemistry. Part I. Chemical Dynamics (translated by R. A. Lehfeldt), Edward Arnold, London, 224–229, 1898.

Verkaik, E. and Braakhekke, W. G.: Kauri trees (*Agathis australis*) affect nutrient, water and light availability for their seedlings, New Zealand J. Ecol., 31, 39–46, 2007.

Verkaik, E., Jongkind, A. G., and Berendse, F.: Short-term and long-term effects of tannins on nitrogen mineralisation and litter decomposition in kauri (*Agathis australis* (D. Don) Lindl.) forests, Plant Soil, 287, 337–345, 2006.

Verkaik, E., Gardner, R. O., and Braakhekke, W. G.: Site conditions affect seedling distribution below and outside the crown of Kauri trees (*Agathis australis*), New Zealand J. Ecol., 31, 13–21, 2007.

Vitousek, P., Chadwick, O. A., Matson, P., Allison, S., Derry, L., Kettley, L., Luers, A., Mecking, E., Monastra, V., and Porder S.: Erosion and the rejuvenation of weathering-derived nutrient supply in an old tropical landscape, Ecosystems, 6, 762–772, 2003.

Warren, J. M., Iversen, C. M., Garten Jr., C. T., Norby, R. J., Childs, J., Brice, D., Evans, R. M., Gu, L., Thornton, P., and Weston, D. J.: Timing and magnitude of C partitioning through a young loblolly pine (*Pinus taeda* L.) stand using ^{13}C labeling and shade treatments, Tree Physiol., 32, 799–813, 2012.

Weir, B. S., Paderes, E. P., Anand, N., Uchida, J. Y., Pennycook, S. R., Bellgard, S. E. and Beever, R. E.: A taxonomic revision of phytophthora clade 5 including two new species, *Phytophthora agathidicida* and *P. cocois*, Phytotaxa, 205, 21–38, 2015.

Whitlock, J. S.: Soil development in a kauri forest succession: Huapai scientific reserve, unpublished Master thesis, University of Auckland, Auckland, 1985.

Wunder, J., Perry, G. L. W., and McCloskey, S. P. J.: Structure and composition of a mature kauri (*Agathis australis*) stand at Huapai Scientific Reserve, Waitakere Range, New Zealand Tree-Ring Site Rep., 33, 1–19, 2010.

Wyse, S. V. and Burns, B. R.: Effects of *Agathis australis* (New Zealand kauri) leaf litter on germination and seedling growth differs among plant species, New Zealand J. Ecol., 37, 178–183, 2013.

Wyse, S. V., Burns, B. R., and Wright, S. D.: Distinctive vegetation communities are associated with the long-lived conifer *Agathis australis* (New Zealand kauri, Araucariaceae) in New Zealand rainforests, Austral Ecol., 39, 388–400, 2014.

Xu, M. and Qi, Y.: Spatial and seasonal variations of Q_{10} determined by soil respiration measurements at a Sierra Nevadan forest, Global Biogeochem. Cy., 15, 687–697, 2001.

Yoo, K., Amundson, R., Heimsath, A. M., and Dietrich, W. E.: Erosion of upland hillslope soil organic carbon: Coupling field measurements with a sediment transport model, Global Biogeochem. Cy., 19, GB3003, doi:10.1029/2004GB002271, 2005.

Zalamea, M., Gonzalez, G., and Gould, W.: Comparing litterfall and standing vegetation: assessing the footprint of litterfall traps, in: Tropical Forests, edited by: Sudarshana, P., Nageswara-Rao, M., and Soneji, J. R., Intech, 21–36, available at: http://www.intechopen.com/books/tropical-forests (last access: 28 January 2015), 2012.

Zang, U., Goisser, M., Häberle, K., Matyssek, R., Matzner, E. and Borken, W.: Effects of drought stress on photosynthesis, rhizosphere respiration, and fine-root characteristics of beech saplings: A rhizotron field study, J. Plant Nutr. Soil Sci., 177, 168–177, 2014.

12

Decision support for the selection of reference sites using ^{137}Cs as a soil erosion tracer

Laura Arata[1], Katrin Meusburger[1], Alexandra Bürge[1], Markus Zehringer[2], Michael E. Ketterer[3], Lionel Mabit[4], and Christine Alewell[1]

[1]Environmental Geosciences, Department of Environmental Sciences, University of Basel, Basel, Switzerland
[2]State Laboratory Basel-City, Basel, Switzerland
[3]Chemistry Department, Metropolitan State University of Denver, Colorado, USA
[4]Soil and Water Management & Crop Nutrition Laboratory, FAO/IAEA Agriculture & Biotechnology Laboratory, Vienna, Austria

Correspondence to: Katrin Meusburger (katrin.meusburger@unibas.ch)

Abstract. The classical approach of using ^{137}Cs as a soil erosion tracer is based on the comparison between stable reference sites and sites affected by soil redistribution processes; it enables the derivation of soil erosion and deposition rates. The method is associated with potentially large sources of uncertainty with major parts of this uncertainty being associated with the selection of the reference sites. We propose a decision support tool to Check the Suitability of reference Sites (CheSS). Commonly, the variation among ^{137}Cs inventories of spatial replicate reference samples is taken as the sole criterion to decide on the suitability of a reference inventory. Here we propose an extension of this procedure using a repeated sampling approach, in which the reference sites are resampled after a certain time period. Suitable reference sites are expected to present no significant temporal variation in their decay-corrected ^{137}Cs depth profiles. Possible causes of variation are assessed by a decision tree. More specifically, the decision tree tests for (i) uncertainty connected to small-scale variability in ^{137}Cs due to its heterogeneous initial fallout (such as in areas affected by the Chernobyl fallout), (ii) signs of erosion or deposition processes and (iii) artefacts due to the collection, preparation and measurement of the samples; (iv) finally, if none of the above can be assigned, this variation might be attributed to "turbation" processes (e.g. bioturbation, cryoturbation and mechanical turbation, such as avalanches or rockfalls). CheSS was exemplarily applied in one Swiss alpine valley where the apparent temporal variability called into question the suitability of the selected reference sites. In general we suggest the application of CheSS as a first step towards a comprehensible approach to test for the suitability of reference sites.

1 Introduction

Soil erosion is a global threat (Lal, 2003). Recently estimated erosion rates range from low rates of 0.001–2 t ha^{-1} yr^{-1} on flat relatively undisturbed lands (Pimentel, 2006) to high rates under intensive agricultural use of > 50 t ha^{-1} yr^{-1}. In mountainous regions, rates ranging from 1 to 30 t ha^{-1} yr^{-1} have been reported (e.g. Descroix and Mathys, 2003; Frankenberg et al., 1995; Konz et al., 2012) where they often exceed the natural process of soil formation (Alewell et al., 2015). The use of the artificial radionuclide ^{137}Cs as a soil erosion tracer has been increasing during the last decades, and the method has been applied all over the world with success (e.g. Mabit et al., 2013; Zapata, 2002). The use of ^{137}Cs as a soil erosion tracer allows for an integrated temporal estimate of the total net soil redistribution rate per year since the time of the main fallout, including all erosion processes by water, wind and snow during summer and winter seasons (Meusburger et al., 2014).

^{137}Cs was released into the atmosphere during nuclear bomb tests and as a consequence of nuclear power plant (NPP) accidents, such as Chernobyl in April 1986. It reaches

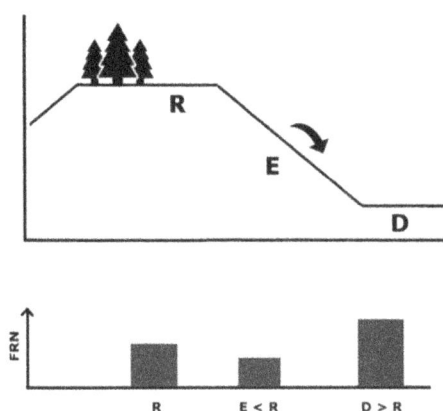

Figure 1. Concept of the fallout radionuclide (FRN) traditional method, in which the FRN content of a reference site located in a flat and undisturbed area (R) is compared to the FRN content of disturbed sites (E and D). If the FRN at the site under investigation is lower than at the reference site, the site has experienced erosion processes (E). If the FRN content is greater than at the reference site, the site has experienced deposition processes (D).

the land surface by dry and wet fallouts and once deposited on the ground, it is strongly bound to fine particles at the soil surface. Due to its low vertical migration rates, it moves predominantly in association with fine soil particles through physical processes and provides an effective track of soil and sediment redistribution processes (Mabit et al., 2008). The traditional approach to using the ^{137}Cs method is based on the comparison between the inventory (total radionuclide activity per unit area) at a given sampling site and that of a so-called reference site located in a flat and undisturbed stable area. The method indicates the occurrence of erosion processes at sites with a lower ^{137}Cs inventory compared to the reference site and sediment deposition processes at sites with a greater ^{137}Cs inventory (Fig. 1a). Specific mathematical conversion models allow for the derivation of quantitative estimates of soil erosion and deposition rates from the latter comparison (IAEA, 2014).

The efficacy of the method relies on an accurate selection of representative reference sites (Mabit et al., 2008; Owens and Walling, 1996; Sutherland, 1996). The measured total ^{137}Cs inventory at the reference sites represents the baseline fallout (i.e. reference inventory), a fundamental parameter for the qualitative and quantitative assessment of soil redistribution rates (Loughran et al., 2002). It is used for the comparison with the total ^{137}Cs inventories of the sampling sites and therefore determines if and how strongly a site is eroding or accumulating sediments. Moreover, the depth profile of the ^{137}Cs distribution in the soil at the reference site plays a very important role, as the shape of this profile is used in the conversion models to convert changes in ^{137}Cs inventory to quantitative estimates of soil erosion rates (Walling et al., 2002). Recent studies have demonstrated the sensitivity of conversion models to uncertainties or even biases in the ref-

erence inventory (e.g. Arata et al., 2016; Iurian et al., 2014; Kirchner, 2013).

The close proximity of a reference site to the area under investigation is required to meet the assumption that both have experienced similar initial fallout. The latter is particularly important if the study area was strongly affected by Chernobyl fallout, which, aside from global fallout from nuclear weapons testing, is the major ^{137}Cs input in many regions of Europe. Because of different geographical situations and meteorological conditions at the time of passage of the radioactive cloud, the contamination associated with Chernobyl fallout was very inhomogeneous (Chawla et al., 2010; Alewell et al., 2014). Therefore, in some areas a significant small-scale variability in ^{137}Cs distribution may be expected. As already pointed out by Lettner et al. (1999) and Owens and Walling (1996), this might impede the comparison between reference and sampling sites. To adequately consider the spatial variability in the FRN fallout, multiple reference sites should be selected and the variability within the sites properly addressed (Kirchner, 2013; Mabit et al., 2013; Pennock and Appleby, 2002). In addition, the reference site should not have experienced any soil erosion or deposition processes since the main ^{137}Cs fallout (which generally requires that it was under continuous vegetation cover, such as perennial grass). Different forms of turbation, including animal turbation, anthropogenic turbation and cryoturbation or snow processes, may also affect the ^{137}Cs soil depth distribution at the reference site. Finally, the collection of the samples, the preparation process and gamma analysis might introduce a certain level of uncertainty, which should be carefully considered. For instance, Lettner et al. (1999) estimated that preparation and measuring processes contribute 12.2 % to the overall variability in the reference inventory. Guidance in the form of independent indicators (e.g. stable isotopes as suggested by Meusburger et al., 2013) for the suitability of reference sites might assist with the selection of reference sites.

All in all the suitability or unsuitability of references site is crucial; it may even be the most crucial step in all FRN-based erosion assessments. The general suitability of ^{137}Cs-based erosion assessment has been recently and controversially discussed (Parsons and Forster, 2011, 2013; Mabit et al., 2013). We would like to propose that the FRN community agree on general concepts and sampling strategies to test the suitability of reference sites in order to improve the method and establish trust in this useful erosion assessment method. Up to now, the variability among spatial replicate samples at reference sites has commonly been the sole criterion to decide on the suitability of a reference value. We propose an extended method to Check the Suitability of reference Sites (CheSS) using a repeated sampling strategy and an assessment of the temporal variability of reference sites. The suitability of reference sites for an accurate application of ^{137}Cs as a soil erosion tracer is tested at Urseren Valley (Canton Uri, Swiss Central Alps).

2 CheSS (Check the Suitability of reference Sites): a concept to assess the suitability of reference sites for the application of ^{137}Cs as a soil erosion tracer

2.1 Repeated sampling strategy and calculation of inventories

The time period for the repeated sampling of reference sites needed for the application of ^{137}Cs as a soil erosion tracer will be site- and case-specific and depends on the initial small-scale spatial variability and the depth distribution of the reference inventory. The time span should be of sufficient length to cause an inventory change that is larger than the uncertainty related to the inventory assessment, e.g. larger than 35 %. In our study site, which is affected by anthropogenic disturbance and snow erosion of several millimetres per winter, 2 years can be considered sufficient (Meusburger et al., 2014). Several spatial repetitions following the suggestion of Sutherland (1996) are necessary and should be analysed separately to investigate the small-scale variability in ^{137}Cs in the area. As we detected measurement differences between different detectors (see below), all samples should ideally be measured for ^{137}Cs activity using the same analytical facilities. Finally, ^{137}Cs activity needs to be decay corrected to the same date (either the period of the first sampling campaign or the second) considering the half-life of ^{137}Cs (30.17 years).

The decay-corrected ^{137}Cs activities (act, Bq kg^{-1}) of each soil layer in the depth profile are converted into inventories (Inv, Bq m^{-2}) with the following equation:

$$\text{Inv} = \text{act} \times xm, \tag{1}$$

where xm is the measured mass depth of fine soil material (< 2 mm fraction; kg m^{-2}) in the respective soil sample. The depth profile of each reference site is then displayed as inventory (Bq m^{-2}) against the depth of each layer (cm). The repeated sampling inventory change (Inv$_{\text{change}}$) can then be defined as

$$\text{Inv}_{\text{change}} = \frac{\text{Inv}_{t0} - \text{Inv}_{t1}}{\text{Inv}_{t0}} \times 100, \tag{2}$$

where t_0 and t_1 are the dates of the first and second sampling campaigns, respectively, Inv$_{t1}$ is the ^{137}Cs inventory (Bq m^{-2}) at t_1 and Inv$_{t0}$ is the ^{137}Cs inventory at $t0$. Positive values of Inv$_{\text{change}}$ indicate erosion, whereas negative values stand for deposition.

2.2 A decision tree to assess the suitability of reference sites

We evaluated the suitability of the reference sites by analysing, in addition to the spatial variability, the temporal variation in the ^{137}Cs inventory. Given the assumption that no additional deposition of ^{137}Cs occurred at the sites during the investigated time window (which is valid worldwide except for the areas affected by the Fukushima Daiichi fallout), any temporal variation in the ^{137}Cs content should be attributable to different forms of soil disturbance or artefacts in the preparation or measurement of the samples. The potential causes of the spatial and temporal variation in the ^{137}Cs total inventories and depth profiles are examined through a decision tree which includes three main nodes (Fig. 2).

2.2.1 Node 1: spatial variation in FRN total inventory

Firstly, the spatial variation in the ^{137}Cs total inventory at each reference site is tested. Ideally, several replicates have been collected. If the coefficient of variation (CV) exceeds 35 % as suggested by Sutherland (1996), this could be a sign of unsuitability of the reference site, but it leaves the possibility of (i) increasing sampling numbers, (ii) analysing the causes of the spatial variation (see CheSS A to D) and (iii) moving to nodes 2 and 3 in CheSS.

2.2.2 Node 2: variation in the ^{137}Cs depth profile

Secondly, whether there is a significant variation between the ^{137}Cs depth profiles measured as spatial or temporal (in $t0$ and $t1$) replicates is tested. In theory, at a stable site the shape of the depth profile should not change between replicates. Consequently, a regression between the FRN activity depth profiles collected as spatial or temporal replicates should follow a 1 : 1 line, and the variability should lie within the range of the observed spatial uncertainty (node 1). A deviation of the linear regression coefficient from the 1 : 1 line in combination with high residues and low R^2 values (< 0.5 R^2) indicates an immediate and significant change in the profile, which is typically caused by anthropogenic disturbance. For the FRN application at ploughed sites, the reference site might still be considered appropriate if the total inventory is not affected because conversion models used for ploughed sites are less sensitive to the shape of the FRN depth distribution. For unploughed soils, again the analysis of causes A to D might help in understanding the causes of the variability. Alternative options would be to take temporal replicates to evaluate the stability and thus the suitability of the reference site (node 3).

2.2.3 Node 3: temporal variation in FRN total inventory

If the CV of all replicates taken in $t0$ and $t1$ is < 35 %, the reference site might be used for the FRN method. The longer the time period between the first and second sampling, the more reliable the yielded assessments. A suitable test for significant differences should confirm or reject the hypothesis of ^{137}Cs total inventory stability over time. If the potential causes of variation (A to D) do not apply, the site is not suitable for the traditional FRN approach, but a repeated sampling approach could still be used to assess soil redistribution rates based on FRN methods (Porto et al., 2014; Kachanoski and de Jong, 1984).

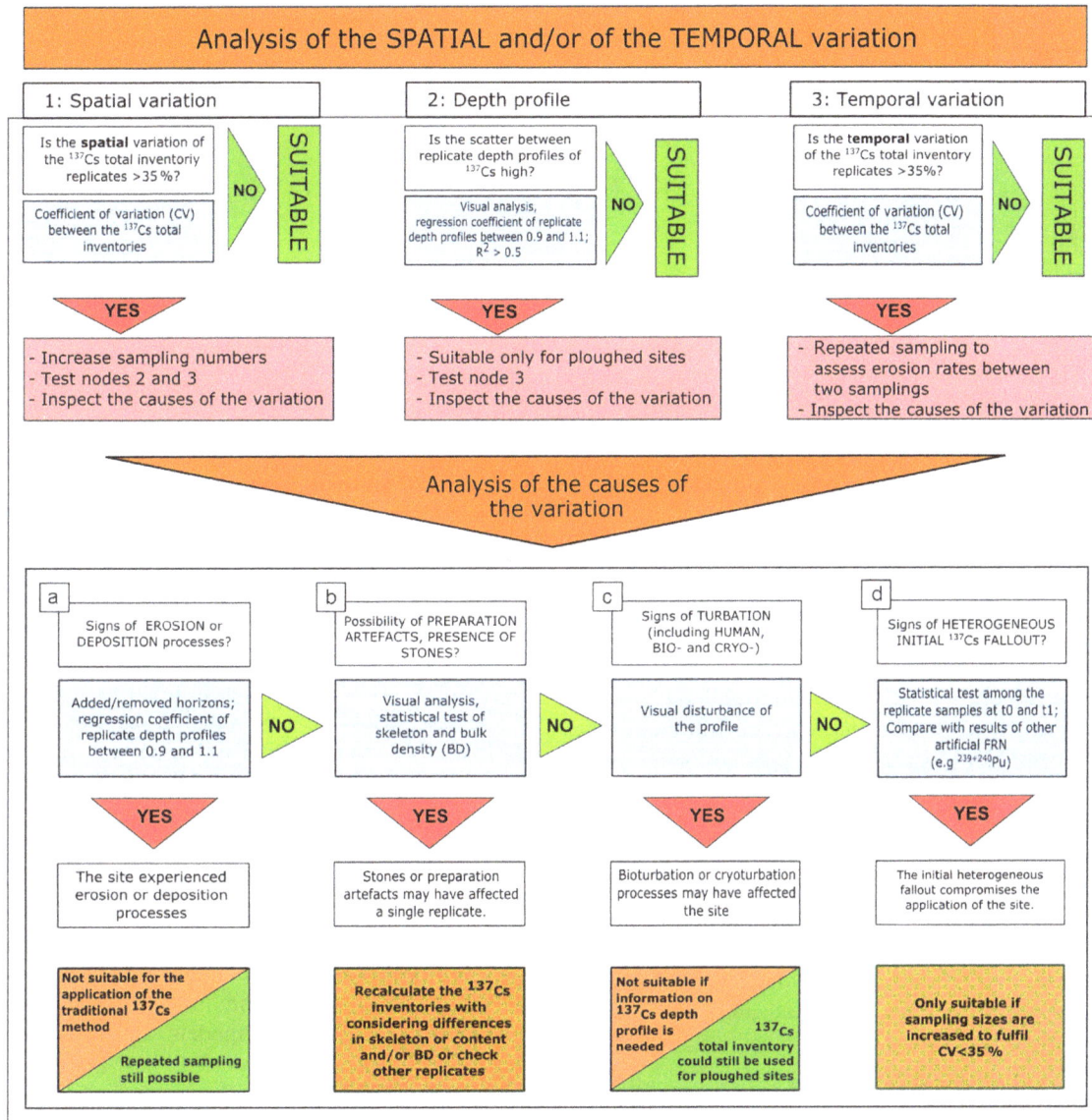

Figure 2. The CheSS decision tree to evaluate the suitability of a reference site for using ^{137}Cs as a soil erosion tracer.

2.2.4 Signs of disturbance associated with erosion and deposition processes (A)

A variation in the ^{137}Cs depth profile may have been caused by soil movement processes affecting the site (Fig. 2a). If the site experienced a loss of soil due to erosion, we expect to observe a removal of the top soil layers of the profile measured, for instance during the second sampling campaign (Fig. 3; red values below the reference profile). Further, the regression coefficient of the reference site that was affected by erosion will tend to be < 0.9 when plotted against a suitable reference profile or (for node 3) the reference profile before the disturbance (Fig. 3). In the case of deposition, a sedimentation layer should be found on the top of the reference depth profile, assuming that no ploughing operations affected

the site (Fig. 3; red values above the reference profile). In this case, the regression coefficient will be > 1.1. Information on the depth distribution of another FRN might provide additional reliable confirmation. If redistribution processes are confirmed, the site is not suitable as a reference site and another location or a repeated FRN sampling approach to estimate erosion rates between the two sampling campaign should be considered (Kachanoski and de Jong, 1984).

2.2.5 Sampling or preparation artefacts (B)

One very common artefact which might bias the comparison between the samples collected at different sites or at $t0$ and $t1$ is the difference in the skeleton content (the percentage of soil fractions > 2 mm; Fig. 2b). The presence of stones might

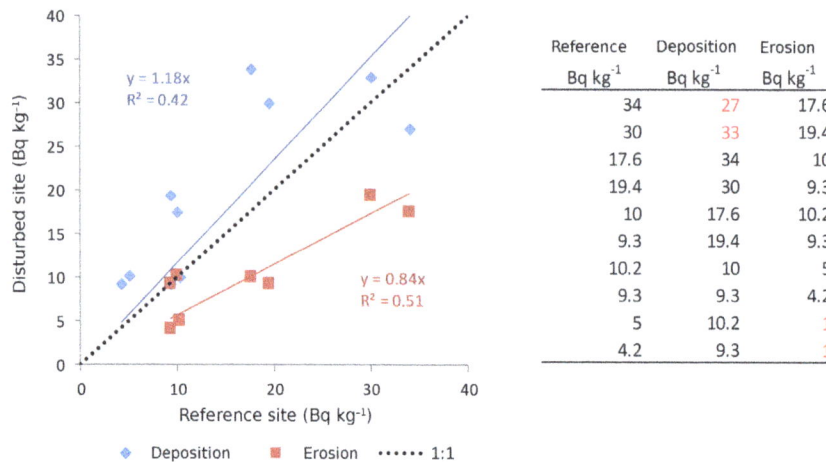

Reference Bq kg^{-1}	Deposition Bq kg^{-1}	Erosion Bq kg^{-1}
34	27	17.6
30	33	19.4
17.6	34	10
19.4	30	9.3
10	17.6	10.2
9.3	19.4	9.3
10.2	10	5
9.3	9.3	4.2
5	10.2	1
4.2	9.3	1

Figure 3. Hypothetical signs of sheet erosion (red) and deposition (blue) on a depth profile compared to an undisturbed site.

determine passways for water and very fine particles and solutes in the soil and thus influence the accumulation and migration of ^{137}Cs through the soil layers. As ^{137}Cs reaches the soil by fallout from the atmosphere, the common shape of the ^{137}Cs distribution along the undisturbed depth profile can be described by an exponential function with the highest ^{137}Cs concentrations located in the uppermost soil layers (Mabit et al., 2008; Walling et al., 2002). This is particularly the case for soils with a low skeleton content (Fig. 4a) since the presence of stones may affect ^{137}Cs depth distribution either through (i) impeding the ^{137}Cs downward migration (^{137}Cs activity could then be concentrated in the layer above the stone; Fig. 4b) or (ii) creating macropores and micropores, favouring the ^{137}Cs associated with fine particles to "migrate" to deeper layers (Fig. 4c) or causing lateral movement which will induce a lower ^{137}Cs content in our samples.

As such, the seemingly spatial or temporal variation in the depth profile might indeed be a spatial variation induced by differences in skeleton content and/or bulk densities. Higher bulk densities will result in higher increment inventories even if ^{137}Cs activities at the layers are comparable. Thus, a thorough control (eventually through a statistical test, such as a paired t test) of whether skeleton content and bulk densities are comparable between replicates is suggested. Finally, sampling, preparation artefacts and measuring processes may produce various sources of error between different sites and years. The latter is especially the case if different people prepare the samples. An estimation of possible errors might be considered, for example through a simulation of different increment assignment along the profile. If different detectors or different calibration sources and/or geometry are used in the two sampling campaigns, a comparability check of the measurements is advisable. For instance, a subset of samples could be measured with the two different detectors, and any potential discrepancy in the results should be properly reported.

2.2.6　Signs of soil disturbance (C)

Different forms of disturbance, such as bioturbation, cryoturbation or even human-induced soil perturbation (e.g. tillage, seedbed preparation or digging), might have influenced the ^{137}Cs depth distribution between different sites and $t0$ and $t1$ (Fig. 2c). Occurrences of turbation are often difficult to identify prior to sampling but might eventually be detected by using other tracing approaches, such as the δ^{13}C depth distribution (Meusburger et al., 2013; Schaub and Alewell, 2009). In the case of turbation, the shape of the depth profile will be highly variable and should not be considered in the estimation of soil redistribution rates for unploughed soils. Nonetheless, the total inventory of ^{137}Cs at a ploughed site could still be used in combination with simple and basic mathematical conversion models, such as the proportional model (Ritchie and McHenry, 1990; IAEA, 2014), which require information only about the total reference inventory of ^{137}Cs and do not need detailed information about the ^{137}Cs depth distribution.

2.2.7　Signs of a heterogeneous initial fallout of ^{137}Cs over the area (D)

Finally, a significant difference between reference replicates may be caused by high small-scale spatial variability in ^{137}Cs distribution at the site due to heterogeneous initial fallout over the study area (Fig. 2d). In Europe, significant small-scale variability in ^{137}Cs distribution is known to be due to the Chernobyl fallout, which was characterized by high ^{137}Cs deposition associated with few rain events. Compared to nuclear bomb test fallout, the Chernobyl fallout was significantly more heterogeneous (e.g. Alewell et al., 2014). Therefore, in the areas affected by the Chernobyl fallout, sites sampled closely to each other may present very different ^{137}Cs contents. It is therefore necessary to investigate the small-scale spatial variability (e.g. the same scale as distance be-

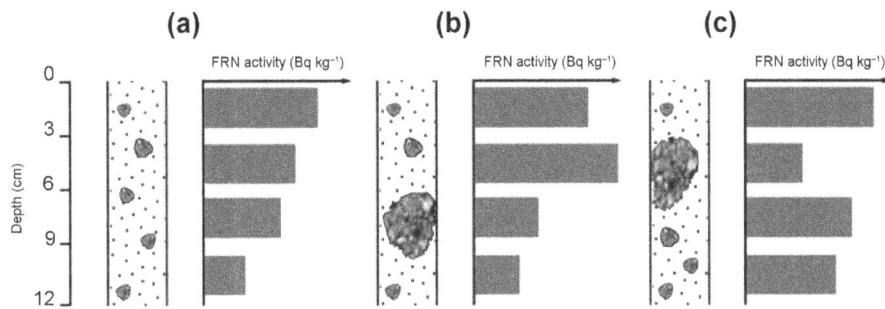

Figure 4. Possible influence of stones on the FRN depth distribution.

tween reference site replicates) measured at both or at least one sampling campaign by looking at the CV again, as presented in the previous sections, or through a statistical test (for example, the analysis of variance, ANOVA). If the spatial variability is highly significant, the site should not be envisaged as a reference site for the application of the ^{137}Cs method unless the number of samples collected for the determination of the reference baseline is large enough (at least 10) to counterweight the small-scale variability within the site (Mabit et al., 2012; Sutherland, 1996; Kirchner, 2013). A possible validation of this cause of heterogeneity might be a comparison with the spatial distribution of another FRN, such as $^{239+240}$Pu or ^{210}Pb$_{ex}$ (Porto et al., 2013; Fig. 2d). As the fallout deposition of $^{239+240}$Pu after the Chernobyl accident was confined to a restricted area in the vicinity of the nuclear power plant (Ketterer et al., 2004), the origin of plutonium fallout in the rest of Europe is linked to the past nuclear bomb tests only. Consequently, the Pu fallout distribution was more homogeneous (Alewell et al., 2014; Ketterer et al., 2004; Zollinger et al., 2015). If the $^{239+240}$Pu depth profiles do not vary significantly between the two sampling years, there should be no disturbance (e.g. turbation, erosion) or measurement artefacts. As such, it might be concluded that the heterogeneous deposition of ^{137}Cs at the time of the fallout prejudices the use of Cs at this site.

3   The application of the CheSS decision tree

3.1   Study area

To test the methodology described above, we used a data set from an alpine study area, the Urseren Valley (30 km^2) in Central Switzerland (Canton Uri), which has an elevation ranging from 1440 to 3200 m a.s.l. At the valley bottom (1442 m a.s.l.), the average annual air temperature for the years 1980–2012 is around 4.1 ± 0.7 °C and the mean annual precipitation is 1457 ± 290 mm with 30 % falling as snow (MeteoSwiss). The U-shaped valley is snow covered from November to April. On the slopes, pasture is the dominant land use, whereas hayfields are prevalent near the valley bottom.

3.2   Sampling design

Supportive information was provided by the local landowners to select the reference sites in both valleys. Sites used for ploughing and grazing activities were excluded. A first sampling campaign was undertaken in autumn 2010 for $^{239+240}$Pu and in 2013 for ^{137}Cs. Six reference sites (REF1 to REF6) were identified in flat and undisturbed areas along the valley. At each site, three cores (40 cm of depth) 1 m apart from each other were sampled. The cores were cut in 3 cm increments to derive information on the ^{137}Cs depth profile. The three cores from each site were bulked to provide one composite sample per site. During the second sampling campaign in spring 2015, all six reference sites were resampled. Considering the typical and high soil redistribution dynamics of the valley of > 1 cm per year caused by snow-induced soil removal (Meusburger et al., 2014), the time span is sufficiently long to ensure the possibility to observe changes in the depth profiles if soil erosion and deposition processes affected the area. At each site, we collected three replicates, which were analysed separately, to investigate the small-scale variability in the FRN content. All cores were air-dried (40 °C for 72 h) and sieved (< 2 mm) to remove coarse particles; the skeleton content and the bulk density (BD) were determined.

3.3   Measurement of anthropogenic FRN activities and inventories

The measurements of the ^{137}Cs activity (Bq kg^{-1}) were performed with high-resolution HPGe detectors. The ^{137}Cs activity (Bq kg^{-1}) from 2013 was analysed at the Institute of Physics at the University of Basel using a coaxial high-resolution germanium lithium detector (Princeton Gamma Tech) with a relative efficiency of 19 % (at 1.33 MeV; ^{60}Co). Counting time was set to 24 h per sample. Samples collected in 2015 were analysed at the State Laboratory Basel-City using coaxial high-resolution germanium detectors with 25 to 50 % relative efficiencies (at 1.33 MeV; ^{60}Co). Counting times were set to provide a precision of less than ± 10 % for ^{137}Cs at the 95 % level of confidence.

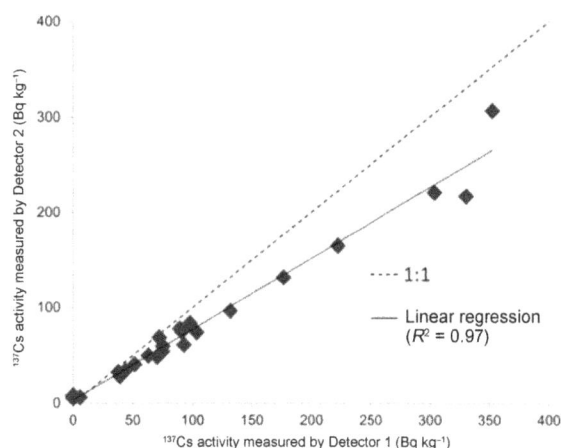

Figure 5. The comparison between the ^{137}Cs measurements of a subset of samples ($n = 16$) performed with two different HPGe detectors; detector 1 is housed by the Physics Department at the University of Basel (CH) and detector 2 is housed by the State Laboratory Basel-City (CH).

All soil samples were counted in sealed discs (65 mm diameter, 12 mm height, 32 cm^3) and the measurements were corrected for sample density and potential radioactivity background. The detectors located at the State Laboratory Basel-City were calibrated with a reference solution of the same geometry. The reference contained ^{152}Eu and ^{241}Am (2.6 kBq rsp. 7.7 kBq) to calibrate the detectors from 60 to 1765 keV. It was obtained from the Czech Metrology Institute, Prague. This solution was bound in silicon resin with a density of 1.0. The efficiency functions were corrected for coincidence summing of the ^{152}Eu lines using a Monte Carlo simulation program (Gespecor). The ^{137}Cs was counted at 662 keV with an emission probability of 0.85 and a (detector) resolution of 1.3 to 1.6 keV (FWHM). All measurements and calculations were performed with the gamma software Interwinner 7. The ^{137}Cs activity measurements were all decay corrected to the year 2015.

To compare the ^{137}Cs results to another artificial FRN, all samples were also measured for $^{239+240}$Pu activity. The determination of plutonium isotopes from both valleys and for both sampling years was performed using a Thermo X Series II quadrupole ICP-MS at Northern Arizona University, USA. A detailed description of the ICP-MS specifications and sample preparation procedure can be found in Alewell et al. (2014). The activities of ^{137}Cs and $^{239+240}$Pu (act, Bq kg^{-1}) were converted into inventories (Bq m^{-2}) according to Eq. (1).

3.4   Application of the CheSS decision support tool to the reference sites

Because the ^{137}Cs activity of the samples was measured with different detectors for the two sampling years, we investigated the potential variability between the two detectors. A

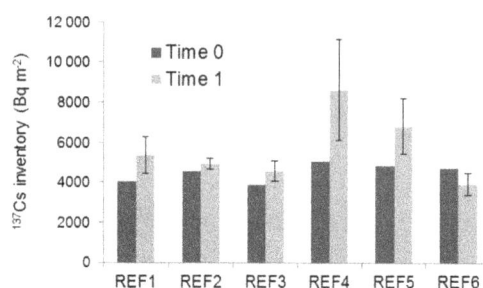

Figure 6. Temporal variation between the total ^{137}Cs inventories measured at the reference sites in the Urseren Valley; time 0 = 2013 and time 1 = 2015. The error bars indicate the standard deviations of the inventories among the replicates collected at each reference site in 2015.

selected subset of samples ($n = 24$) was analysed using both detectors (one located at the Institute of Physics at the University of Basel and the other located at the State Laboratory Basel-City). The results highlight a strong correspondence between the measurements by the two analytical systems ($R^2 = 0.97$; $p < 0.005$); however, the detector at the State Laboratory Basel-City returned slightly lower ^{137}Cs activities (Fig. 5). Thus, the ^{137}Cs activities of the samples measured in 2013 were corrected to the values of the detector at the State Laboratory Basel-City (which has a higher efficiency) to allow for comparability between the different data sets.

Total ^{137}Cs inventories (decay corrected to the year 2015) of the six reference sites collected in the Urseren Valley in 2013 range from 3858 to 5057 Bq m^{-2} with a mean value of 4515 Bq m^{-2} and a standard deviation (SD) of 468 Bq m^{-2}. Data from 2015 range between 3925 and 8619 Bq m^{-2} with a mean value of 5701 Bq m^{-2} and a SD of 1730 Bq m^{-2} (Fig. 6).

When following the CheSS decision tree, we investigated the variation in the ^{137}Cs total inventories at each reference site (node 1). The replicate samples were analysed separately only during the second sampling campaign ($t1$), while during the first sampling campaign ($t0$) only composite samples were analysed. Reference sites REF3, REF5 and REF6 presented signs of high small-scale variability, as expressed by a CV of 48 %. Such variability excluded them from any further application as reference sites without subsequent additional sampling. For sites REF1, REF2 and REF4, the CV was between 19 and 31 %.

Passing to node 2 of the CheSS decision tree, the analysis focuses on the variation in the shape of the ^{137}Cs depth profile (Fig. 7). Here we examined the regression between the reference depth profiles in $t0$ and $t1$. For the three sites with acceptable spatial variability (i.e. reference sites 1, 2 and 4), the site REF4 shows signs of deposition with a regression coefficient between $t0$ and $t1 = 1.34$. The deposition was confirmed by field observations of construction work conducted between the two samplings. After this disturbance the site is

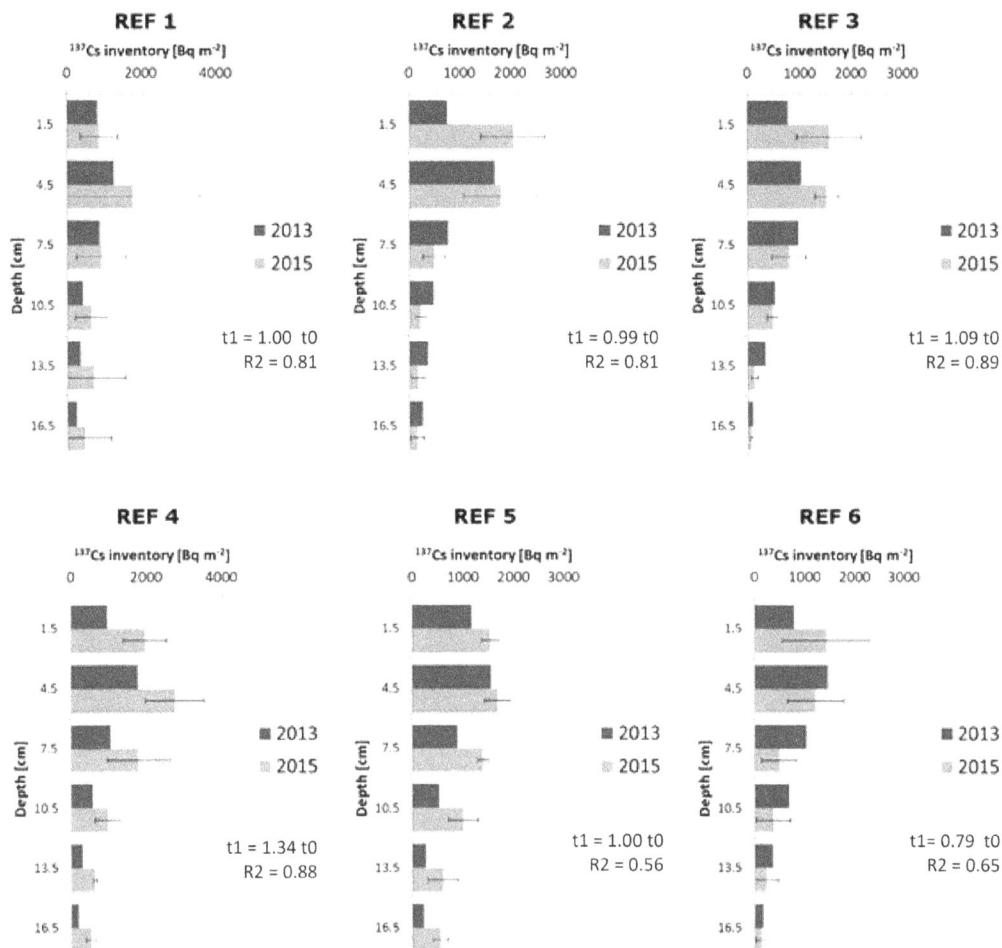

Figure 7. The ^{137}Cs depth profiles of the six investigated reference sites in the Urseren Valley for the two different sampling campaigns. The error bars indicate the standard deviations of the inventories among the replicates collected at each reference site 2015. The regression equation between the depth profile at $t0$ and $t1$ is displayed together with the R^2.

no longer a suitable reference site. Among the sites with high spatial variability, the site REF6 showed signs of erosion with a regression coefficient between $t0$ and $t1 = 0.79$.

In node 3 the temporal differences in total inventories between $t0$ and $t1$ were assessed. Here only the site REF4 showed a significant difference in the total ^{137}Cs inventories between $t0$ and $t1$, thus confirming the unsuitability of the site after the construction work.

To further investigate the causes of the spatial variation, $^{239+240}$Pu inventories measured at the three replicates of each site were analysed for $t0 = 2010$ and $t1 = 2015$ (Fig. 8). Clearly, deposition for REF4 and erosion processes for REF6 were confirmed with an increase of 46% and a decrease of 27% in the total $^{239+240}$Pu inventory between $t0$ and $t1$, respectively.

Further, we looked at the differences in the skeleton content of the three replicate samples collected at $t1$ (Fig. 2b). For site REF1, an ANOVA test showed a significant difference (p value of 0.025), and thus a difference in the pres-

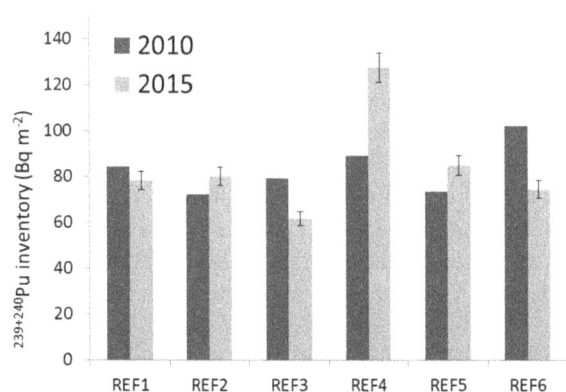

Figure 8. Temporal variation between the total $^{239+240}$Pu inventories measured at the reference sites in the Urseren Valley; time $0 = 2010$ and time $1 = 2015$. The error bars indicate the standard deviations of the inventories among the replicates collected at each reference site in 2015.

ence of stones in the three soil cores might have affected the FRN depth distribution. In particular, a Tukey's HSD (honest significant difference) post-hoc pairwise comparison identified the replicate number 3 at REF1 as a potential outlier. To validate the suitability of REF1, more replicates should be collected and measured in order to compare their ^{137}Cs depth profiles to the results obtained during the first sampling campaign. In summary, only the reference site REF2 appeared to be suitable for ^{137}Cs-based studies. For the site REF4, the construction work precluded its suitability for further application as a reference site. Form visual inspection of the soil profile, B could exclude cause C and consequently the final cause of heterogeneous fallout with high spatial variability (D) applies for the sites REF3 and REF5. These sites may be suitable for other FRNs or for ^{137}Cs if more samples are collected to constrain the spatial heterogeneity that was introduced by the ^{137}Cs Chernobyl fallout.

4 Conclusion

With the decision tree CheSS, a support tool to verify the suitability of reference sites for a ^{137}Cs-based soil erosion assessment is presented. Great attention has to be given to analysis of the small-scale variability in ^{137}Cs distribution in the reference areas, especially in regions affected by nuclear accident fallout. To cope with small-scale variability, sampling numbers might be increased or the temporal variation in ^{137}Cs or another radionuclide, such as $^{239+240}$Pu, might be analysed. The CheSS test in the Urseren Valley indicated that the heterogeneity and disturbance of ^{137}Cs distribution prejudiced the suitability of some reference sites. Additionally, the presence of stones affected the shapes of the depth profiles in at least one replicate sample at reference site 1. Including unsuitable reference sites, the application of the traditional ^{137}Cs approach, based on a spatial comparison between reference and sampling sites, is compromised. To derive soil redistribution rates, a ^{137}Cs repeated sampling approach should be preferred. This approach is based on a temporal comparison of the FRN inventories measured at the same site in different times (Kachanoski and de Jong, 1984). It does not require the selection of reference sites because the inventory documented by the initial sampling campaign is used as the reference inventory for that point (Porto et al., 2014).

Accurate soil erosion assessment is crucially needed to validate soil erosion modelling, which can help prevent and mitigate soil losses on larger spatial scales. In this context, FRN could play a decisive role if we are able to overcome its potential pitfalls, especially related to the selection of suitable reference sites. The decision tree CheSS provides a concept for objective and comparable reference site testing, which enables the exclusion of sites which present signs of uncertainty. We are convinced that this can contribute to improving the reliability of FRN-based soil erosion assessments.

Author contributions. LA, KM, LM and CA designed the concept of the method and analysed the data. AB contributed to the collection and preparation of the soil samples and to the analysis of the data. MZ measured the ^{137}Cs activity in the soil samples and analysed the results. MEK measured the $^{239+240}$Pu activity in the soil samples. LA prepared the paper with contributions from all co-authors.

Competing interests. The authors declare that they have no conflict of interest.

Acknowledgements. The authors would like to thank Annette Ramp, Gregor Juretzko, Simon Tresch, Carmelo La Spada and Axel Birkholz for support during fieldwork. This work was financially supported by the Swiss National Science Foundation (SNF; project no. 200021-146018) and was finalized in the framework of the IAEA Coordinated Research Project (CRP) "Nuclear techniques for a better understanding of the impact of climate change on soil erosion in upland agro-ecosystems" (D1.50.17).

Edited by: Olivier Evrard

References

Alewell, C., Meusburger, K., Juretzko, G., Mabit, L., and Ketterer, M. E.: Suitability of $^{239+240}$Pu and ^{137}Cs as tracers for soil erosion assessment in mountain grasslands, Chemosphere, 103, 274–280, 2014.

Alewell, C., Egli, M., and Meusburger, K.: An attempt to estimate tolerable soil erosion rates by matching soil formation with denudation in Alpine grasslands, J. Soil. Sediment., 15, 1383–1399, 2015.

Arata, L., Meusburger, K., Frenkel, E., A'Campo-Neuen, A., Iuran, A. R., Ketterer, M. E., Mabit, L., and Alewell, C.: Modelling Deposition and Erosion rates with RadioNuclides (MODERN) – Part 2: A comparison of different models to convert $^{239+240}$Pu inventories into soil redistribution rates at unploughed sites, J. Environ. Radioactiv., 162, 97–106, 2016.

Chawla, F., Steinmann, P., Pfeifer, H. R., and Froidevaux, P.: Atmospheric deposition and migration of artificial radionuclides in Alpine soils (Val Piora, Switzerland) compared to the distribution of selected major and trace elements, Sci. Total Environ., 408, 3292–3302, 2010.

Descroix, L. and Mathys, N.: Processes, spatio-temporal factors and measurements of current erosion in the French southern Alps: a review, Earth Surf. Proc. Land., 28, 993–1011, 2003.

Frankenberg, P., Geier, B., Proswitz, E., Schütz, J., and Seeling, S.: Untersuchungen zu Bodenerosion und Massenbewegungen im Gunzesrieder Tal/Oberallgäu, Forstwissenschaftliches Centralblatt vereinigt mit Tharandter forstliches Jahrbuch, 114, 214–231, 1995.

IAEA (International Atomic Energy Agency): Guidelines for using Fallout radionuclides to assess erosion and effectiveness of soil conservation strategies. IAEA-TECDOC-1741, IAEA publication, Vienna, Austria, 213 pp., 2014.

Iurian, A. R., Mabit, L., and Cosma, C.: Uncertainty related to input parameters of ^{137}Cs soil redistribution model for undisturbed fields, J. Environ. Radioactiv., 136, 112–120, 2014.

Kachanoski, R. G. and De Jong, E.: Predicting the temporal relationship between soil cesium-137 and erosion rate, J. Environ. Qual., 13, 301–304, 1984.

Ketterer, M. E., Hafer, K. M., Jones, V. J., and Appleby, P. G.: Rapid dating of recent sediments in Loch Ness: ICPMS measurements of global fallout Pu, Sci. Total Environ., 322, 221–229, 2004.

Kirchner, G.: Establishing reference inventories of ^{137}Cs for soil erosion studies: methodological aspects, Geoderma, 211, 107–115, 2013.

Konz, N., Prasuhn, V., and Alewell, C.: On the measurement of alpine soil erosion, Catena, 91, 63–71, 2012.

Lal, R.: Soil erosion and the global carbon budget, Environ. Int., 29, 437–450, 2003.

Lettner, H., Bossew, P., and Hubmer, A. K.: Spatial variability of fallout Caesium-137 in Austrian alpine regions, J. Environ. Radioact., 47, 71–82, 1999.

Loughran, R. J., Pennock, D. J., and Walling, D. E.: Spatial distribution of caesium-137, in: Handbook for the Assessment of Soil Erosion and Sedimentation Using Environmental Radionuclides, Springer, Dordrecht, 97–109, 2002.

Mabit, L., Benmansour, M., and Walling, D. E.: Comparative advantages and limitations of the fallout radionuclides ^{137}Cs, ^{210}Pb$_{ex}$ and ^{7}Be for assessing soil erosion and sedimentation, J. Environ. Radioact., 99, 1799–1807, 2008.

Mabit, L., Chhem-Kieth, S., Toloza, A., Vanwalleghem, T., Bernard, C., Amate, J. I., de Molina, M. G., and Gómez, J. A.: Radioisotopic and physicochemical background indicators to assess soil degradation affecting olive orchards in southern Spain, Agr. Ecosyst. Environ., 159, 70–80, 2012.

Mabit, L., Meusburger, K., Fulajtar, E., and Alewell, C.: The usefulness of ^{137}Cs as a tracer for soil erosion assessment: A critical reply to Parsons and Foster (2011), Earth-Sci. Rev., 127, 300–307, 2013.

Meusburger, K., Mabit, L., Park, J.-H., Sandor, T., and Alewell, C.: Combined use of stable isotopes and fallout radionuclides as soil erosion indicators in a forested mountain site, South Korea, Biogeosciences, 10, 5627–5638, https://doi.org/10.5194/bg-10-5627-2013, 2013.

Meusburger, K., Leitinger, G., Mabit, L., Mueller, M. H., Walter, A., and Alewell, C.: Soil erosion by snow gliding – a first quantification attempt in a subalpine area in Switzerland, Hydrol. Earth Syst. Sci., 18, 3763–3775, https://doi.org/10.5194/hess-18-3763-2014, 2014.

Owens, P. N. and Walling, D. E.: Spatial variability of caesium-137 inventories at reference sites: an example from two contrasting sites in England and Zimbabwe, Appl. Radiat. Isot., 47, 699–707, 1996.

Parsons, A. J. and Foster, I. D. L.: What can we learn about soil erosion from the use of ^{137}Cs?, Earth Sci. Rev., 108, 101–113, https://doi.org/10.1016/j.earscirev.2011.06.004, 2011.

Parsons, A. J. and Foster, I. D. L.: The assumptions of science A reply to Mabit et al. (2013), Earth-Sci. Rev., 127, 308–310, 2013.

Pennock, D. J. and Appleby, P. G.: Site selection and sampling design, in: Handbook for the Assessment of Soil Erosion and Sedimentation Using Environmental Radionuclides, Springer, Dordrecht, 15–40, 2002.

Pimentel, D.: Soil erosion: A food and environmental threat, Environment Development and Sustainability, 8, 119–137, 2006.

Porto, P., Walling, D. E., and Callegari, G.: Using ^{137}Cs and ^{210}Pb$_{ex}$ measurements to investigate the sediment budget of a small forested catchment in southern Italy, Hydrol. Process., 27, 795–806, 2013.

Porto, P., Walling, D. E., Alewell, C., Callegari, G., Mabit, L., Mallimo, N., Meusburger, K., and Zehringer, M.: Use of a ^{137}Cs re-sampling technique to investigate temporal changes in soil erosion and sediment mobilisation for a small forested catchment in southern Italy, J. Environ. Radioact., 138, 137–148, 2014.

Ritchie, J. C. and McHenry, J. R.: Application of radioactive fallout Cesium-137 for measuring soil-erosion and sediment accumulation rates and patterns – a review, J. Environ. Qual., 19, 215–233, 1990.

Schaub, M. and Alewell, C.: Stable carbon isotopes as an indicator for soil degradation in an alpine environment (Urseren Valley, Switzerland), Rapid Commun. Mass Sp., 23, 1499–1507, 2009.

Sutherland, R. A.: Caesium-137 soil sampling and inventory variability in reference locations: A literature survey, Hydrol. Process., 10, 43–53, 1996.

Walling, D. E., He, Q., and Appleby, P. G.: Conversion models for use in soil-erosion, soil-redistribution and sedimentation investigations, in: Handbook for the assessment of soil erosion and sedimentation using environmental radionuclides, edited by: Zapata, F., Kluwer, Dordrecht, the Netherlands, 111–164, 2002.

Zapata, F. (Ed.): Handbook for the assessment of soil erosion and sedimentation using environmental radionuclides (Vol. 219), Dordrecht, Kluwer Academic Publishers, 2002.

Zollinger, B., Alewell, C., Kneisel, C., Meusburger, K., Brandová, D., Kubik, P., Schaller M., Ketterer M., and Egli, M.: The effect of permafrost on time-split soil erosion using radionuclides (^{137}Cs, $^{239+240}$Pu, meteoric ^{10}Be) and stable isotopes (δ^{13}C) in the eastern Swiss Alps, J. Soil. Sediment., 15, 1400–1419, 2015.

13

Soil conservation in the 21st century: why we need smart agricultural intensification

Gerard Govers[1], Roel Merckx[1], Bas van Wesemael[2], and Kristof Van Oost[2]

[1]KU Leuven, Department of Earth and Environmental Sciences, Celestijnenlaan 200E, 3001 Leuven, Belgium
[2]Université Catholique de Louvain, Earth and Life Institute, 3 Place Louis Pasteur,
1348 Louvain-la-Neuve, Belgium

Correspondence to: Gerard Govers (gerard.govers@ees.kuleuven.be)

Abstract. Soil erosion severely threatens the soil resource and the sustainability of agriculture. After decades of research, this problem still persists, despite the fact that adequate technical solutions now exist for most situations. This begs the question as to why soil conservation is not more rapidly and more generally implemented. Studies show that the implementation of soil conservation measures depends on a multitude of factors but it is also clear that rapid change in agricultural systems only happens when a clear economic incentive is present for the farmer. Conservation measures are often more or less cost-neutral, which explains why they are often less generally adopted than expected. This needs to be accounted for when developing a strategy on how we may achieve effective soil conservation in the Global South, where agriculture will fundamentally change in the next century. In this paper we argue that smart intensification is a necessary component of such a strategy. Smart intensification will not only allow for soil conservation to be made more economical, but will also allow for significant gains to be made in terms of soil organic carbon storage, water efficiency and biodiversity, while at the same time lowering the overall erosion risk. While smart intensification as such will not lead to adequate soil conservation, it will facilitate it and, at the same time, allow for the farmers of the Global South to be offered a more viable future.

1 Introduction

The terrestrial land surface provides critical services to humanity and this is largely possible because soils are present. Humanity uses ca. 15 million km^2 of the total Earth's surface as arable farmland (Ramankutty et al., 2008). Besides this, ca. 30 million km^2 is being used as grazing lands: on all these lands, plants grow which are either directly (as food) or indirectly (as feed, fibre or fuel) used by humans for nutrition and a large range of economic activities. Agricultural areas, especially areas used as arable land, have often been selected because they have soils that make them suitable for agriculture. But it is not only the soils on agricultural land that provide humanity with essential services. Also, on non-agricultural land soils provide the necessary rooting space for plants, store the water necessary for their growth and provide nutrients in forms that plants can access. On both agricul-

tural and non-agricultural land, soils are host to an important fauna whose diversity is, by some measures, larger than that of its aboveground counterpart (De Deyn and Van der Putten, 2005); furthermore, on both land types, soils store massive amounts of organic carbon, the total amount of which (ca. 2500 Gt; Batjes, 1996; Hiederer and Köchyl, 2012) is much larger than the amount of carbon present in the atmosphere (ca. 800 Gt). Importantly, organic carbon storage per unit area is generally much higher on non-agricultural land (Poeplau et al., 2011; Hiederer and Köchyl, 2012). By allowing plants to grow, soils significantly contribute to the terrestrial carbon sink, which removes an amount equal to 30–40 % of the carbon annually emitted by humans from the atmosphere (Le Quere et al., 2009). Soils, both those on agricultural and those on non-agricultural lands, are therefore a vital part of humanity's global life support system, just like the atmosphere and the oceans. An Earth without soils would

Table 1. Conditions and trends with respect to soil erosion as assessed by experts (data from FAO and ITPS, 2015).

Region	Condition	Trend
Asia	Poor	Negative
Latin America	Poor	Negative
Middle East and North Africa	Very poor	Negative
Sub-Saharan Africa	Poor	Negative
Europe and Eurasia	Fair	Positive
North America	Fair	Positive
Southwest Pacific	Fair	Positive

be fundamentally different from the Earth as we know it and would, in all likelihood, not be able to support human life as we know it.

No further arguments should be necessary to protect soils from the different threats posed to them by modern agriculture and other human activities. Yet, as is the case with many other natural resources, soils are under intensive pressure. Organic carbon loss, salinisation, compaction and sealing all threaten the functioning of soils to different extents in different areas of the world. One of the most important and perhaps the ultimate threat posed to soils is accelerated erosion due to agricultural disturbance. When soils are used for farming their natural vegetation cover is removed and they are often disturbed by tillage. The result is that, under conventional tillage, erosion rates by water on arable land are, on average, up to 2 orders of magnitude higher than those observed under natural vegetation. This acceleration creates a major imbalance as soil production is outstripped by soil erosion by a factor 10–100 so that soil is effectively mined (Johnson, 1987; Montgomery, 2007; Vanacker et al., 2007b). Eroded soil is, in many cases, truly lost and cannot be restored (although there are exceptions to this rule), which explains why land prices in areas heavily affected by erosion may remain lower than expected, even when excessive erosion has been halted for several decades (Hornbeck, 2012).

It is rather surprising that agricultural soil erosion is still such an important problem. Pre-industrial societies such as the Inca already understood that erosion threatened agricultural productivity and used soil conservation techniques such as terracing for centuries (Krajick, 1998). In France, environmental degradation by excessive water erosion of mountain hillslopes literally ruined the livelihood of entire mountain communities at the end of the 19th century (Robb, 2008). A similar situation developed in Iceland, where excessive wind and water erosion forced entire villages to be abandoned in the same period. In both countries overexploitation of the natural environment by subsistence farmers through excessive deforestation and overgrazing were key factors. Both countries responded to this situation: in Iceland the first soil conservation service of the world was founded in 1907 (Arnalds, 2005), while France started an extensive pro-

gramme to restore its mountain environments (RTM) as early as 1860 (Lilin, 1986). In the United States, the Dust Bowl years (1930s) moved the erosion problem high up the political agenda: President Franklin Roosevelt not only erected the Soil Conservation Service but also, famously, said "A nation that destroys its soils destroys itself" (FAO and ITPS, 2015).

One might therefore expect that, by now, detailed information would exist on the status of the global soil resource and the necessary measures would have been taken to stop soil degradation due to human action and/or mitigate the consequences. Yet, this is clearly not the case: recent estimates of human-induced agricultural erosion amount to 25–40 for water erosion, ca. 5 $Gt\,yr^{-1}$ for tillage erosion and 2–3 $Gt\,yr^{-1}$ for wind erosion (Van Oost et al., 2007; Govers et al., 2014). Measured soil production rates are, on average, ca. $0.036 \pm 0.04\,mm\,yr^{-1}$ (Montgomery, 2007) and are even lower on most agricultural soils because agricultural soils have a certain thickness and soil production rates decrease with increasing soil depth (Stockmann et al., 2014). Thus, over all agricultural land (arable and pasture) total soil formation would amount to maximum ca. $2\,Gt\,yr^{-1}$, which implies that the global soil reservoir is depleted by erosion at a rate which is ca. 20 times higher than the supply rate. Although these numbers are only an approximation (for instance, they do not account for the fact that eroded soil may be re-deposited on agricultural land) they clearly illustrate that we are still far away from a sustainable situation: the rate at which the soil resource is being depleted is, over the longer term, a clear threat to agricultural productivity (FAO and ITPS, 2015). The loss of mineral soil is not the only issue: soil erosion also mobilises 23–42 $Tg\,yr^{-1}$ of nitrogen and 14–26 $Tg\,yr^{-1}$ of phosphorus (Quinton et al., 2010). These numbers may be compared with the annual application rate of mineral fertilisers, which are ca. 122 $Tg\,yr^{-1}$ for N and ca. 18 $Tg\,yr^{-1}$ of mineral P respectively. At 2013 USA mineral fertiliser prices of ca. USD 1.35 $(kg\,N)^{-1}$ and ca. USD 4.75 $(kg\,P)^{-1}$ (http://www.ers.usda.gov/data-products/fertilizer-use-and-price.aspx) the annual amount of fertilisers mobilised by soil erosion is equivalent to ca. USD 35 billion for N and ca. USD 80 billion for P: this is a significant financial loss, even if one considers that the total global agricultural food production is currently valued at ca. USD 4000 billion (http://faostat.fao.org/site/613/DesktopDefault.aspx?PageID=613#ancor). Most of these soil and nutrient losses take place in the hilly and mountain areas in the so-called Global South: a recent scientific appraisal by FAO and the ITPS (the Intergovernmental Technical Panel on Soils) showed that erosion problems are still increasing in Africa, Latin America and Asia (FAO and ITPS, 2015). The situation is perceived to be improving in Europe and North America (FAO and ITPS, 2015), even though soil losses in these regions are also often still above the tolerable level (Verheijen et al., 2009). Thus, it is especially the agriculture in the Global South (Latin America, Africa, the developing nations of Asia and the Middle East), where it is often one of

Figure 1. The presence of a dense network of rills and of significant deposition at the footslope (here in Huldenberg, Belgium, in July 2006) is as such sufficient proof for excessive soil erosion (in this case erosion exceeded $100\,t\,ha^{-1}$ in a single event).

the main economic activities, which suffers excessively from these losses.

In this paper we reflect on why, despite these clear facts, effective soil conservation is not yet a done deal and what might be done about this. We argue that there is a need for a novel vision on soil conservation in the Global South, shifting the focus away from not only the technical issues of soil conservation but also soil conservation as such. Soil conservation efforts need to be framed into a general vision on how agriculture will develop in the Global South: this vision needs to account for soil protection, but must also guarantee food security and allow the development of an agricultural system that does provide a sufficient income to farmers. We will first assess possible reasons as to why soils do not yet get the protection they deserve. Thereafter, we will discuss the building blocks of a vision on future soil conservation.

2   The status of soil conservation

2.1   Do we have the necessary data to guide soil conservation?

Investing in the application of soil conservation measures is only meaningful when erosion rates are higher than acceptable. This can most easily be established when erosion rates can reliably be quantified. Quantitative information is indeed available for North America and Europe (Cerdan et al., 2010; NRCS, 2010). However, the quality of our estimates of soil erosion rates by water for other areas on the globe is often poor. Sometimes, estimates are based on a limited number of data which are simply extrapolated to larger areas: this often leads to bias, simply because erosion rates are generally measured at locations where erosion intensity is much higher than average (Boardman, 1998; Cerdan et al., 2010).

Also, when models are used to make an extrapolation, estimates are often incorrect. There are two reasons for this: (i) the models that are used are often improperly calibrated, i.e. model parameters are set to values that are not appropriate for the location under consideration, and (ii) the model parameterisation may be correct but the spatial data used to drive the model are inappropriate. A typical example of the latter is when slope lengths are directly derived from a digital elevation model so that the impact of slope breaks such as field borders is not accounted for (e.g. Yang et al., 2003). This can lead to a considerable overestimation of erosion rates (Desmet and Govers, 1996; Cerdan et al., 2010; Quinton et al., 2010). Erroneous predictions do not only make it difficult to identify the most vulnerable areas in which conservation measures are most urgent: they may also invalidate the cost–benefit evaluations of soil conservation programmes and lead to disinformation of the general public about the extent and severity of the problem.

Although there is a clear need for better, quantitative data on erosion rates, the lack of such data is not the most important explanation as to why excessive soil erosion often still goes unchecked. While it may indeed be difficult to quantify erosion rates correctly, it is much easier to identify those areas where intense soil erosion is indeed a problem and where action is necessary, whatever the exact erosion rates are. This is, after all, what institutions such as the soil conservation services of Iceland and the United States did long before accurate erosion measurements were available. Simple visual observations on the presence of rills and gullies or wind deflation areas are clear indications that the implementation of conservation measures is necessary (Fig. 1). Another reason why an exact quantification is not always necessary is that conservation measures generally are not proportional: their implementation is most often of a yes/no type – one can decide whether or not to implement conservation tillage, but not by how much.

2.2   Do we have the necessary technology for soil conservation?

There is no doubt that soil conservation technology has matured over the last decades: we now have the tools to effectively reduce erosion rates to acceptable levels in many, if not all, agricultural systems. Conservation tillage is the tool of choice in many areas, especially in the Americas. This is hardly surprising: erosion plot research has consistently shown that water erosion rates under conservation tillage are reduced by 1 to 2 orders of magnitude in comparison to conventional systems (Montgomery, 2007; Leys et al., 2010). Moreover, the effectiveness of conservation tillage as calculated by plot studies is likely to be underestimated: for various reasons the effectiveness of conservation does increase if the slope length increases (Leys et al., 2010). As a consequence, water erosion rates under conservation tillage on moderate slopes are generally very low ($< 1\,t\,ha^{-1}\,yr^{-1}$) and often comparable to those occurring under natural vegetation (Montgomery, 2007). Conservation tillage may also be used to drastically control wind erosion not only because residue cover does reduce the shear stress to which soil particles are exposed but also because the presence of residue helps to keep the surface soil layer moist, thereby increasing its shear strength.

Conservation tillage is not always the best tool. It may be difficult or impossible to apply with certain crops, such as potatoes grown on ridges, and/or difficult to introduce into specific agricultural systems as it may affect the overall workload or the gender balance of the workload (Giller et al., 2009). It may also not be sufficient to implement conservation tillage as processes such as gully erosion may not be effectively controlled and may in some cases even be enhanced by conservation tillage as the latter is much more effective in reducing erosion than in reducing surface runoff (Leys et al., 2010). However, in such cases technological solutions also exist: they can consist of infrastructural measures such as stone bunds and terrace building in combination or vegetation measures such as grassed waterways, but proper land use allocation can also make a significant difference. Water and wind erosion rates can often be reduced to acceptable levels through the use of such measures in combination with modifications of tillage techniques and crop rotations (Sterk, 2003; Valentin et al., 2008; Nyssen et al., 2009).

Not only arable land can be affected by excessive erosion. Grazing lands may suffer from a drastic reduction in vegetation cover due to overgrazing and compaction, again resulting in excessive water and/or wind erosion with rates up to 2 orders of magnitude higher than those observed under natural conditions (Vanacker et al., 2007b). Reduction of grazing pressure (at least in a first stage) and the introduction of controlled grazing are key strategies (i) to restore the vegetation cover and (ii) to allow these lands to become productive again so that they can be sustainably used (Mekuria et al., 2007). Such measures can be further supported by the planting of trees (Sendzimir et al., 2011). Reforestation may also be a solution as it reduces erosion rates to near-natural levels but it has evident implications for the type of agriculture that can be supported (Vanacker et al., 2007b). Thus, as is the case on arable land, the key to erosion reduction on grasslands is in most cases the maintenance or restoration of a good vegetation cover, possibly supported by technical measures.

Erosion in agricultural areas is often directly related to not only agricultural activities but also the infrastructure related to these activities such as roads and field boundaries. Unpaved roads on sloping surfaces are important sources of sediment not only in many agricultural areas (Rijsdijk et al., 2007; Vanacker et al., 2007a) but also in cities (Imwangana et al., 2015). Water is often concentrated at field boundaries, thereby leading to gully formation (Poesen et al., 2003). Again, the necessary technological know-how to control such erosion phenomena is available: check dams, better water drainage infrastructure, the implementation of field buffer zones and a better landscape organisation all help to reduce sediment production on road networks and in built-up areas.

2.3   Why, then, is soil conservation not more generally adopted?

Thus, neither the lack of conservation technology nor the lack of data on the erosion hazard can fully explain why efficient soil conservation measures are still not implemented on most agricultural land, especially in the Global South. It has indeed long been clear that several factors other than (the lack of) scientific knowledge or data hamper the adoption of conservation tillage. These factors include the training level of the farmer, the farm size and work organisation as well as access to information. However, a thorough analysis by Knowler and Bradshaw (2007) showed that the effect of these variables was often ambiguous (when different studies are compared) and that few, if any, variables showed a consistent effect. One might conclude from this that changing farming practices must be inherently difficult, as our understanding of controlling factors is relatively poor and many barriers to the adoption of novel technology need to be overcome. This is not only a problem in the Global South: in Europe the adoption of conservation tillage is also slow in many countries due to a multitude of factors, including the fact that soil tillage is deeply rooted in the culture of many farmers (Lahmar, 2010).

Clearly, farming systems are, to some extent, "locked in": they rely on well-tried technology, division of labour and crop types and are therefore difficult to change. There are, nevertheless, also cases where farming systems change rapidly and conservation technology is quickly adopted. Once the necessary technology was available, conservation tillage spread very rapidly through most of Argentina and Brazil: in Argentina, it took ca. 20 years (from 1990 to 2010) to bring ca. 80 % of the arable land under no-till (Peiretti and Dumanski, 2014), thereby effectively halting excessive soil erosion on most of the arable land of the country. In Brazil, more than 25 million ha of land was under no-tillage in 2006, whereas the technique was virtually unused before 1990 (Derpsch et al., 2010). Rapid changes in agricultural systems are not limited to the adoption of conservation tillage. When subsistence farmers in remote areas gain access to profitable markets, very rapid transformations can occur, even in areas where existing technology is poor: such changes can have very negative effects in terms of soil degradation rates as a switch to cash cropping may introduce crops with which a much higher erosion risk is associated (Valentin et al., 2008). Thus, while cultural and technological barriers to change certainly do exist, farmers are most certainly capable of rapid change. Whether such rapid change occurs critically depends on whether farmers think change will bring them a personal gain.

This is where the problem lies. Under some conditions, the adoption of conservation technology is indeed clearly economically beneficial to the farmer: this appears to be true for large farming operations in (sub)tropical regions growing cash crops such as soy beans (Peiretti and Dumanski, 2014). But in most other cases the direct benefits of the implementation of conservation agriculture and/or other soil conservation measures are small, if they exist at all. This appears to be the case for both large-scale mechanised agriculture in the temperate zone as well as for subsistence hillslope farming in developing countries (Knowler et al., 2001). In both scenarios, potential savings are offset by additional costs: in mechanised systems the cost of machinery and agrochemicals offsets savings in fuel costs (Zentner et al., 1996; Janosky et al., 2002), while in traditional hillslope farming extra work hours are needed to maintain conservation structures and some land has to be sacrificed to implement these structures, thereby reducing overall yields (Nyssen et al., 2007; Quang et al., 2014). Importantly and contrary to common belief, crop yields do not rise significantly in conservation systems if no additional inputs are provided: this is true for advanced technological systems (Van den Putte et al., 2010; Pittelkow et al., 2015) as well as for tropical smallholder farming (Brouder and Gomez-Macpherson, 2014). As a consequence, farmers often do not have direct incentives to implement soil conservation measures and change becomes difficult to put into effect.

One may argue that benefits should be considered not only at the level of the individual farmer but also at the societal level, where soil conservation may generate co-benefits. Often, carbon storage and biodiversity protection under conservation systems are mentioned as important ecosystem services for which farmers could be paid. Research in the last decade has consistently shown that carbon storage gains in conservation systems are lower than was anticipated two decades ago and are generally well below $1\,\mathrm{t\,C\,ha^{-1}\,yr^{-1}}$ (Oorts et al., 2007; Angers and Eriksen-Hamel, 2008; Christopher et al., 2009; Eagle et al., 2012; Govers et al., 2013). Furthermore, paying farmers to store carbon would only be viable at much higher carbon prices than the current market prices, which are around USD 10–$15\,\mathrm{t^{-1}}$ (Grace et al., 2012; Govers et al., 2013). Paying farmers at current market prices can only generate a relatively small economic benefit for the farmer and prices would have to rise significantly for soil carbon storage to become an important element on the farmers' balance sheet. On the other hand, soil conservation generally has a positive impact on (soil) biodiversity on the farm land as soils are less frequently disturbed (Mader et al., 2002; Verbruggen et al., 2010). Where agriculture is interspersed with densely populated areas, additional co-benefits may consist of a reduction of flooding and/or siltation of sewage systems and water treatment plants, which are important problems in many areas in Europe (Boardman et al., 1994). These benefits, however, are difficult to convert to financial income for the farmer. This is not only because the economic value of increased biodiversity on farmland is difficult to quantify but

also because such on-farm benefits in biodiversity have to be weighed against possible off-farm losses (see below). The reduction in flooding risk, on the other hand, will generally not be considered as a benefit by society but rather as damage repair: the problems were caused by agriculture in the first place.

3 The way forward

How, then, should we proceed to stimulate a more rapid adoption of soil conservation measures to protect the world's soil resource? The answer to this question will obviously depend on the characteristics of the local agro-ecological system. Agricultural systems show a large variety so that not only the factors impeding the adoption of conservation tillage vary locally (Knowler and Bradshaw, 2007) but also the tools that societies have at their disposal to reduce it.

Western societies with highly developed information systems tackle the problem by means of a policy combining regulation (e.g. by forbidding the cultivation of certain crops on land that is very erosion-prone) and subsidies or compensations in combination with well-guided campaigns to inform farmers on the potential benefits and risks for themselves as well as for the broader society. Such combined approaches have had demonstrable success in various parts of Europe and North America, where farmers are not only well trained and highly specialised but also dependent to a large extent on subsidies, giving the administrations the necessary financial leverage to stimulate or even coerce farmers (Napier et al., 1990). As a result erosion rates in North America have gone down considerably over the last decades and are still declining (Kok et al., 2009). One may therefore assume that in these societies erosion rates can be reduced to tolerable levels provided that the necessary policies are maintained and/or strengthened. Countries with a strong central government that can impose decisions on land use and soil conservation, as is the case in China, can successfully reduce erosion: the excessive erosion rates on the Chinese Loess Plateau were strongly reduced through massive government programmes implementing erosion control measures (Chen et al., 2007; Zhao et al., 2016)

These approaches are, at present, not possible in most countries of the Global South. Many governments in the Global South are not able to implement a successful soil conservation policy as they do not have at their disposal the necessary data and/or the necessary political and societal instruments to do so. At first sight it may therefore appear unlikely that soils will become effectively protected in most of the developing world within a foreseeable time span. Yet this conclusion foregoes the fact that agriculture in the Global South, and especially in sub-Saharan Africa, will see fundamental changes in the next decades. At least three fundamental tendencies can be identified that will change the nature of agriculture in the Global South in the 21st century: these should be accounted for when developing a vision on soil conservation.

In many areas where soils are most seriously threatened, the human population will continue to grow strongly. In the next decades, the locus of world population growth will shift in an unprecedented manner. Population growth in the Global North has stopped and many regions in the Global South will follow suit in the next decades: Asia is expected to reach its maximum population around 2050. China's population will peak around 2030 and that of India no later than 2070. Latin America will follow around 2060 (http://esa.un.org/unpd/wpp/, Lutz and KC, 2010; Gerland et al., 2014). Sub-Saharan Africa is a different matter: here the demographic transition started only after the Second World War and the population will continue to grow rapidly during most of the 21st century. As a result of these diverging tendencies the distribution of the world's population will have changed beyond recognition in 2100: Europe's share in the global population will have fallen from its maximum of ca. 22 % in 1950 down to ca. 6 % in 2100, while the share of Africa will rise from ca. 9 % in 1950 to ca. 39 % in 2100 (http://esa.un.org/unpd/wpp/).

The population in the Global South will also become more urban. By 2050 ca. two-thirds of the global population is expected to live in cities (as compared to ca. 55 % at this moment). Urbanisation rates are especially high in Africa, where the fraction of urban population is expected to increase from 40 in 2014 to 55 % in 2050, and in Asia, where urbanisation will increase from ca. 47.5 to ca. 65 % over the same period (United Nations, 2014). There is no alternative for this evolution: despite all their problems, cities are the engines of modern economic development as they allow a population to create the added value that is so desperately needed through advantages of scale, intense interaction and exchange (Glaeser, 2011). This is the fundamental reason of the attractiveness of cities and the major factor explaining rural to urban migration: poor rural populations perceive the city as a place of opportunity and moving there as an opportunity to improve their own lives or at least those of their children (Perlman, 2006; Saunders, 2011). A consequence of this massive migration movement is that rural populations rapidly age and that the average farm worker is significantly older than the average non-farm worker (40 vs. 34 years in Africa, http://www.gallup.com/poll/168593/one-five-african-adults-work-farms.aspx). Clearly, the evolution sketched above is a generalisation: local dynamics depend, amongst other things, on the presence of attractive labour opportunities in the cities and the local availability of land (Ellis-Jones and Sims, 1995).

It is not overly optimistic to expect that, while population growth continues, these populations will at the same time gain in purchase power. While incomes in southern Asia and especially sub-Saharan Africa are presently much smaller than those in the Global North, their growth rates are, fortunately, much bigger. For example, Ethiopia's economy has,

over the last decade, consistently been growing at 8 to 10 % per year, leading to a rise of the per capita gross national income from USD 110 (2015 dollars) in 2004 to USD 550 in 2015 (http://data.worldbank.org/country/ethiopia).

Combined, these tendencies will lead to an increased market demand for food. Furthermore, diets will move away from a diet largely based on cereals towards a more varied (but not necessarily healthier) food palate in which meat is likely to have a larger share than is currently the case. Global estimates therefore sometimes predict that global food production (in terms of kcal) will more or less double in the first half of the 21st century (Tilman et al., 2011), but an increase in demand by 60–70 % is more likely (Alexandratos and Bruinsma, 2012). As (relatively) more people will live in cities, there will be relatively fewer people working on the land to produce the food that is necessary. Furthermore, as most of future population growth will take place in sub-Saharan Africa, food demand will rise most rapidly in this area.

Thus, agriculture in the Global South will be fundamentally different from what it is now in less than a century. More food will have to be produced with less people and the increasingly urban population will more and more rely on markets to obtain the food it needs. This begs a basic question: how can we make sure that the soils necessary to produce all this food are sustainably managed and preserved for future generations?

4   Soil conservation in a changing global context

Two contrasting pathways can be followed to meet the expected increase in food demand in the Global South. More food can be produced either by extending the area over which current food production systems are applied or by agricultural intensification, i.e. by increasing the amount of food produced per unit of land.

Both pathways are, in principle, possible: until present, Africa has followed the first path. Over the last five decades, the increasing food demand of African populations has mainly been met by increasing the area used for farming, while yields per unit of surface area remained stable and very low (Henao and Baanante, 2006). This evolution sharply contrasts with the one observed in most parts of Asia: here agricultural production was mainly increased through intensification (Henao and Baanante, 2006). In Asia, the Green Revolution led to a dramatic rise in agricultural yields through the combination of new crop varieties, better farming technology and the increased use of fertilisers. As a consequence, Asia now manages to feed its population much better than it did in 1970: the amount of available kcal per person rose from ca. 2000 to ca. 2400 kcal (South Asia) or even 3000 kcal (East Asia) in 2005 (Alexandratos and Bruinsma, 2012) despite the fact that the amount of land used for agriculture only marginally increased (Henao and Baanante,

2006) and despite the fact that the population in these regions increased from 0.98 to 1.53 billion (East Asia) and from 1.06 to 2.20 billion (South Asia) over the same period (http://esa.un.org/unpd/wpp/).

While the challenge for African agriculture is not dissimilar to that of Asia in the 1960s, Africa does not necessarily have to go down the same route. In principle, it could continue to follow the areal extension strategy policy for some time to come. At present, ca. 290 million ha of agricultural land is in use in Africa, but another 400 million ha of African land is suitable (good) or very suitable (prime) for agriculture (Alexandratos and Bruinsma, 2012). Therefore, there is scope for a strategy whereby significantly more land would be used for agriculture than is the case at present although this would pose important problems: a large fraction of the suitable land is located in politically unstable countries and/or far from existing markets (Chamberlin et al., 2014).

An extension strategy may, at first sight, be attractive from the point of view of soil conservation. One might indeed argue that this would be based on agricultural technology that has been in use for decades, and may therefore be best suited to increase agricultural production without causing excessive soil degradation. Indeed, the occurrence of erosion in mechanised, intensive agricultural systems is often attributed to the loss of traditional soil conservation methods (Bocco, 1991). Averting intensification and aiming at area extension may therefore seem a suitable solution to avoid excessive soil degradation as traditional farming methods can be maintained and optimised to be as environmentally friendly as possible. Many organisations do indeed stress environmental protection and sustainability as key issues to be addressed in the further development of African agriculture and explicitly state that Africa should indeed follow a path different from the Asian Green Revolution (De Schutter, 2011).

While it is evident that we should learn from agricultural developments in Asia and avoid the dramatic negative effects the Asian Green Revolution had in some places, we argue here that tropical smallholder farming does need intensification for soil conservation to become successful. This intensification should be smart: it is not sufficient that intensification is sustainable so that the capability of the natural resources to meet the needs of future generations is not jeopardised. Intensification strategies should also maximise the opportunities of current and future farmers to generate an acceptable income by providing them with access to profitable markets and supplying them with the necessary knowledge and technology to produce for these markets. Smart intensification requires an approach that does not focus on the conservation of natural resources alone but also on the creation of added value using a future-oriented perspective and the quantity and quality of food production and supply. Clearly, improving the livelihood of the farmers and farming communities should be a key element. However, the capability of this farming community to provide the necessary agricultural supplies to an ever growing non-farming population

also needs to be taken into account. Thus, it is important to consider not only the current socio-economic conditions but also how demographic and socio-economic conditions are likely to change in the future. We argue that smart intensification will not only make soil conservation more achievable but that it would also allow for reaping additional environmental benefits that may be lost when a less intensive or less future-oriented development path is chosen. As is the case for "smart cities", we do not believe a single, all-encompassing definition of smart intensification can be formulated. However, we summarised the components that we consider to be essential in Fig. 2. In the rest of the paper we focus the discussion on how soil conservation may benefit from smart intensification.

Smart intensification will allow for the most erosion-prone land to be spared from agriculture, thereby reducing landscape-scale erosion rates. When farmers select land for arable production, they will select the most suitable land that is available. In general this means that, for obvious reasons, flatter land is preferred over steeper land (Van Rompaey et al., 2001; Bakker et al., 2005). Steep lands are generally much more difficult to cultivate than flatter areas and yields can be expected to be lower in comparison to yields (for the same amount of inputs) on flat land because soils are intrinsically less productive and/or because soil productivity is negatively affected by accelerated erosion (Stone et al., 1985; Ellis-Jones and Sims, 1995; Lu and van Ittersum, 2004). The combination of both effects (more labour required and lower yields) invariably implies that the net returns of arable farming decrease with increasing terrain steepness. The total amount of erosion as well as the amount of erosion per unit of crop yield will therefore necessarily increase when area expansion is preferred over intensification (Figs. 3, 5).

Increasing agricultural production in Africa through areal extension alone would therefore imply that overall soil losses would increase much more rapidly than agricultural production would. If, on the other hand, agricultural yields on good agricultural land would be improved, it may be possible to set aside some of the marginal land that is currently used for arable farming. The somewhat counterintuitive result of this will be that, even if erosion rates on the arable land that remains in production were to increase due to intensification, the overall soil loss (at the landscape scale) would still decrease (Fig. 3).

Smart intensification will conserve soil carbon, which will, in turn, reduce erosion risks. Over the last decades, a significant body of scientific literature has emerged on the potential of agricultural land to store additional soil organic carbon through the use of appropriate management techniques. While studies do suggest that some gains are indeed possible, most studies report modest gains at best. Reported average sequestration rates under conservation tillage in Canada are between 0 and $0.14\,t\,C\,ha^{-1}\,yr^{-1}$ in Canada (VandenBygaart et al., 2010), while an average sequestration rate of $0.12\,t\,C\,ha^{-1}\,yr^{-1}$ has been calculated for the USA (Eagle

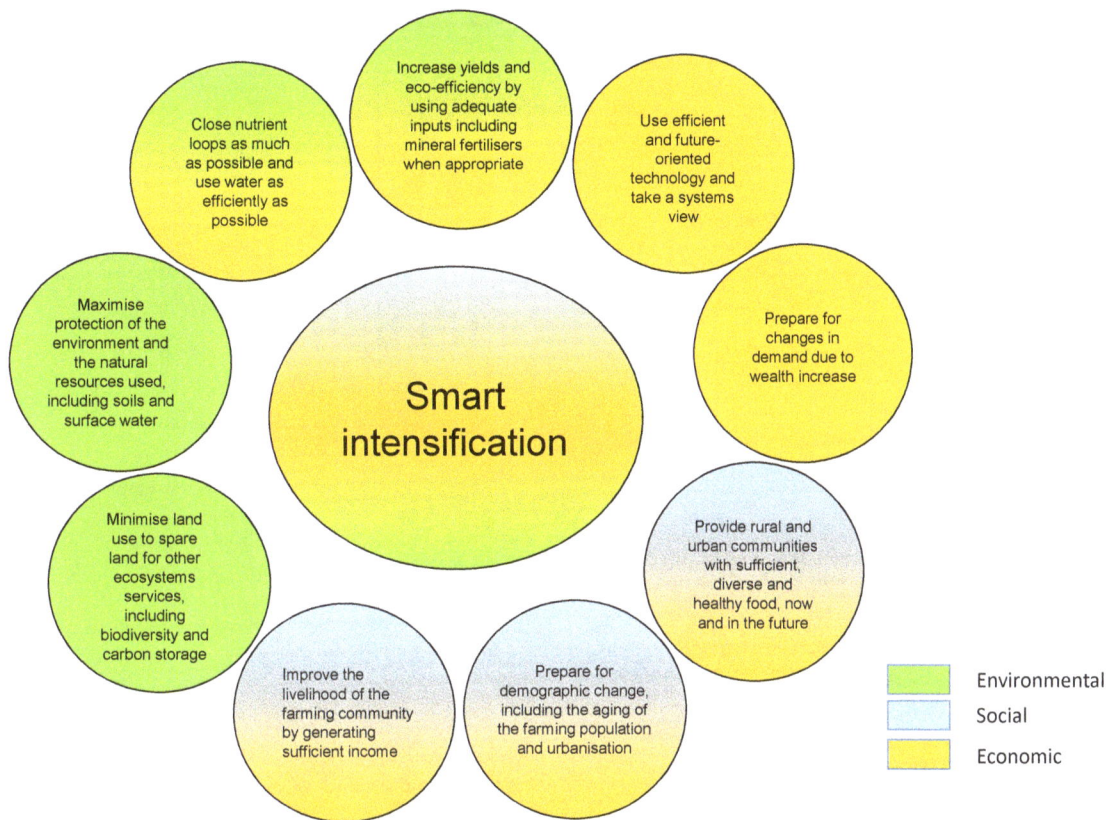

Figure 2. Different aspects of smart agricultural intensification. Colouring refers to main reason as to why each aspect is important.

et al., 2012). In a study covering 12 study sites in three Midwestern states of the USA, Christopher et al. (2009) did not find any significant increase in soil organic carbon storage under no-till in real farming conditions. Experimental studies also showed that under agroforestry gains in soil organic carbon are small, with an average of $0.25 \, \text{t} \, \text{C} \, \text{ha}^{-1} \, \text{yr}^{-1}$ (Govers et al., 2013). These findings contrast not only with claims in the literature (Ramachandran Nair et al., 2009) but also with the observation that soil carbon stocks on natural (or undisturbed) land are generally much higher (often more than three times higher) than those observed on arable land (e.g. Poeplau et al., 2011; Hiederer and Köchyl, 2012).

The latter is related to two main factors: (i) biomass is not removed from natural land, which results in larger organic carbon inputs, and (ii) these lands are not mechanically disturbed which reduces carbon respiration rates. Thus, more soil carbon will be conserved when the extent of agricultural land is reduced and more land is preserved under or restored towards natural conditions. An additional beneficial effect of the latter is that soil organic carbon stocks may increase on agricultural land with increasing agricultural yields, provided that the residual biomass is adequately managed (VandenBygaart et al., 2010; Minasny et al., 2012): this, in turn, will reduce the erosion and degradation risk (Torri and Poesen, 1997). Thus, intensification will allow for more carbon to

be preserved than areal extension (Figs. 3, 5). The fact that intensification is beneficial for soil carbon conservation has also been demonstrated at the global level: agricultural intensification has allowed for avoidance of ca. 161 Gt of carbon emissions from the soil to the atmosphere between 1960 and 2005 (Burney et al., 2010).

Smart intensification will help to make agriculture in the Global South more water-efficient. Agriculture is by far the largest global consumptive user of blue water (water extracted from rivers and groundwater): at the global scale, over 80 % of all consumptive water use is related to agricultural activities (e.g. Döll et al., 2009). As the amount of available water will not significantly increase in the future, a more efficient water use is a prerequisite to increase agricultural production in the Global South. Less productive systems are often more water-intensive, i.e. more units of water are needed for each unit of crop that is produced. Striving towards higher yields will remedy this problem as it allows for the amount of crop produced per unit of water to increase (Rockström et al., 2007). Higher yields are therefore a means to increase water conservation and to make sure that more water is available for the functioning of non-agricultural ecosystems. Clearly, the realisation of this potential requires other measures as well such as a realistic pricing of water and water use monitoring in areas where water scarcity is a problem so that in-

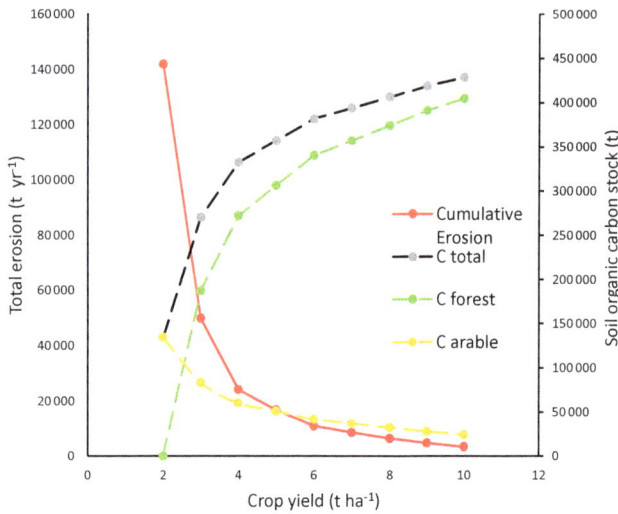

Figure 3. Modelled total erosion (t yr^{-1} left axis) and soil organic carbon stocks (t, right axis) vs. crop yield per hectare for a hypothetical test area of 2900 ha and assuming a total cereal production of 5000 t. We assumed that slope gradients (sin θ) were uniformly distributed between 0.02 and 0.58, i.e. an area of 100 ha in each 0.02 slope class. The crop yield shown is the crop yield on a zero slope and relative crop yield (P) is assumed to vary with slope: $P = 1 - (sin\theta)^{0.5}$. Erosion ($E$, t ha^{-1} yr^{-1}) is assumed to vary with slope gradient according to the slope function derived by Nearing (1997): $E \sim -1.5 + 17/[1 + \exp(2.3–6.1 \sin\theta)]$, and an erosion rate of 10 t ha^{-1} yr^{-1} is assumed on a 0.09 slope. Soil organic carbon stocks per unit area are assumed to be 40 on arable land and 170 t ha^{-1} under forest (Poeplau et al., 2011). The total soil organic carbon stock (C total) in the area strongly increases with increasing crop yield because the gain in soil organic carbon stocks on forested land (C forest) is much more important than the loss on arable land (C arable).

efficient use of this scarce resource can be prevented. Again, the implementation of such systems will be far more efficient in high-yield systems as the return per unit of capital cost will be higher.

Smart intensification is beneficial for biodiversity at the landscape scale. Environments where intensive agriculture is dominant are often very poor in terms of biodiversity. One might therefore suggest that, in order to preserve biodiversity, intensification should be avoided and a certain biodiversity on agricultural lands maintained. Again, such a strategy would necessarily imply that more land would be needed to produce the same amount of agricultural goods. Recent studies have consistently shown that such a strategy is not beneficial for biodiversity at a larger scale: the biodiversity gained on agricultural land is, in general, not sufficient to compensate for the additional biodiversity loss due to agricultural land expansion (e.g. Phalan et al., 2011b; De Beenhouwer et al., 2013; Schneider et al., 2014). Thus, land sparing and concentrating intensive agriculture on designated areas is generally a better strategy than land sharing with low-intensity

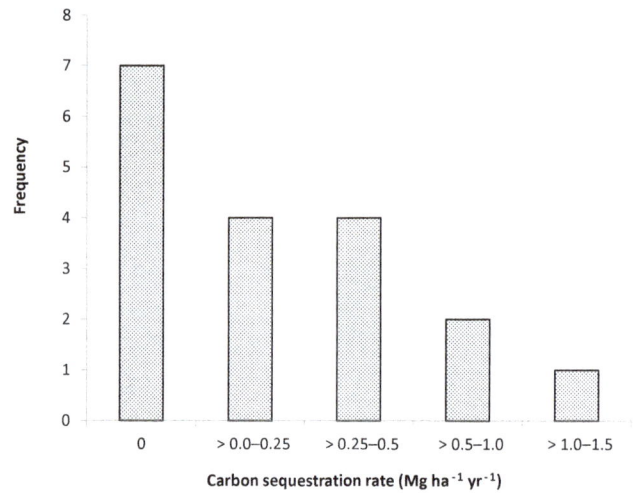

Figure 4. Frequency distribution of experimentally observed carbon sequestration rates under agroforestry. Data from 18 paired field studies in both (sub)tropical and temperate climates (details and references of studies in Govers et al., 2013). The average soil organic carbon sequestration rate reported over all 18 studies is 0.25 ± 0.33 t ha^{-1} yr^{-1}.

agriculture that will occupy a much larger fraction of the available land (Fig. 5). Sparing will not always be the best strategy as this will depend on local conditions: for instance, wildlife-friendly agriculture may be the best solution in the buffer zones around wildlife reserves.

Smart intensification will increase the added value of the land used for agricultural production and hence make the implementation of conservation measures economically sound. Clearly the economic value of a good such as arable land depends on the economic return that can be gained from the use of it. Intensification will allow for these returns to increase. This is especially true for sub-Saharan Africa, where yields are still extremely low (Neumann et al., 2010). While there are many reasons for this, a key factor is that African soils are chronically underfertilised (Henao and Baanante, 2006; Keating et al., 2010). The amount of fertiliser used per unit of surface are of agricultural land in Africa is only 10 % of what is being used in Europe or the United States: the consequence is that, in many cases, the nutrient balance of many African agricultural systems is negative, i.e. more nutrients are removed through harvesting than there are supplied by fertilisation (Smaling et al., 1993; Henao and Baanante, 2006). This negative balance is further aggravated by soil erosion, which annually mobilises more nutrients than are applied in sub-Saharan Africa (Quinton et al., 2010). Even a modest increase in fertiliser use may therefore allow for a significant boost to agricultural yields in sub-Saharan Africa, at least if this increase were accompanied by other measures such as the introduction of high-yield varieties and the necessary training for the farmers (Sanchez, 2010; Twomlow et al., 2010; Mueller et al., 2012).

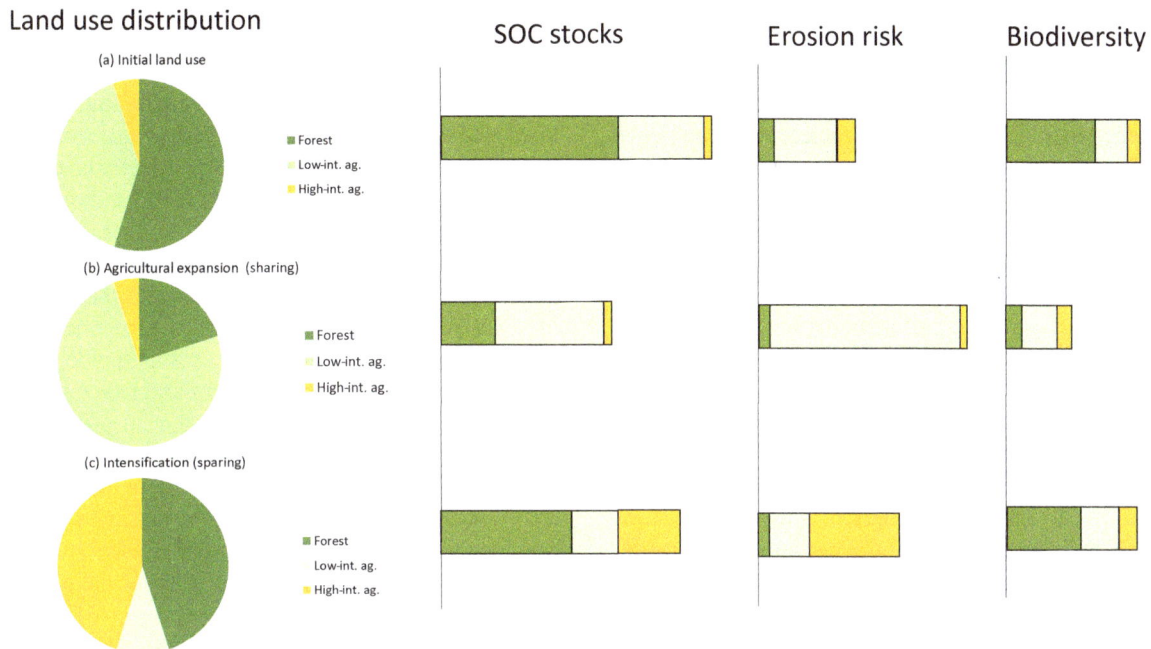

Figure 5. Semi-quantitative illustration of the effects of a significant increase in agricultural production through smart intensification (sparing land) vs. agricultural expansion (sharing land) on soil organic carbon stocks, the erosion risk and biodiversity. We assume that in a given area the required increase in agricultural production is such that, if yields are not increased, the entire area that is potentially suitable for agriculture (80 % of the total area) has to be used for agriculture and that smart intensification would reduce the area needed to ca. 55 % of the total area. The bar graphs give a semi-quantitative assessment, at the landscape scale, of the impact of these alternatives according to current scientific insights. Smart intensification is beneficial with respect to soil organic carbon storage because soil organic stocks under natural forest are much higher than under arable land (e.g. Poeplau et al., 2011). Smart intensification will reduce total soil erosion because less marginal (sloping) land needs to be taken into production (e.g. Van Rompaey et al., 2002). Finally, smart intensification is beneficial for biodiversity because more forest is preserved and the biodiversity of undisturbed forests is much higher than that of land used for agriculture (e.g. Phalan et al., 2011a).

Higher agricultural yields will increase the added value that may be produced per unit of agricultural land and hence its value. A consequence of this is that the economic stimulus to implement conservation measures on this land will increase as land will become a more precious resource. Furthermore, intensification will also reduce the overall conservation investment that has to be made as the acreage that needs to be treated will be smaller, which will allow for the available resources to be concentrated on a smaller area. Finally, many conservation strategies are based on the use of crop residue (i) to return nutrients and carbon to the soil and (ii) to reduce the soil erosion risk. Such strategies are likely to be more successful when more residue per unit of area is available. Case studies have repeatedly shown that the mechanisms described above can indeed lead to more effective soil conservation under increasing intensification and population pressure (e.g. Tiffen et al., 1994; Boyd and Slaymaker, 2000).

Smart intensification will help to create the market opportunities needed for sustainable agriculture. The dramatic increase in population that will occur in the Global South over the next century, in combination with rapid urbanisation and economic growth, make the transition towards a market-oriented agriculture inevitable. This is not a bad thing: all too often we have a far too rosy view on the potential of subsistence agriculture. The truth is that subsistence farming does not generate the necessary financial means for the farmers to get out of poverty, although improvements in agricultural technology may contribute to increased food security (Harris and Orr, 2014). Only when farmers have access to markets they can generate an income that allows them to fully participate in society so that they can not only benefit from the material perks of modern life but also provide a high-quality education to their children and the necessary health care to those who need it: soil conservation as such cannot achieve this (Posthumus and Stroosnijder, 2010). Case studies support that a symbiosis between the development of a market-oriented agriculture and soil conservation is indeed likely as market access provides farmers with the economic incentives to implement soil conservation measures (Boyd and Slaymaker, 2000). Again, the transition from a subsistence to a market-oriented system will almost inevitably have to be accompanied by intensification as the latter will allow a better return on both capital and input investment.

Smart intensification will not be sufficient to achieve adequate soil conservation (but it will help). The points raised above illustrate that adequate soil conservation is much more likely to be achieved if more intensive agricultural systems are developed in the Global South as the economic and environmental stimuli to implement soil conservation measures will be much larger. Yet, the experiences in Europe and North America illustrate that this may not be sufficient to achieve adequate soil conservation and that government stimulation (through financial measures) and/or coercion may be necessary to further reduce soil degradation. It is, however, the magnitude of such efforts and their effectiveness that should be considered. The societal efforts and costs that will be needed to achieve adequate soil conservation will be far smaller when less land is used for agriculture as much less land will need treatment. Furthermore, one may also imagine that efforts to convince farmers to adopt conservation measures will be more successful in an intensive, market-oriented agricultural system as they will, generally, be more open to changes and both governments and other stakeholders will have more leverage in discussions on how the agricultural system needs to be organised. This is, obviously, no guarantee for success as potential direct financial benefits may seduce the stakeholders to neglect the necessary investments to achieve long-term sustainability. The latter is a problem that occurs everywhere where environmental and economic concerns conflict and, while general principles to resolve such problems have been formulated (Ostrom, 2009), specific policies to deal with this conflict will depend on local conditions.

5 Conclusions

All too often, soil conservation is discussed in isolation, whereby much attention is given to the effectiveness of technical solutions in reducing excessive soil and water losses at a given location. Agriculture, however, is a system wherein lateral connections at different scales are very important: actions at a specific location will necessarily have implications at other locations. Agricultural systems are also subject to constant change as they respond to changes in population numbers, population distribution, economic wealth and cultural preferences. A coherent vision on the development of soil conservation in 21st century needs to account for this context and needs to consider both the spatial and temporal dynamics of agricultural systems.

While it is certainly true that conservation technology can be further developed other considerations may be more important for the successful implementation of soil conservation programmes. In our view, smart intensification is an essential ingredient of any strategy seeking efficient soil conservation while at the same time meeting the growing food demands of a strongly increasing, more urbanised global population. Smart intensification will help to reduce the land

surface area exposed to a high soil degradation risk, while it will, at the same time, increase the return on the soil conservation measures that will still be necessary. Smart intensification will also allow for additional environmental benefits to be reaped in terms of soil organic carbon storage, biodiversity and water availability. It will also be directly beneficial to the farmers, allowing them to produce food for more people and to achieve an acceptable income. It is therefore no surprise that, when considering these other angles, other researchers have reached similar conclusions, stating that agriculture in the Global South and particularly in Africa needs to intensify and that the exclusive focus on smallholders as engines for growth needs to change (Collier and Dercon, 2009).

Intensification is not a panacea that magically solves all problems. Striving towards higher crop yields will require the use of more external inputs, including the use of mineral fertilisers. This is often assumed to be detrimental to the environment: yet this only will be true if fertilisers are used excessively, as is the case now in many areas of the world (Sattari et al., 2012; Lassaletta et al., 2014; Zhang et al., 2015). If correctly used, the environmental benefits of judicious mineral fertiliser use will more often than not outweigh their potential negative impacts by reducing the amount of land needed for agricultural production (Tilman et al., 2011). Furthermore, intensification will require higher energy and capital inputs per unit of surface area: these extra investments will partly be compensated for by the fact that a smaller area of land needs to be cultivated, but access to markets will often be essential to make intensification profitable.

Smart intensification as such will not be sufficient to reduce soil loss to acceptable levels: in intensive systems, soil losses are also often higher than is tolerable and conflicts between (long-term) environmental and (short-term) economic goals will be present. Yet, these problems will be easier to tackle when we give smart intensification adequate consideration in any plan on future agricultural development in the Global South.

6 Data availability

All data used in this paper are in the public domain and can be accessed through the URL links provided.

Competing interests. The authors declare that they have no conflict of interest.

Acknowledgements. This paper greatly benefited from the critical comments of Martin Van Ittersum and an anonymous referee. The financial support of STAP (Scientific and Technical Advisory Panel of the Global Environmental Facility) and the IUAP project SOGLO (The Soil System under Global Change, IUAP P7/24) is gratefully acknowledged.

Edited by: J. Wallinga

References

Alexandratos, N. and Bruinsma, J.: World agriculture towards 2030/2050: The 2012 revision, ESA Working paper, 2012.

Angers, D. A. and Eriksen-Hamel, N. S.: Full-inversion tillage and organic carbon distribution in soil profiles: A meta-analysis, Soil Sci. Soc. Am. J., 72, 1370–1374, doi:10.2136/sssaj2007.0342, 2008.

Arnalds, A.: Approaches to Landcare–a century of soil conservation in Iceland, Land Degrad. Dev., 16, 113–125, 2005.

Bakker, M. M., Govers, G., Kosmas, C., Vanacker, V., van Oost, K., and Rounsevell, M.: Soil erosion as a driver of land-use change, Agr. Ecosyst. Environ., 105, 467–481, doi:10.1016/j.agee.2004.07.009, 2005.

Batjes, N. H.: Total carbon and nitrogen in the soils of the world, Eur. J. Soil Sci., 47, 151–163, doi:10.1111/j.1365-2389.1996.tb01386.x, 1996.

Boardman, J.: An average soil erosion rate for Europe: Myth or reality?, J. Soil Water Conserv., 53, 46–50, 1998.

Boardman, J., Ligneau, L., de Roo, A., and Vandaele, K.: Flooding of property by runoff from agricultural land in northwestern Europe, Geomorphology, 10, 183–196, 1994.

Bocco, G.: Traditional knowledge for soil conservation in central Mexico, J. Soil Water Conserv., 46, 346–348, 1991.

Boyd, C. and Slaymaker, T.: Re-examining the "more people less erosion" hypothesis: Special case or wider trend, Natural Resource Perspective, 63, 1–6, 2000.

Brouder, S. M. and Gomez-Macpherson, H.: The impact of conservation agriculture on smallholder agricultural yields: A scoping review of the evidence, Agr. Ecosyst. Environ., 187, 11–32, 2014.

Burney, J. A., Davis, S. J., and Lobell, D. B.: Greenhouse gas mitigation by agricultural intensification, P. Natl. Acad. Sci. USA, 107, 12052–12057, doi:10.1073/pnas.0914216107, 2010.

Cerdan, O., Govers, G., Le Bissonnais, Y., Van Oost, K., Poesen, J., Saby, N., Gobin, A., Vacca, A., Quinton, J., Auerswald, K., Klik, A., Kwaad, F., Raclot, D., Ionita, I., Rejman, J., Rousseva, S., Muxart, T., Roxo, M. J., and Dostal, T.: Rates and spatial variations of soil erosion in Europe: A study based on erosion plot data, Geomorphology, 122, 167–177, doi:10.1016/j.geomorph.2010.06.011, 2010.

Chamberlin, J., Jayne, T. S., and Headey, D.: Scarcity amidst abundance? Reassessing the potential for cropland expansion in Africa, Food Policy, 48, 51–65, doi:10.1016/j.foodpol.2014.05.002, 2014.

Chen, L., Wei, W., Fu, B., and Lü, Y.: Soil and water conservation on the Loess Plateau in China: review and perspective, Prog. Phys. Geogr., 31, 389–403, doi:10.1177/0309133307081290, 2007.

Christopher, S. F., Lal, R., and Mishra, U.: Regional study of no-till effects on carbon sequestration in the midwestern United States, Soil Sci. Soc. Am. J., 73, 207–216, doi:10.2136/sssaj2007.0336, 2009.

Collier, P. and Dercon, S.: African agriculture in 50 years: Smallholders in a rapidly changing world?, FAO Expert meeting on how to feed the world in 2050 (June 2009), FAO, 13 pp., 2009.

De Beenhouwer, M., Aerts, R., and Honnay, O.: A global meta-analysis of the biodiversity and ecosystem service benefits of coffee and cacao agroforestry, Agr. Ecosyst. Environ., 175, 1–7, 2013.

De Deyn, G. B. and Van der Putten, W. H.: Linking aboveground and belowground diversity, Trends Ecol. Evol., 20, 625–633, doi:10.1016/j.tree.2005.08.009, 2005.

Derpsch, R., Friedrich, T., Kassam, A., and Li, H.: Current status of adoption of no-till farming in the world and some of its main benefits, International Journal of Agricultural and Biological Engineering, 3, 1–25, 2010.

De Schutter, O.: Agroecology and the Right to Food, Report presented at the 16th Session of the United Nations Human Rights Council, Geneva, Switzerland, United Nations Human Rights Council, 2011.

Desmet, P. J. J. and Govers, G.: A GIS procedure for automatically calculating the USLE LS factor on topographically complex landscape units, J. Soil Water Conserv., 51, 427–433, 1996.

Döll, P., Fiedler, K., and Zhang, J.: Global-scale analysis of river flow alterations due to water withdrawals and reservoirs, Hydrol. Earth Syst. Sci., 13, 2413–2432, doi:10.5194/hess-13-2413-2009, 2009.

Eagle, A. J., Henry, L. R., Olander, L. P., Haugen-Kozyra, K., Millar, N., and Robertson, G. P.: Greenhouse Gas Mitigation Potential of Agricultural Land Management in the United States: A synthesis of the literature, Nicholas Institute for Environmental Policy Solutions, Duke University, 2012.

Ellis-Jones, J. and Sims, B.: An appraisal of soil conservation technologies on hillside farms in Honduras, Mexico and Nicaragua, Project Appraisal, 10, 125–134, 1995.

FAO and ITPS: Status of the World's Soil Resources – Main Report, FAO and ITPS, 608 pp., 2015.

Gerland, P., Raftery, A. E., Ševčíková, H., Li, N., Gu, D., Spoorenberg, T., Alkema, L., Fosdick, B. K., Chunn, J., Lalic, N., Bay, G., Buettner, T., Heilig, G. K., and Wilmoth, J.: World population stabilization unlikely this century, Science, 346, 234–237, doi:10.1126/science.1257469, 2014.

Giller, K. E., Witter, E., Corbeels, M., and Tittonell, P.: Conservation agriculture and smallholder farming in Africa: The heretics' view, Field Crop. Res., 114, 23–34, doi:10.1016/j.fcr.2009.06.017, 2009.

Glaeser, E.: Triumph of the City: How Our Greatest Invention Makes Us Richer, Smarter, Greener, Healthier and Happier, Pan Macmillan, 2011.

Govers, G., Merckx, R., Van Oost, K., and van Wesemael, B.: Managing Soil Organic Carbon for Global Benefits: a STAP Technical Report, Global Environmental Facility, Washington, D.C., 70 pp., 2013.

Govers, G., Van Oost, K., and Wang, Z.: Scratching the Critical Zone: The Global Footprint of Agricultural Soil Erosion, Procedia Earth and Planetary Science, 10, 313–318, doi:10.1016/j.proeps.2014.08.023, 2014.

Grace, P. R., Antle, J., Aggarwal, P. K., Ogle, S., Paustian, K., and Basso, B.: Soil carbon sequestration and associated economic costs for farming systems of the Indo-Gangetic Plain: A meta-analysis, Agr. Ecosyst. Environ., 146, 137–146, doi:10.1016/j.agee.2011.10.019, 2012.

Harris, D. and Orr, A.: Is rainfed agriculture really a pathway from poverty?, Agr. Syst., 123, 84–96, 2014.

Henao, J. and Baanante, C.: Agricultural production and soil nutrient mining in Africa: implications for resource conservation and policy development, IFDC-AN International Center for Soil Fertility and Agricultural Development, 2006.

Hiederer, R. and Köchyl, M.: Global soil organic carbon estimates and the harmonized world soil database, Luxembourg EUR 25225EN, 79 pp., 2012.

Hornbeck, R.: The enduring impact of the American dust bowl: Short-and long-run adjustments to environmental catastrophe, The American Economic Review, 1477–1507, 2012.

Imwangana, F. M., Vandecasteele, I., Trefois, P., Ozer, P., and Moeyersons, J.: The origin and control of mega-gullies in Kinshasa (DR Congo), Catena, 125, 38–49, 2015.

Janosky, J. S., Young, D. L., and Schillinger, W. F.: Economics of conservation tillage in a wheat–fallow rotation, Agron. J., 94, 527–531, 2002.

Johnson, L. C.: Soil loss tolerance: Fact or myth, J. Soil Water Conserv., 42, 155–160, 1987.

Keating, B. A., Carberry, P. S., Bindraban, P. S., Asseng, S., Meinke, H., and Dixon, J.: Eco-efficient agriculture: Concepts, challenges, and opportunities, Crop Science, 50, S-109–S-119, doi:10.2135/cropsci2009.10.0594, 2010.

Knowler, D. and Bradshaw, B.: Farmers' adoption of conservation agriculture: A review and synthesis of recent research, Food Policy, 32, 25–48, doi:10.1016/j.foodpol.2006.01.003, 2007.

Knowler, D., Bradshaw, B., and Gordon, D.: The economics of conservation agriculture, Land and Water Division, FAO, Rome, 2001.

Kok, H., Papendick, R., and Saxton, K. E.: STEEP: Impact of long-term conservation farming research and education in Pacific Northwest wheatlands, J. Soil Water Conserv., 64, 253–264, 2009.

Krajick, K.: Green farming by the Incas?, Science, 281, 322–322, 1998.

Lahmar, R.: Adoption of conservation agriculture in Europe: lessons of the KASSA project, Land Use Policy, 27, 4–10, 2010.

Lassaletta, L., Billen, G., Grizzetti, B., Anglade, J., and Garnier, J.: 50 year trends in nitrogen use efficiency of world cropping systems: the relationship between yield and nitrogen input to cropland, Environ. Res. Lett., 9, 105011, doi:10.1088/1748-9326/9/10/105011, 2014.

Le Quere, C., Raupach, M. R., Canadell, J. G., Marland, G., Bopp, L., Ciais, P., Conway, T. J., Doney, S. C., Feely, R. A., Foster, P., Friedlingstein, P., Gurney, K., Houghton, R. A., House, J. I., Huntingford, C., Levy, P. E., Lomas, M. R., Majkut, J., Metzl, N., Ometto, J. P., Peters, G. P., Prentice, I. C., Randerson, J. T., Running, S. W., Sarmiento, J. L., Schuster, U., Sitch, S., Takahashi, T., Viovy, N., van der Werf, G. R., and Woodward, F. I.: Trends in the sources and sinks of carbon dioxide, Nat. Geosci., 2, 831–836, doi:10.1038/ngeo689, 2009.

Leys, A., Govers, G., Gillijns, K., Berckmoes, E., and Takken, I.: Scale effects on runoff and erosion losses from arable land under conservation and conventional tillage: The role of residue cover, J. Hydrol., 390, 143–154, doi:10.1016/j.jhydrol.2010.06.034, 2010.

Lilin, C.: Histoire de la restauration des terrains en montagne au 19ème siècle, Cahiers ORSTOM, Série Pédologie, 22, 139–145, 1986.

Lu, C. H. and van Ittersum, M. K.: A trade-off analysis of policy objectives for Ansai, the Loess Plateau of China, Agr. Ecosyst. Environ., 102, 235–246, doi:10.1016/j.agee.2003.09.023, 2004.

Lutz, W. and Kc, S.: Dimensions of global population projections: what do we know about future population trends and structures?, Philos. T. R. Soc. B, 365, 2779–2791, doi:10.1098/rstb.2010.0133, 2010.

Mader, P., Fliessbach, A., Dubois, D., Gunst, L., Fried, P., and Niggli, U.: Soil fertility and biodiversity in organic farming, Science, 296, 1694–1697, doi:10.1126/science.1071148, 2002.

Mekuria, W., Veldkamp, E., Haile, M., Nyssen, J., Muys, B., and Gebrehiwot, K.: Effectiveness of exclosures to restore degraded soils as a result of overgrazing in Tigray, Ethiopia, J. Arid Environ., 69, 270–284, doi:10.1016/j.jaridenv.2006.10.009, 2007.

Minasny, B., McBratney, A., Hong, S. Y., Sulaeman, Y., Kim, M. S., Zhang, Y. S., Kim, Y. H., and Han, K. H.: Continuous rice cropping has been sequestering carbon in soils in Java and South Korea for the past 30 years, Global Biogeochem. Cy., 26, 1–8, doi:10.1029/2012GB004406, 2012.

Montgomery, D. R.: Soil erosion and agricultural sustainability, P. Natl. Acad. Sci. USA, 104, 13268–13272, doi:10.1073/pnas.0611508104, 2007.

Mueller, N. D., Gerber, J. S., Johnston, M., Ray, D. K., Ramankutty, N., and Foley, J. A.: Closing yield gaps through nutrient and water management, Nature, 490, 254–257, 2012.

Napier, T. L., Boardman, J., Foster, I., and Dearing, J.: The evolution of US soil-conservation policy: from voluntary adoption to coercion, Soil erosion on agricultural land, Proceedings of a workshop sponsored by the British Geomorphological Research Group, Coventry, UK, January 1989, 627–644, 1990.

Nearing, M. A.: A single, continuous function for slope steepness influence on soil loss, Soil Sci. Soc. Am. J., 61, 917–919, 1997.

Neumann, K., Verburg, P. H., Stehfest, E., and Müller, C.: The yield gap of global grain production: A spatial analysis, Agr. Syst., 103, 316–326, 2010.

NRCS: 2007 National Resources Inventory – Soil Erosion on Cropland, 29, 2010.

Nyssen, J., Poesen, J., Gebremichael, D., Vancampenhout, K., D'Aes, M., Yihdego, G., Govers, G., Leirs, H., Moeyersons, J., Naudts, J., Haregeweyn, N., Haile, M., and Deckers, J.: Interdisciplinary on-site evaluation of stone bunds to control soil erosion on cropland in Northern Ethiopia, Soil Till. Res., 94, 151–163, doi:10.1016/j.still.2006.07.011, 2007.

Nyssen, J., Clymans, W., Poesen, J., Vandecasteele, I., De Raets, S., Haregeweyn, N., Naudts, J., Hadera, A., Moeyersons, J., Haile, M., and Deckers, J.: How soil conservation affects the catchment sediment budget – a comprehensive study in the north Ethiopian highlands, Earth Surf. Proc. Land., 34, 1216–1233, doi:10.1002/Esp.1805, 2009.

Oorts, K., Bossuyt, H., Labreuche, J., Merckx, R., and Nicolardot, B.: Carbon and nitrogen stocks in relation to organic matter fractions, aggregation and pore size distribution in no-tillage and conventional tillage in northern France, Eur. J. Soil Sci., 58, 248–259, doi:10.1111/j.1365-2389.2006.00832.x, 2007.

Ostrom, E.: A general framework for analyzing sustainability of social-ecological systems, Science, 325, 419–422, doi:10.1126/science.1172133, 2009.

Peiretti, R. and Dumanski, J.: The transformation of agriculture in Argentina through soil conservation, Int. Soil Water Conserv. Res., 2, 14–20, doi:10.1016/S2095-6339(15)30010-1, 2014.

Perlman, J. E.: The metamorphosis of marginality: four generations in the favelas of Rio de Janeiro, Ann. Am. Acad. Polit. SS., 606, 154–177, 2006.

Phalan, B., Balmford, A., Green, R. E., and Scharlemann, J. R. P. W.: Minimising the harm to biodiversity of producing more food globally, Food Policy, 36, S62–S71, 2011a.

Phalan, B., Onial, M., Balmford, A., and Green, R. E.: Reconciling food production and biodiversity conservation: Land sharing and land sparing compared, Science, 333, 1289–1291, doi:10.1126/science.1208742, 2011b.

Pittelkow, C. M., Linquist, B. A., Lundy, M. E., Liang, X., Van Groenigen, K. J., Lee, J., Van Gestel, N., Six, J., Venterea, R. T., and Van Kessel, C.: When does no-till yield more? A global meta-analysis, Field Crop. Res., 183, 156–168, 2015.

Poeplau, C., Don, A., Vesterdal, L., Leifeld, J., Van Wesemael, B., Schumacher, J., and Gensior, A.: Temporal dynamics of soil organic carbon after land-use change in the temperate zone – carbon response functions as a model approach, Glob. Change Biol., 17, 2415–2427, doi:10.1111/j.1365-2486.2011.02408.x, 2011.

Poesen, J., Nachtergaele, J., Verstraeten, G., and Valentin, C.: Gully erosion and environmental change: importance and research needs, Catena, 50, 91–133, 2003.

Posthumus, H. and Stroosnijder, L.: To terrace or not: the short-term impact of bench terraces on soil properties and crop response in the Peruvian Andes, Environment, Development and Sustainability, 12, 263–276, 2010.

Quang, D. V., Schreinemachers, P., and Berger, T.: Ex-ante assessment of soil conservation methods in the uplands of Vietnam: An agent-based modeling approach, Agr. Syst., 123, 108–119, doi:10.1016/j.agsy.2013.10.002, 2014.

Quinton, J. N., Govers, G., Van Oost, K., and Bardgett, R. D.: The impact of agricultural soil erosion on biogeochemical cycling, Nat. Geosci., 3, 311–314, doi:10.1038/Ngeo838, 2010.

Ramachandran Nair, P. K., Mohan Kumar, B., and Nair, V. D.: Agroforestry as a strategy for carbon sequestration, J. Plant Nutr. Soil Sci., 172, 10–23, doi:10.1002/jpln.200800030, 2009.

Ramankutty, N., Evan, A. T., Monfreda, C., and Foley, J. A.: Farming the planet: 1. Geographic distribution of global agricultural lands in the year 2000, Global Biogeochem. Cy., 22, GB1003, doi:10.1029/2007GB002952, 2008.

Rijsdijk, A., Bruijnzeel, L. A. S., and Sutoto, C. K.: Runoff and sediment yield from rural roads, trails and settlements in the upper Konto catchment, East Java, Indonesia, Geomorphology, 87, 28–37, doi:10.1016/j.geomorph.2006.06.040, 2007.

Robb, G.: The discovery of France, Pan Macmillan, 2008.

Rockström, J., Lannerstad, M., and Falkenmark, M.: Assessing the water challenge of a new green revolution in developing countries, P. Natl. Acad. Sci. USA, 104, 6253–6260, doi:10.1073/pnas.0605739104, 2007.

Sanchez, P. A.: Tripling crop yields in tropical Africa, Nat. Geosci., 3, 299–300, 2010.

Sattari, S. Z., Bouwman, A. F., Giller, K. E., and van Ittersum, M. K.: Residual soil phosphorus as the missing piece in the global phosphorus crisis puzzle, P. Natl. Acad. Sci. USA, 109, 6348–6353, 2012.

Saunders, D.: Arrival City: How the Largest Migration in History Is Reshaping Our World, Knopf Doubleday Publishing Group, 2011.

Schneider, M. K., Lüscher, G., Jeanneret, P., Arndorfer, M., Ammari, Y., Bailey, D., Balázs, K., Báldi, A., Choisis, J.-P., and Dennis, P.: Gains to species diversity in organically farmed fields are not propagated at the farm level, Nat. Commun., 5, 4151, doi:10.1038/ncomms5151, 2014.

Sendzimir, J., Reij, C. P., and Magnuszewski, P.: Rebuilding resilience in the Sahel: regreening in the Maradi and Zinder regions of Niger, Ecol. Soc., 16, 1, doi:10.5751/ES-04198-160301, 2011.

Smaling, E. M. A., Stoorvogel, J. J., and Windmeijer, P. N.: Calculating soil nutrient balances in Africa at different scales, Fert. Res., 35, 237–250, doi:10.1007/bf00750642, 1993.

Sterk, G.: Causes, consequences and control of wind erosion in Sahelian Africa: a review, Land Degrad. Dev., 14, 95–108, 2003.

Stockmann, U., Minasny, B., and McBratney, A. B.: How fast does soil grow?, Geoderma, 216, 48–61, 2014.

Stone, J. R., Gilliam, J. W., Cassel, D. K., Daniels, R. B., Nelson, L. A., and Kleiss, H. J.: Effect of erosion and landscape position on the productivity of Piedmont soils, Soil Sci. Soc. Am. J., 49, 987–991, 1985.

Tiffen, M., Mortimore, M., and Gichuki, F.: More People, Less Erosion: Environmental Recovery in Kenya, John Wiley & Sons Ltd, 1994.

Tilman, D., Balzer, C., Hill, J., and Befort, B. L.: Global food demand and the sustainable intensification of agriculture, P. Natl. Acad. Sci. USA, 108, 20260–20264, doi:10.1073/pnas.1116437108, 2011.

Torri, D. and Poesen, J.: Predictability and uncertainty of the soil erodibility factor using a global dataset, Catena, 31, 1–22, 1997.

Twomlow, S., Rohrbach, D., Dimes, J., Rusike, J., Mupangwa, W., Ncube, B., Hove, L., Moyo, M., Mashingaidze, N., and Mahposa, P.: Micro-dosing as a pathway to Africa's Green Revolution: evidence from broad-scale on-farm trials, Nutr. Cycl. Agroecosys., 88, 3–15, 2010.

United Nations: World Urbanization Prospects 2014: Highlights, United Nations Publications, 2014.

Valentin, C., Agus, F., Alamban, R., Boosaner, A., Bricquet, J.-P., Chaplot, V., De Guzman, T., De Rouw, A., Janeau, J.-L., and Orange, D.: Runoff and sediment losses from 27 upland catchments in Southeast Asia: Impact of rapid land use changes and conservation practices, Agr. Ecosyst. Environ., 128, 225–238, 2008.

Vanacker, V., Molina, A., Govers, G., Poesen, J., and Deckers, J.: Spatial variation of suspended sediment concentrations in a tropical Andean river system: The Paute River, southern Ecuador, Geomorphology, 87, 53–67, doi:10.1016/j.geomorph.2006.06.042, 2007a.

Vanacker, V., von Blanckenburg, F., Govers, G., Molina, A., Poesen, J., Deckers, J., and Kubik, P.: Restoring dense vegetation can slow mountain erosion to near natural benchmark levels, Geology, 35, 303–306, doi:10.1130/G23109a.1, 2007b.

VandenBygaart A. J., Bremer, E., McConkey, B. G., Janzen, H. H., Angers, D. A., Carter, M. R., Drury, C. F., Lafond, G. P., and McKenzie, R. H.: Soil organic carbon stocks on long-term agroecosystem experiments in Canada, Can. J. Soil Sci., 90, 543–550, doi:10.4141/cjss10028, 2010.

Van den Putte, A., Govers, G., Diels, J., Gillijns, K., and Demuzere, M.: Assessing the effect of soil tillage on crop growth: A meta-regression analysis on European crop yields under conservation agriculture, Eur. J. Agron., 33, 231–241, doi:10.1016/j.eja.2010.05.008, 2010.

Van Oost, K., Quine, T. A., Govers, G., De Gryze, S., Six, J., Harden, J. W., Ritchie, J. C., McCarty, G. W., Heckrath, G., Kosmas, C., Giraldez, J. V., da Silva, J. R. M., and Merckx, R.: The impact of agricultural soil erosion on the global carbon cycle, Science, 318, 626–629, doi:10.1126/science.1145724, 2007.

Van Rompaey, A. J. J., Govers, G., Van Hecke, E., and Jacobs, K.: The impacts of land use policy on the soil erosion risk: a case study in central Belgium, Agr. Ecosyst. Environ., 83, 83–94, 2001.

Van Rompaey, A. J. J., Govers, G., and Puttemans, C.: Modelling land use changes and their impact on soil erosion and sediment supply to rivers, Earth Surf. Proc. Land., 27, 481–494, doi:10.1002/Esp.335, 2002.

Verbruggen, E., Roling, W. F. M., Gamper, H. A., Kowalchuk, G. A., Verhoef, H. A., and van der Heijden, M. G. A.: Positive effects of organic farming on below-ground mutualists: large-scale comparison of mycorrhizal fungal communities in agricultural soils, New Phytol., 186, 968–979, doi:10.1111/j.1469-8137.2010.03230.x, 2010.

Verheijen, F. G., Jones, R. J., Rickson, R., and Smith, C.: Tolerable versus actual soil erosion rates in Europe, Earth-Sci. Rev., 94, 23–38, 2009.

Yang, D. W., Kanae, S., Oki, T., Koike, T., and Musiake, K.: Global potential soil erosion with reference to land use and climate changes, Hydrol. Process., 17, 2913–2928, doi:10.1002/hyp.1441, 2003.

Zentner, R., McConkey, B., Campbell, C., Dyck, F., and Selles, F.: Economics of conservation tillage in the semiarid prairie, Can. J. Plant Sci., 76, 697–705, 1996.

Zhang, X., Davidson, E. A., Mauzerall, D. L., Searchinger, T. D., Dumas, P., and Shen, Y.: Managing nitrogen for sustainable development, Nature, 528, 51–59, doi:10.1038/nature15743, 2015.

Zhao, J., Van Oost, K., Chen, L., and Govers, G.: Moderate topsoil erosion rates constrain the magnitude of the erosion-induced carbon sink and agricultural productivity losses on the Chinese Loess Plateau, Biogeosciences, 13, 4735–4750, doi:10.5194/bg-13-4735-2016, 2016.

14

Leaf waxes in litter and topsoils along a European transect

Imke K. Schäfer[1], **Verena Lanny**[2], **Jörg Franke**[1], **Timothy I. Eglinton**[2], **Michael Zech**[3,4],
Barbora Vysloužilová[5,6], **and Roland Zech**[1]

[1]Institute of Geography and Oeschger Centre for Climate Change Research, University of Bern,
3012 Bern, Switzerland
[2]Department of Earth Science, ETH Zurich, 8092 Zurich, Switzerland
[3]Landscape- & Geoecology, Faculty of Environmental Sciences, Technical University of Dresden,
01062 Dresden, Germany
[4]Institute of Agronomy and Nutritional Sciences, Soil Biogeochemistry, Martin Luther University
Halle-Wittenberg, 06120 Halle, Germany
[5]Institute of Archaeology of Academy of Science of the Czech Republic, Letenská 4,
11801 Prague 1, Czech Republic
[6]Laboratoire Image, Ville, Environnement, UMR7362, CNRS/Université de Strasbourg,
67083 Strasbourg CEDEX, France

Correspondence to: Imke K. Schäfer (imke.schaefer@giub.unibe.ch)

Abstract. Lipid biomarkers are increasingly used to reconstruct past environmental and climate conditions. Leaf-wax-derived long-chain n-alkanes and n-alkanoic acids may have great potential for reconstructing past changes in vegetation, but the factors that affect the leaf wax distribution in fresh plant material, as well as in soils and sediments, are not yet fully understood and need further research. We systematically investigated the influence of vegetation and soil depth on leaf waxes in litter and topsoils along a European transect. The deciduous forest sites are often dominated by the n-C_{27} alkane and n-C_{28} alkanoic acid. Conifers produce few n-alkanes but show high abundances of the C_{24} n-alkanoic acid. Grasslands are characterized by relatively high amounts of C_{31} and C_{33} n-alkanes and C_{32} and C_{34} n-alkanoic acids. Chain length ratios thus may allow for distinguishing between different vegetation types, but caution must be exercised given the large species-specific variability in chain length patterns. An updated endmember model with the new n-alkane ratio $(n$-$C_{31} + n$-$C_{33})/(n$-$C_{27} + n$-$C_{31} + n$-$C_{33})$ is provided to illustrate, and tentatively account for, degradation effects on n-alkanes.

1 Introduction

To improve our understanding of ongoing environmental changes and to predict consequences of future climate change more precisely, it is important to investigate the magnitude of, and interactions between, climate and environmental variations in the past. Lipid biomarkers are well preserved in many geological archives and are increasingly used for palaeoclimate and palaeoenvironmental reconstructions (Eglinton and Eglinton, 2008). Long-chain n-alkanes ($> C_{25}$) and n-alkanoic acids ($> C_{20}$), for example, are essential constituents of epicuticular leaf waxes and thus serve as specific biomarkers for higher terrestrial plants (Eglinton et al., 1962; Eglinton and Hamilton, 1967; Otto and Simpson, 2005).

Leaf wax n-alkanes typically show an odd-over-even predominance (OEP; Eglinton and Hamilton, 1967). The relative odd homologue abundance may be useful to discriminate between different vegetation types: C_{27} and C_{29} have been reported to be predominant in leaf waxes of trees and shrubs, whereas C$_{31}$ and C$_{33}$ mostly derive from grasses and herbs (Maffei, 1996; Maffei et al., 2004; Rommerskirchen et al., 2006; Zech et al., 2009; Lei et al., 2010; Kirkels et al., 2013). From this distribution, various n-alkane ratios have been proposed that allow estimating the main contributing vegetation type to a palaeosample (Schwark et al., 2002; Zech et al., 2009; Lei et al., 2010; Schatz et al., 2011; Wiesenberg et al., 2015). Where pollen grains are preserved, leaf wax and pollen records are in good agreement (e.g. Brincat et al., 2000; Schwark et al., 2002; Zech et al., 2010; Tarasov et al., 2013). In contrast, a recent study that summarizes n-alkane patterns in modern plants from all over the world showed no discrimination power for vegetation reconstruction at a global scale (Bush and McInerney, 2013), so regional calibration studies may be more appropriate. Additionally, the accuracy of n-alkane records remains somewhat uncertain due to several potential pitfalls:

i. Leaf wax production and concentration can vary widely between species (e.g. Diefendorf et al., 2011; Bush and McInerney, 2013). Moreover, species abundance in an ecosystem controls leaf wax signals in soils and sediment.

ii. Several studies reported long-chain n-alkanes not only in leaves but also in other plant parts (roots, stems, blossoms; e.g. Wöstmann, 2006; Jansen et al., 2006; Kirkels et al., 2013; Gocke et al., 2013), but in these studies the patterns show preferential synthesis of shorter chains ($< n$-C$_{25}$) with low OEPs, as well as much lower n-alkane concentrations (3 to 10 times) than in leaves.

iii. Leaf waxes are affected by mineralization and degradation (e.g. Zech et al., 2009, 2011a; Nguyen Tu et al., 2011). As OEP values become lower during degradation, Zech et al. (2009) and Buggle et al. (2010) proposed procedures to quantify and correct n-alkane ratios for degradation using the OEP.

iv. Apart from vegetation type, many environmental parameters may influence leaf wax patterns, for example temperature and precipitation (e.g. Poynter et al., 1989; Sachse et al., 2006; Tipple and Pagani, 2013; Bush and McInerney, 2015), as well as radiation, nutrient and water availability, salinity, mechanical stress and pollution (e.g. Shepherd and Wynne Griffiths, 2006; Guo et al., 2014). Sachse et al. (2006) and Duan and He (2011) found longer chain lengths at lower latitudes, which could indicate (i) enhanced loss of shorter n-alkanes with increasing evaporation and (ii) preferential production of long-chain n-alkanes, providing better protection against evaporation at higher temperatures and radiation. Schefuß et al. (2003) reported a higher n-C$_{31}$ vs. n-C$_{29}$ abundance in dust samples from drier regions along the West African margin and suggested that humidity may be the driving factor. Bush and McInerney (2015) showed a correlation between average chain

length (ACL) and temperature along a transect throughout the central United States and concluded that temperature is directly responsible for the synthesis of longer chain length.

So far, only few studies have investigated homologue long-chain n-alkanoic acid patterns in plants, soils and sediments (e.g. Almendros et al., 1996; Marseille et al., 1999; Bull et al., 2000b; Zocatelli et al., 2012; Feakins et al., 2014; Wiesenberg et al., 2015). Leaf wax n-alkanoic acids have a distinct even-over-odd predominance (EOP; Eglinton and Hamilton, 1967). Zocatelli et al. (2012) found n-C$_{26}$ to be predominant in grassland soil relative to forest soil, whereas the forest soils contained n-C$_{22}$ to n-C$_{28}$ in greater amounts. Since n-alkane patterns alone seem to not always allow a reliable conclusion about the dominant vegetation type, there is an urgent need to more systematically investigate and understand the factors controlling the homologue n-alkanoic acid patterns.

In order to contribute to a better understanding of the various factors controlling leaf wax patterns, we collected and analysed litter and topsoil samples in deciduous and coniferous forests and from grasslands along a transect in central Europe.

Specifically, our study aims to evaluate (1) the role of different vegetation types for n-alkane and n-alkanoic acid chain length patterns and (2) the effects of degradation on leaf wax patterns.

Sampling of litter and topsoil samples rather than individual plants or plant parts has the advantage of integrating over the whole ecosystem and implicitly taking into account variable production and concentration of leaf waxes between species, individual plants and plant parts. Such an approach also implicitly accounts for the fact that vegetation types generally do not totally dominate a specific ecosystem but mostly co-occur to a variable degree. For example, we defined forests with a dominance $> 80\%$ of deciduous or coniferous trees as deciduous or coniferous forests, respectively, and forests generally have some grass and herb understorey as well. Our study will thus provide leaf wax patterns for ecosystems dominated by specific vegetation types, and those patterns will be a sound basis for comparing and interpreting leaf wax patterns from palaeosols or sediments. Changes in leaf wax patterns from litter to the topsoil are expected to indicate effects of degradation and microbial reworking of organic material in the topsoil. Differences in climatic conditions along the transect are not very pronounced, so this paper cannot focus on potential direct climatic effects on leaf wax patterns.

2 Material and methods

2.1 Geographical setting and sampling

In 2012 and 2014, we collected litter (L) and topsoil samples (Ah1: 0–3 cm; Ah2: 3–10 cm) from 26 locations along a transect in central Europe (Fig. 1). Samples BRO, HUB, HUG, HUM, HUR and KOC were kindly provided by B. Vyslouřilová; more information about sampling and the regional setting of these locations can be found in Vyslouřilová et al. (2015). The study area is in general characterized by relatively mild temperatures and moderate rainfall. The mean annual air temperature along the transect ranges from 5.5 to 11.0 °C, and mean annual precipitation ranges from 470 to 1700 mm (see Table S1 in the Supplement for individual data and data source). Altitudes range from 16 to 899 m a.s.l. (above sea level). The natural vegetation consists of grasslands in the south of the transect and a higher amount of deciduous broadleaf and mixed forests (varying percentages of deciduous trees and conifers) in the north. Also, the proportion of evergreen conifers increases northwards.

We sampled soils in forests with a dominance of deciduous and coniferous trees, as well as soils below grasslands (referred to as "dec", "con" and "grass" in the following text and figures). Photographs of the sampling sites and descriptions of the dominant vegetation are provided in Tables S2 and S1, respectively. For the dec and con sites, we were able to collect litter samples, but the grass sites had virtually no litter. Sampling sites were chosen in forests with large old trees, indicating a stable environment for more than approximately 30 years. This limits the risk that a vegetation change might have influenced the leaf wax signal in the soil, as leaf waxes are stable over long time periods (e.g. Derenne and Largeau, 2001). Please note that our grass sites can be dominated by grasses, herbs or heaths. Soil and litter samples at each site were composites from three sampling points approximately 5 to 7 m apart from each other.

2.2 Lipid analysis

Lipids were extracted from 1–6 g freeze-dried and ground samples by microwave extraction with 15 mL of dichloromethane (DCM)/methanol (MeOH) (9 : 1) at 100 °C for 1 h. Each total lipid extract was passed over a pipette column filled with aminopropyl silica gel as the stationary phase (Supelco, 45 μm). The apolar fraction (including n-alkanes) was eluted with hexane, more polar compounds (e.g. alcohols) with DCM/MeOH (1 : 1), and acids (including n-alkanoic acids) with 5 % acetic acid in diethyl ether. The n-alkanes were purified by passing the apolar fraction over a pipette column filled with activated AgNO$_3$ impregnated silica gel (to retain unsaturated compounds) and another pipette column filled with zeolite (Geokleen). After drying, the zeolite (containing straight-chain compounds) was dissolved in HF and the n-alkanes were recovered by liquid–liquid extraction with hexane. For quantification, the n-alkane fractions were spiked with a known amount of the 5α-androstane and analysed with an Agilent 7890 gas chromatograph (GC) equipped with a VF1 column (30 m length × 0.25 mm i.d., 0.25 μm film thickness) and a flame ionization detector (FID).

Figure 1. Sample locations (black dots) along the transect (map source: US National Park Service, Esri, HERE, DeLorme, MapmyIndia, OpenStreetMap contributors and the GIS user community).

The n-alkanoic acids were converted to fatty acid methyl esters (FAMEs) with MeOH/HCl (95/5; 70 °C, 8 h). The FAMEs were recovered by liquid–liquid extraction with hexane and cleaned over silica, AgNO$_3$ and zeolite columns as described above before quantification with GC-FID. For quantification of the FAMEs, 5α-androstane was again used as an internal standard. Unfortunately, due to some problems during FAME preparation, not all samples were available for methylation (missing samples: BRO, HUB, HUG, HUM, HUR, KOC).

2.3 Leaf wax proxies

Total n-alkane and n-alkanoic acid concentrations (c_{tot}) were calculated as the sum of C$_{25}$ to C$_{35}$ and C$_{20}$ to C$_{34}$ (odd as well as even ones), respectively, and given in μg g^{-1} dry weight (dw).

Changes in the average chain length (ACL) of n-alkyl lipids can show changes in the input of vegetation type. The ACL was determined by modifying the equation of Poynter et al. (1989). We used odd chain lengths only for n-alkanes (Eq. 1) and even chain lengths for n-alkanoic acids (Eq. 2).

$$\text{ACL}(n\text{-alkanes})$$
$$= \frac{27 \times n\text{-C}_{27} + 29 \times n\text{-C}_{29} + 31 \times n\text{-C}_{31} + 33 \times n\text{-C}_{33}}{n\text{-C}_{27} + n\text{-C}_{29} + n\text{-C}_{31} n\text{-C}_{33}} \quad (1)$$

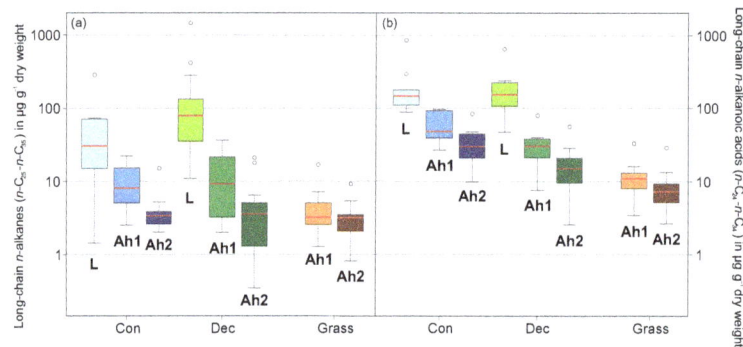

Figure 2. Total concentrations of (**a**) n-alkanes and (**b**) n-alkanoic acids in $\mu g\,g^{-1}$ dry weight. Abbreviations: con, coniferous forest sites ($n = 9$); dec, deciduous forest sites ($n = 14$); grass, grassland sites ($n = 22$); L, litter; Ah1, topsoil 1 (0–3 cm); Ah2, topsoil 2 (3–10 cm). Box plots show median (red line), interquartile range (IQR) with upper (75 %) and lower (25 %) quartiles, lowest datum still within $1.5 \times$ IQR of lower quartile, and highest datum still within $1.5 \times$ IQR of upper quartile. Note that the y axis is logarithmic.

ACL(n-alkanoic acids)

$$= \frac{24 \times n\text{-}C_{24} + 26 \times n\text{-}C_{26} + 28 \times n\text{-}C_{28} + 30 \times n\text{-}C_{30} + 32 \times n\text{-}C_{32}}{n\text{-}C_{24} + n\text{-}C_{26} + n\text{-}C_{28} + n\text{-}C_{30} + n\text{-}C_{32}} \qquad (2)$$

The OEP of the n-alkanes (Eq. 3) and the EOP of the n-alkanoic acids (Eq. 4) can be used as a proxy for degradation and were determined after Hoefs et al. (2002):

$$\text{OEP} = \frac{n\text{-}C_{27} + n\text{-}C_{29} + n\text{-}C_{31} + n\text{-}C_{33}}{n\text{-}C_{26} + n\text{-}C_{28} + n\text{-}C_{30} + n\text{-}C_{32}}, \qquad (3)$$

$$\text{EOP} = \frac{n\text{-}C_{24} + n\text{-}C_{26} + n\text{-}C_{28} + n\text{-}C_{30} + n\text{-}C_{32}}{n\text{-}C_{23} + n\text{-}C_{25} + n\text{-}C_{27} + n\text{-}C_{29} + n\text{-}C_{31}}. \qquad (4)$$

High OEP values are characteristic of fresh plant material, while the OEP values decrease with ongoing soil organic matter degradation in the topsoil (e.g. Buggle et al., 2010; Zech et al., 2009, 2011b).

2.4 Statistical analysis

First, we tested whether the data were normally distributed (Shapiro and Wilk, 1965) and whether variances of the samples were equal (Levene, 1960). In the case of normality and equal variances, we conducted an analysis of variance (ANOVA) test or otherwise a Kruskal–Wallis test to check for significant differences ($\alpha = 0.05$) between depths horizons within the same vegetation type or between vegetation types within the same horizon, respectively. If the ANOVA/Kruskal–Wallis test indicated significant differences in the means, we applied a "post hoc" test to identify which of the means differ, accounting for the effect of multiple testing. The appropriate post hoc test after ANOVA was selected as recommended by Field (2013): for samples with equal size and equal variance, we applied the Tukey's honest significance test (Tukey, 1949). In the case of equal variance and unequal sample size, we used the Hochberg test (Hochberg, 1988) and for unequal variances the Games–Howell test (Games and Howell, 1976). After Kruskal–Wallis tests, we performed the non-parametric

Conover–Iman post hoc test with a Bonferroni adjustment of p values (Conover and Iman, 1979; Conover, 1999). This test is similar to the well-known Dunn test (Dunn, 1964) but is based on the t distribution instead of the z distribution. It is statistically more powerful than the Dunn test and better suited for our small sample size.

3 Results

3.1 Leaf wax n-alkane abundances and chain length patterns

All samples show a dominance of long ($> C_{25}$) odd-chain n-alkanes, characteristic for epicuticular leaf waxes (e.g. Eglinton et al., 1962; Eglinton and Hamilton, 1967; Rieley et al., 1991; Collister et al., 1994). Total n-alkane concentrations (C_{tot}) range from 0.4 to 1468 $\mu g\,g^{-1}$ dw (Table S3). Such concentrations and huge variability are in agreement with published data from fresh plant material (e.g. Diefendorf et al., 2011; Hoffmann et al., 2013) and from soil and sediments (e.g. Marseille et al., 1999; Freeman and Colarusso, 2001; Liebezeit and Wöstmann, 2009). Differences exist depending on the vegetation type and litter/soil horizon (Fig. 2a) but are mostly not significant (Table S4).

Differences in chain length patterns between deciduous forests, coniferous forests and grasslands are illustrated in Fig. 3 for litter (a), Ah1 (b) and Ah2 (c). The deciduous forest samples are strongly dominated by n-C_{27}, although its relative abundance decreases from litter to Ah2. However, most of our sampling sites are dominated by beech trees (L11, L13, L14, L16, L17, L18, L20 and L23) that are known to produce mostly n-C_{27} (Bush and McInerney, 2013, and references therein). To check whether the observed n-C_{27} dominance could be explained only with the presence of beech we performed the same correlations and plots as mentioned above, excluding the beech-dominated sites. Figure S5 in the Supplement shows that the dominance of n-C_{27} is less pronounced here and disappears in Ah2. Grass sites are dom-

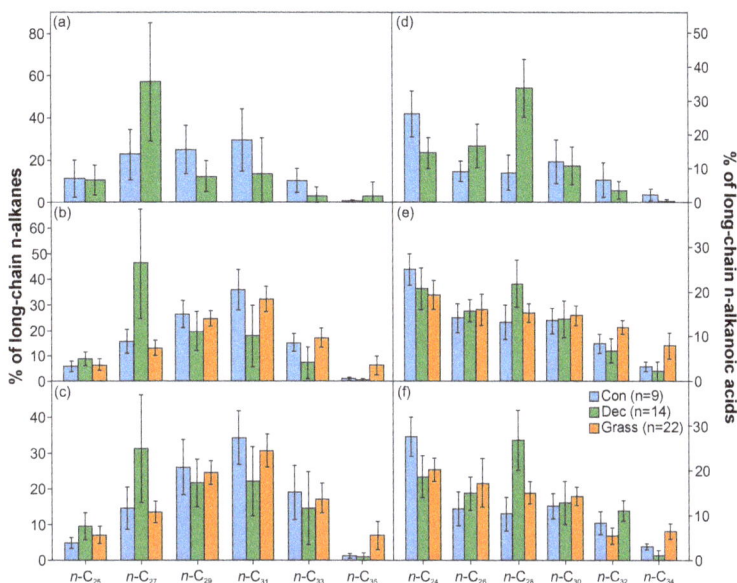

Figure 3. Chain length patterns for odd long-chain n-alkanes in **(a)** litter, **(b)** Ah1 and **(c)** Ah2, as well as long- and even-chain n-alkanoic acids in **(d)** litter, **(e)** Ah1 and **(f)** Ah2.

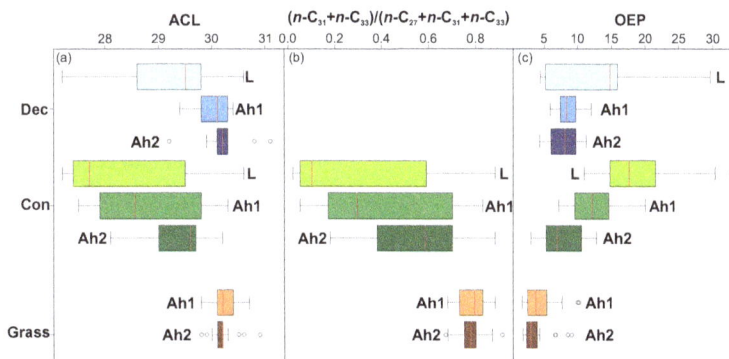

Figure 4. Box plots of **(a)** n-alkane ACL, **(b)** $(n\text{-}C_{31} + n\text{-}C_{33}) / (n\text{-}C_{27} + n\text{-}C_{31} + n\text{-}C_{33})$ ratio, and **(c)** OEP. Con: coniferous forest sites ($n = 9$); dec: deciduous forest sites ($n = 14$); grass: grassland sites ($n = 22$); L: litter; Ah1: topsoil 1 (0–3 cm); Ah2: topsoil 2 (3–10 cm). Box plots show median (red line), interquartile range (IQR) with upper (75 %) and lower (25 %) quartiles, lowest datum still within $1.5 \times$ IQR of lower quartile, and highest datum still within $1.5 \times$ IQR of upper quartile.

inated by $n\text{-}C_{31}$ and are also characterized by high abundances of $n\text{-}C_{33}$ compared to deciduous sites. The con sites show a pattern very similar to that of our grass sites.

The ACLs of the grass and con sites are significantly higher than those of the dec sites (Fig. 4a; grass: Ah1 = 30.5; Ah2 = 30.3; dec: Ah1 = 28.6; Ah2 = 29.6; see Table S6 for p values). Without the beech-dominated sites, the dec sites' ACL shifts to higher values, but they are still significantly lower than mean ACLs of the grass sites (Fig. S7a). However, significant differences disappear in Ah2 (Table S6). The dec litter samples have lower ACLs (27.5) than the Ah1 (28.6) and Ah2 (29.6). Our grass samples show almost no decrease in ACL from Ah1 (30.5) to Ah2 (30.3), and the con sites likewise show no significant decrease from L to Ah2 (Table S6).

To study past changes in dec vs. grass vegetation, various n-alkane ratios have been proposed and used (e.g. Zhang et al., 2006; Lei et al., 2010; Bush and McInerney, 2013; Zech et al., 2013a, b). We tested several n-alkane ratios (i.e. $n\text{-}C_{33} / (n\text{-}C_{27} + n\text{-}C_{33})$, $(n\text{-}C_{31} + n\text{-}C_{33}) / (n\text{-}C_{27} + n\text{-}C_{31} + n\text{-}C_{33})$, and $(n\text{-}C_{31} + n\text{-}C_{33}) / (n\text{-}C_{27} + n\text{-}C_{29} + n\text{-}C_{31} + n\text{-}C_{33})$) and found the largest differences between grass and dec samples for the ratio $(n\text{-}C_{31} + n\text{-}C_{33}) / (n\text{-}C_{27} + n\text{-}C_{31} + n\text{-}C_{33})$. This ratio is low in the dec samples and high in the grass samples (Fig. 4b; dec L: 0.08; Ah1: 0.29; Ah2: 0.56; grass Ah1: 0.79; Ah2: 0.78). Differences between dec and grass are significant in Ah1 and Ah2 (Table S8). Without the beech-dominated sites, the ratio becomes higher for dec but still shows significant differences between dec and grass (Fig. S7b, Table S8).

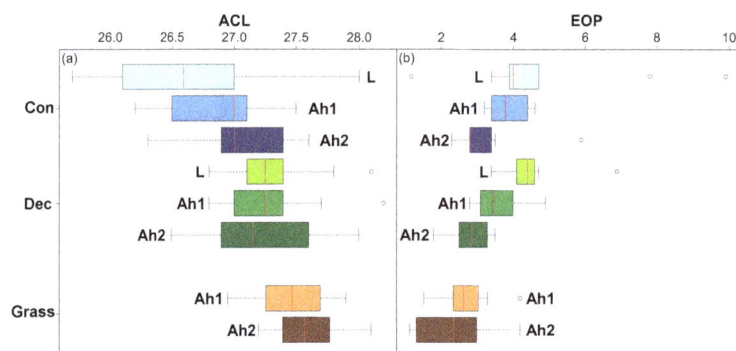

Figure 5. Box plots for **(a)** n-alkanoic acid ACL and **(b)** EOP. Con: coniferous forest sites ($n = 9$); dec: deciduous forest sites ($n = 14$); grass: grassland sites ($n = 14$); L: litter; Ah1: topsoil 1 (0–3 cm); Ah2: topsoil 2 (3–10 cm). Box plots show median (red line), interquartile range (IQR) with upper (75 %) and lower (25 %) quartiles, lowest datum still within $1.5 \times$ IQR of lower quartile, and highest datum still within $1.5 \times$ IQR of upper quartile.

The OEP (or CPI, carbon preference index, which is very similar to the OEP) is often regarded as a proxy for the preservation status of the leaf-wax-derived n-alkanes (e.g. Huang et al., 1996; Tipple and Pagani, 2010; Vogts et al., 2012; Wang et al., 2014, and references therein). High OEPs are characteristic of fresh plant material and modern soils (Kirkels et al., 2013; Diefendorf et al., 2011; Collister et al., 1994), whereas low OEPs indicate degradation of n-alkanes during pedogenesis and early diagenesis (Marseille et al., 1999; Freeman and Colarusso, 2001; Buggle et al., 2010; Zech et al., 2011a; Wang et al., 2014). OEPs in the samples range from 3 to 32.8, typical for fresh plant material and soils (Table S3). Values significantly decrease from litter (18.4) to Ah1 (12.1) and Ah2 (6.8) for the dec sites, and a minor decrease can be observed in the grass sites (Fig. 4c, Table S9). Significant differences occur between dec and grass sites in all horizons (Table S9).

3.2 Leaf wax n-alkanoic acid abundances and chain length patterns

All samples show high abundances of long ($> C_{20}$) even-chain n-alkanoic acids (Table S10), characteristic of epicuticular leaf waxes (Eglinton and Hamilton, 1967). Many samples also have large amounts of C_{16} and C_{18}, yet those are ubiquitous and cannot be considered as leaf wax biomarkers. Total n-alkanoic acid concentrations (c_{tot} refers here to the sum of C_{20} to C_{34}) range from 3 to 854 µg g^{-1} dw, consistent with previous studies (e.g. Marseille et al., 1999; Jandl et al., 2002). As for the n-alkanes, total n-alkanoic acid concentrations vary between the vegetation types and horizons (Fig. 2b). In general, c_{tot} decreases from litter to Ah1 and Ah2. In contrast to the n-alkanes, the highest n-alkanoic acid concentrations occur in our con samples. Decrease from litter to Ah1 and Ah2 are only significant in the dec samples (Table S11).

The n-alkanoic acid chain length patterns show differences between the different vegetation types (Fig. 3d: litter; 3e: Ah1; 3f: Ah2). While the dec samples are dominated by n-C_{28}, the con samples show a maximum for the shorter homologue n-C_{24} in the litter, Ah1 and Ah2. The grass sites have high abundances of n-C_{24} to n-C_{30}, but when compared to dec and con, the relative high n-C_{32} and n-C_{34} abundances are distinct.

Significant differences in the ACL between the three vegetation types exist and reflect the above-mentioned predominance of various homologues (Fig. 5a, Table S12). While the grass sites show a tendency towards higher ACLs, the con sites tend to have the lowest values. Differences stay significant, even in Ah1 and Ah2 (Table S12).

On the basis of the above-mentioned significant differences between the ACL and the three vegetation types we propose the following three indices, referred to as CDG indices (Eqs. 5–7) for coniferous forests (C), deciduous forests (D) and grasslands (G):

$$\text{Index C} = \frac{n\text{-}C_{24}}{n\text{-}C_{24} + n\text{-}C_{28} + n\text{-}C_{32} + n\text{-}C_{34}}, \tag{5}$$

$$\text{Index D} = \frac{n\text{-}C_{28}}{n\text{-}C_{24} + n\text{-}C_{28} + n\text{-}C_{32} + n\text{-}C_{34}}, \tag{6}$$

$$\text{Index G} = \frac{n\text{-}C_{32} + n\text{-}C_{34}}{n\text{-}C_{24} + n\text{-}C_{28} + n\text{-}C_{32} + n\text{-}C_{34}}. \tag{7}$$

Index C ranges from 0.25 to 0.91, with the highest values for the con sites (Fig. 6a), and shows significant differences between our con and dec samples, as well as between the con and grass in all horizons, but not between the dec and grass locations (Table S13). Index D ranges from 0.03 to 0.62 and has highest values for the dec sites (Fig. 6b). It differs significantly between all vegetation types in L, Ah1 and Ah2, except for con and grass sites in Ah2. It also significantly decreases from L to Ah1 and Ah2 in the dec sites (Table S14). Index G ranges from 0.00 to 0.37 and discriminates between forest and grass sites, with systematically higher values for

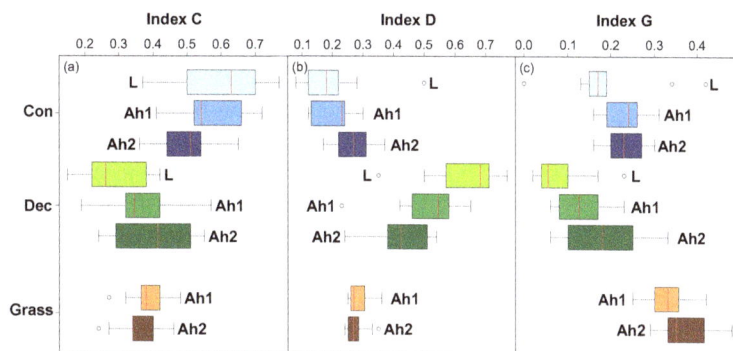

Figure 6. Box plots for **(a)** indices C, **(b)** D and **(c)** G. Con: coniferous forest sites ($n = 9$); dec: deciduous forest sites ($n = 14$); grass: grassland sites ($n = 14$); L: litter; Ah1: topsoil 1 (0–3 cm); Ah2: topsoil 2 (3–10 cm). Box plots show median (red line), interquartile range (IQR) with upper (75 %) and lower (25 %) quartiles, lowest datum still within $1.5 \times$ IQR of lower quartile, and highest datum still within $1.5 \times$ IQR of upper quartile.

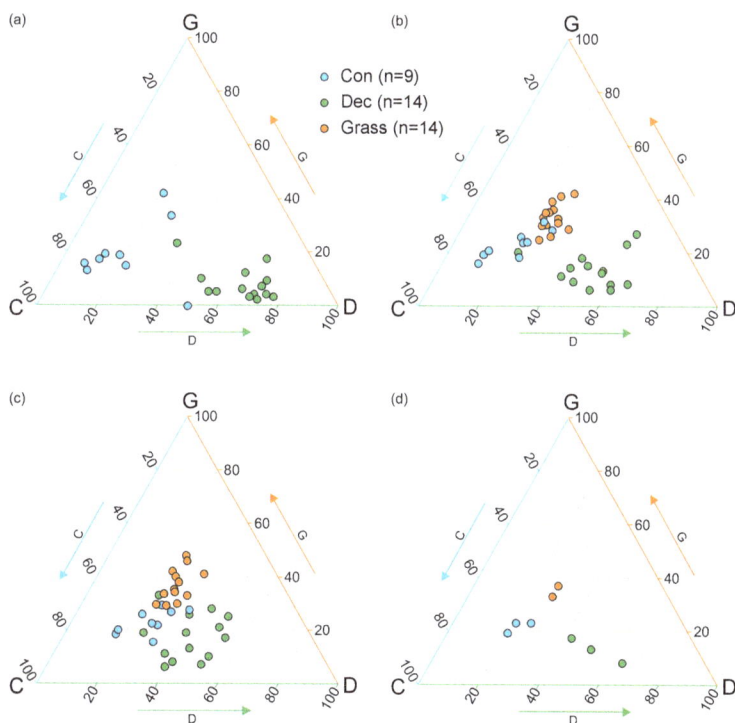

Figure 7. Ternary plots for the CDG indices: **(a)** litter, **(b)** Ah1 and **(c)** Ah2, as well as **(d)** means for vegetation types. Each point represents the mean of litter, Ah1 and Ah2, with regard to con, dec and grass.

the grass sites (Fig. 6c). Like index D, index G shows significant differences between all three vegetation types in all horizons, except for the con and dec locations in Ah1. The index likewise shows a significant decrease from L to Ah2 in the dec samples (Table S15). The CDG indices can conveniently be plotted in ternary diagrams, which illustrate the clusters for the different vegetation types and the scatter within the cluster (Fig. 7).

EOPs in our samples range from 1.2 to 9.9 (Table S16), typical for n-alkanoic acids that originate from epicuticular leaf waxes and are found in soils (Killops and Killops, 2005). The EOP decreases significantly from litter to Ah1 and Ah2 in the dec samples (L: 4.3; Ah1: 3.46; Ah2: 2.85; Fig. 5b, Table S16), and without significance in the con samples (L: 4.0; Ah1: 3.8; Ah2: 2.81).

4 Discussion

4.1 *n*-alkane pattern in litter and topsoil

The lower c_{tot} values of the con litter samples compared to the dec litter (median con: 30.9 µg g^{-1} dw; dec: 80.4 µg g^{-1} dw) are in good agreement with findings of much lower *n*-alkane abundances in conifer needles than deciduous leaves (e.g. Sachse et al., 2006; Zech et al., 2009; Diefendorf et al., 2011; Norris et al., 2013; Tarasov et al., 2013) and in forest soils below conifers (e.g. Almendros et al., 1996). Given these low reported *n*-alkane concentrations in conifer needles, we ascribe the *n*-alkane patterns in the con litter and topsoil samples to the *n*-alkane input from the understorey, and we focus in the following discussion on the differences between grass and dec sites. Our grass soils have very low *n*-alkane abundances (median Ah1, 3.5 µg g^{-1} dw), which we interpret to be an artefact of (former) plowing and admixture with inorganic soil material. Based on the data we therefore cannot infer a low *n*-alkane production in grass sites.

4.1.1 *n*-alkanes to distinguish between vegetation types

The domination of n-C_{27} in our dec samples in all horizons, and the relatively high concentration of n-C_{31} and n-C_{33} in the grass samples, implies that the established source-specific compounds, at least along the transect, allow for conclusions to be made regarding the vegetation type that generated them. However, Fig. S5 proves that the pattern is less specific when the beech-dominated sites are excluded. This supports former results and shows that n-C_{27} is strongly produced by beech trees (Bush and McInerney, 2013, and references therein). The ACL and our proposed *n*-alkane ratio of $(n$-$C_{31} + n$-$C_{33})/(n$-$C_{27} + n$-$C_{31} + n$-$C_{33})$ show significant differences between the dec and the grass sites in Ah1 and Ah2, even when the beech-dominated sites are excluded (Tables S6 and S8). Although n-C_{27} is not the dominant long odd-chain *n*-alkane in Ah2 at the dec locations that are not dominated by beeches, its percentage in the dec samples is still higher than at the grass sites (Fig. S5), whereas the percentage of n-C_{31} and n-C_{33} is the highest in the grass sites in Ah1 as well as in Ah2. Therefore, our results corroborate, at least for the studied transect, that the ACL and our proposed *n*-alkane ratio of $(n$-$C_{31} + n$-$C_{33})/(n$-$C_{27} + n$-$C_{31} + n$-$C_{33})$ allow for differentiation between the input of dec and grass vegetation. Care has to be taken when interpreting palaeovegetation changes solely on the dominance of one *n*-alkane compound over the others (e.g. proxies like C_{max}; Wiesenberg et al., 2015), because this might lead to an underestimation of the deciduous tree input, at least when beech trees were not the main contributors to the soil. Nevertheless, we strongly emphasize that the observed patterns are very likely a regional phenomenon and our results should not be transmitted to other regions with different climate and vegetation types, because *n*-alkane patterns do not work on a global scale (Bush and McInerney, 2013). Thus, our results underline the need for regional calibrations for the *n*-alkane pattern, because they corroborate its potential for palaeovegetation reconstruction on a regional base.

Although significant differences occur between the dec and grass sites OEP in Ah1 and Ah2, we would not recommend using the OEP as a proxy to distinguish between the two vegetation types, because it can very likely show significant decreases with increasing soil depth, as it does in the dec samples (Table S9), so it is probably strongly influenced by degradation and microbial reworking.

4.1.2 Influence of soil depth on the *n*-alkane pattern

Although we observed no significant decreases in C_{tot} from L to Ah1 and Ah2 for the dec and con sites, as well as from Ah1 to Ah2 for all three vegetation forms (Table S4), the slightly decreasing trends in Fig. 2a are most likely due to lipid degradation and admixture with inorganic soil material.

As stated above, the significant decrease in OEP from L to Ah2 in the dec samples (Fig. 4c, Table S9) is probably due to degradation effects. Therefore, despite the wide range of OEPs in modern plants (Bush and McInerney, 2013, and references therein), the OEP can serve as a degradation proxy along our transect. The OEPs in the grass samples show no significant decrease from Ah1 (7.2) to Ah2 (6.5), but the degradation effects here are probably biased by plowing and mixing of Ah1 and Ah2.

The decrease in the relative percentage of the dominant *n*-alkane(s) in the dec and grass samples from L to Ah1 and from Ah1 to Ah2 (Fig. 3a–c) as well as the significant increase in ACL from L to Ah2 in the dec samples (Fig. 4a, Table S6) is probably another indication of the effect of degradation on the *n*-alkane pattern. Our grass samples show an insignificant but decreasing trend in ACL from Ah1 (30.5) to Ah2 (30.3). These observed changes are consistent with the notion that the more abundant homologues are preferentially degraded and lost during pedogenesis. This affects *n*-alkane patterns in soils and sediments (Zech et al., 2009, 2013a, b). Degradation also affects our $(n$-$C_{31} + n$-$C_{33})/(n$-$C_{27} + n$-$C_{31} + n$-$C_{33})$ ratio, which is expressed in the significant increases in the dec samples from litter (0.08) to Ah1 (0.3) to Ah2 (0.56) and in the slightly decreasing trend in the grass samples from 0.84 to 0.80 (Fig. 4b, Table S8).

In order to illustrate and correct for degradation effects, Zech et al. (2009) proposed an endmember model, which was later modified by Zech et al. (2013a, b). We added the dec and grass samples to the dataset for Europe, provided by Zech et al. (2013b). Figure 8 shows the new endmember plot and illustrates that our *n*-alkane ratio differs between grass and dec samples and that it changes depending on the OEP, i.e. with degradation. As already described above, the *n*-alkane ratio is wider for grass samples and lower values are more typical for dec. With increasing degradation, differences seem to become less: the trend lines, or "degradation lines", for

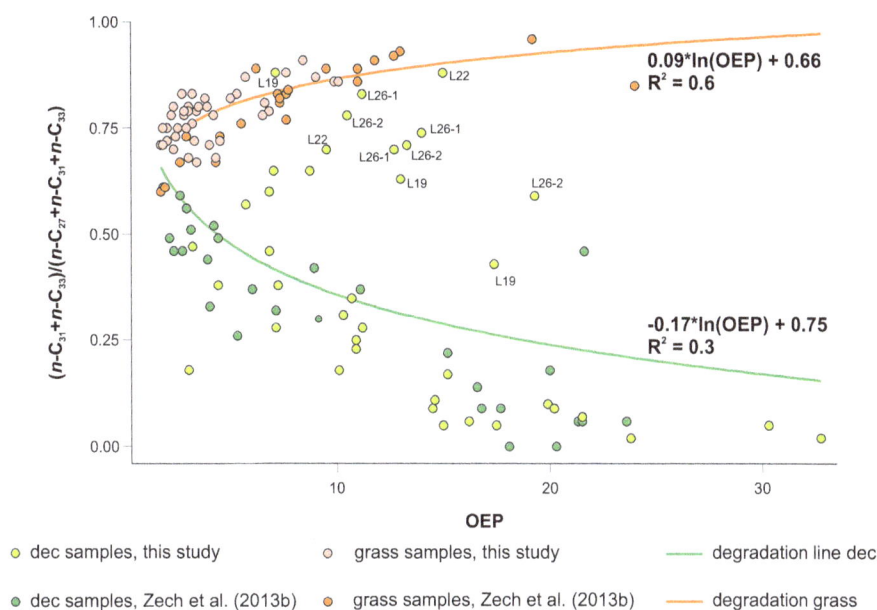

Figure 8. Endmember plot modified after Zech et al. (2013b). Degradation lines refer to the complete dataset. Samples that deviated markedly from the degradation lines are labelled and discussed in the text.

grass and dec converge. In principle, the endmember model allows for degradation effects to be tentatively corrected for and for the contribution of grasses versus deciduous trees in (palaeo)samples to be quantified: the equations in Fig. 8 are used to calculate the grass and tree endmember for a specific OEP, and following the rule of proportion % grass can then be estimated as

$$\% \text{ grass} = \frac{n\text{-alkane ratio}_{\text{sample}} - \text{equation}_{\text{degradation line trees}}}{\text{equation}_{\text{degradation line grass}} - \text{equation}_{\text{degradation line trees}}}. \quad (8)$$

Again, we tested whether the endmember plot can be explained only by the presence of beech by applying the same model and plot without the beech-dominated sites. The dec degradation line shifts upward, closer to the grass degradation line (Fig. S17), and R^2 drops from 0.3 to only 0.1, but it still shows a separation of the dec samples from the grass samples. Four sites plot particularly high above the dec degradation line and deserve a closer look. Sample location L19 is a birch forest surrounded by fields, and L26-2 dec is an open forest with birch and oak trees, with a larger number of grasses in the understorey. For both sites, the n-alkane ratios for Ah1 and Ah2 plot much closer to the grass degradation line than the litter samples, and we speculate that both sites may have been grasslands in the past that were only recently reforested. Since turnover times of n-alkanes are in the order of decades (e.g. Amelung et al., 2008, and references therein; Wiesenberg et al., 2004) we would expect to see the n-alkane pattern prior to reforestation in the upper soil. Unfortunately, we do not have information about former land use at the study locations to verify this speculation. The sample locations L22 (acer, elder, ash, poplar) and L26-1

(acer, oak, beech, fir) are characterized by litter samples that plot close to the grass degradation line. We cannot exclude the possibility that these litter samples and sites are affected by n-alkane input from grasses, but likely the data simply reflect the large species-specific variability in n-alkane patterns reported repeatedly in the literature (e.g. Diefendorf et al., 2011; Bush and McInerney, 2013).

In summary, our results show that (1) n-alkane patterns are systematically different between the investigated dec and grass sites, (2) soil depth/degradation affects the homologue patterns, and (3) endmember modelling is a useful tool for palaeovegetation reconstruction along the transect, but one needs to be aware of the uncertainties related mainly to the large species-specific variability in the n-alkane patterns. However, the fact that coniferous trees produce only a few n-alkanes makes respective palaeovegetation reconstructions "blind" for coniferous trees.

4.2 n-alkanoic acid pattern in vegetation and topsoil

To the best of our knowledge, this is the first study which systematically investigates long-chain n-alkanoic acid patterns in litter and topsoil along a transect that encompasses a range of environmental conditions and vegetation types. Since differences in con concentration compared to dec are much more pronounced in the topsoil than in the litter, we infer that better preservation of n-alkanoic acids in soils under coniferous forests is the reason for the observed differences, and not higher alkanoic acid production by conifers. This is further consistent with studies showing better preservation of alkanoic acids in soils with low pH typical for coniferous forests, while n-alkanes are better preserved in soils with a high pH, more typical for deciduous forests (Bull et al., 2000a; Zocatelli et al., 2012). We again attribute the low c_{tot} in the grass sites mainly to plowing and admixture with inorganic soil material.

4.2.1 n-alkanoic acids to distinguish between vegetation types

The n-alkanoic acid distribution in the vegetation types implies that specific compounds can be used to characterize them (Fig. 3d–f). The n-C_{24} alkanoic acid can represent the input of conifers in L, Ah1 and Ah2; n-C_{28} shows the contribution of deciduous trees in all horizons; and the relative amount of n-C_{32} and n-C_{34} can be used to estimate the grass contribution. From that, we suggest the CDG indices. They show strong differences between the three vegetation types (Fig. 6, Tables S13–S15), which are significant in nearly all horizons, apart from index C, which does not allow a distinction between dec and grass sites (Table S13). The ternary plots of the three indices visualize the discrimination potential by showing clusters for the different vegetation types, although we must emphasize that outliers exist (Fig. 7a–c). Index C based on the dominance of the shorter chain n-alkanoic acid C_{24}, which might be more strongly affected by microbial degradation and reworking compared to the longer-chain counterparts n-C_{28}, n-C_{32} and n-C_{34} that are included in the D and G indices (Sect. 4.2.2). The ACL of the n-alkanoic acids, on the other hand, may not be a particularly useful proxy for palaeovegetation because the observed differences are small (although they are significant, Table S12) and mixing a con and grass signal could falsely yield a dec signal.

4.2.2 Influence of soil depth on the n-alkanoic acid pattern

The significant decrease in C_{tot} of the dec and con samples from L to Ah2 (Table S11) is most likely attributed to enhanced degradation effects on the acids with increasing soil depth.

The preferential loss of n-C_{28} (n-C_{24}) in the dec samples (con samples) can be visualized by comparing the homologue patterns of litter, Ah1 and Ah2 (Fig. 3d–f). This degradation effect is not documented in a significant change in the ACL, and only the con sites show a slight but non-significant increasing trend with increasing soil depth (Fig. 5a, Table S12). The degradation effect of a certain homologue on the CDG indices is illustrated in Figs. 6 and 7. Index D, for example, is high for the dec litter samples but significantly decreases from litter to Ah1 and Ah2 (Table S14). The same applies for index C and G with regard to the con and grass samples, respectively, although the changes are not significant. With the preferential loss of the most abundant compound (n-C_{28} for dec and n-C_{24} for con), the respective characteristic index decreases; however, the other two indices un-

avoidably increase. This is illustrated by the clusters moving closer together (Fig. 7b–d). Nevertheless, they still allow for discrimination between the three vegetation types as at least index D and G show significant differences between all three vegetation types in all horizons (Tables S14 and S15). All three indices show a significant decrease (index D) or increase (index C and index G) with soil depth in the dec samples, which implies that they are more prone to degradation under the more alkaline deciduous forest soils. The grass samples do not show changes in ACL or in the indices from Ah1 to Ah2, which we again ascribe to plowing.

The significant decrease in EOP from litter to Ah1 and Ah2 in the dec samples (dec litter: 4.3; Ah1: 3.46; Ah2: 2.85; Fig. 5b) resembles the decrease in the OEP for the n-alkanes and suggests that the most abundant (even-numbered) compounds are preferentially degraded during pedogenesis. The decrease in EOP in the con samples is not significant, which we again ascribe to a better preservation of n-alkanoic acids in the acidic soils under conifers. Nevertheless, Fig. 5b indicates a decreasing trend in the EOP in the con samples from L to Ah2, probably due to slight degradation effects on the acids in the con Ah1 and Ah2 samples. Grass samples do not show a trend in EOP, which is most likely because of the plowing that affected these sites. Like the OEP, the EOP might thus serve as a proxy for degradation.

Although our results demonstrate that the leaf-wax-derived n-alkanoic acids in soils under coniferous forests are less prone to degradation compared to soils under deciduous forests, the risk still exists that the leaf wax contribution from coniferous trees to soil and sedimentary archives might be underestimated when the alkanoic acid pattern is not corrected for degradation. The same applies for the deciduous forests and probably also for the grass sites.

Overall, our results show that

i. n-alkanoic acid patterns are significantly different between the investigated dec, con and grass sites;

ii. the specific CDG indices might be valuable proxies for palaeovegetation;

iii. degradation affects the homologue patterns and CDG indices, at least in the dec samples, so that procedures to correct for degradation need to be developed and tested.

5 Conclusions

We have systematically investigated leaf-wax-derived long-chain n-alkane and n-alkanoic acid patterns in litter and top soils along a European transect. Our findings are as follows:

1. Both compound classes show distinct differences depending on the type of vegetation. The vegetation signal is not only found in the litter; it can also be preserved to some degree in the topsoil. The grass sites contain more n-C_{31} and n-C_{33} alkanes than the dec sites but less

n-C_{27}. The ratio $(n\text{-}C_{31} + n\text{-}C_{33}) / (n\text{-}C_{27} + n\text{-}C_{31} + n\text{-}C_{33})$ seems to be most suitable to distinguish between those two vegetation types in our study area. Litter and soil samples in coniferous forests are probably biased by the understorey, so vegetation reconstructions solely based on the n-alkane pattern are blind for coniferous trees. Nevertheless, the n-alkanes show a great potential for palaeovegetation reconstruction along our transect, but the species-specific absolute and relative variability in the homologue abundances need to be taken into account.

We propose three n-alkanoic acid indices to distinguish contributions from the three investigated vegetation types: index C is the relative abundance of the C_{24} n-alkanoic acid and represents the input of coniferous trees. Index D is the relative abundance of the C_{28} n-alkanoic acid and is particularly high in litter and in topsoil of deciduous forests. The relative abundance of the C_{32} and C_{34} n-alkanoic acids is expressed as index G and shows the contribution from grasses and herbs.

2. The homologue patterns of leaf waxes change from litter to Ah1 and Ah2. Although we cannot completely rule out effects related to possible land use and vegetation change in the past, the overall consistent trends imply that degradation plays an important role. Degradation not only lowers the OEP and EOP of n-alkanes and n-alkanoic acids, respectively, but also reduces the vegetation-specific differences of the homologue patterns. An updated endmember model is suggested to account for degradation effects on n-alkanes, but similar procedures still need to be developed and tested for the n-alkanoic acids before their potential for palaeovegetation reconstructions can be fully exploited.

Overall, our findings suggest that combined investigations of n-alkane and n-alkanoic acid distributions on a regional scale have great potential for palaeovegetation reconstruction, although degradation effects need to be taken into account. In particular, with regard to the n-alkanoic acids, more research is needed to gain a better understanding of those effects.

6   Data availability

The dataset we used in this paper is accessible via the Supplement. For the endmember model we combined our dataset with the dataset published in Zech et al. (2013b) at doi:10.1016/j.palaeo.2013.07.023.

Acknowledgements. We thank P. Neitzel, who contributed in large part to the work in the field and in the laboratory at ETH Zurich, and Q. Lejeune for support in the field, as well as C. Magill for scientific discussions. C. Diebold helped with the laboratory work at the University of Bern. We also acknowledge L. Wüthrich and M. Bliedtner for helpful discussions. The research was funded by the Swiss National Science Foundation (PP00P2 150590).

Edited by: R. Zornoza

References

Almendros, G., Sanz, J., and Velasco, F.: Signatures of lipid assemblages in soils under continental Mediterranean forests, Eur. J. Soil Sci., 47, 183–196, doi:10.1111/j.1365-2389.1996.tb01389.x, 1996.

Amelung, W., Brodowski, S., Sandhage-Hofmann, A., and Bol, R.: Combining biomarker with stable isotope analyses for assessing the transformation and turnover of soil organic matter, in: Advances in Agronomy, edited by: Sparks, D. L., Elsevier Inc Academic Press, Burlington, 155–250, doi:10.1016/S0065-2113(08)00606-8, 2008.

Brincat, D., Yamada, K., Ishiwatari, R., Uemura, H. and Naraoka, H.: Molecular-isotopic stratigraphy of long-chain n-alkanes in Lake Baikal Holocene and glacial age sediments, Org. Geochem., 31, 287–294, doi:10.1016/S0146-6380(99)00164-3, 2000.

Buggle, B., Wiesenberg, G. L., and Glaser, B.: Is there a possibility to correct fossil n-alkane data for postsedimentary alteration effects?, Appl. Geochem., 25, 947–957, doi:10.1016/j.apgeochem.2010.04.003, 2010.

Bull, I. D., v. Bergen, P. F., Nott, C. J., Poulton, P. R., and Evershed, R. P.: Organic geochemical studies of soils from the Rothamsted classical experiments – V. The fate of lipids in different long-term experiments, Org. Geochem., 31, 389–408, doi:10.1016/S0146-6380(00)00008-5, 2000a.

Bull, I. D., Nott, C. J., van Bergen, P. F., Poulton, P. R., and Evershed, R. P.: Organic geochemical studies of soils from the Rothamsted classical experiments – VI. The occurrence and source of organic acids in an experimental grassland soil, Soil Biol. Biochem., 32, 1367–1376, doi:10.1016/S0038-0717(00)00054-7, 2000b.

Bush, R. T. and McInerney, F. A.: Influence of temperature and C4 abundance on n-alkane chain length distributions across the central USA, Org. Geochem., 79, 65–73, doi:10.1016/j.gca.2013.04.016, 2015.

Bush, R. T. and McInerney, F. A.: Leaf wax n-alkane distributions in and across modern plants: Implications for paleoecology and chemotaxonomy, Geochim. Cosmochim. Ac., 117, 161–179, doi:10.1016/j.gca.2013.04.016, 2013.

Collister, J. W., Rieley, G., Stern, B., Eglinton, G., and Fry, B.: Compound-Specific Delta-C-13 Analyses of Leaf Lipids from Plants with Differing Carbon-Dioxide Metabolisms, Org. Geochem., 21, 619–627, doi:10.1016/0146-6380(94)90008-6, 1994.

Conover, W. J.: Practical Nonparametric Statistics, 3rd Edn., Wiley, Hoboken, NJ, 1999.

Conover, W. J. and Iman, R. L.: On multiple-comparisons procedures, Technical Report LA-7677-MS, Los Alamos Scientific Laboratory, Los Alamos, 1979.

Derenne, S. and Largeau, C.: A review of some important families of refractory macromolecules: composition, origin and fate in soils and sediments, Soil Sci., 166, 833–847, 2001.

Diefendorf, A. F., Freeman, K. H., Wing, S. L., and Graham, H. V.: Production of n-alkyl lipids in living plants and implications for the geologic past, Geochim. Cosmochim. Ac., 75, 7472–7485, doi:10.1016/j.gca.2011.09.028, 2011.

Duan, Y. and He, J.: Distribution and isotopic composition of n-alkanes from grass, reed and tree leaves along a latitudinal gradient in China, Geochem. J., 45, 199–207, doi:10.2343/geochemj.1.0115, 2011.

Dunn, O. J.: Multiple comparisons using rank sums, Technometrics, 6, 241–252, 1964.

Eglinton, G. and Hamilton, R. J.: Leaf epicuticular waxes, Science, 156, 1322–1335, doi:10.1126/science.156.3780.1322, 1967.

Eglinton, G., Hamilton, R. J., Raphael, R. A., and Gonzalez, A. G.: Hydrocarbon Constituents of the Wax Coatings of Plant Leaves: A Taxonomic Survey, Nature, 193, 739–742, doi:10.1038/193739a0, 1962.

Eglinton, T. I. and Eglinton, G.: Molecular proxies for paleoclimatology, Earth Planet. Sc. Lett., 275, 1–16, doi:10.1016/j.epsl.2008.07.012, 2008.

Feakins, S. J., Kirby, M. E., Cheetham, M. I., Ibarra, Y., and Zimmerman, S. R. H.: Fluctuation in leaf wax D/H ratio from a southern California lake records significant variability in isotopes in precipitation during the late Holocene, Org. Geochem., 66, 48–59, doi:10.1016/j.orggeochem.2013.10.015, 2014.

Field, A. P.: Discovering statistics using IBM SPSS Statistics, 4th Edn., Sage publications, London, 2013.

Freeman, K. H. and Colarusso, L. A.: Molecular and isotopic records of C4 grassland expansion in the late Miocene, Geochim. Cosmochim. Ac., 65, 1439–1454, doi:10.1016/S0016-7037(00)00573-1, 2001.

Games, P. A. and Howell, J. F.: Pairwise Multiple Multiple Comparison Comparison Procedures Procedures with Unequal Unequal N's and/or Variances: A Monte Carlo Study, J. Educ. Stat., 1, 113–125, 1976.

Gocke, M., Kuzyakov, Y., and Wiesenberg, G. B.: Differentiation of plant derived organic matter in soil, loess and rhizoliths based on n-alkane molecular proxies, Biogeochemistry, 112, 23–40, doi:10.1007/s10533-011-9659-y, 2013.

Guo, N., Gao, J., He, Y., Zhang, Z., and Guo, Y.: Variations in leaf epicuticular n-alkanes in some Broussonetia, Ficus and Humulus species, Biochem. Syst. Ecol., 54, 150–156, doi:10.1016/j.bse.2014.02.005, 2014.

Hochberg, Y.: A sharper Bonferroni procedure for multiple tests of significance, Biometrika, 75, 800–803, 1988.

Hoefs, M. J. L., Rijpstra, W. I. C., and Sinninghe Damsté, J. S.: The influence of oxic degradation on the sedimentary biomarker record I: evidence from Madeira Abyssal plain turbidites, Geochim. Chosmochim. Ac., 66, 2719–2735, doi:10.5194/bg-11-2455-2014, 2002.

Hoffmann, B., Kahmen, A., Cernusak, L. A., Arndt, S. K., and Sachse, D.: Abundance and distribution of leaf wax n-alkanes in leaves of Acacia and Eucalyptus trees along a strong humidity gradient in northern Australia, Org. Geochem., 62, 62–67, doi:10.1016/j.orggeochem.2013.07.003, 2013.

Huang, Y., Bol, R., Harkness, D. D., Ineson, P., and Eglinton, G.: Post-glacial variations in distributions, 13C and 14C contents of aliphatic hydrocarbons and bulk organic matter in three types of British acid upland soils, Org. Geochem., 24, 273–287, doi:10.1016/0146-6380(96)00039-3, 1996.

Jandl, G., Schulten, H.-R. and Leinweber, P.: Quantification of long-chain fatty acids in dissolved organic matter and soils, J. Plant Nutr. Soil Sci., 165, 133–139, doi:10.1002/1522-2624(200204)165:2<133::AID-JPLN133>3.0.CO;2-T, 2002.

Jansen, B., Nierop, K. G. J., Hageman, J. A., Cleef, A. M., and Verstraten, J. M.: The straight-chain lipid biomarker composition of plant species responsible for the dominant biomass production along two altitudinal transects in the Ecuadorian Andes, Org. Geochem., 37, 1514–1536, doi:10.1016/j.orggeochem.2006.06.018, 2006.

Killops, S. and Killops, V.: Chemical Stratigraphic Concepts and Tools, Introduction to Organic Geochemistry, 2nd Edn., Blackwell Publishing Ltd., 166–245, 2005.

Kirkels, F. M., Jansen, B., and Kalbitz, K.: Consistency of plant-specific n-alkane patterns in plaggen ecosystems: A review, Holocene 23, 1355–1368, doi:10.1177/0959683613486943, 2013.

Lei, G., Zhang, H., Chang, F., Pu, Y., Zhu, Y., Yang, M., and Zhang, W.: Biomarkers of modern plants and soils from Xinglong Mountain in the transitional area between the Tibetan and Loess Plateaus, Quatern. Int., 218, 143–150, doi:10.1016/j.quaint.2009.12.009, 2010.

Levene, H.: Robust tests for equality of variances, in: Contributions to Probability and Statistics: Essays in Honor of Harold Hotelling, edited by: Olkin, I., Ghurye, S. G., Hoefding, W., Madow, W. G., and Mann, H. B., Stanford University Press, Stanford, 278–292, 1960.

Liebezeit, G. and Wöstmann, R.: n-Alkanes as Indicators of Natural and Anthropogenic Organic Matter Sources in the Siak River and its Estuary, E Sumatra, Indonesia, Bull. Environ. Contam. Toxicol., 83, 403–409, doi:10.1007/s00128-009-9734-4, 2009.

Maffei, M.: Chemotaxonomic significance of leaf wax alkanes in the gramineae, Biochem. Syst. Ecol., 24, 53–64, doi:10.1016/0305-1978(95)00102-6, 1996.

Maffei, M., Badino, S. and Bossi, S.: Chemotaxonomic significance of leaf wax n-alkanes in the Pinales (Coniferales), J. Biol. Res., 1, 3–19, 2004.

Marseille, F., Disnar, J. R., Guillet, B., and Noack, Y.: n-Alkanes and free fatty acids in humus and A1 horizons of soils under beech, spruce and grass in the Massif-Central (Mont-Lozère), France, Eur. J. Soil Sci., 50, 433–441, doi:10.1046/j.1365-2389.1999.00243.x, 1999.

Norris, C. E., Dungait, J. A. J., Joynes, A., and Quideau, S. A.: Biomarkers of novel ecosystem development in boreal forest soils, Org. Geochem., 64, 9–18, doi:10.1016/j.orggeochem.2013.08.014, 2013.

Nguyen Tu, T. T., Egasse, C., Zeller, B., Bardoux, G., Biron, P., Ponge, J.-F., David, B., and Derenne, S.: Early degradation of plant alkanes in soils: A litterbag experiment using 13C-labelled leaves, Soil Biol. Biochem., 43, 2222–2228, doi:10.1016/j.soilbio.2011.07.009, 2011.

Otto, A. and Simpson, M.: Degradation and Preservation of Vascular Plant-derived Biomarkers in Grassland and Forest Soils from Western Canada, Biogeochemistry, 74, 377–409, doi:10.1007/s10533-004-5834-8, 2005.

Poynter, J. G., Farrimond, P., Robinson, N., and Eglinton, G.: Aeolian-Derived Higher Plant Lipids in the Marine Sedimentary Record: Links with Palaeoclimate, in: Paleoclimatology and

Paleometeorology: Modern and Past Patterns of Global Atmospheric Transport, edited by: Leinen, M. and Sarnthein, M., Springer Netherlands, 435–462, 1989.

Rieley, G., Collier, R. J., Jones, D. M., Eglinton, G., Eakin, P. A., and Fallick, A. E.: Sources of Sedimentary Lipids Deduced from Stable Carbon Isotope Analyses of Individual Compounds, Nature, 352, 425–427, doi:10.1038/352425a0, 1991.

Rommerskirchen, F., Plader, A., Eglinton, G., Chikaraishi, Y., and Rullkötter, J.: Chemotaxonomic significance of distribution and stable carbon isotopic composition of long-chain alkanes and alkan-1-ols in C4 grass waxes, Org. Geochem., 37, 1303–1332, doi:10.1016/j.orggeochem.2005.12.013, 2006.

Sachse, D., Radke, J., and Gleixner, G.: δD values of individual n-alkanes from terrestrial plants along a climatic gradient – Implications for the sedimentary biomarker record, Org. Geochem., 37, 469–483, doi:10.1016/j.orggeochem.2005.12.003, 2006.

Schatz, A.-K., Zech, M., Buggle, B., Gulyás, S., Hambach, U., Marković, S. B., Sümegi, P., and Scholten, T.: The late Quaternary loess record of Tokaj, Hungary: Reconstructing palaeoenvironment, vegetation and climate using stable C and N isotopes and biomarkers, Quatern. Int., 240, 52–61, doi:10.1016/j.quaint.2010.10.009, 2011.

Schefuß, E., Ratmeyer, V., Stuut, J.-B. W., Jansen, J. H. F., and Sinninghe Damsté, J. S.: Carbon isotope analyses of n-alkanes in dust from the lower atmosphere over the central eastern Atlantic, Geochim. Cosmochim. Ac., 67, 1757–1767, doi:10.1016/S0016-7037(02)01414-X, 2003.

Schwark, L., Zink, K., and Lechterbeck, J.: Reconstruction of postglacial to early Holocene vegetation history in terrestrial Central Europe via cuticular lipid biomarkers and pollen records from lake sediments, Geology, 30, 463-466, doi:10.1130/0091-7613(2002)030<0463:ropteh>2.0.co;2, 2002.

Shapiro, S. S. and Wilk, M. B.: An analysis of variance test for normality (complete samples), Biometrika, 52, 591–611, doi:10.1093/biomet/52.3-4.591, 1965.

Shepherd, T. and Wynne Griffiths, D.: The effects of stress on plant cuticular waxes, New Phytol., 171, 469–499, doi:10.1111/j.1469-8137.2006.01826.x, 2006.

Tarasov, P. E., Müller, S., Zech, M., Andreeva, D., Diekmann, B., and Leipe, C.: Last glacial vegetation reconstructions in the extreme-continental eastern Asia: Potentials of pollen and n-alkane biomarker analyses, Quatern. Int., 290–291, 253–263, doi:10.1016/j.quaint.2012.04.007, 2013.

Tipple, B. J. and Pagani, M.: A 35 Myr North American leaf-wax compound-specific carbon and hydrogen isotope record: Implications for C4 grasslands and hydrologic cycle dynamics, Earth Planet. Sc. Lett., 299, 250–262, doi:10.1016/j.epsl.2010.09.006, 2010.

Tipple, B. J and Pagani, M.: Environmental control on eastern broadleaf forest species' leaf wax distributions and D/H ratios, Geochim. Cosmochim. Ac., 111, 64–77, doi:10.1016/j.gca.2012.10.042, 2013.

Tukey, J.: Comparing Individual Means in the Analysis of Variance, Biometrics, 5, 99–114, 1949.

Vogts, A., Schefuß, E., Badewien, T., and Rullkötter, J.: n-Alkane parameters from a deep sea sediment transect off southwest Africa reflect continental vegetation and climate conditions, Org. Geochem., 47, 109–119, doi:10.1016/j.orggeochem.2012.03.011, 2012.

Vysloužilová, B. Ertlen, D.; Šefrna, L., Novak, T., Viragh, K., Rué, M., Campaner, A., Dreslerová, D., and Schwartz, D.: Investigation of vegetation history of buried chernozem soils using near-infrared spectroscopy (NIRS), Quatern. Int., 365, 203–211, doi:10.1016/j.quaint.2014.07.035, 2015.

Wang, N., Zong, Y., Brodie, C. R., and Zheng, Z.: An examination of the fidelity of n-alkanes as a palaeoclimate proxy from sediments of Palaeolake Tianyang, South China, Quatern. Int., 333, 100–109, doi:10.1016/j.quaint.2014.01.044, 2014.

Wiesenberg, G. L. B., Schwarzbauer, J., Schmidt, M. W. I., and Schwark, L.: Source and turnover of organic matter in agricultural soils derived from n-alkane/n-carboxylic acid compositions and C-isotope signatures, Org. Geochem., 35, 1371–1393, 2004.

Wiesenberg, G. L. B., Andreeva, D. B., Chimitdorgieva, G. D., Erbajeva, M. A., and Zech, W.: Reconsruction of environmental changes during the late glacial and Holocene reflected in a soil-sedimentary sequence from the lower Selenga River valley, Lake Baikal region, Siberia, assessed by lipid molecular proxies, Quatern. Int., 365, 190–202, 2015.

Wöstmann, R.: Biomarker in torfbildenden Pflanzen und ihren Ablagerungen im nordwestdeutschen Küstenraum als Indikatoren nacheiszeitlicher Vegetationsänderungen, Fakultät für Mathematik und Naturwissenschaft, Carl von Ossietzky Universität, Oldenburg, p. 233, 2006.

Zech, M., Buggle, B., Leiber, K., Markovic, S., Glaser, B., Hambach, U., Huwe, B., Stevens, T., Sümegi, P., Wiesenberg, G. and Zöller, L.: Reconstructing Quaternary vegetation history in the Carpathian Basin, SE Europe, using n-alkane biomarkers as molecular fossils – Problems and possible solutions, potential and limitations, Quaternary Sci. J., 58, 148–155, 2009.

Zech, M., Andreev, A., Zech, R., Müller, S., Hambach, U., Frechen, M., and Zech, W.: Quaternary vegetation changes derived from a loess-like permafrost palaeosol sequence in northeast Siberia using alkane biomarker and pollen analyses, Boreas, 39, 540–550, doi:10.1111/j.1502-3885.2009.00132.x, 2010.

Zech, M., Zech, R., Buggle, B., and Zöller, L.: Novel methodological approaches in loess research – interrogating biomarkers and compound-specific stable isotopes, Eiszeitalter Gegenwart-Quatern. Sci. J., 60, 170–187, 2011a.

Zech, M., Pedentchouk, N., Buggle, B., Leiber, K., Kalbitz, K., Markovic, S., and Glaser, B.: Effect of leaf-litter decomposition and seasonality on D/H isotope ratios of n-alkane biomarkers, Geochim. Cosmochim. Ac., 75, 4917–4928, 2011b.

Zech, M., Krause, T., Meszner, S. and Faust, D.: Incorrect when uncorrected: Reconstructing vegetation history using n-alkane biomarkers in loess-paleosol sequences – A case study from the Saxonian loess region, Germany, Quatern. Int., 296, 108–116, doi:10.1016/j.quaint.2012.01.023, 2013a.

Zech, R., Zech, M., Marković, S., Hambach, U., and Huang, Y.: Humid glacials, arid interglacials? Critical thoughts on pedogenesis and paleoclimate based on multi-proxy analyses of the loess–paleosol sequence Crvenka, Northern Serbia, Palaeogeogr. Palaeocl., 387, 165–175, doi:10.1016/j.palaeo.2013.07.023, 2013b.

Zhang, Z., Zhao, M., Eglinton, G., Lu, H., and Huang, C.-Y.: Leaf wax lipids as paleovegetational and paleoenvironmental proxies for the Chinese Loess Plateau over the last 170 kyr, Quaternary Sci. Rev., 25, 575–594, doi:10.1016/j.quascirev.2005.03.009, 2006.

15

Soil organic carbon stocks are systematically overestimated by misuse of the parameters bulk density and rock fragment content

Christopher Poeplau, Cora Vos, and Axel Don

Thünen Institute of Climate-Smart Agriculture, Bundesallee 50, 38116 Braunschweig, Germany

Correspondence to: Christopher Poeplau (christopher.poeplau@thuenen.de)

Abstract. Estimation of soil organic carbon (SOC) stocks requires estimates of the carbon content, bulk density, rock fragment content and depth of a respective soil layer. However, different application of these parameters could introduce a considerable bias. Here, we explain why three out of four frequently applied methods overestimate SOC stocks. In soils rich in rock fragments (> 30 vol. %), SOC stocks could be overestimated by more than 100 %, as revealed by using German Agricultural Soil Inventory data. Due to relatively low rock fragments content, the mean systematic overestimation for German agricultural soils was 2.1–10.1 % for three different commonly used equations. The equation ensemble as re-formulated here might help to unify SOC stock determination and avoid overestimation in future studies.

1 Introduction

Size and changes in the soil organic carbon (SOC) pool are major uncertainties in global earth system models used for climate predictions. Accurate estimation of SOC stocks is vital to understanding the links between atmospheric and terrestrial carbon (Friedlingstein et al., 2014). Estimates of global SOC stocks are based on soil inventories from regional to continental scale, involving multiplication of measured carbon content by soil bulk density (BD, oven-dry mass of soil per unit volume) and the depth of the respective soil layer (Batjes, 1996). The content of elements such as carbon and nitrogen in soils is usually determined in an aliquot sample of the fine soil, which is defined as the part of the soil that passes through a 2 mm sieve (Corti et al., 1998). Coarse mineral fragments > 2 mm, in the following referred to as rock fragments (Poesen and Lavee, 1994), are considered free of SOC (Perruchoud et al., 2000), although this may not be completely true as shown by Corti et al. (2002). Furthermore, living root fragments > 2 mm are not considered part of SOC, but usually as part of plant biomass. It is thus widely accepted that accurate estimates of SOC stocks

should account in some way for the presence of fragments > 2 mm (Rytter, 2012; Throop et al., 2012).

The accuracy of SOC estimates depends in the first instance on the available data and their quality. Soil organic carbon content of the fine soil is usually measured with high throughput and precision in elemental analysers, while BD and rock fragments content are often only assessed in plot-scale studies due to much more elaborate sampling requirements (Don et al., 2007). In regional-scale studies or national soil inventories, BD is therefore often approximated using pedotransfer functions and the fraction of rock fragments is often ignored (Wiesmeier et al., 2012). Stoniness is therefore regarded as the greatest uncertainty in SOC stock estimates (IPCC, 2003). However, even when all parameters are recorded, considerable difference in SOC stocks can arise from varying use of the parameters in equations. Apart from the methodological bias caused by using different methods for determining BD and rock fragment content (Beem-Miller et al., 2016; Blake, 1965), the different calculation approaches could lead to systematically different SOC stock estimates if soils contain rock fragments. Several of the approaches commonly used to calculate SOC stocks are not correct and inflate SOC stocks. The aim of this study was

(i) to reveal the conceptual differences in widely used methods for SOC stock calculation, (ii) to quantify the methodological bias in SOC stocks in a regional-scale soil inventory, (iii) to identify the most affected soil layers and finally (iv) to suggest the most adequate method for unified and unbiased SOC stock calculation.

2 Materials and methods

In a preliminary literature review we selected a total of 100 publications for which the method used to calculate SOC stocks was recorded. The search was restricted to publications listed in ISI Web of Knowledge, where "soil carbon stocks" was used as the search term. We ordered the 4915 search results by "relevance", excluded reviews and modelling studies and avoided redundant senior authors (Table S1 in the Supplement). In the literature we identified four different methods, which vary in use of the parameters BD and rock fragments content (Henkner et al., 2016; Lozano-García and Parras-Alcántara, 2013; Poeplau and Don, 2013; Wang and Dalal, 2006):

In method one (M1), a certain volume of soil is sampled, dried and weighed to determine BD. Thereby, no separation into fine soil and coarse soil (rock fragments, roots) fraction is made, while C concentration is determined in a sieved fine soil sample (usually < 2 mm). Soil organic carbon stocks are then calculated as follows:

M1:

$$BD_{sample} = \frac{mass_{sample}}{volume_{sample}} \qquad (1)$$

$$SOCstock_i = SOCcon_{fine\ soil} \times, BD_{sample} \times depth_i, \qquad (2)$$

where BD_{sample} is the bulk density of the total sample, $mass_{sample}$ is the total mass of the sample, $volume_{sample}$ is the total volume of the sample, $SOCstock_i$ is the SOC stock of the investigated soil layer (i) (Mg ha^{-1}), $SOCcon_{fine\ soil}$ is the content of SOC in the fine soil (%) and $depth_i$ is the depth of the respective soil layer (cm). This method does not account for rock fragments at all. In method two (M2), a certain volume of soil is sampled, dried and weighed. However, after sieving, the mass and volume of rock fragments and coarse roots are determined. In the following, we simplify the equations by omitting coarse roots, which is also "common practice", although the volume occupied by roots can be considerably high. This source of error is not further discussed in this study. By approximating a rock fragments density ($\rho_{rock\ fragments}$) of 2.6 g cm^{-3} (Don et al., 2007) (root density is usually assumed to be close to 1 g cm^{-3}), BD of the fine soil is subsequently calculated as

M2:

$$BD_{fine\ soil} = \frac{mass_{sample} - mass_{rock\ fragments}}{volume_{sample} - \frac{mass_{rock\ fragments}}{\rho_{rock\ fragments}}}, \qquad (3)$$

$$SOCstock_i = SOCcon_{fine\ soil} \times BD_{fine\ soil} \times depth_i. \qquad (4)$$

Thus in M2, coarse soil content is accounted for in Eq. (3), not in Eq. (4). The opposite is true for the next method (M3), in which the *rock fragments fraction* (vol. %/100) is determined, but only applied to reduce the soil volume (Eq. 5), and not to determine $BD_{fine\ soil}$:

M3: Eq. (1),

$$SOCstock_i = SOCcon_{fine\ soil} \times BD_{sample} \\ \times depth_i \times (1 - rock\ fragments\ fraction). \qquad (5)$$

In method four (M4), the coarse soil fraction is accounted for in both equations, i.e. to calculate $BD_{fine\ soil}$ (Eq. 3) and the volume of the fine soil (Eq. 6)

M4: Eq. (3),

$$SOCstock_i = SOCcon_{fine\ soil} \times BD_{fine\ soil} \times depth_i \\ \times (1 - rock\ fragments\ fraction) \qquad (6)$$

It has to be noted that when the term *rock fragments fraction* in Eq. (5) corresponds to the mass fraction of rock fragments and not to the volume fraction, results of M3 resembles results of M4.

In the German Agricultural Soil Inventory, more than 3000 agricultural soils (cropland and grassland) have been sampled as described by Grüneberg et al. (2014). To date, a total of 2515 sites were sampled and analysed for all relevant parameters (rock fragments content, fine soil mass, carbon content of the fine soil) in five different depth increments: 0–10, 10–30, 30–50, 50–70 and 70–100 cm. Here, we excluded soils with a SOC content > 8.7 %, which are not considered mineral soils anymore (Ad-Hoc-Ag Boden, 2005), giving a total of 2350 sites and 11 514 soil samples. The most common soil types sampled were Cambisols (24 %), Anthrosols (16 %), Stagnosols (13 %) and Albeluvisols (11 %) and the parent material was at 93 % of all sites loose sediments of varying origins. We expected the strongest effects in soils with high stoniness and therefore stratified the dataset by rock fragments content (vol. %). Therefore, we additionally calculated the method-induced potential deviation in SOC stocks as a function of rock fragments content (0–70 vol. %) for the average $BD_{fine\ soil}$ of the inventory dataset (1.4 g cm^{-3}). Due to the fact that method-induced deviations were systematic, we did not conduct statistics. As soon as the rock fragments content is not 0, there is always a significant difference between calculation methods, no matter how small the differences between methods would be. Data analysis and plotting was performed in the R 3.1.2 environment (R Development Core Team, 2010).

Volumetric stone content classes

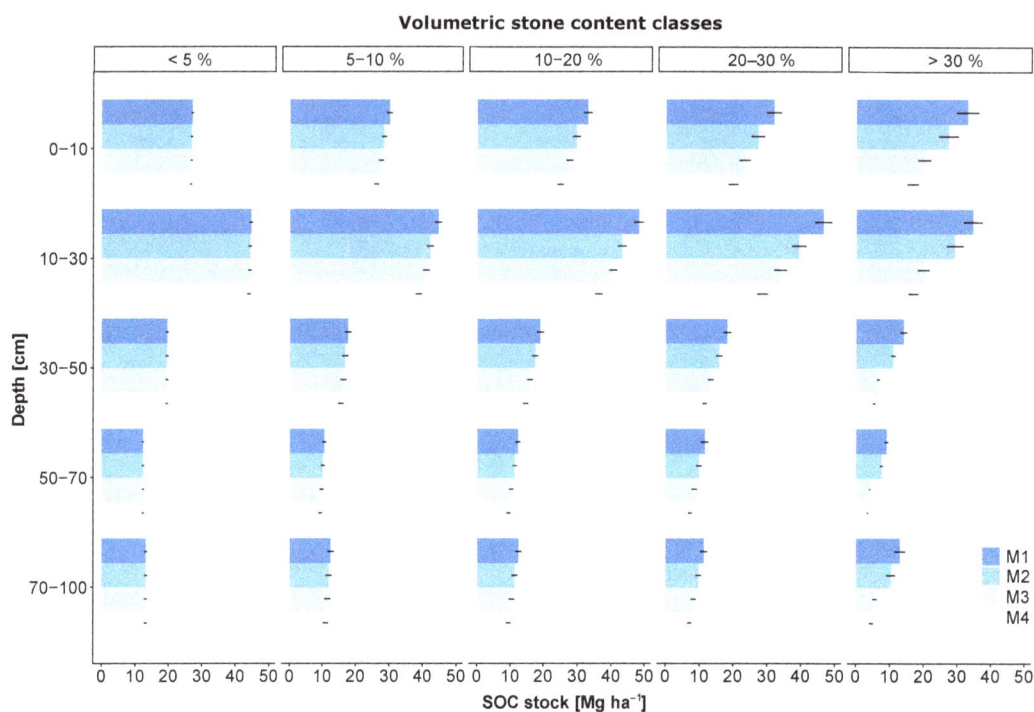

Figure 1. Average soil organic carbon stocks of the German Agricultural Soil Inventory in different depth increments calculated by different calculation methods (M1–M4) for five volumetric rock fragment content classes. Error bars indicate standard errors.

3 Results and discussion

3.1 Bias of three calculation methods to estimate SOC stocks

Three out of the four SOC calculation methods produced systematically overestimated SOC stocks. These deviations are systematic errors (bias) that cannot be reduced with optimized methods to determine the parameters SOC content, BD and rock fragments content but reduce the accuracy of SOC stock estimates. As expected, the differences in SOC stocks between calculation methods increased with rock fragments content (Fig. 1). This is in line with findings by Rytter (2012), who observed that the method of BD estimation is most important in very stony soils. While differences between methods for soils with a rock fragment content of less than 5 vol. % were small to almost negligible, M1–M3 deviated strongly from M4 in soils with > 30 % rock fragments (Fig. 1). Since M4 is the closest approximation to reality, the systematic bias was expressed as relative deviation from M4 (Table 1). In soils with > 30 % rock fragments, M1 caused the highest bias of all three calculation methods, overestimating SOC stocks by on average 144 %, i.e. more than doubling the real SOC stocks. Methods M2 and M3 also produced biased SOC stocks with 98 and 21 % overestimations for the highest rock fragment content class (> 30 % rock fragment content).

Using the average $BD_{fine\,soil}$ of $1.4\,g\,cm^{-3}$, we plotted the deviation from M4 as a function of volumetric rock fragment

content for M1-M3 (Fig. 2). Thereby, M1 and M2 showed exponential responses, while M3 showed a linear response. These responses would increase with decreasing bulk density of the fine soil. The literature review revealed that M1, M2, M3 and M4 were used by 52, 5, 30, and 13 studies respectively. In 19 out of 30 studies using M3, it was unclear whether the correction term (1 − rock fragment fraction) referred to the volumetric or gravimetric rock fragment fraction. Thus, in 68–87 % of all studies reviewed, SOC stocks were systematically overestimated assuming a rock fragment fraction > 0. More than half of the studies reviewed did not account for the rock fragment fraction at all. Cropland was the land-use type in which rock fragment were most often completely ignored. Eighty-five percent of all reviewed cropland studies used M1 to calculate SOC stocks (Table S2). In contrast, 54 % of all studies that used M4 were conducted in forest soils. This might be related to the fact that rock fragment are more abundant in forest soils and that SOC investigations in cropland soils are often restricted to the surface layer with low rock fragment fraction. However, only 17 % of all assessed forest studies used method M4, while M1 was the most often applied (41 %).

The number of soils with high rock fragment contents in the German dataset is limited due to the dominance of parent material from glacio-fluvial deposits (Table 1). Thus, the majority of soils (67–78 %, depending on soil depth increment) had a volumetric rock fragment content of < 5 %. As a consequence, the average SOC stocks were only moderately

Table 1. Fraction of total observations for different volumetric rock fragment content classes in the German Agricultural Soil Inventory and average soil organic carbon stock deviations (%) from M4 for the calculation methods M1–M3 in different depth increments.

Depth	Fraction of total observations					Average relative deviation from M4		
	< 5 %	5–10 %	10–20 %	20–30 %	> 30 %	M1	M2	M3
0–10	78.4	12.9	5.7	1.8	1.2	6.1	3.6	2.2
10–30	72.4	14.0	6.4	3.1	4.2	7.3	4.3	2.5
30–50	68.4	10.3	6.4	4.1	10.7	8.4	5.3	2.2
50–70	67.5	9.4	6.4	4.1	12.6	8.8	5.8	2.1
70–100	68.4	9.3	5.7	3.3	13.3	10.1	6.5	2.3

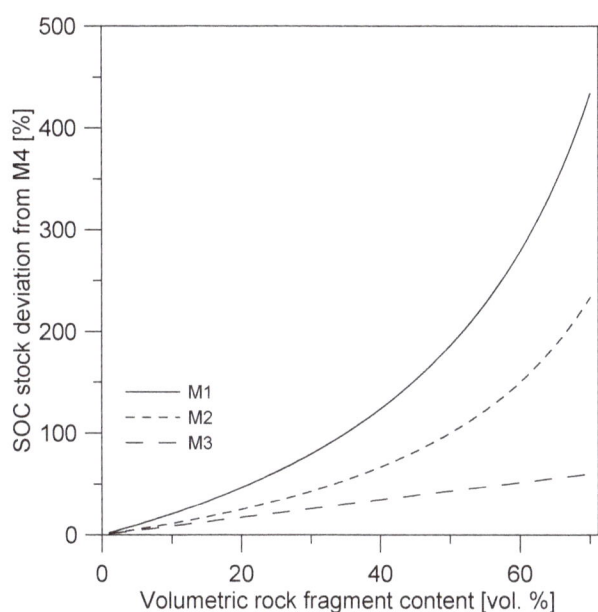

Figure 2. Systematic deviations in SOC stock from calculation method M4 for methods M1–M3 as a function of volumetric rock fragment content. Bulk density of the fine soil was set to $1.4\,\mathrm{g\,cm^{-3}}$ in this example.

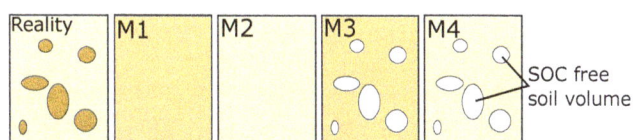

Figure 3. Schematic overview on the four methods applied to estimate the mass of soil needed to calculate soil organic carbon stocks. Different shades of brown are used to indicate different densities: the rock fragment fraction (ellipsoids) has the darkest brown and the fine soil fraction the lightest brown.

influenced by the calculation method (2.1–10.1 % deviation, Table 1). For forests, which are usually found on soils less suitable for agriculture, e.g. due to high stoniness, the bias would be stronger. Overall, the results highlight the importance of a correct use of the parameters BD and rock fragment fraction when calculating SOC stocks.

3.2　Evaluation of the four different calculation methods

Since all four methods use the same $SOCcon_{fine\,soil}$ due to equal preparation of the fine soil, differences between the calculation methods arise from differences in use of the parameters BD and rock fragment content. The individual bias of each method is visualized in Fig. 3. In M1, BD of the soil containing SOC (fine soil) is overestimated due to inclusion of rock fragment in the BD estimate. The volume of soil which contains SOC present in the respective soil layer is also overestimated, since the rock fragment fraction is not subtracted from the total soil volume (Eq. 2). Thus, M1 "fills" the space occupied by rock fragments with fine soil with an overestimated BD. In the German Agricultural Soil Inventory, only 9 % of all sampled layers were found to be free of rock fragments. Thus, for most soils M1 is not the correct way to calculate SOC stocks. Similarly, M2 overestimates $SOCstock_i$ by filling the volume of rock fragments with fine soil. However, BD is calculated and used correctly leading to a smaller systemic overestimation of SOC compared to M1. Finally, M3 correctly accounts for the rock fragment fraction that can be assumed to be SOC-free. However, in M3 an overestimated BD is applied as in M1, i.e. BD_{sample} and not the $BD_{fine\,soil}$. Methods to estimate BD and rock fragment content vary, primarily owing to size and abundance of the latter and may have large uncertainty (Blake, 1965; Parfitt et al., 2010; Rytter, 2012). However, the presented difference between calculation methods is independent of the method of determination of the these parameters with one exception: if the sampled soil layer contains no gravel, but only fine soil and rock fragments that exceed the diameter of a soil ring used to determine BD_{sample}, and this ring is placed at a position (in the profile wall) which is completely free of rock fragments, while the rock fragment content is estimated with a different method and accounted for, then M3 does resemble M4. Bulk density is often determined with soil rings with a volume between 100 and $500\,\mathrm{cm^2}$ or soil probes (Walter et al., 2016). In the German Agricultural Soil Inventory, $250\,\mathrm{cm^2}$ soil rings are used to determine BD. In 91 % of all soils inventoried, small rock fragments were detected which end up in the soil ring and have to be corrected for. Thus, method M3 is rarely a correct method to estimate SOC stocks. It is erroneously often cited as the IPCC default method. However, while the equations given in IPCC resemble M3, IPCC provides a footnote that is most likely often overlooked, which states that BD estimates should be corrected for the proportion of "coarse fragments" (IPCC, 2003). Even if the rock fragment fraction might store a certain amount of organic carbon (Corti et al., 2002), which might lead to slight underestimation of SOC stocks in M4, we suggest use of this method in future studies.

3.3　Proposed equations to calculate SOC stocks

Bulk density might be of interest as an important soil property. However, for the calculation of SOC stocks alone it is not needed, while it is the fine soil stock of the investigated soil layer (FSS_i, Mg ha^{-1}) that is of interest since it contains the SOC. Thus, the equations in M4 could be reformulated as

$$FSS_i = \frac{mass_{fine\,soil}}{volume_{sample}} \times depth_i, \tag{7}$$

$$SOCstock_i = SOCcon_{fine\,soil} \times FSS_i. \tag{8}$$

This has implications for sample preparation: for $BD_{fine\,soil}$ the volume of coarse fragments has to be estimated by weighing rock fragments and coarse roots separately, while FSS_i would only need the total mass of the fine soil contained in the known volume of sample. When using soil probes to sample soil cores with a known volume, FSS_i calculation can further be simplified to

$$FSS_i = \frac{mass_{fine\,soil}}{surface_{sample}}, \tag{9}$$

where $surface_{sample}$ is the surface area (cm^2) of the sampling probe.

4　Conclusions

We show here that substantially different methods are used for the calculation of SOC stocks. These methods differ in use of the parameters bulk density and rock fragment content, which causes systematic overestimation of SOC stocks in three out of four, more or less frequently applied methods, or in 68–87 of 100 publications reviewed. We showed that this overestimation can exceed 100 % in stony soils. For future studies, we suggest to calculate the fine soil stock of a certain soil layer which is to be multiplied with its SOC content to derive unbiased SOC stock estimates. If rock fragments were measured, SOC stocks of existing datasets could also be recalculated, e.g. in the case of resamplings.

Competing interests. The authors declare that they have no conflict of interest.

Acknowledgements. This study was funded by the German Federal Ministry of Food and Agriculture in the framework of the German Agricultural Soil Inventory.

Edited by: B. van Wesemael

References

Ad-Hoc-Ag Boden: Bodenkundliche Kartieranleitung, E. Schweizerbart'sche Verlagsbuchhandlung, Hannover, 2005.

Batjes, N.: Total carbon and nitrogen in the soils of the world, Eur. J. Soil Sci., 47, 151–163, 1996.

Beem-Miller, J. P., Kong, A. Y. Y., Ogle, S., and Wolfe, D.: Sampling for Soil Carbon Stock Assessment in Rocky Agricultural Soils, Soil Sci. Soc. Am. J., 80, 1411–1423, 2016.

Blake, G.: Bulk density, Methods of Soil Analysis, Part 1, Physical and Mineralogical Properties, Including Statistics of Measurement and Sampling, 1965, 374–390, 1965.

Corti, G., Ugolini, F. C., and Agnelli, A.: Classing the Soil Skeleton (Greater than Two Millimeters): Proposed Approach and Procedure, Soil Sci. Soc. Am. J., 62, 1620–1629, 1998.

Corti, G., Ugolini, F., Agnelli, A., Certini, G., Cuniglio, R., Berna, F., and Fernández Sanjurjo, M.: The soil skeleton, a forgotten pool of carbon and nitrogen in soil, Eur. J. Soil Sci., 53, 283–298, 2002.

Don, A., Schumacher, J., Scherer-Lorenzen, M., Scholten, T., and Schulze, E. D.: Spatial and vertical variation of soil carbon at two grassland sites – Implications for measuring soil carbon stocks, Geoderma, 141, 272–282, 2007.

Friedlingstein, P., Meinshausen, M., Arora, V. K., Jones, C. D., Anav, A., Liddicoat, S. K., and Knutti, R.: Uncertainties in CMIP5 Climate Projections due to Carbon Cycle Feedbacks, J. Clim., 27, 511–526, 2014.

Grüneberg, E., Ziche, D., and Wellbrock, N.: Organic carbon stocks and sequestration rates of forest soils in Germany, Glob. Change Biol., 20, 2644–2662, 2014.

Henkner, J., Scholten, T., and Kühn, P.: Soil organic carbon stocks in permafrost-affected soils in West Greenland, Geoderma, 282, 147–159, 2016.

IPCC: Good practice guidance for land use, land-use change and forestry, Good practice guidance for land use, land-use change and forestry, http://www.ipcc-nggip.iges.or.jp/public/gpglulucf/gpglulucf.html, 2003.

Lozano-García, B. and Parras-Alcántara, L.: Land use and management effects on carbon and nitrogen in Mediterranean Cambisols, Agr. Ecosyst. Environ., 179, 208–214, 2013.

Parfitt, R. L., Ross, C., Schipper, L. A., Claydon, J. J., Baisden, W. T., and Arnold, G.: Correcting bulk density measurements made with driving hammer equipment, Geoderma, 157, 46–50, 2010.

Perruchoud, D., Walthert, L., Zimmermann, S., and Lüscher, P.: Contemporary carbon stocks of mineral forest soils in the Swiss Alps, Biogeochemistry, 50, 111–136, 2000.

Poeplau, C.: SOIL data, doi:10.17605/OSF.IO/N8W9J, last access: 9 March 2017.

Poeplau, C. and Don, A.: Sensitivity of soil organic carbon stocks and fractions to different land-use changes across Europe, Geoderma, 192, 189–201, 2013.

Poesen, J. and Lavee, H.: Rock fragments in top soils: significance and processes, Catena, 23, 1–28, 1994.

R Development Core Team: R: A language and environment for statistical computing, R Foundation for Statistical Computing, Vienna, Austria, 2010.

Rytter, R.-M.: Stone and gravel contents of arable soils influence estimates of C and N stocks, Catena, 95, 153–159, 2012.

Throop, H. L., Archer, S. R., Monger, H. C., and Waltman, S.: When bulk density methods matter: Implications for estimating soil organic carbon pools in rocky soils, J. Arid Environ., 77, 66–71, 2012.

Walter, K., Don, A., Tiemeyer, B., and Freibauer, A.: Determining Soil Bulk Density for Carbon Stock Calculations: A Systematic Method Comparison, Soil Sci. Soc. Am. J., 80, 579–591, 2016.

Wang, W. and Dalal, R.: Carbon inventory for a cereal cropping system under contrasting tillage, nitrogen fertilisation and stubble management practices, Soil Till. Res., 91, 68–74, 2006.

Wiesmeier, M., Sporlein, P., Geuss, U., Hangen, E., Haug, S., Reischl, A., Schilling, B., von Lutzow, M., and Kogel-Knabner, I.: Soil organic carbon stocks in southeast Germany (Bavaria) as affected by land use, soil type and sampling depth, Glob. Change Biol., 18, 2233–2245, 2012.

Permissions

All chapters in this book were first published in SOIL, by Copernicus Publications; hereby published with permission under the Creative Commons Attribution License or equivalent. Every chapter published in this book has been scrutinized by our experts. Their significance has been extensively debated. The topics covered herein carry significant findings which will fuel the growth of the discipline. They may even be implemented as practical applications or may be referred to as a beginning point for another development.

The contributors of this book come from diverse backgrounds, making this book a truly international effort. This book will bring forth new frontiers with its revolutionizing research information and detailed analysis of the nascent developments around the world.

We would like to thank all the contributing authors for lending their expertise to make the book truly unique. They have played a crucial role in the development of this book. Without their invaluable contributions this book wouldn't have been possible. They have made vital efforts to compile up to date information on the varied aspects of this subject to make this book a valuable addition to the collection of many professionals and students.

This book was conceptualized with the vision of imparting up-to-date information and advanced data in this field. To ensure the same, a matchless editorial board was set up. Every individual on the board went through rigorous rounds of assessment to prove their worth. After which they invested a large part of their time researching and compiling the most relevant data for our readers.

The editorial board has been involved in producing this book since its inception. They have spent rigorous hours researching and exploring the diverse topics which have resulted in the successful publishing of this book. They have passed on their knowledge of decades through this book. To expedite this challenging task, the publisher supported the team at every step. A small team of assistant editors was also appointed to further simplify the editing procedure and attain best results for the readers.

Apart from the editorial board, the designing team has also invested a significant amount of their time in understanding the subject and creating the most relevant covers. They scrutinized every image to scout for the most suitable representation of the subject and create an appropriate cover for the book.

The publishing team has been an ardent support to the editorial, designing and production team. Their endless efforts to recruit the best for this project, has resulted in the accomplishment of this book. They are a veteran in the field of academics and their pool of knowledge is as vast as their experience in printing. Their expertise and guidance has proved useful at every step. Their uncompromising quality standards have made this book an exceptional effort. Their encouragement from time to time has been an inspiration for everyone.

The publisher and the editorial board hope that this book will prove to be a valuable piece of knowledge for researchers, students, practitioners and scholars across the globe.

List of Contributors

Leonor Rodrigues, Mareike Trauerstein and Heinz Veit
Institute of Geography, University of Berne, Hallerstrasse 12, 3012 Bern, Switzerland

Umberto Lombardo
University of Pompeu Fabra, Ramon Trias Fargas 25–27, Mercè Rodoreda, 08005 Barcelona, Spain

Perrine Huber and Sandra Mohr
Independent researcher, 3012 Bern, Switzerland

Juliane Filser
Center for Environmental Research and Sustainable Technology, University of Bremen, General and Theoretical Ecology, Leobener Str. – UFT, 28359 Bremen, Germany

Jack H. Faber
Wageningen Environmental Research (Alterra), P.O. Box 47, 6700 AA Wageningen, the Netherlands

Alexei V. Tiunov and Alexei V. Uvarov
Laboratory of Soil Zoology, Institute of Ecology&Evolution, Russian Academy of Sciences, Leninsky prospekt 33, 119071 Moscow, Russia

Lijbert Brussaard and Gerlinde De Deyn
Dept. of Soil Quality, Wageningen University, P.O. Box 47, 6700 AA Wageningen, the Netherlands

Jan Frouz
Institute for Environmental Studies, Charles University in Prague, Faculty of Science, Benátská 2, 128 43 Praha 2, Czech Republic

Matty P. Berg
Vrije Universiteit Amsterdam, Department of Ecological Science, De Boelelaan 1085, 1081 HV Amsterdam, the Netherlands

Patrick Lavelle
Université Pierre et Marie Curie, Centre IRD Ile de France, 32, rue H. Varagnat, 93143 Bondy CEDEX, France

Michel Loreau
Centre for Biodiversity Theory and Modelling, Station d'Ecologie Théorique et Expérimentale, UMR5321 – CNRS&Université Paul Sabatier, 2, route du CNRS, 09200 Moulis, France

Diana H. Wall
School of Global Environmental Sustainability&Dept. Biology, Colorado State University, Fort Collins, CO 80523-1036, USA

Pascal Querner
University of Natural Resources and Life Sciences, Department of Integrated Biology and Biodiversity Research, Institute of Zoology, Gregor-Mendel-Straße 33, 1180 Vienna, Austria

Herman Eijsackers
Wageningen University and Research Centre, P.O. Box 9101, 6700 HB Wageningen, the Netherlands

Juan José Jiménez
ARAID, Soil Ecology Unit, Department of Biodiversity Conservation and Ecosystem Restoration, IPE-CSIC, Avda. Llano de la Victoria s/n, 22700 Jaca (Huesca), Spain

Marshall D. McDaniel and A. Stuart Grandy
Department of Natural Resources and the Environment, University of New Hampshire, Durham, NH, USA acurrent address: Department of Agronomy, Iowa State University, 2517 Agronomy Hall, 716 Farm House Lane, Ames, IA 50011, USA

Christopher Shepard, Marcel G. Schaap, and Craig Rasmussen
Department of Soil, Water and Environmental Science, The University of Arizona, Tucson, AZ 85721-0038, USA

Jon D. Pelletier
Department of Geosciences, The University of Arizona, Tucson, AZ 85721-0077, USA

Samuel N. Araya
Environmental Systems Graduate Group, University of California, Merced, CA 95343, USA

Marilyn L. Fogel and Asmeret Asefaw Berhe
Environmental Systems Graduate Group, University of California, Merced, CA 95343, USA
Life and Environmental Sciences Unit, University of California, Merced, CA 95343, USA

Eléonore Beckers and Aurore Degré
Université de Liège, Gembloux Agro-Bio Tech, UR Biosystems Engineering, Passage des déportés 2, 5030 Gembloux, Belgium

Mathieu Pichault
Université de Liège, Gembloux Agro-Bio Tech, UR Biosystems Engineering, Passage des déportés 2, 5030 Gembloux, Belgium
Université de Liège, Gembloux Agro-Bio Tech, UR TERRA, Passage des déportés 2, 5030 Gembloux, Belgium

Sarah Garré
Université de Liège, Gembloux Agro-Bio Tech, UR TERRA, Passage des déportés 2, 5030 Gembloux, Belgium

Wanwisa Pansak
Naresuan University, Department of Agricultural Science, 65000 Phitsanulok, Thailand

Florian Wilken
Institute for Geography, Universität Augsburg, Augsburg, Germany
Chair of Soil Protection and Recultivation, Brandenburg University of Technology Cottbus-Senftenberg,Cottbus, Germany
Institute of Soil Landscape Research, Leibniz Centre for Agricultural Landscape Research ZALF e.V., Müncheberg, Germany

Michael Sommer
Institute of Soil Landscape Research, Leibniz Centre for Agricultural Landscape Research ZALF e.V., Müncheberg, Germany
University of Potsdam, Institute of Earth and Environmental Sciences, Potsdam, Germany

Kristof Van Oost
Earth & Life Institute, TECLIM, Université catholique de Louvain, Louvain-la-Neuve, Belgium

Oliver Bens
Helmholtz Centre Potsdam GFZ German Research Centre for Geosciences, Potsdam, Germany

Peter Fiener
Institute for Geography, Universität Augsburg, Augsburg, Germany

Sami Touil
Superior National School of Agronomy, El Harrach, Algiers, Algeria

Gembloux Agro-Bio Tech, Biosystem Engineering, Soil–Water–Plant Exchanges, University of Liege, Passage des Déportés, Gembloux, Belgium
Laboratory of Crop Production and Sustainable Valorization of Natural Resources, University of Djilali Bounaama Khemis Miliana, Ain Defla, Algeria

Aurore Degre
Gembloux Agro-Bio Tech, Biosystem Engineering, Soil–Water–Plant Exchanges, University of Liege, Passage des Déportés, Gembloux, Belgium

Mohamed Nacer Chabaca
Superior National School of Agronomy, El Harrach, Algiers, Algeria

Johan Bouma
Wageningen University, Wageningen, the Netherlands

Jean-Christophe Calvet, Noureddine Fritz, Christine Berne, Bruno Piguet, William Maurel and Catherine Meurey
CNRM, UMR 3589 (Météo-France, CNRS), Toulouse, France

Luitgard Schwendenmann
School of Environment, University of Auckland, Private Bag 92019, 1142 Auckland, New Zealand

Cate Macinnis-Ng
School of Biological Sciences, University of Auckland, Private Bag 92019, 1142 Auckland, New Zealand

Laura Arata, Katrin Meusburger, Alexandra Bürge and Christine Alewell
Environmental Geosciences, Department of Environmental Sciences, University of Basel, Basel, Switzerland

Markus Zehringer
State Laboratory Basel-City, Basel, Switzerland

Michael E. Ketterer
Chemistry Department, Metropolitan State University of Denver, Colorado, USA

Lionel Mabit
Soil and Water Management & Crop Nutrition Laboratory, FAO/IAEA Agriculture & Biotechnology Laboratory, Vienna, Austria

Gerard Govers and Roel Merckx
KU Leuven, Department of Earth and Environmental Sciences, Celestijnenlaan 200E, 3001 Leuven, Belgium

Bas van Wesemael, and Kristof Van Oost
Université Catholique de Louvain, Earth and Life Institute, 3 Place Louis Pasteur, 1348 Louvain-la-Neuve, Belgium

Imke K. Schäfer, Jörg Franke and Roland Zech
Institute of Geography and Oeschger Centre for Climate Change Research, University of Bern, 3012 Bern, Switzerland